AIR, WATER AND SOIL POLLUTION SCIENCE AND TECHNOLOGY

Additional books in this series can be found on Nova's website under the Series tab.

Additional e-books in this series can be found on Nova's website under the e-book tab.

AIR, WATER AND SOIL POLLUTION SCIENCE AND TECHNOLOGY

MICROBES IN SOIL AND THEIR AGRICULTURAL PROSPECTS

AIR, WATER AND SOIL POLLUTION SCIENCE AND TECHNOLOGY

MICROBES IN SOIL AND THEIR AGRICULTURAL PROSPECTS

KAUSHAL KISHORE CHOUDHARY, PH.D.
AND
DOLLY WATTAL DHAR, PH.D.
EDITORS

nova publishers
New York

Library of Congress Cataloging-in-Publication Data

ISBN: 978-1-63482-824-6

Published by Nova Science Publishers, Inc. † New York

CONTENTS

PREFACE

Microbes are the most lucrative and unique creature of the nature on the earth surface. They constitute a large percentage of Earth's biodiversity. They constitute a structurally and functionally diverse group of organisms that include archaea, bacteria, cyanobacteria, and single-celled eukaryotic organisms. Microbes are the pioneer colonizer of the earth surface, and had created a platform for origin, survival and persistence of other biosystems. They produced a layer of soil crust by secreting acidic substances and bring about weathering of rocks forming soil fertile layer. Nevertheless, cyanobacteria is one of the major and most vital microorganisms evolved oxygen and makes the environment oxygenic opening the way for origin of diverse and advanced form of present day organisms. They synthesize complex compounds from simple ones and *vice versa*, resulting in the continuation of nutrient cycles, which is essential for persistence of life forms on the earth surface. Moreover, they can transform unavailable sources of nitrogen into available forms of nitrogenous compounds, mobilize organic phosphates into inorganic phosphates, synthesize growth promoting hormones, and most importantly cope with fluctuation in physic-chemical behavior of soil and anthropogenic climate change. This book is a compilation of articles dealing with microbial functioning and its interaction with plants and other soil microbial communities. Their implication for sustainable agriculture is also discussed.

Current understanding of harmful effects chemical fertilizer has on soil fertility agriculture practices, it has become more important and more demanding for agriculturalists, ecophysiologists and molecular biologists to understand the biology of essential microbes that are playing significant role in maintaining productivity and fertility of soil. Moreover, proper and potential exploitation of microbes for production of secure food requires a basic understanding of microbial behavior in its natural habitat. This book contains contributions from leading researchers on aspects such as: social networking and molecular signaling in soil, plant-microbe interaction and agriculturally useful aspects of microbes. The book discusses the basic understanding of biological functioning of microbes under natural conditions in soil and its applied aspects with due consideration to biotechnology, particularly with reference to exploitation of microbes for supplementary sources of fertilizer. Concerted efforts have been made to include most of the microbial groups, their interactions with plants and their significance in agriculture. Each chapter in this book has been contributed by eminent scholars and scientists in their own field of expertise. The articles describe the advanced state of knowledge without compromising with the basic understanding of the subject. Articles have been grouped into two sections viz. a. Plant-Microbes interactions/symbiosis and stress tolerance and, b. Agricultural Prospects.

The first section under the heading of 'Plant-Microbes interactions' and stress tolerance' covers the article from eminent scholars, scientists and academicians on molecular signaling in gram negative bacteria and their ecological significance, how does microbe support plants in various abiotic stress tolerance, physiological, biochemical and molecular processes involved in establishment of symbiotic association with plants, methods of analyzing microbial diversity and UV-B protection mechanism in cyanobacteria that might be of significant value in future agricultural practices. The article discussing molecular signaling in gram-negative bacteria covers the topic like intracellular communication of gram-negative bacteria with plants, responses of bacterial quorum sensing and their biotechnological significance. Articles on microbe-symbiotic association (Rhizobium-legume symbiosis, Frankia association with different plants) discuss the mechanisms involved in the effect of environmental factors on association and their significance to host plants. Most importantly, the articles also cover the biochemical and molecular mechanism of evolution of life on the earth surface i.e., nitrogen fixation. Another important aspect of this book is to include articles on physiological, biochemical and molecular mechanisms involved in the interaction between plants and microbes that increases tolerance level of plants under various abiotic stresses and supports growth and development. Plants secrete root exudates during changing soil environment. Hence, one article has been included that discusses the role of root exudates on microbial community structure. Subsequent chapters discuss methods for analyzing microbial communities and UV-protection mechanisms in cyanobacteria and their possible future application in sustainable agriculture.

The second sections have been aimed towards focusing on technological advances in biofertilizers and the role of microbes in agriculture. This section covers the possible future contributions of diverse microorganism groups and their product viz. gram-negative and methylotrophic bacteria, Frankia, Burkholederia, cyanobacteria, Arbuscular Mycorrhizal (AM) fungi, bacteriophages, Azolla-Anabaena symbiosis, Trichoderma, biosurfactant etc. in augmentation of nitrogen, phosphorus and growth promoting hormones to soil and ultimately soil productivity for sustainable agriculture practices.

We have tried our best to incorporate most of the topics on physiological and biochemical aspects involved in regulating metabolic machinery of microbes to adopt diverse environments that is of significant value for continuing sustainable agronomic practices. But, we take undiluted responsibility for mistakes that might have crept into the text inadvertently.

We are grateful to our authors for their support without which the compilation of book would not possible. We express our heartiest gratitude to Nova Science Publishers (USA) for continuous support in delivering suggestions during compilation of the book.

Editors:

Dr. Kaushal Kishore Choudhary, PhD
Dr. Jagannath Mishra College
Affiliated to Babasaheb BhimRao Ambedkar Bihar University
Muzaffarpur – 842001, Bihar, India
Email: kkc1970@gmail.com

Dr. Dolly Wattal Dhar, Ph. D.
Division of Microbiology,
Indian Agricultural Research Institute,
New Delhi – 110012, New Delhi, India
Email: dollywattaldhar@yahoo.com

SECTION A. PLANT-MICROBE INTERACTIONS AND STRESS TOLERANCE

In: Microbes in Soil and Their Agricultural Prospects
Editors: K. Kishore Choudhary and D. Wattal Dhar
ISBN: 978-1-63482-824-6

Chapter 1

MOLECULAR SIGNALING IN GRAM NEGATIVE RHIZOBACTERIA: BIOCONTROL PERSPECTIVES AND ECOLOGICAL IMPLICATIONS

***Ankit Kumar*[1,4,*], *Sandeep Saini*[1,3], *Anil Prakash*[2] and *B. N. Johri*[1]**
[1]Department of Biotechnology, Barkatullah University, Bhopal, Madhya Pradesh, India
[2]Department of Microbiology, Barkatullah University, Bhopal, Madhya Pradesh, India
[3]National Bureau of Agriculturally Impotant Microorganisms (ICAR), Uttar Pradesh, India
[4]Division of Entomology, Indian Agricultural Research Institute, Pusa, New Delhi, India

ABSTRACT

Most bacterial species use cellular communication mediated through diffusible signal molecules to monitor their population density; they respond by changing their behaviour to coordinate gene regulation through a process known as quorum sensing (QS). The most common signal molecules found in Gram negative bacteria are acylated homoserine lactones (HSLs). Their derivatives regulate many important physiological functions, including virulent traits, motility, conjugation, and sporulation which regulate competence development as well as some plant growth promoting traits in rhizobacteria. Plants also respond to the presence of signal molecules produced from bacteria and consequently produce a variety of molecules that mimic the bacterial QS signals to confuse the pathogen. Alternatively, plants might detect the QS signals and trigger various defense responses depending on the type and concentration of signals detected. Since QS regulatory systems are often associated with pathogenesis, interference with QS signaling may be exploited as an alternative approach to control bacterial diseases of plants. This chapter focusses over the intra- and inter-cellular communication among the Gram negative rhizobacteria associated with plants, various signal molecules produced by rhizobacteria, how plants respond to the bacterial QS, the physiological attributes of QS, quorum quenching (QQ) and its biotechnological applications.

---

* Corresponding author: Division of Entomology, Indian Agricultural Research Institute (IARI), Pusa, New Delhi-110012, India. E-mail: ankitmehta1707@gmail.com, ankit1707@ rediffmail.com.

Keywords: Acyl-homoserine lactone, biocontrol, quorum quenching, quorum sensing, rhizobacteria.

INTRODUCTION

During the course of co-evolution of plant-microbe interactions, bacteria have developed very sophisticated mechanisms of gene expression at population and community levels through biosynthesis, regulation and perception of diffusible compounds. Bacteria sense their population density and coordinate the expression of target genes, including the virulence factors in Gram negative bacteria, by the N-acylhomoserine lactones (AHLs) dependent mechanism known as QS. As the level of signal produced is a function of bacterial cell density, the regulatory pathway was termed as QS [1]. The nature of QS signals is highly diverse and has been widely studied in Gram negative bacteria, such as *Pseudomonas* [2, 3]. Most QS dependent reactions are unproductive due to a single bacterial cell. However, they may become beneficial when undertaken simultaneously by a group of bacterial population. This enables bacteria to behave as multi-cellular organisms. Different bacterial species employ QS to coordinate their gene expression depending upon their own population density (Table 1). QS allows bacteria to monitor the environment and consequently change their behaviour on population wide scale in response to the changes in the number and species that occur in a community.

Table 1. Some popular AHL dependent quorum sensing (QS) systems in Gram negative bacteria

Organism	QS genes	Major Signal molecule(s)	Regulated function	Reference(s)
Agrobacterium tumefaciens	*traI/traR*	3-oxooctanoyl-HSL	Conjugal transfer of plasmid	Zhang et al. 2002; Hoedecoeur and Faure, 2010; Lang and Faure, 2014
Erwinia carotovora sspp. *carotovora*	(i) *expI/expR* (ii) *carI/carR*	3-oxohexanoyl-HSL	(i) Extracellular enzyme secretion (ii) Carbapenem antibiotic production	McGowan et al., 1995, 2005; Burr et al. 2006; Pollumaa et al., 2012
Pantoea (*Erwinia*) *stewartii* subspp. *Stewartii*	*esaI/esaR*	N-(3-oxohexanoyl)-HSL 3-oxooctanoyl-HSL	Exopolysachharide expression, biofilm formation and colonization	von Bodman et al., 2003; Koutsoudis et al., 2006
Pseudomonas aeruginosa	(i) *lasI/lasR* (ii) *rhlI/rhlR*	(i) 3-oxododecanoyl-HSL (ii) N-butyryl-HSL	Exoenzyme production, biofilm formation Rhamnolipid production	Smith and Iglewski, 2003; Juhas et al., 2005; Antunes et al., 2010; Karlsson et al., 2012
Pseudomonas aureofaciens	(i) *phzI/phzR* (ii) *csaI/csaR*	N-hexanoyl-HSL	Phenazine production Colony morphology, aggregation	Pierson et al., 1994; Zhang and Pierson, 2001; Maddula et al., 2006

Organism	QS genes	Major Signal molecule(s)	Regulated function	Reference(s)
Pseudomonas chlororaphis	*phzI/phzR*	N-hexanoyl-HSL	Phenazine-1-carboxamide biosynthesis	Chin et al., 2001, 2005; Maddula et al., 2006
Pseudomonas fluorescens	(i) *mpuI/mpuR* (ii) *hdtS*	Long acyl-chain-AHL 3-oxododecanoyl-HSL	Mupirocin biosynthesis Unknown	Laue et al., 2000; El-Sayed et al., 2001; Hothersall et al., 2011; Maeyer et al., 2011
Pseudomonas putida	*ppuI/ppuR*	3-oxododecanoyl-HSL	Biofilm formation	Steidle et al., 2002; Bertani and Venturi, 2004; Maeyer et al., 2011; Chen et al., 2013
Pseudomonas syringae	*ahlI/ahlR*	3-oxohexanoyl-HSL	Colony morphology, ecological fitness	Dumenyo et al., 1998; Quinones et al., 2004
Ralstonia solanacearum	*solI/solR*	N-hexanoyl-HSL, N-octanoyl-HSL	Not defined	Flavier et al., 1997; Cha et al., 1998; Schell, 2000
Rhizobium leguminosarum	(i) *rhiI/rhiR* (ii) *cinI/cinR*	N-hexanoyl-HSL N-(-3-hydroxy-7-cis-tetradecenoyl)-HSL	Rhizospheric competence Quorum sensing regulated cascade	Daniels et al., 2002; Edwards et al., 2009; Downie, 2010; Ahlgren et al., 2011; Nievas et al., 2012
Serratia liquefaciens	(i) *swrI/swrR* (ii) *smaI/ smaR*	N-butanoyl-HSL N-butanoyl-HSL, N-hexanoyl-HSL	Swarming movement, biosurfactant production Carbapenem antibiotic biosynthesis	Eberl et al., 1999; Thomson et al., 2000; Daniels et al., 2004; Liu et al., 2011; Bakkiyaraj et al., 2012
Xanthomonas campestris	*rpf* regulon	Diffusible factor (DF), Diffusible signal factor (DSF)	Exoenzyme and exopolysachharide production	Barber et al., 1997; Vojnov et al., 2001, He et al., 2010; Zhao et al., 2011; Guo et al., 2012

Cellular communication involves the exchange of low mol wt., diffusible signal molecules among the members of a localized population. If concentration of a signal molecule produced by the population is greater than its loss by diffusion or inactivation, the signal molecules start accumulating and as soon as a threshold level is attained, it activates the cognate receptor proteins. Consequently, it may trigger widespread changes in gene expression within members of the population. This cell density dependent response is believed to have been evolved as a means to provide advantages to a particular group of cells, for example, improved access to environmental niches or improved defense capabilities against other microbes or eukaryotic host-defense mechanisms. Most importantly, QS has changed our traditional perception that bacteria are just individual cells, unable to interact with each other and fail to collectively respond to environmental stimuli; a trait typical to the multi-cellular organisms. The very first example and paradigm of Gram-negative quorum sensing is the *luxI-luxR* QS system of *Vibrio fischeri*, which is involved in the regulation of bioluminescence in a density dependent manner [4]. *V.fischeri* colonizes the light organ of the Hawaiin squid *Euphrymna scolopes* where this bacterium grows to high cell density, thereafter inducing the expression of genes required for bioluminescence. Genetically the bioluminescent gene cluster of *V.fischeri* is composed of eight *lux* genes viz., *luxA-E*, *luxG*, *luxI* and *luxR* [5]. While LuxI is the autoinducer synthase enzyme and takes part in the production of an homoserine lactone (HSL), N-(3-oxohexanoyl)-HSL (3-oxo-C6HSL), *luxR* helps to bind the autoinducer and to activate the transcription of bioluminscence operon [6]. As the population density of *V.fischeri* grows, it produces and releases an autoinducer hormone (AHL) into extra-cellular environment. As soon as a threshold level is attained, the signal molecule binds to the *luxR* transcriptional activator leading to the expression of bioluminescence operon. Detection of the autoinducer by *V. fischeri* elicits a signaling cascade that culminates in the emission of light [7].

QS systems have been shown to be key regulators of virulence in both Gram-negative and Gram-positive pathogens [8, 9]. However, the regulated genes encode other proteins also, including those that are involved in basic metabolic processes in addition to the classic virulence factors [10]. This suggests that a significant portion of a bacterial genome (4–10%) and proteome (20% or more) can be influenced by quorum signaling, and that QS is a mechanism used by pathogenic bacteria not only to modulate virulence factor production but also to adapt to the metabolic demands of living in communities [11, 12]. In fact, since the discovery of cell density dependent regulation of bioluminescence in marine bacteria, cell to cell communication in bacteria, now a days, is considered to play an inevitable role in the life cycle of various bacterial species studied. This is evident by large volumes of reports that came out in past few years suggesting that bacteria not only form well organized communities but also exchange information with other members of the community in order to coordinate their activities [13-18]. Given a diversity of biological functions are regulated by QS systems and the widespread effects these have upon human health and agriculture, it is not surprising that the field has attracted significant interest in past few years [19]. Worthy of particular note is the possibility that quorum sensing systems may offer novel therapeutic targets for the treatment of a variety of bacterial infections [20]. Thus, extra-cellular signaling seems to provide a new basis for control of molecular and cellular processes as well as population behaviour, probably in a way more reliable over that of native physiology. QS study in bacteria might work as a foundation upon which highly complicated intracellular communication as found in higher organisms has evolved. Considering it as a foundation stone of signaling architecture the aim of this chapter is to outline the recent breakthroughs in the areas of QS with signaling molecules and pathways studied.

SIGNALING MOLECULES

The chemical signal molecules produced are termed as pheromones or sometimes autoinducers (AI), the concentration of which is a function of bacterial population/cell density. Chemically, these can be categorized into two main classes: (i) amino acids and short chain peptides and, (ii) derivatives of fatty acids [18]. However, molecules such as 3-hydroxypalmitic acid methyl ester (3-OH-PAME), 3,4-dihydroxy-2-heptylquinoline and a furanosyl borate diester can also act as QS signals in Gram negative bacterial population [21, 22]. Bacteria, belonging to genera from a wide variety of environmental niches ranging from marine and freshwater environments to soil, plants and animals, including many pathogens, symbionts, extremophiles and plant-growthpromoting bacteria are known to produce AHLs [23]. Figure 1 shows some representative signaling molecules and their structures produced by different bacterial species.

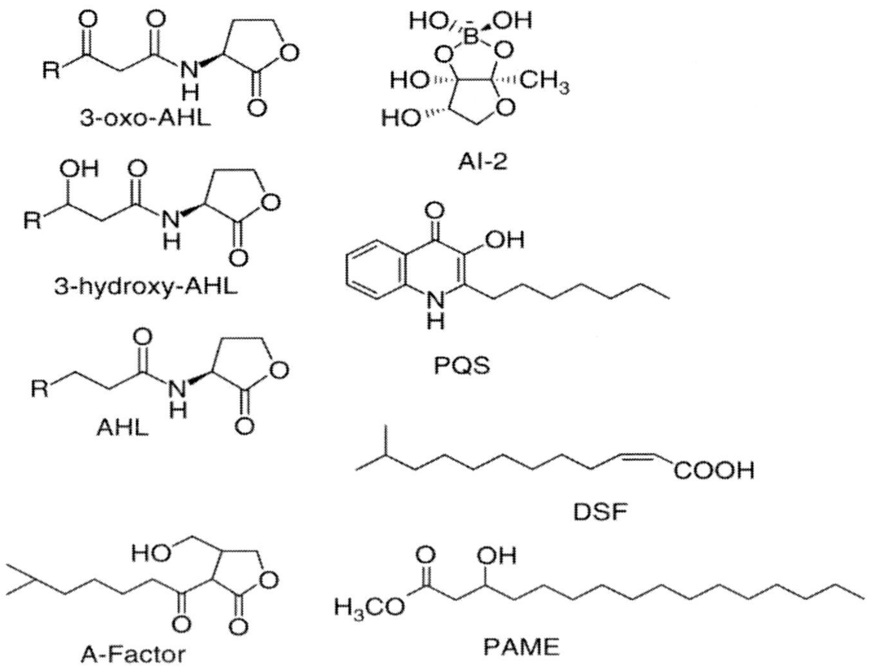

Figure 1. Some popular signalling molecules and their structures. 3-oxo-AHL, N-(3-oxoacyl)homoserine lactone; 3-hydroxy-AHL, N-(3-hydroxyacyl)homoserine lactone; AHL, N-acylhomoserine lactone wherein R ranges from C1 to C15; A factor (double bond in acyl side chain), 2-isocapryloyl-3-hydroxymethyl-γ-butyrolactone; AI-2, autoinducer-2 (furanosyl borate ester); PQS (quinolone signal from *Pseudomonas* spp.),2-heptyl-3-hydroxy-4(1H)-quinolone; DSF, diffusible signal factor, methyl dodecenoic acid; PAME, hydroxyl-palmitic acid methyl ester.

While AHLs and other substituted γ-butyrolactones are synthesized by Gram negative bacteria, certain oligopeptides and substituted γ-butyrolactones are the primary signal molecules studied in Gram positive bacteria [24]. The signal molecules used by Gram-negative bacteria may vary from AHL signals in different bacterial species to 3-OH PAME of *Ralstonia solanacearum*, AI-2 from *Vibrio harveyi*, bradyoxetin from *Bradyrhizobium japonicum* and 2-heptyl-3-hydroxy-4-quinolone from *Pseudomonas aeruginosa* [25]. However, *Salmonella*, which does not synthesize AHLs, nevertheless has a receptor (designated as SdiA) for them that enables it to detect and respond to nearby AHL-producing bacteria [26]. Although, AI-2 is produced by some Gram-positive bacteria also, generally they produce linear, modified or cyclic peptides as signal molecules, such as autoinducing peptides (AIPs) produced by staphylococci [23]. AI-2 is known to be produced by numerous bacterial species, and later the gene (designated as *luxS*) concerned with AI-2 biosynthesis was demonstrated to be present in over 70 bacterial species including *V.harveyi*, *E.coli*, and *S.typhimurium* [27, 28]. Based on these findings, it was proposed that AI-2 is a nonspecies-specific autoinducer that mediates the intra- and inter-species communication between Gram-positive and Gram-negative bacteria and is thus, a universal signaling molecule for interspecies communication [22, 29]. AI-3 is another signal molecule known that activates transcription of virulence genes and controls virulence in enterohaemorrhagic *E.coli* O157:H7. This bacterial signal-response system was originally defined through a role in interkingdom signaling as it is required for bacterial responses to the eukaryotic hormones epinephrine and norepinephrine [30]. However, the most widely studied signal molecules involved in QS are the AHLs [15, 18].

About 10 to 20% of the cultivable bacterial populations from bulk soil as well as from the rhizosphere are believed to be AHL producing [31]. Cha et al. successfully demonstrated that majority of plant-associated bacteria produce AHL signal molecules, especially, they showed that almost all isolates from the genera *Agrobacterium*, *Pantoea* and *Rhizobium* as well as nearly 50% of erwinias and pseudomonads could synthesize detectable levels of AHLs [32]. Among them, QS is involved in the regulation of antibiotic biosynthesis, extracellular enzymes, anti-fungal production, plasmid conjugation, biofilm formation, virulence factors and rhizosphere gene expression [33, 34]. Given that a large proportion of the root colonizing bacteria produce AHL signals, it is not surprising that these pheromones not only function as population density sensors of just one species but also participate in cellular communication between different species [22]. Structurally, AHLs may be saturated or unsaturated and mainly vary with respect to the length (4-14 carbons) and the substituent (H, O or OH) at the third carbon of the acyl-side chain [35]. However, AHLs comprising of 16 to 18 carbons have been isolated from *Rhodobacter capsulatus* and *Sinorhizobium meliloti* [35, 36].

QS IN THE RHIZOBACTERIA

QS in *Pseudomonas* spp.

P.aeruginosa is the model organism for QS study in Gram negative bacteria and most research is focused on this opportunistic human pathogen; environmental isolates can infect humans, undergo rapid adaptation, and cause nosocomial pneumonia, sepsis in burn wounds, urinary-tract infections and chronic pulmonary inflammation in hosts rendered susceptible by cystic fibrosis [37]. Two closely linked QS systems operational in *P.aeruginosa* are the *las* and the *rhl* systems (Figure 2) [38]. The *las* system is comprised of AHL synthase LasI that directs the synthesis of signal molecule, N-3-oxo-dodecanoyl-homoserine lactone (3-oxo-C12HSL) and transcriptional activator LasR. Similarly, *rhl* system consists of AHL synthase Rhll which encodes for synthesis of N-butanoyl-homoserine lactone (C4HSL) and transcriptional activator RhlR [13, 39]. A third LuxR homologue termed QscR has been identified, which is shown to regulate the transcription of both *lasI* and *rhlI* [40]. LasR is known to positively regulate the expression of extracellular virulence factors and is functionally homologous with LuxR in *V.fischerii* [13]. Similarly *rhl* system affects the expression of a broad spectrum of genes. Some of the QS genes are regulated via both *las* and *rhl* QS systems while others are controlled exclusively by either *las* or *rhl* system. However the two systems do not operate independently since LasR-HSL complex has been found to positively regulate the transcription of r*hl*R and r*hl*l [41]. Interestingly, the LasI encoded signal molecule 3-oxo-C12HSL, has been shown to prevent the binding of RhlI encoded signal molecule, N-butanoyl-homoserine lactone, with its cognate receptor RhlR. Therefore, it is speculated that regulation of *RhlI-RhlR* circuit by *LasI-LasR* system ensures the two systems operate sequentially in an appropriate order.

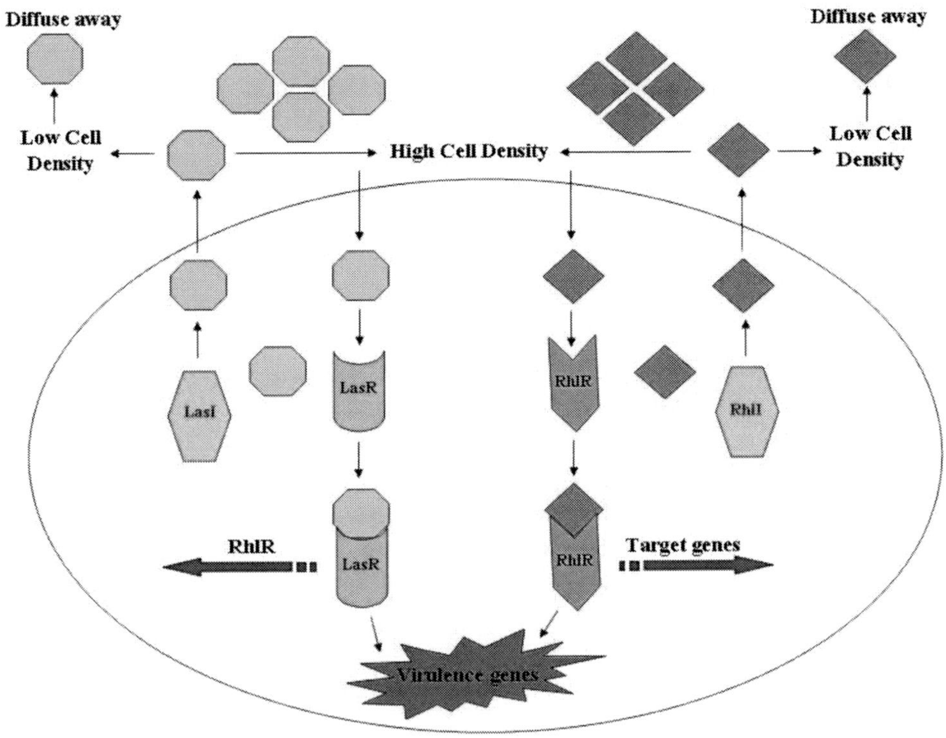

Figure 2. The LasI/LasR-RhlI/RhlR quorum sensing system of *Pseudomonas aeruginosa*. The two quorum sensing systems regulate a variety of functions in *P.aeruginosa*. LasI is a LuxI homologue protein which produces signal molecule, N-(3-oxododecanoyl)-homoserine lactone. After a minimum threshold level is achieved, the signal molecule binds with the LasR protein. This LasR-signal complex binds with the promoter elements of virulent genes and initiates their transcription. Moreover, LasI-signal complex also induces the transcription of rhlR and stimulates another QS system. Binding of RhlI induced signal molecule with the RhlR leads to transcription of a subset of LasR regulated virulent genes. Additionally, RhlR-signal complex also controls expression of several target genes which are not directly controlled by LasR. However, LasI encoded signal molecule prevents the binding of RhlI generated signal with its cognate receptor RhlR.

The other pseudomonads studied for QS include *P.chlororaphis*, *P.aureofaciens*, *P.fluorescens*, and *P.putida*; some of these are well known for their ability to colonize the plant related niches, such as rhizosphere whereby they act as plant growth promoting rhizobacteria (PGPR) by antagonizing the plant pathogens and through production of certain molecules that influence the plant disease resistance and growth in direct or indirect manner [42-44]. *P.fluorescens* NCIMB 10586 produces a mixture of antibacterial pseudomonic acids (PAs) mediated through polyketide synthases (PKSs), of which muciporin is an important component [45]. Muciporin blocks protein synthesis by targeting isoleucyltRNA synthetase of Gram-positive bacteria including methicillin-resistant *Staphylococcus aureus*. Biosynthesis of muciporin depends on QS-dependent regulation via the LuxI/LuxR homologues MupI/MupR. Hothersall et al. investigated the factors which affect the mup gene expression and identified the signal molecule produced by MupI as 3-oxo-C12HSL [46]. However, unlike other cases wherein exogenous addition of QS signaling molecule has been previously used to increase antibiotic production, such as 2,4-DAPG production by *P.fluorescens* S272, no muciporin production was observed when 3-oxo-C12HSL was supplied exogenously. Mutation studies confirmed that *mupR* expression in trans to wild type and mutants can increase production of antibiotic several folds in presence of an amidase/hydrolase, which can degrade 3-oxo-C12HSL. In pseudomonads, phenazines are the most extensively studied group of QS regulated biocontrol related antibiotics. de Maeyer et al. evaluated forty strains of fluorescent *Pseudomonas* isolated from white and red cocoyam roots (*Xanthosoma segittifolium* (L.) Schott) for their ability to synthesize acyl-HSLs [44]. The two varieties differ in their susceptibility to the cocoyam root pathogen, *Pythium myriotylum*, former being highly susceptible while latter being somewhat tolerant to this disease. It was found that only isolates from red cocoyam rhizosphere which synthesized phenazines and antagonized the pathogenic fungus produced acyl-HSLs suggesting that acyl-HSL production is related to the antagonistic activity of the strains. Chromatographic studies coupled with mass spectrometric analysis (LC-MS/MS) revealed the chemical nature of signal molecules and confirmed the presence of two QS systems for phenazine production in strain *Pseudomonas* CMR12a. Similarly, Wei and Zhang identified a QS system consisting of PcoR and PcoI of the LuxR–LuxI family in *P.fluorescens* strain 2P24 that was isolated from wheat take-all decline soil in China [47]. Molecular studies showed that strain 2P24 employs a QS system to regulate its biocontrol activity as mutants deficient in *pcol* were significantly defective in biofilm formation, colonization on wheat rhizosphere and biocontrol ability against wheat take-all, however complementation of *pcoI* restored the biocontrol activity to the wild-type level. Later work revealed that the PcoI/PcoR QS system in strain 2P24 is regulated by PhoP/PhoQ two-component system [48]. More recently, Alavi and coworkers have demonstrated that *Stenotrophomonas maltophilia* strain R551-3 requires a diffusible signal factor (DSF) to exert beneficial effects on host plants [49]. A comparative study of the wild-type strain with a mutant deficient in the *rpfF* (regulation of pathogenicity factors) gene that is essential for the synthesis of DSF revealed that numerous genes known to play a role in plant colonization (e.g., chemotaxis, cell motility, biofilm formation, multidrug efflux pumps) are controlled by the rpf/DSF system in *S.maltophilia*.

AHL mediated QS has also been studied in plant pathogenic *P.syringae* pv. *syringae* that causes brown spot in bean [50]. *P.syringae* strain B728a produces AHL that interacts with AhlR and this whole AHL system plays an important role in cellular aggregation and ecological fitness of the organism during *in planta* growth and disease. A *luxI* homolog derived from *P.syringae* has been identified and mutants with impaired *luxI* homolog resulted in loss of acyl HSL production thereby alteration in colony morphology and loss of viability on plant surfaces. In *P.chlororaphis* strain PCL1391 that inhibits the tomato foot and root rot caused by *Fusarium oxysporum* f.sp. *radicislycopersici*, production of antifungal compound, phenazine-1-carboxamide (PCN) is regulated by AHL QS system in a density dependent manner via PhzI and PhzR system analogous to LuxI/LuxR [51]. A broad survey carried out by Elasri et al. showed heterogeneity in AHL QS systems in soil pseudomonads and suggested that AHL production was more common among plant associated bacteria than free-living soil isolates, indicating the importance of AHL QS in plant-bacterial interactions [52]. Certain *P.putida* strains not only produce secondary metabolites with antagonistic activity but also exhibit plant growth promotory activities and degradation of toxic organic compounds [53]. These properties make *P.putida* an attractive candidate for agricultural and environmental applications.

TraI/TraR* Virulence System in *Agrobacterium Tumefaciens

A.tumefaciens is a pathogenic soil α-proteobacterium which induces crown gall tumors in a broad range of dicotyledonous plants via transfer of oncogenic DNA fragment, the T-DNA of its tumor inducing (Ti) plasmid into the nuclear genome of host plants [54, 55]. The T-DNA genes encode for synthesis of plant growth regulators (auxin and cytokinins) that results in uncontrolled cell proliferation and development of a plant tumor, wherein intercellular spaces are colonized by the pathogen. Further, T-DNA genes also encode for synthesis of specific compounds in transformed plant cells, called opines, which are used by the pathogen as a source of carbon and nitrogen. Additionally, some opines are known to stimulate the synthesis of a bacterial QS signal, 3-oxo-octanoylhomoserine lactone (3-oxo-C8HSL), which in turn controls amplification of the copy number of the Ti plasmid in *A.tumefaciens*, and its horizontal transfer by conjugation [56]. Thus, it is envisaged that QS contributes to *A.tumefaciens* aggressiveness leading to aggravated virulence symptoms in plant hosts [57]. The Ti plasmid also confers to the non-Agrobacterial hosts the capacity to assimilate opine and, in some instances, to induce tumours on the plant hosts; it also remains transferable to other bacteria [58].

In *A.tumefaciens* QS is controlled by TraR, a LuxR type AHL receptor which responds very strongly towards 3-oxo-C8HSL for its biological activity [59]. This AHL is synthesized by TraI that is homologous to LuxI [60]. Both TraI and TraR, the regulatory components of QS in *A.tumefaciens* are situated on the transmissible Ti plasmid [61]. Two sensory signals, a host opine signal and the host HSL autoinducer 3-oxo-C8HSL which is a metabolic product of bacterial enzyme coded by TraI are required for conjugal transfer of plasmid between *A.tumefaciens* cells [59]. Transfer of octopine type Ti plasmid is induced by octopine while transfer of nopaline type Ti plasmid is induced by agrocinopines [33]. Like other LuxI/LuxR type QS systems, *A.tumefaciens* QS also comprises another component that negatively modulates the activity of TraR and 3-oxo-C8HSL and this component is the Ti plasmid-encoded protein TraM which can suppress TraR transcriptional activity. For a long time, it was believed that TraM proteins are not related to any other proteins found in the databases, but recent studies on the *P.aeruginosa* QslA protein contradicted this view, suggesting that TraM-type functions might be relatively common in bacteria [62]. Further studies strengthened the negative regulatory functions exerted by TraM on QS. For example, TraM was later found to block TraR activity even after the transcription factor was bound to DNA. More recently TraM has also been demonstrated to promote TraR proteolysis [63, 64].

Haudecoeur and Faure elaborated a study on fine control of QS communication in *A.tumefaciens* strain C58. Strain C58 expresses two lactonases, AttM (now BlcC) and AiiB, that cleave 3-oxo-C8HSL and are potential modulators of QS [65]. In *A. tumefaciens* strain C58, two lactonase-encoding genes, *attM* and *aiiB* are borne on At and Ti plasmids, respectively [66]. Both AttM and AiiB cleave the gamma-butyrolactone ring of a wide spectrum of QS signals [67]. Purified AiiB, which is highly similar to QS-signal lactonase AiiA from *Bacillus thuringiensis*, preferentially cleaves the QS signals with an acyl chain longer than four carbons, such as hexanoyl-, octanoyl- and decanoylhomoserine lactones, while AttM efficiently cleaves the simplest gamma-butyrolactone (GBL) into gamma-hydroxybutyric acid (GHB) that can be used as a carbon source by *A.tumefaciens* [68]. Other studies have shown the direct contribution of lactonases AttM and AiiB in the control of 3-oxo-C8HSL level and QS-regulated functions such as conjugation of Ti plasmid and seriousness of plant symptoms [69]. However, expression of the two lactonases is also regulated by different plant signals, such as gamma-aminobutyrate (GABA) and salicylic acid suggesting a putative role of plant derived signal molecules in communication between plants and bacteria [57, 70].

It seems in *A.tumefaciens*, the TraI/TraR system and other QS-regulated genes are well conserved in all nopaline-and octopine-type strains studied, indicating that this regulatory mechanism has been anciently selected [55]. The target genes of *A.tumefaciens* QS are involved in the dissemination of Ti plasmids (both by replication and conjugation) and also in positive and negative feedback controls with the 3-oxo-C8HSL-synthesis TraI enzyme and the TraM antiactivator. Different studies have shown that TraM plays a critical role in implementation of the QS, even if it is not clear whether TraM is more relevant in delaying QS activation or in stabilizing and limiting QS activity.

The *CarI/CarR* and *ExpI/ExpR* Virulence/Antibiotic System in *Erwinia* spp.

Species of genus *Erwinia* have been recognized primarily as plant pathogens and belong to one of the three principal groups: *Erwinia carotovora* subspp. *amylovora* (Eca), *E.carotovora* subspp. *carotovora* (Ecc) and *E.carotovora* subspp. *herbicola* (Ech); these cause various diseases of economically important crops viz., potato, onion, cucumber, carrot and other plant hosts. Ecc is a pathogenic bacterium which causes soft-rot in potato. The pathogenicity of this bacterium is attributed mainly to the secretion of plant cell wall degrading enzymes (PCWDEs) such as cellulase, proteases, polygalacturonases and several pectinases that macerate plant cell walls [71, 72]. Extra-cellular secretion of these enzymes is crucial to the pathogenic success of bacterium as evident in some reports wherein disruption of genes encoding individual exo-enzymes has resulted in significant reduction of virulence *in planta* [73, 74]. The exo-enzyme production in *E.carotovora* is also cell density dependent and relies upon synthesis of HSL, through a pair of cognate receptors CarI/CarR homologous to LuxI/LuxR [2, 74]. In addition to PCWDEs, certain strains of *E.carotovora* have been found to produce detectable amounts of 1-carbapen-2-em-3-carboxylic acid (carbapenem), an antibiotic of β-lactam group [2, 75]. Available evidence suggests that antibiotic production by *Erwinia* spp. is imperative for competitive exclusion of sensitive strains of *Erwinia* spp. at the site of infection [76].

The *Ecc* and *Eca* strains mainly synthesize two different types of AHSLs and have been divided into two groups based on these [77]. Class I, including *Ecc* strains SCC3193 and EC153, synthesize mainly 3-oxo-C8HSL, while class II, including *Ecc* strains ATCC39048, GS101, ATTn10 and MS1, Ecc71 and *Eca* strain SCRI1043, synthesize predominantly 3-oxo-C6HSL [77, 78]. However, the primary AHSL, 3-oxo-C8HSL, is produced by LuxI like signal synthase CarI. Mutants deficient in *carI* are unable to produce carbapenem, pectinases, endoglucanases and other hydrolases, thus they are completely non-pathogenic [79, 80]. The Car QS system regulates the gene expression for production of antibiotic carbapenem through *carA-H* biosynthetic operon [81]. Molecular studies revealed that mutants defective in either *carR* or *carI* were unable to synthesize carbapenem, while mutations in *carI*, but not in *carR* affected extracellular production of enzymes [75, 76]. However, disruption of *carR* had no effect on extracellular enzyme production by *E.carotovora* suggesting that *carR* is not responsible for HSL mediated regulation of enzymes [76]. There are several other species of

Erwinia which produce various acyl HSLs viz., *E.herbicola, E.chrysanthemi, E.carotovora* subspecies *atroseptica* and *betavascolorum* [2, 33]. Nasser et al. reported that *E.chrysanthemi* strain 3937 produces three different AHLs, N-(3-oxoctanoyl)-homoserine lactone (OHL), N-hexanoyl-L-homoserine lactone (HHL) and N-decanoyl-L-homoserine lactone through a homologous set of linked *expI* and *expR* genes [82]. Interestingly, disruption of either *expI* or *expR* has very little effect on extracellular production of pectolytic enzymes *in vitro* in *E.chrysanthemi* strain 3937 [82]. Therefore it is not clear, what function ExpI/R QS regulatory system may have in *E.chrysanthemi*? The QS system of *Erwinia stewartii* subspecies *stewartii*, now known as *Pantoea stewartii* subspecies *stewartii* also consists of LuxI homologue, EsaI, that directs the synthesis of OHL, and LuxR homologue EsaR [83, 84]. The bacterium causes Stewart's wilt in sweetcorn and leaf blight in maize and produces a virulent factor called as stewartan (an exopolysaccharide) in a cell density dependent manner. In large quantities, this compound can cause blockage in xylem vessels of plants, with subsequent wilting. Mutants deficient in *esaI* failed to produce stewartan and showed reduced virulence *in planta* [84]. However, mutants with impaired *esaR* gene, showed overproducing supermucoid phenotypes, suggesting that function of EsaR is to suppress production of stewartan when population density of *P.stewartii* subspecies *stewartii* is low [84]. As expected, at higher cell densities this suppression was relieved.

In *E.carotovora* both carbapenem and exoenzymes are produced simultaneously, so it is presumed that exoenzymes produced, damage the plant cell wall while antibiotic keeps the competing bacteria away that attempt to invade the plants by taking advantage of the wound produced by the exoenzymes of *E.carotovora*. Furthermore, several plants are known to mount a defense response when they sense breakdown products of their own cell wall [85, 86]. Therefore, it is suggested that QS controls the biosynthesis of extracellular enzymes in a way that it may serve to conceal the *E.carotovora* populations by preventing the elicitation of host response until a threshold population density is attained.

QS in *Rhizobium* spp.

Rhizobium species are best known for symbiotic association with leguminous plants and their ability of nitrogen fixation in the root nodules of legumes [87, 88]. The symbiosis is characterized by exchange of signals between rhizobia and host legume [88]. The rhizospheric environment of legumes exert a strong attractive power on rhizobia due to the presence of abundant polycyclic aromatic compounds such as flavonoids, sugars, dicarboxylic acids and amino acids which trigger a chemotectic response and directs the bacteria towards their compatible host [89]. Indeed, signal exchange between rhizobia and host legumes in the rhizosphere relies on the production of signal molecules such as AHLs which allow the whole bacterial population to set off a concerted action as soon as a threshold population density is achieved [90-92]. In fact, the chemical molecules responsible for this symbiotic association are encoded by the genes present on, so called symbiotic (Sym) plasmids carried by *Rhizobium* species. Loh et al. reported a new class of QS molecules in *Bradyrhizobium japonicum* involved in symbiotic gene regulation [93]. Recent studies have shown the production of two new signaling molecules by bacteria of the *Bradyrhizobium* genus: cinnamoyl-homoserine lactone (an aryl-HSL) in photosynthetic stem-nodulating bradyrhizobia and isovaleryl-homoserine lactone (a branched-chain fatty HSL) in the soybean symbiont *Bradyrhizobium japonicum* USDA 110 [94, 95]. More recently, Nievas et al. identified and characterized QS signals produced in 53 peanut-nodulating bradyrhizobial strains along with biosensor strains *A.tumefaciens* NTL4 (pZLR4) and *Chromobacterium violaceum* CV026 and they also studied their effects on bacterial motility and processes involving cell-cell interaction, such as aggregation and biofilm formation [96]. Similarly, *R.etli* strain CNPAF512 has been found to produce seven different signal molecules as part of QS system, of which some are produced by *cinI/cinR* system homologous to *luxI* and *luxR* [97]. Gene *cinI* encodes for the enzyme AHL synthase that directs the synthesis of signal molecule, N-(3-hydroxy-7-cis-tetradecenoyl)-HSL while *cinR* codes for transcriptional regulator that binds with AHL. Expression of both the genes *cinI* and *cinR*, is a function of population density and are expressed under symbiotic conditions. Plants nodulated with mutant strains deficient in gene *cin* were found to be limited in nitrogen fixation capacity, most likely because of arrested bacteriod differentiation [90].

Nevertheless, the most widely studied species is *R.leguminosarum* and information about the QS system of this bacterium is limited even today. However, from the available evidence, there seems to be present several LuxI/LuxR homologues. A plasmid encoded *luxI* homologue, *rhiI* has been characterized which is regulated by *cinR/I* [98]. The enzyme coded by *rhiI* directs the biosynthesis of signal molecules, such as HHL, OHL, and an acyl HSL molecule which is supposed to contain a seven carbon acyl moiety as side chain [99]. Regulation of *rhiI* is under the control of *rhiR* encoded by the Sym plasmid. The acyl HSLs produced by *rhiI* in conjunction with some other uncharacterized acyl HSL molecules activate the RhiR in *R.leguminosarum* [98]. *rhiR* regulates the gene expression within *rhi*ABC operon [98, 100]. Intriguingly, the gene products from this operon are expressed in the rhizosphere very efficiently, but fail to express in differentiated bacteroids of root nodules and thus their function is unclear [100, 101].

Sinorhizobium meliloti is another member of group rhizobia, frequently observed as a symbiotic partner of *Medicago sativa* and *Medicago truncatula* [24]. *S.meliloti* has been found to produce a diverse array of compounds with acyl-HSL like activity [32]. At high cell densities the QS signal molecules, long chain AHLs accumulate and at threshold concentration triggers responses in the population affecting root colonization and nodule invasion efficiency of the bacterium [102]. However, it has been observed that AHLs produced by *S.meliloti* exert a strong impact on the proteome of *M.truncatula,* affecting many physiological functions such as protein processing, host defense, transcriptional regulation and other related activities [103, 104]. But the plant itself has been found to produce QS mimics which transform the normal bacterial response in the rhizosphere [105]. Unexpectedly, exposure of the plant roots to bacterial AHL molecules has been found to alter the types of AHLs secreted by *M.truncatula* [104].

Even today, it is not certain which cellular processes are regulated by QS system in *R.leguminosarum*. For example, there are certain reports that show the inhibition of nodule formation by RhiI, as mutants with impaired *rhiI* gene result in very slight increase in nodule formation compared to the wild type [98]. In another case, the Sym plasmid of an isolate of *Rhizobium* was found to carry open reading frames which shared a high level of sequence homology with the conjugal transfer genes of *A.tumefaciens* suggesting that regulation of conjugal transfer of Sym plasmid by *Rhizobium* spp. may be similar to that of conjugal transfer of Ti plasmids in *A.tumefaciens* [106]. Certainly, the acyl HSL production is widespread in other rhizobial species and a rich diversity of AHLs is produced by *Rhizobium* species such that many signal molecules are common to different species [32].

Cell to Cell Communication in *Serratia* spp.

The genus *Serratia* includes species, many of which are opportunistic pathogens, colonizing a variety of ecological habitats such as water, soil, plants, insects, fishes and humans [107]. *S.liquefaciens* MG1 is a motile bacterium, which forms swarming colony on solid surfaces such as agar medium. Swarming movement of the colony is due to the swarming action of specialized cells, called as swarmer cells present on the edge of the bacterial colonies. According to a report, at least 28 genes are under the control of *swrI/R* QS system in *S.liquefaciens* [108]. A lipodepsipentapeptide biosurfactant, known as serrawettin has been found essential for the swarming movement of the bacterium. Production of serrawettin is under the control of SwrA, a multidomain enzyme complex, which is QS regulated [109]. Gene *swrI* is a *luxI* homologue that encodes for an enzyme AHL synthase which synthesizes N-butanoyl-L-homoserinelactone (BHL) while *swrR* is a *luxR* type transcriptional activator. At high cell densities, when the AHL concentration reaches a threshold level, the SwrR/BHL complex activates the transcription of target genes such as, *swrA*, leading to the synthesis of biosurfactant which is necessary for swarming motility of the organism [110, 111]. It has been found that disruption of *swrI* resulted in substantial loss of swarming ability of the strain [112]. However, this swarming ability could be restored by the addition of exogenous AHLs. Further, inactivation of *swrI* neither affected the growth rate or swimming motility nor the development of hyperflagellation and cell elongation [112]. These findings clearly suggest that in *S.liquefaciens* MG1 production of biosurfactant serrawttin is regulated by QS system; mutants deficient in acyl-HSL production are therefore incapable of exhibiting swarming movement. More recently, Liu et al. characterized two AHL-mediated QS systems in the wheat stem endophyte *Serratia plymuthica* G3 and showed that QS network was involved in the global regulation of biocontrol related traits in the endophytic strain G3 [113]. Treatment with *Bacillus* A24AiiA lactonase in strain G3 resulted in altered adhesion, biofilm formation and antifungal activity. These results showed that in strain G3 the antifungal activity and production of exoenzymes was positively regulated through QS, however indol-3-acetic acid (IAA) production was negatively regulated.

Cox et al. have reported the production of an antibiotic molecule, carbapenem from *Serratia* spp. ATCC39006 [114]. Just as in case of *Erwinia carotovora* subspecies *carotovora* (Ecc), the production of carbapenem in *Serratia* spp. ATCC39006 is also believed to be regulated, by an acyl-HSL independent LuxR homologue [2]. But further analysis of the different acyl HSLs produced by this organism revealed the presence of at least two acyl HSLs, BHL and HHL, which are products of gene locus *smaI* [115]. Therefore, it is apparent that expression and regulation of carbapenem by *Serratia* spp. ATCC39006 is under the control of QS system. Further, genetic studies showed that carbapenem synthesis and resistance genes of *Serratia* spp. ATCC39006 and car genes of Ecc shared a high level of sequence homology [115]. Interestingly, the two organisms employ different acyl HSLs to activate their respective CarR enzymes. Thus it is of particular interest to analyze the transcriptional activators of these organisms in order to understand how LuxR homologues differentially respond to their cognate acyl HSLs.

Bacterial Cross Talk in *Ralstonia* spp.

Ralstonia solanacearum is a plant pathogenic bacterium that causes vascular wilt of a wide range of plant species including tobacco, tomato, potato and bananas [116]. The pathogen infects the plant through wounds or other openings. It penetrates the cortex, transcends through the endodermis and reaches the vascular system. Once inside the vascular system, it starts multiplying and colonizes the entire plant. Inside the cortex, the pectolytic enzymes produced by the pathogen disrupt the middle lamella, thus creating channels for distribution of bacteria to other parts and helps in the release of nutrients from host cells [117, 118]. Eventually, the plant undergoes wilting due to obstruction caused by accumulation of bacteria and EPS in the xylem. Expression of these virulent factors is regulated by a Phc (phenotype conversion) regulatory system in a cell density dependent manner [117, 118]. The LysR type transcriptional regulator, PhcA is central to the regulation of pathogenicity factors (EPS and extracellular enzymes) in *R.solanacearum* and its activity is modulated by a volatile signaling molecule, 3-OH PAME [119, 120]. 3-OH PAME acts as a signal molecule for a two component regulatory system comprising of a membrane bound sensor-kinase, PhcS and a response regulator, PhcR [121]. The gene products of PhcS and PhcR, post-transcriptionally modulate the activity of PhcA. However, PhcR is considered as unusual in a way that its putative output domain resembles the histidine kinase domain of a sensor protein [2]. Studies have shown that in absence of 3-OH PAME, PhcS and PhcR, operate together to negatively regulate the gene expression of PhcA regulated genes [119]. It is believed that at low population density coupled with low conc. of 3-OH PAME, PhcS phosphorylates PhcR, which represses the expression of phcA. But, as soon as the concentration of 3-OH PAME reaches a minimum threshold level, it suppresses the ability of PhcS to phosphorylate the PhcR, thereby resulting in increased expression of PhcA and subsequent production of PhcA governed virulence factors [122]. This indicates that unphosphorylated PhcR acts as a negative regulator of Phc phenotype at subthreshold concentration of 3-OH PAME and phosphorylation of PhcR subsequently inactivates it. *phcA* mutants are incapable of producing EPS, pectin methyl esterase and endoglucanase but they are hypermotile and show increased production of polygalacturonase and siderophore [2, 33]. As a result *phcA* mutants of *R.solanacearum* are almost avirulent. This is because PhcA directly regulates the production of endoglucanase and pectin methyl esterase.

Homologues of *luxI* and *luxR* have also been identified in *Ralstonia* spp. and are designated as *solI* and *solR*. Disruption of *solI* inhibits the biosynthesis of acyl HSLs, however production of EPS and exoenzymes remains unaffected, so is the pathogenicity of the bacterium [119]. Intriguingly, the expression of *solI/solR* is regulated by PhcA and RpoS [123]. In case of *R.solanacearum*, QS seems to play an important role since it helps in making transition from a free-living state in soil to a parasitic form. At low concentration of 3-OH PAME, the bacterium exhibits reduced production of EPS and exoenzymes but increased motility and siderophore production, a trait typical of free-living form in soil [124]. However, increased level of 3-OH PAME promotes the PhcA activity, hence increased biosynthesis of EPS and exoenzymes, traits desirable for pathogenic life. Conversely, this reduces motility and siderophore production. Thus, Phc regulon, in a way acts as a regulatory switch that determines the transition from behaviour suited for free living form in soil to development of traits for pathogenesis. However, AHL-acylase (Aac), a quorum quenching enzyme has also been reported from *R.solanacearum* GMI 1000 that can control AHL-mediated pathogenicity [125].

Cellular Communication in *Xanthomonas Campestris*

X.campestris pv. *campestris* (Xcc) is a plant pathogen that causes black rot of cruciferous plants [126]. The pathogen enters into the vascular system through hydathodes and multiplies in the xylem ultimately leading to the vascular wilt due to the obstruction in the xylem caused by bacterial cells and xanthan gum, the main component of bacterial EPS. Just as in case of *R.solanacearum*, expression of virulent traits (exoenzymes and EPS) in Xcc is regulated in a cell density dependent manner via two QS-like systems through diffusible signaling molecules [127, 128]. Besides AHL-type of QS signals and AI-2, the diffusible signal factor (DSF) family signals have also been detected in a variety of bacterial species, including Xcc, *X.oryzae* pv. *oryzae*, *Xylella fastidiosa*, *Stenotrophomonas maltophilia*, and *Burkholderia cenocepacia* [129-133]. However, distinct DSF family molecules can be produced by different bacterial species. For example, a DSF from Xcc has been characterized as *cis*-11-methyl-2-dodecenoic acid while a DSF-like signal molecule, designated BDSF, has been characterized as *cis*-2-dodecenoic acid in *B.cenocepacia* [132, 134]. Noticeably, multiple DSF family molecules can also be produced by one organism; for example, *X.oryzae* pv. *oryzae* is known to produce three DSF family signals: DSF, BDSF, and cis,cis-11-methyldodeca-2,5-dienoic acid (CDSF) in rich media [133]. QS mediated through DSF (cis-11-methyl-2-dodecenoic acid) has been linked to the regulation of virulence, motility, toxin production, aerobic respiration, biofilm dispersal, extracellular enzyme and extracellular polysaccharide (EPS) production [129, 135]. It has been found that *rpf* gene cluster is responsible for DSF-dependent QS system, core genes of which are *rpfF*, *rpfC* and *rpfG* [127, 135, 136]. *rpfF* encodes a protein similar to enoyl-CoA hydratase that catalyzes the synthesis of DSF [137]. DSF is sensed by a unique two-component signal-transduction system consisting of RpfC and RpfG, which also regulates the expression of downstream genes [120, 138].

Production of DSF seems to be restricted to *Xanthomonas* species; largely to *X.campestris* strains and the only other xanthomonad known to have Rpf system for production and detection of DSF is *X.oryzae* pv. *oryzae* [139]. Unfortunately, the role of DSF signaling pathway and putative proteins regulated by the DSF-dependent QS system is not completely understood. However, high-throughput identification of DSF-mediated genes has become feasible by microarray technology or 2-DE at the whole-genome/proteome level respectively. For example, microarray analysis has led to identification of 165 genes in Xcc, whose expressions were significantly influenced by DSF [127]. In another study, Schuster et al. identified 315 positively and 38 negatively regulated genes by AHL via microarray technology in *P.aeruginosa* [12]. Similarly, Zhao et al. studied the proteomic analysis of the regulatory function of DSF-dependent quorum sensing in *Xanthomonas oryzae* pv. *oryzicola* (Xoc) and showed that DSF might play an important role in virulence and growth of Xoc by mediating expression of proteins [140]. These authors observed that the *rpfF* mutant lost the ability to produce DSF molecules, and exhibited a significant reduction of virulence in rice compared to the wildtype strain. Further, the mutation of *rpfF* impaired EPS production, and led to Xoc cell aggregation.

Plant Response to Bacterial QS

In response to the host-microbe interactions, plants also have evolved complex and versatile mechanisms to detect and accordingly respond to the presence of bacteria. Besides, they also respond to the bacteria-bacteria communication mediated through QS signals, but the chemistry behind such action is still unclear [104, 141, 142]. However, molecular patterns associated with LPS, peptidoglycan, proteins like flagellin are perceived through specific receptors known as pattern recognition receptors (PRRs) and this perception triggers the primary immune response of plants [24]. Only two PRRs are known in *Arabidopsis,* despite the fact that genome of this plant has been found to possess several hundreds potential PRRs [143, 144]. Intriguingly, rhizosphere of plants offers a spatially structured habitat densely colonized by the diverse bacterial communities. Available reports suggest that higher plants, including pea, rice, tomato, soybean and *M.truncatula*, secrete a variety of compounds which mimic the bacterial QS signals [105]. These signal mimicing molecules may provide plants with important tools to interrupt or control QS regulation in bacteria associated with them. Mathesius et al. reported that *M.truncatula* controls production of over 150 proteins in response to AHLs secreted by *S.meliloti* and *P.aeruginosa*, the two model organisms extensively used for QS studies [104]. These compounds are produced in response to AHLs released by bacteria. Similarly, *Pisum sativum* (pea) is known to produce AHL mimics that govern the AHL regulated behaviour in different bacterial strains [105].

A large number of microorganisms with recognized acyl HSL producing ability are well known for their capacity to associate with higher organisms either in pathogenic or symbiotic relationship. In nature, bacteria and plants as well as their genetically modified derivatives can produce QS signal biomimics including signal modifying enzymes which can interfere in normal QS [145]. Certain signal mimic compounds are known which can inhibit the bacterial responses to an added AHL signal. However, in contrast many of the signal mimics from higher plants also stimulate AHL inducing behaviour [105]. But chemical isolation of these plant-associated compounds has not been successful as yet. Interestingly, the QS signals used by bacteria while they attack plant hosts, may exhibit unintentional side effects that make their hosts alert to impending invasion and subsequently trigger defense responses. This response was found in *M.truncatula* to added AHLs wherein roots of *M.truncatula* were exposed to nanomolar or picomolar concentration of 3-oxo-N-(tetrahydro-2-oxo-3-furanyl)-hexadecenamide (3-oxo-C16:1 HL) from *S.meliloti* resulting in accumulation of more than 150 different proteins with diverse functions [104]. Additionally, secretion of AHL and other signal mimic molecules from root was also altered suggesting that detection of bacterial QS signals by host plants and subsequent triggering of diverse host responses could be a result of interactions in many of the pathogenic and symbiotic associations.

The concentration of AHL mimicking molecules at the plant surface may be high enough that it affects the AHL regulated gene expression in bacteria under natural conditions [105]. For example, the plant may prevent bacteria from concentrating in sufficient numbers to attach to the host successfully by stimulating swarming motion in bacterial colonies [146]. Though the chemical structure of AHL mimics is not clear, it is presumed that compounds which mimic AHL signal molecules are chemically different from bacterial AHLs [105]. A study was carried out on *Serratia liquefaciens* to evaluate the effect of AHL signal mimics and swarming activity of the bacterium was found to be greatly influenced by the compounds secreted by pea seedlings [105]. Moreover, the methanolic extract of seedling exudates stimulated the swarming mobility in *swrI* mutants which are unable to synthesize AHLs of their own. However, pea seedlings as control were unable to stimulate the *swrA* mutant, impaired in its biosurfactant production, suggesting that pea did not secrete a biosurfactant; rather, a signal molecule which induced swarming mobility in the bacterium.

Biocontrol Perspectives and Ecological Implications

Quorum Quenching – Biotechnological Applications

Contrary to the use of acyl HSLs for their own advantage, available evidence indicates that other bacteria may attempt to interrupt cell to cell communication by actively destroying the message through a process known as quorum quenching (QQ) [20,145]. Given the pivotal role QS plays in pathogenic interactions, inactivation or degradation of AHLs presents an attractive target for designing innovative disease control strategies from agriculture application point of view [18,93,145]. Several plant and microbial molecules have been reported with such bioactivity against AHLs, however their function is still speculative [147]. For example, halogenated furanones from marine red algae, *Delisea pulchra* have been found to inhibit the bioluminescence and virulence of *V.harveyi* by competitively binding with the transcriptional factor LuxR [148]. Pea and crown vetch seedlings were found to exude chemical compounds which mimic bacterial AHLs and interfere with AHL induced biosynthesis of violacein in *Chromobacterium violaceum* [105]. Thus, QQ can be used under anti-virulence/anti-disease strategies to develop novel medical and animal therapies or novel biological control strategies for phyto-pathogens [16, 20, 145]. Starting from the generation of signal molecule till its journey to signal receptors, three target sites implying the degradation of QS signals have been proposed: one, targeting the precursors of signal molecules to prevent the generation of signal, second to prevent the diffusion of signal molecule from the site of generation and, third, to stop the binding of signal with the target receptor. Signal generation can be inhibited by using natural/synthetic analogues of precursors of signal molecules. Several bacteria and plants along with their genetically modified descendents have been constructed for research and applications in biotechnology to produce QS-signal analogues or molecules that interfere with QS, including some QS-signal modifying enzymes [16,18,24,145]. However, their function is speculative and needs further evaluation to be implemented for agricultural applications [147].

Preventing the dissemination of signal generated is another way that can be exploited for inhibition of QS. For example, some plants infected by pathogenic *E.carotovora* tend to increase pH at the site of infection resulting in lactonolysis of AHL in which the biological activity of the molecule is lost while there are many bacteria which do so by producing lactonolysing enzymes, AiiA, for example *B.thuringiensis*, *B.cereus* [149,150]. Consequently, the signal molecule fails to escape from the site of generation thereby resulting in inhibition of QS. Targeting the QS signal receptors by QS antagonists is another promising way and several synthetic analogues have been designed which bind with the signal receptors (e.g., LuxR in case of AHLs) leading to the formation of an inactive complex that fails to express the virulence [147]. The most promising strategy for QS disruption at present appears to be the blockage of AHL receptor protein preventing its binding with the signal molecule [13]. Givskov and collaborators reported that certain molecules, identified as halogenated furanones, produced from macroalgae *Delisea pulchra* interfere with the AHL regulated functions in *P.aeruginosa* [151]. It was found that these compounds not only blocked the cellular communication in *P.aeruginosa* biofilm but also caused the sloughing of the biofilm upon prolonged incubation. However the exact mechanism of action of furanones remains to be elucidated.

Several species of soil bacteria have been reported to cause signal interference by enzymatic degradation of AHLs including members of α-proteobacteria (*Agrobacterium, Bosea, Ochrobactrum* and *Sphingopyxis*), β-proteobacteria (*Comamomonas, Delftia, Ralstonia*, and *Variovorax*), and γ-proteobacteria (*Acinetobacter* and *Pseudomonas*) [24]. Dong et al. first showed acylhomoserine lectonase activity (AiiA) in a *Bacillus* isolate from soil that hydrolyzes the lactone ring of AHLs [152]. The first attempt of disease control by disruption of QS involved introduction of aiia gene cloned from *Bacillus* species against *E.carotovora* into transgenic tobacco and potato plants [149]. Expression of this gene and production of AHL-lactonase in transgenic plants resulted in inactivation of QS systems in pathogen and led to significant reduction in disease incidence. Two types of enzymes that inactivate AHLs have been identified in several species/genera of bacteria: the AHL-lactonase which causes lactonolysis (opening of the γ-butyrolactone ring) resulting in acyl-homoserine with reduced biological activity and other is AHL-acylase which breaks the amide linkage of AHLs to produce homoserine lactone and fatty acids with no biological activity [153-155].

Molina et al. demonstrated the efficacy of using wild type soil bacterium with AHL degradation potential for biological control of plant diseases [156]. They isolated a *Bacillus* spp. strain A24 which showed a broad spectrum antagonistic activity and could degrade AHLs produced from pathogenic bacteria *E.carotovora* and *A.tumefaciens*, causal agents of potato soft rot and crown gall of tomato respectively. The disease incidence was significantly reduced and the results obtained were comparable or even better than those obtained using antibiotic producing strains of *P.chlororaphis* PCL1391, *Pseudomonas* spp. PITR2 and *Pseudomonas* spp. Q2-87. Their experiment clearly showed that AHL degradation could be employed as a means to control bacterial diseases and could be effective or be better than that observed with phenazine and 2,4-diacetylphloroglucinol producing bacterial strains. Even when the pathogen was given time to establish and initiate disease symptoms, subsequent application of an AHL degrading strain prevented further disease development thus exhibiting a curative disease control. Similarly, Dong et al. showed that heterologous expression of an *aiia* gene cloned from *Bacillus* spp. strain 240B1 in pathogenic bacterium *E.carotovora* decreased pathogen's own virulence [152]. Thus heterologous expression of AHL degrading genes in plants may confer significant protection against bacterial diseases providing a curative and preventive biocontrol activity. However, biological control based on the application of AHL degradation ability which relies on QS for disease development has its own limits when applied to control fungal pathogens which do not utilize AHL signals. This was confirmed in a study of AHL degrading strains of *Bacillus* spp. A24 and *P.fluorescens* P3/pME6863 which were ineffective against fusarial wilt of tomato [156, 157].

Ecological Significance

It has been clear in many instances that QS plays a critical role in regulation of bacterial pathogenicity; interest in past few years has been increased in designing and implementation of novel antimicrobial strategies which specifically targets the QS [71]. In this series, drugs that antagonize autoinduction are being developed but it is more likely that eukaryotes susceptible to bacterial infection mediated through QS would have already evolved natural ways to impede bacterial invasion through interfering QS regulated functions [18]. Human cells also possess QQ ability. For example primary and immortalized epithelial cell lines of human origin have been shown to specifically inactivate *P.aeruginosa* 3OC12-HSL autoinducer produced by LasI but not C4-HSL, a product of RhlI [158]. This activity opens up new vistas in current scenario of the development of anti-*P.aeruginosa* therapies for treatment of CF. In this way, naturally occurring QQ processes may be targeted to develop novel antimicrobial therapies. For example, over-expression of aiiA in tobacco and potato plants conferred resistance to *E.carotovora* wherein expression of virulence factors is mediated through AHL [149]. In another study, mice treated with synthetic antagonists of bacterial AI developed resistance to bacterial infection [159, 160]. These examples elegantly suggest that QQ can be used as an alternative to the traditional antibiotics since conventional antibiotics kill or inhibit bacterial growth by interfering with essential bacterial housekeeping functions such as DNA, RNA or protein synthesis thus imposing a selection pressure resulting in the emergence of antibiotic resistant microbial pathogens [145].

Indeed bacterial cooperation provides bacteria with advantages and makes those processes feasible which otherwise were not attainable by a bacterium alone. However, it also raises certain questions regarding evolution of cell to cell communication in bacteria, the cost bacteria pay for communication, and how the fidelity is maintained in communication in QS systems [34]. The ecological and evolutionary implications of QS in bacteria are only beginning to be addressed and continued research will possibly provide new insights of the ecological and evolutionary significance of group behaviour of bacterial communities.

CONCLUSION

During the past decade, QS has been studied intensively indicating the complexity of the system more than it was thought initially. QS is believed to play a crucial role in bacterial physiology including regulation of rhizospheric competence factors like antibiotic production, horizontal gene transfer and control of those functions which are directly or indirectly related to plant-microbe interactions. Indeed, plant response to the associated microorganisms or vice versa is controlled by an array of diverse signal molecules in the rhizosphere. However, decoding the intercellular network of communication mediated through signals appears to be a future challenge. Apparently, understanding the cellular communication in bacterial world is fundamental to the science of microbiology including industrial and clinical microbiology. Although QQ has been proposed as potential supplementary means of biocontrol for agricultural developments, ecological consequences of QQ enzymes under *in situ* conditions remains to be assessed. One can easily ask whether the regulation acts under natural conditions where QS is possible? QQ is a generalized mechanism to inhibit QS regulated functions irrespective of the nature of producing species. One can not exclude the possibility that the organism targeted may itself act as a quencher instead of one which is supposed to act like quencher, although such studies are rare. For example, in a study *P.ultimum* caused interference in the pathways involved in biosynthesis of cellular components of the biocontrol bacterium *P.fluorescens* strain F113. The signals released by the pathogenic oomycete *P.ultimum* have been found to down regulate the genes of *P.fluorescens* strain F113, essential for competitive colonization of roots, thus preventing the strain F113 to attain a threshold population density high enough to result in biocontrol of *P.ultimum* in the rhizosphere [161].

The studies presented in this chapter exemplify the intercellular communication among the Gram negative rhizobacteria of same or different species. QS regulation seems to be an important mechanism which controls the transition from normal vegetative/non-pathogenic state to pathogenic form in many bacterial pathogens. For example, in case of *R.solanacearum*, 3-OH PAME-dependent Phc pathway determines the switch over from motile, free living, low virulence state of the bacterium to EPS and exocellular enzyme producing high virulence/pathogenic state. Expression of virulent traits in such cases regulated via QS is crucial to the survival of pathogen since early expression of virulence factors may prove detrimental to the pathogen itself. This is partly due to elicitation of host defense systems of the plants which may get triggered up as the breakdown products (oligogalacturonide) of cell wall released by pectolytic enzymes can also induce the defense system of the plants before the pathogen attains threshold density necessary to overwhelm the local defense barrier and host defense responses. Thus although biochemical mechanisms beneath the AHL mediated QS have been well understood in culture, modes of action of these signal molecules under natural conditions remains to be elucidated and needs the help of advanced molecular tools of transcriptomics, proteomics, metabolomics and metagenomics. Further, the emergence of antibiotic resistance in microbial pathogens highlights the need to develop new strategies in order to prevent/control of infectious diseases. Earlier reports, particularly the research progress over the past few years have shown how individual cells of bacterial pathogens use QS to synchronize their activities among themselves so as to take advantage of microbe–microbe and pathogen–host interactions. Moreover, due to rapid progress in understanding and discovery of novel QQ mechanisms which interfere effectively with microbial QS have been consecutively found in a wide range of organisms, including both prokaryotes and eukaryotes. These naturally occurring QQ mechanisms act by blocking the key steps of QS, such as signal generation, signal accumulation or signal reception. They have promising potential in both basic research and biotechnological applications.

ACKNOWLEDGMENTS

This work was partly supported by the grants received from Madhya Pradesh Council of Science & Technology (MPCST), Bhopal in the form of Silver Jubilee Fellowship (No. 4788/CST/SJRF/2008) to BNJ.

REFERENCES

[1] Fuqua, W. C., Winans, S. C. & Greenberg, E. P. (1994). Quorum sensing in bacteria: the LuxR/LuxI family of cell density responsive transcriptional regulators. *J. Bacteriol. 176*, 269-275.

[2] Whitehead, N. A., Barnard, A. M. L., Slater, H. L. S. N. J. & Salmond, G. P. C. (2001). Quorum sensing in Gram negative bacteria. *FEMS Microbiol. Rev. 25*, 365-404.

[3] Schaefer, A. L., Greenberg, E. P., Oliver, C. M., Oda, Y., Huang, J. L., Bittan-Banin, G., Peres, C. M., Schmidt, S., Juhaszona, K., Sufrin, J. R. & Harwood, C. S. (2008). A new class of homoserine lactone quorum sensing signals. *Nature*, *454*, 595-599.

[4] Nealson, K. H. & Hastings, J. W. (1979). Bacterial bioluminescence: its control and ecological significance. *Microbiol. Rev. 43*, 496-518.

[5] Mok, K. C., Wingreen, N. S. & Bassler, B. L. (2003). *Vibrio harveyi* quorum sensing: a coincidence detector for two autoinducers controls gene expression. *EMBO J. 22*, 870-881.

[6] Milton, D. L., Chalker, V. J., Kirke, D., Hardman, A., Camara, M. & Williams, P. (2001). The Luxm homologue VanM from *Vibrio anguillarum* directs the synthesis of N-(3-hydroxyhexanoyl) homoserine lactone and N-hexanoylhomoserine lactone. *J. Bacteriol. 183*, 3537-3547.

[7] Waters, C. M. & Bassler, B. L. (2006). The *Vibrio harveyi* quorum-sensing system uses shared regulatory components to discriminate between multiple autoinducers. *Genes Dev. 20*, 2754-2767.

[8] Khan, M. S. A., Zahin, M., Hasan, S., Husain, F. M. & Ahmad, I. (2009). Inhibition of quorum sensing regulated bacterial functions by plant essential oils with special reference to clove oil. *Lett. Appl. Microbiol. 49*, 354-360.

[9] Chu, Y.-Y., Nega, M., Wolfle, M., Plener, L., Grond, S., Jung, K. & Gotz, F. (2013). A new class of quorum quenching molecules from *Staphylococcus* species affects communication and growth of gram-negative bacteria. *PLOS Pathog. 9*, e1003654.

[10] Sifri, C. D. (2008). Quorum sensing: bacteria talk sense. *Clin. Infect. Dis. 47*, 1070-1076.

[11] Dunman, P. M., Murphy, E., Haney, S., Palacios, D., tucker-Kellogg, G., Wu, S., Brown, E. L., Zagursky, R. J., Shlaes, D. & Projan, S. J. (2001). Transcription profiling-based identification of *Staphylococcus aureus* genes regulated by the agr and/or sarA loci. *J. Bacteriol. 183*, 7341–7453.

[12] Schuster, M., Lostroh, C. P., Ogi, T. & Greenberg, E. P. (2003). Identification, timing, and signal specificity of *Pseudomonas aeruginosa* quorum- controlled genes: A transcriptome analysis. *J. Bacteriol. 185*, 2066-2079.

[13] Juhas, M., Eberl, L. & Tummler, B. (2005). Quorum sensing: the power of cooperation in the world of *Pseudomonas*. *Environ. Microbiol. 7*, 459-471.

[14] Armbuster, C. E., Wenzhou, H., Pang, B., Weimer, K. E. D., Juneau, R. A., Turner, J. & Swords, W. E. (2010). Indirect pathogenicity of haemophilus influenza and Moraxella catarrhalis in polymicrobial otitis media occurs via interspecies quorum signaling. *mBio 1*, e00102-10.

[15] Galloway, W. R. J. D., Hodgkinson, J. T., Bowden, S. D., Welch, M. & Spring, D. R. (2011). Quorum sensing in gram-negative bacteria: small-molecule modulation of AHL and AI-2 quorum sensing pathways. *Chem. Rev. 111*, 28-67.
[16] Kalia, V. C. & Purohit, H. J. (2011). Quenching the quorum sensing system: potential antibacterial drug targets. *Crit. Rev. Microbiol. 37*, 121–140.
[17] Rutherford, S. T. & Bassler, B. L. (2012). Bacterial quorum sensing: its role in virulence and possibilities for its control. *Cold Spring* Nov. 1, *2*(11). pii, a012427.
[18] Kalia, V. C. (2013). Quorum sensing inhibitors: an overview. *Biotechnol. Adv. 31*, 224-245.
[19] Geske, G. D., O'Neill, J. C., Miller, D. M., Mattmann, M. E. & Blackwell, H. E. (2007). Modulation of bacterial quorum sensing with synthetic ligands: systematic evaluation of N-acylated homoserine lactones in multiple species and new insights into their mechanisms of action. *J. Am. Chem. Soc. 129*, 13613-13625.
[20] Rasmussen, T. B. & Givskov, M. (2006). Quorum sensing inhibitors: a bargain of effects. *Microbiol. 152*, 895-904.
[21] Chen, X., Schauder, S., Potier, N., Van Dorsselaer, A., Peclzer, I., Bassler, B. L. & Hughson, F. M. (2002). Structural identification of a bacterial quorum sensing signal containing boron. *Nature*, *415*, 545-549.
[22] Thiel, V., Vilchez, R., Sztajer, H., Wagner-Dobler, I. & Schulz, S. (2009). Identification, quantification, and determination of the absolute configuration of the bacterial quorum-sensing signal autoinducer-2 by gas chromatography-mass spectrometry. *Chem. Bio. Chem. 10*, 479-485.
[23] Williams, P. (2007). Quorum sensing, communication and cross-kingdom signalling in the bacterial world. *Microbiol.153*, 3923–3938.
[24] Faure, D., Vereecke, D. & Leveau, J. H. J. (2009). Molecular communication in the rhizosphere. *Plant Soil*, *321*, 279-303.
[25] Dirix, G., Monsieurs, P., Dombrecht, B., Daniels, R., Marchal, K., Vanderleyden, J. & Michiels, J. (2004). Peptide signal molecules and bacteriocins in Gram-negative bacteria: a genome-wide in silico screening for peptides containing a double-glycine leader sequence and their cognate transporters. *Peptides*, *25*, 1425–1440.
[26] Michael, B., Smith, J. N., Swift, S., Heffron, F. & Ahmer, B. M. (2001). SdiA of *Salmonella enterica* is a LuxR homolog that detects mixed microbial communities. *J. Bacteriol. 183*, 5733-5742.
[27] Lowery, C. A., Dickerson, T. J. & Janda, K. D. (2008). Interspecies and interkingdom communication mediated by bacterial quorum sensing. *Chem. Soc. Rev. 37*, 1337-1346.
[28] Pereira, C. S., Thompson, J. & Xavier, K. B. (2013). AI-2-mediated signalling in bacteria. *FEMS Microbiol. Rev. 37*, 156–181.
[29] Vendeville, A., Winzer, K., Heurlier, K., Tang, C. M. & Hardie, K. R. (2005). Making 'sense' of metabolism: autoinducer-2, LuxS and pathogenic bacteria. *Nat. Rev. Microbiol. 3*, 383-396.
[30] Ryan, R. P. & Dow, J. M. (2008). Diffusible signals and interspecies communication in bacteria. *Microbiol. 154*, 1845–1858.
[31] D'Angelo-Picard, C., Faure, D., Carlier, A., Uroz, S., Raffoux, A., Fray, R. & Dessaux, Y. (2004). Bacterial populations in the rhizosphere of tobacco plants producing the quorum-sensing signals hexanoyl-homoserine lactone and 3-oxo-hexanoyl-homoserine lactone. *FEMS Microbiol. Ecol. 51*, 19–29.

[32] Cha, C., Gao, P., Chen, Y.-C., Shaw, P. D. & Farrand, S. K. (1998). Production of acyl-homoserine lactone quorum-sensing signals by gram-negative plant-associated bacteria. *Mol. Plant Microbe. Interact. 11*, 1119–1129.

[33] von Bodman, S. B., Bauer, W. D. & Coplin, D. L. (2003). Quorum sensing in plant-pathogenic bacteria. Ann. Rev. *Phytopathol. 41*, 455–482.

[34] Waters, C. M. & Bassler, B. L. (2005). Quorum Sensing: Cell-to-Cell Communication in Bacteria. *Ann. Rev. Cell Dev. Biol. 21*, 319-346.

[35] Marketon, M. M., Gronquist, M. R., Eberhard, A. & Gonzalez, J. E. (2002). Characterization of the *Sinorhizobium meliloti* sinR/sinI locus and the production of novel N-acyl homoserine lactones. *J. Bacteriol. 184*, 5686-5695.

[36] Schaefer, A. L., Taylor, T. A., Beatty, J. T. & Greenberg, E. P. (2002). Long-chain acyl-homoserine lactone quorum-sensing regulation of *Rhodobacter capsulatus* gene transfer agent production. *J. Bacteriol. 184*, 6515-6521.

[37] Venturi, V. (2006). Regulation of quorum sensing in *Pseudomonas*. *FEMS Microbiol. Rev. 30*, 274–291.

[38] Bratu, S., Gupta, J. & Quale, J. (2006). Expression of the las and rhl quorum-sensing systems in clinical isolates of *Pseudomonas aeruginosa* does not correlate with efflux pump expression or antimicrobial resistance. *J. Antimicrob. Chem. 58*, 1250–1253.

[39] Lazdunski, A. M., Ventre, I. & Sturgis, J. N. (2004). Regulatory circuits and communication in Gram negative bacteria. *Nat. Rev. Microbiol. 7*, 581-592.

[40] Chugani, S. A., Whiteley, M., Lee, K. M., D'Argenio, D., Manoil, C. & Greenberg, E. P. (2001). QscR, a modulator of quorum-sensing signal synthesis and virulence in *Pseudomonas aeruginosa*. *Proc. Natl. Acad. Sci.* USA. *98*, 2752-2757.

[41] Pesci, E. C., Pearson, J. P., Seed, P. C. & Iglewski, B. H. (1997). Regulation of las and rhl quorum sensing in *Pseudomonas aeruginosa*. *J. Bacteriol. 179*, 3127-3232.

[42] Bertani, I. & Venturi, V. (2004). Regulation of the N-acylhomoserine lactone-dependent quorum-sensing system in rhizosphere *Pseudomonas putida* WCS358 and cross-talk with the stationary-phase RpoS sigma factor and the global regulator GacA. *Appl. Environ. Microbiol. 70*, 5493–5502.

[43] Ferluga, S., Steindler, L. & Venturi, V. (2008). N-acyl homoserine lactone quorum sensing in Gram-negative rhizobacteria, In: *Secondary Metabolites in Soil Ecology* (Karlovsky, P., Ed.), Springer: Heidelberg, Germany, 69-92.

[44] de Maeyer, K., D'Aes, J., Hua, G. K., Perneel, M., Vanhaecke, L., Noppe, H. & Hofte, M. (2011). N-Acylhomoserine lactone quorum-sensing signalling in antagonistic phenazine-producing *Pseudomonas* isolates from the red cocoyam rhizosphere. *Microbiol. 157*(Pt 2), 459–472.

[45] El-Sayed, A. K., Hothersall, J., Cooper, S. M., Stephens, E., Simpson, T. J. & Thomas, C. M. (2003). Characterization of the mupirocin biosynthesis gene cluster from *Pseudomonas fluorescens* NCIMB 10586. *Chem. Biol. 10*, 419–430.

[46] Hothersall, J., Murphy, A. C., Iqbal, Z., Campbell, G., Stephens, E. R., Wu, J., Cooper, H., Atkinson, S., Williams, P., Crosby, J., Willis, C. L., Cox, R. J., Simpson, T. J. & Thomas, C.M. (2011). Manipulation of quorum sensing regulation in *Pseudomonas fluorescens* NCIMB 10586 to increase mupirocin production. *Appl. Microbiol. Biotechnol. 90*, 1017–1026.

[47] Wei, H. L. & Zhang, L.Q. (2006). Quorum-sensing system influences root colonization and biological control ability in *Pseudomonas fluorescens* 2P24. *Anton. Van Leeuw. 89*, 267–280.

[48] Yan, Q., Gao, W., Wu, X-G. & Zhang, L.-Q. (2009). Regulation of the PcoI/PcoR quorum-sensing system in *Pseudomonas fluorescens* 2P24 by the PhoP/PhoQ two-component system. *Microbiol. 155*, 124–133.

[49] Alavi, P., Muller, H., Cardinale, M., Zachow, C., Sanchez, M. B., Martınez, J. L. & Berg, G. (2013). The DSF quorum sensing system controls the positive influence of *Stenotrophomonas maltophilia* on plants. *PLoS One 8*, e67103.

[50] Quinones, B., Pujol, C. J. & Lindow, S. E. (2004). Regulation of AHL production and its contribution to epiphytic fitness in *Pseudomonas syringae*. *Mol. Plant Microbe. Interact. 17*, 521–531.

[51] Chin, A. W. T. F., van den Broek, D., de Voer, G., van der Drift, K. M., Tuinman, S., Thomas-Oates, J. E., Lugtenberg, B. J. & Bloemberg, G. V. (2001). Phenazine-1-carboxamide production in the biocontrol strain *Pseudomonas chlororaphis* PCL1391 is regulated by multiple factors secreted into the growth medium. *Mol. Plant Microbe. Interact. 14*, 969–979.

[52] Elasri, M., Delorme, S., Lemanceau, P., Stewart, G., Laue, B., Glickmann, E., Oger, P. M. & Dessaux, Y. (2001). Acyl-homoserine lactone production is more common among plant-associated *Pseudomonas* spp. than among soil-borne *Pseudomonas* spp. *Appl. Environ. Microbiol. 67*, 1198–1209.

[53] Weller, D. M. (1988). Biological control of soil borne plant pathogens in the rhizosphere with bacteria. *Ann. Rev. Phytopathol. 26*, 379-407.

[54] Gelvin, S. B. (2000). *Agrobacterium* and plant genes involved in T-DNA transfer and integration. *Ann. Rev. Plant Physiol. Plant Mol. Biol. 51*, 223–256.

[55] Lang, J. & Faure, D. (2014). Functions and regulation of quorum sensing in *Agrobacterium tumefaciens*. *Front. Plant Sci.*, *5*, 1-13.

[56] White, C. E. & Winans, S. C. (2007).The quorum-sensing transcription factor TraR decodes its DNA binding site by direct contacts with DNA bases and by detection of DNA flexibility. *Mol. Microbiol.*, *64*, 245–256.

[57] Chevrot, R., Rosen, R., Haudecoeur, E., Cirou, A., Shelp, B. J., Ron, E. & Faure, D. (2006). GABA controls the level of quorum-sensing signal in *Agrobacterium tumefaciens*. *Proc. Natl. Acad. Sci. USA*, *103*, 7460–7464.

[58] Teyssier-Cuvelle, S., Oger, P., Mougel, C., Groud, K., Farrand, S. K. & Nesme, X. (2004). A highly selectable and highly transferable Ti plasmid to study conjugal host range and Ti plasmid dissemination in complex ecosystems. *Microb. Ecol. 48*, 10–18.

[59] Zhang, L. H., Murphy, P. J., Kerr, A. & Tate, M. E. (1993). *Agrobacterium* conjugation and gene regulation by N-acyl-L-homoserine lactones. *Nature*, *362*, 446–448.

[60] Hwang, I., Li, P. L., Zhang, L., Piper, K. R., Cook, D. M., Tate, M. E. & Farrand, S. K. (1994). TraI, a LuxI homologue, is responsible for production of conjugation factor, the Ti plasmid N-acylhomoserine lactone autoinducer. *Proc. Natl. Acad. Sci. USA*, *91*, 4639–4643.

[61] Fuqua, W. C. & Winans, S. C. (1994). A LuxR-LuxI type regulatory system activates *Agrobacterium* Ti plasmid conjugal transfer in the presence of a plant tumor metabolite. *J. Bacteriol. 176*, 2796–2806.

[62] Seet, Q. & Zhang, L. H. (2011). Antiactivator QsI A defines the quorum sensing threshold and response in *Pseudomonas aeruginosa*. *Mol. Microbiol. 80*, 951–965.

[63] Luo, Z. Q., Qin, Y. & Farrand, S. K. (2000). The antiactivator TraM interferes with the autoinducer – dependent binding of TraR to DNA by interacting with the C- terminal region of the quorum-sensing activator. *J. Biol. Chem. 275*, 7713–7722.

[64] Costa, E. D., Chai, Y. & Winans, S. C. (2012). The quorum-sensing protein TraR of *Agrobacterium tumefaciens* is susceptible to intrinsic and TraM-mediated proteolytic instability. *Mol. Microbiol. 84*, 807–815.

[65] Haudecoeur, E. & Faure, D. (2010). A fine control of quorum-sensing communication in *Agrobacterium tumefaciens*. *Commun. Integrat. Biol. 3*, 84-88.

[66] Carlier, A., Uroz, S., Smadja, B., Fray, R., Latour, X., Dessaux, Y. & Faure, D. (2003). The Ti plasmid of *Agrobacterium tumefaciens* harbors an attM-paralogous gene, aiiB, also encoding N-Acyl homoserine lactonase activity. *Appl. Environ. Microbiol. 69*, 4989-4693.

[67] Zhang, H. B. Wang, L. H. & Zhang, L. H. (2002). Genetic control of quorum- sensing signal turnover in *Agrobacterium tumefaciens*. *Proc. Natl. Acad. Sci. USA*, *99*, 4638–4643.

[68] Chai, Y., Tsai, C. S., Cho, H. & Winans, S. C. (2007). Reconstitution of the biochemical activities of the AttJ repressor and the AttK, AttL, and AttM catabolic enzymes of *Agrobacterium tumefaciens*. *J. Bacteriol. 189*, 3674–3679.

[69] von Bodman, S., Hayman, G. T. & Farrand, S. K. (1992). Opine catabolism and conjugal transfer of the nopaline Ti plasmid pTiC58 are coordinately regulated by a single repressor. *Proc. Natl. Acad. Sci. USA*, *89*, 643–647.

[70] Carlier, A., Chevrot, R., Dessaux, Y. & Faure, D. (2004). The assimilation of γ-butyrolactone in *Agrobacterium tumefaciens* C58 interferes with the accumulation of the N-acyl-homoserine lactone signal. *Mol. Plant Microbe Interact. 17*, 951-957.

[71] Miller, M. B. & Bassler, B. (2001). Quorum sensing in bacteria. *Ann. Rev. Microbiol. 55*, 165-199.

[72] Toth, I. K., Birch, P. R. J. (2005). Rotting softly and stealthily. *Curr. Opin. Plant Biol. 8*, 424-429.

[73] Marits, R., Koiv, V., Laasik, E. & Mae, A. (1999). Isolation of an extracellular protease gene of *Erwinia carotovora* subsp. *carotovora* strain SCC3193 by transposon mutagenesis and the role of protease in phytopathogenicity. *Microbiol. 145*, 1959–1966.

[74] Sjoblom, S., Brader, G., Koch, G. & Palva, E. T. (2006). Cooperation of two distinct ExpR regulators controls quorum sensing specificity and virulence in the plant pathogen *Erwinia carotovora*. *Mol. Microbiol. 60*, 1474-1489.

[75] McGowan, S. J., Barnard, A. M., Bosgelmez, G., Sebaihia, M., Simpson, N. J., Thomson, N. R., Todd, D. E., Welch, M., Whitehead, N. A. & Salmond, G. P. (2005). Carbapenem antibiotic biosynthesis in *Erwinia carotovora* is regulated by physiological and genetic factors modulating the quorum sensing-dependent control pathway. *Mol. Microbiol. 55*, 526–545.

[76] McGowan, S., Sebaihia, M., Jones, S., Yu, B., Bainton, N., Chan, P. F., Bycroft, B., Stewart, G. S., Williams, P. & Salmond, G. P. (1995). Carbapenem antibiotic production in *Erwinia carotovora* is regulated by CarR, a homologue of the LuxR transcriptional activator. *Microbiol. 141*, 541–550.

[77] Chatterjee, A., Cui, Y., Hasegawa, H., Leigh, N., Dixit, V. & Chatterjee, A. K. (2005). Comparative analysis of two classes of quorum-sensing signaling systems that control production of extracellular proteins and secondary metabolites in *Erwinia carotovora* subspecies. *J. Bacteriol. 187*, 8026–8038.

[78] Barnard, A. M. & Salmond, G. P. (2007). Quorum sensing in *Erwinia species. Anal. Bioanal. Chem. 387*, 415–423.

[79] Chatterjee, A., Cui, Y., Liu, Y., Dumenyo, C. K. & Chatterjee, A. K. (1995). Inactivation of *rsmA* leads to overproduction of extracellular pectinases, cellulases, proteases in *Erwinia carotovora* subsp. *carotovora* in the absence of the starvation/cell density-sensing singal, *N*-(3-oxohexanoyl)-L-homoserine lactone. *Appl. Environ. Microbiol. 61*, 1959–1967.

[80] Cui, Y., Madi, L., Mukherjee, A., Dumenyo, C. K. & Chatterjee, A. K. (1996). The RsmA mutants of *Erwinia carotovora* subsp. *carotovora* strain Ecc71 overexpress hrpNEcc and elicit a hypersensitive reaction-like response in tobacco leaves. *Mol. Plant Microbe. Interact. 9*, 565-573.

[81] McGowan, S. J., Sebaihia, M., Porter, L. E., Stewart, G. S., Williams, P., Bycroft, B. W. & Salmond, G. P. (1996). Analysis of bacterial carbapenem antibiotic production genes reveals a novel β-lactam biosynthesis pathway. *Mol. Microbiol. 22*, 415–426.

[82] Nasser, W., Bouillant, M. L., Salmond, G. P. & Reverchon, S. (1998). Characterization of the *Erwinia chrysanthemi* expI-expR locus directing the synthesis of two N-acyl homoserine lactone signal molecules. *Mol. Microbiol. 29*, 1391–1405.

[83] von Bodman, S. & Farrand, S. K. (1995). Capsular polysaccharide biosynthesis and pathogenicity in *Erwinia stewartii* require induction by an N-acylhomoserine lactone autoinducer. *J. Bacteriol. 177*, 5000–5008.

[84] von Bodman, S. B., Majerczak, D. R. & Coplin, D. L. (1998). A negative regulator mediates quorum-sensing control of exopolysaccharide production in *Pantoea stewartii* subsp. *stewartii*. *Proc. Natl. Acad. Sci. USA*, *95*, 7687–7692.

[85] Palva, T. K., Holmstrom, K.-O., Heino, P. & Palva, E. T. (1993). Induction of plant defense response by exoenzymes of *Erwinia carotovora* subsp. *carotovora*. *Mol. Plant Microbe. Interact. 6*, 190-196.

[86] Dixon, R. A., Harrison, M. J. & Lamb, C. J. (1994). Early events in the activation of plant defense responses. *Ann. Rev. Phytopathol. 32*, 479-501.

[87] Oke, V. & Long, S. R. (1999). Bacterial genes induced within the nodule during the *Rhizobium*-legume symbiosis. *Mol. Microbiol. 32*, 837-849.

[88] Perret, X., Staehelin, C. & Broughton, W. J. (2000). Molecular basis of symbiotic promiscuity. *Microbiol. Mol. Biol. Rev. 64*, 180–201.

[89] Reddy, P. M., Rendon-Anaya, M., Soto del Rio, M. D. & Khandual, S. (2007). Flavonoids as signaling molecules and regulators of root nodule development. *Dynamic Soil Dyn. Plant. 1*, 83–94.

[90] Daniels, R., De vos, D. E., Desair, J., Raedschelders, G., Luyten, E., Rosemeyer, V., Verreth, C., Schoeters, E., Vanderleyden, J., Michiels, J. (2002). The cin quorum sensing locus of *Rhizobium etli* CNPAF512 affects growth and symbiotic nitrogen fixation. *J. Biol. Chem. 277*, 462–468.

[91] Pellock, B. J., Teplitski, M., Boinay, R. P., Bauer, W. D. & Walker, G. C. (2002). A LuxR homolog controls production of symbiotically active extracellular polysaccharide II by *Sinorhizobium meliloti*. *J. Bacteriol. 184*, 5067–5076.

[92] Marketon, M. M., Glenn, S. A., Eberhard, A. & Gonzalez, J. E. (2003). Quorum sensing controls exopolysaccharide production in *Sinorhizobium meliloti*. *J. Bacteriol.* *185*, 325–331.

[93] Loh, J., Carlson, R. W., York, W. S. & Stacey, G. (2002). Bradyoxetin, a unique chemical signal involved in symbiotic gene regulation. *Proc. Natl. Acad. Sci. USA*, *99*, 14446–14451.

[94] Ahlgren, N. A., Harwood, C. S., Schaefer, A. L., Giraud, E. & Greenberg, E. P. (2011). Aryl-homoserine lactone quorum sensing in stem-nodulating photosynthetic bradyrhizobia. *Proc. Natl. Acad. Sci. USA*, *108*, 7183-7188.

[95] Lindemann, A., Pessi, G., Schaefer, A. L., Mattmann, M. E., Christensen, Q. H., Kessler, A., Hennecke, H., Blackwell, H. E., Greenberg, E. P. & Harwood, C. S. (2011). Isovaleryl-homoserine lactone, an unusual branched-chain quorum-sensing signal from the soybean symbiont *Bradyrhizobium japonicum*. *Proc. Natl. Acad. Sci. USA*, *108*, 16765–16770.

[96] Nievas, F., Bogino, P., Sorroche, F. & Giordano, W. (2012). Detection, characterization, and biological effect of quorum-sensing signaling molecules in peanut-nodulating bradyrhizobia. *Sensors*, *12*, 2851-2873.

[97] Rosemeyer, V., Michiels, J., Verreth, C. & Vanderleyden, J. (1998). luxI and luxR-homologous genes of *Rhizobium etli* CNPAF512 contribute to synthesis of autoinducer molecules and nodulation of *Phaseolus vulgaris*. *J. Bacteriol.* *180*, 815-821.

[98] Rodelas, B., Lithgow, J. K., Wisniewski-Dye, F., Hardman, A., Wilkinson, A., Economou, A., Williams, P. & Downie, J. A. (1999). Analysis of quorum-sensing-dependent control of rhizosphere-expressed (rhi) genes in *Rhizobium leguminosarum* bv. *viciae*. *J. Bacteriol.* *181*, 3816–3823.

[99] Lithgow, J. K., Wilkinson, A., Hardman, A., Rodelas, B., Wisniewski-Dye, F., Williams, P. & Downie, J. A. (2000). The regulatory locus cinRI in *Rhizobium leguminosarum* controls a network of quorum sensing loci. *Mol. Microbiol.* *37*, 81–97.

[100] Cubo, M. T., Economou, A., Murphy, G., Johnston, A. W. B. & Downie, J. A. (1992). Molecular characterization and regulation of the rhizosphere-expressed genes rhiABCR that can influence nodulation by *Rhizobium leguminosarum* biovar *viciae*. *J. Bacteriol.* *174*, 4026–4035.

[101] Dibb, N. J., Downie, J. A. & Brewin, N. J. (1984). Identification of a rhizosphere protein encoded by the symbiotic plasmid of *Rhizobium leguminosarum*. *J. Bacteriol.* *158*, 621–627.

[102] Hoang, H. H., Gurich, N. & Gonzalez, J. E. (2008). Regulation of motility by the ExpR/Sin quorum sensing system in *Sinorhizobium meliloti*. *J. Bacteriol.* *190*, 861-871.

[103] Gao, M., Teplitski, M., Robinson, J. B. & Bauer, W. D. (2003). Production of substances by *Medicago truncatula* that affect bacterial quorum sensing. *Mol. Plant Microbe. Interact.* *16*, 827-834.

[104] Mathesius, U., Mulders, S., Gao, M., Teplitski, M., Caetano-Anolles, G., Rolfe, B. G. & Bauer, W. D. (2003). Extensive and specific responses of a eukaryote to bacterial quorum-sensing signals. *Proc. Natl. Acad. Sci. USA*, *100*, 1444-1449.

[105] Teplitski, M., Robinson, J. B. & Bauer, W. D. (2000). Plants secrete substances that mimic bacterial N-acyl homoserine lactone signal activities and affect population density-dependent behaviors in associated bacteria. *Mol. Plant Microbe. Interact.* *13*, 637-648.

[106] Freiberg, C., Fellay, R., Bairoch, A., Broughton, W. J., Rosenthal, A. & Perret, X. (1997). Molecular basis of symbiosis between *Rhizobium* and legumes. *Nature*, *387*, 394–401.

[107] Grimont, P. A. D. & Grimont, F. (1978). The genus *Serratia*. *Ann. Rev. Microbiol. 32*, 221–248.

[108] Givskov, M., Olsen, L. & Molin, S. (1988). Cloning and expression in *Escherichia coli* of the gene for extracellular phospholipase A1 from *Serratia liquefaciens*. *J. Bacteriol. 170*, 5855–5862.

[109] Daniels, R., Vanderleyden, J. & Michiels, J. (2004). Quorum sensing and swarming migration in bacteria. *FEMS Microbiol. Rev. 28*, 261–289.

[110] Eberl, L., Molin, S., Givskov, M. (1999). Surface motility of *Serratia liquefaciens* MG1. *J. Bacteriol. 181*, 1703–1712.

[111] Bakkiyaraj, D., Sivasankar, C. & Pandian, S. K. (2012). Inhibition of quorum sensing regulated biofilm formation in *Serratia marcescens* causing nosocomial infections. *Bioorg. Med. Chem. Lett. 22*, 3089–3094.

[112] Eberl, L., Christiansen, G., Molin, S. & Givskov, M. (1996). Differentiation of *Serratia liquefaciens* into swarm cells is controlled by the expression of the flhD master operon. *J. Bacteriol. 178*, 554–559.

[113] Liu, X., Jia, J., Popat, R., Ortori, C. A., Li, J., Diggle, S. P., Gao, K. & Camara, M. (2011). Characterization of two quorum sensing systems in the endophytic *Serratia plymuthica* strain G3: differential control of motility and biofilm formation according to life-style. *BMC Microbiol. 11*, 26.

[114] Cox, A. R. J., Thomson, N. R., Bycroft, B., Stewart, G. S. A. B., Williams, P. & Salmond, G. P. C. (1998). A pheromone-independent CarR protein controls carbapenem antibiotic synthesis in the opportunistic human pathogen *Serratia marcescens*. *Microbiol. 144*, 201-209.

[115] Thomson, N. R., Crow, M. A., McGowan, S. J., Cox, A. & Salmond, G. P. C. (2000). Biosynthesis of carbapenem antibiotic and prodigiosin pigment in *Serratia* is under quorum sensing control. *Mol. Microbiol. 36*, 539-556.

[116] Alvarez, B., Biosca, E. G. & Lopez, M. M. (2010). On the life of *Ralstonia solanacearum*, a destructive bacterial plant pathogen. In: *Current research, technology, and education topics in applied microbiology and microbial biotechnology* (Mendez-Vilas, A., Ed.), Formatex, Badajoz, Spain. vol. *1*. 267-279.

[117] Devine, J. H., Shadel, G. S. & Baldwin, T. O. (1989). Identification of the operator of the lux regulon from the *Vibrio fischeri* strain ATCC7744. *Proc. Natl. Acad. Sci. USA*, *86*, 5688-5692.

[118] Pearson, J. P., Feldman, M., Iglewski, B. H., Prince, A. (2000). *Pseudomonas aeruginosa* cell-to-cell signaling is required for virulence in a model of acute pulmonary infection. *Infect. Immun. 68*, 4331-4334.

[119] Flavier, A. B., Clough, S. J., Schell, M. A. & Denny, T. P. (1997). Identification of 3-hydroxypalmitic acid methyl ester as a novel autoregulator controlling virulence in *Ralstonia solanacearum*. *Mol. Microbiol. 26*, 251–259.

[120] Mole, B. M., Baltrus, D. A., Dangl, J. L. & Grant, S. R. (2007). Global virulence regulation networks in phytopathogenic bacteria. *Trends Microbiol. 15*, 363-371.

[121] Clough, S. J., Lee, K. E., Schell, M. A. & Denny, T. P. (1997). A two-component system in *Ralstonia (Pseudomonas) solanacearum* modulates production of PhcA-

regulated virulence factors in response to 3-hydroxypalmitic acid methyl ester. *J. Bacteriol. 179*, 3639-3648.

[122] Genin, S., Brito, B., Denny, T. P. & Boucher, C. (2005). Control of the *Ralstonia solanacearum* Type III secretion system (Hrp) genes by the global virulence regulator PhcA. *FEBS Lett. 579*, 2077-2081.

[123] Flavier, A. B., Schell, M. A. & Denny, T. P. (1998). An RpoS (¾S) homologue regulates acylhomoserine lactone-dependent autoinduction in *Ralstonia solanacearum. Mol. Microbiol. 28*, 475–486.

[124] Clough, S. J., Flavier, A. B., Schell, M. A. & Denny, T. P. (1997a). Differential expression of virulence genes and motility in *Ralstonia (Pseudomonas) solanacearum* during exponential growth. *Appl. Environ. Microbiol. 63*, 844–850.

[125] Chen, C.-N., Chen, C.-J., Liao, C.-T. & Lee, C.-Y. (2009). A probable aculeacin A acylase from the *Ralstonia solanacearum* GMI1000 is N-acyl-homoserine lactone acylase with quorum-quenching activity. *BMC Microbiol. 9*, 89.

[126] Onsando, J. M. (1992). Black rot of crucifers. In: *Diseases of Vegetables and Oil Seed Crops* (Chaube, H.S., Kumar, J., Mukhopadhyay, A.N., Singh, U.S., Eds.). Prentice Hall, Englewood Cliffs, NJ. 243-252.

[127] He, Y. W., Xu, M., Lin, K., Ng, Y. J., Wen, C. M., Wang, L. H., Liu, Z. D., Zhang, H. B., Dong, Y. H., Dow, J. M. & Zhang, L. H. (2006). Genome scale analysis of diffusible signal factor regulon in *Xanthomonas campestris* pv. *campestris*: identification of novel cell-cell communication-dependent genes and functions. *Mol. Microbiol. 59*, 610–622.

[128] Guo, Y., Zhang, Y., Li, J.-L. & Wang, N. (2012). Diffusible signal factor-mediated quorum sensing plays a central role in coordinating gene expression of *Xanthomonas citri* subsp. *citri. Mol. Plant Microbe. Interact. 25*, 165–179.

[129] Barber, C. E., Tang, J. L., Feng, J. X., Pan, M. Q., Wilson, T. J., Slater, H., Dow, J. M., Williams, P. & Daniels, M. J. (1997). A novel regulatory system required for pathogenicity of *Xanthomonas campestris* is mediated by a small diffusible signal molecule. *Mol. Microbiol. 24*, 555–566.

[130] Fouhy, Y., Scanlon, K., Schouest, K., Spillane, C., Crossman, L., Avison, M. B., Ryan, R. P. & Dow, J. M. (2007). Diffusible signal factor-dependent cell-cell signaling and virulence in the nosocomial pathogen *Stenotrophomonas maltophilia. J. Bacteriol. 189*, 4964-4968.

[131] Colnaghi Simionato, A. V., da Silva, D. S., Lambais, M. R. & Carrilho, E. (2007). Characterization of a putative *Xylella fastidiosa* diffusible signal factor by HRGC-EI-MS. *J. Mass Spect. 42*, 490-496.

[132] Boon, C., Deng, Y., Wang, L. H., He, Y., Xu, J. L., Fan, Y., Pan, S. Q. & Zhang, L. H. (2008). A novel DSF-like signal from *Burkholderia cenocepacia* interferes with *Candida albicans* morphological transition. *ISME J. 2*, 27-36.

[133] He, Y.-W., Wu, J., Cha, J.-S. & Zhang, L.-H. (2010). Rice bacterial blight pathogen *Xanthomonas oryzae* pv. *oryzae* produces multiple DSF-family signals in regulation of virulence factor production. *BMC Microbiol. 10*, 187.

[134] Wang, L. H., He, Y. W., Gao, Y. F., Wu, J. E., Dong, Y. H., He, C., Wang, S. X., Weng, L. X., Xu, J. L., Tay, L., Fang, R. X. & Zhang, L. H. (2004). A bacterial cell-cell communication signal with cross-kingdom structural analogues. *Mol. Microbiol. 51*, 903-912.

[135] Chatterjee, S., Wistrom, C. & Lindow, S. E. (2008). A cell-cell signaling sensor is required for virulence and insect transmission of *Xylella fastidiosa*. *Proc. Natl. Acad. Sci. USA*, *105*, 2670-2675.

[136] Jeong, K. S., Lee, S. E., Han, J. W., Yang, S. U., Lee, B. M., Noh, T. H. & Cha, J. S. (2008). Virulence reduction and differing regulation of virulence genes in rpf mutants of *Xanthomonas oryzae* pv. *oryzae*. *Plant Pathol. J. 24*, 143-151.

[137] Slater, H., Alvarez-Morales, A., Barber, C. E., Daniels, M. J., Dow, J. M. (2000). A two component system involving an HD-GYP domain protein links cell-cell signaling to pathogenicity gene expression in *Xanthomonas campestris*. *Mol. Microbiol. 38*, 986-1003.

[138] He, Y. W. & Zhang, L. H. (2008). Quorum sensing and virulence regulation in *Xanthomonas campestris*. *FEMS Microbiol. Rev. 32*, 842–857.

[139] Chatterjee, S. & Sonti, R. V. (2002). rpfF mutants of *Xanthomonas oryzae* pv. *oryzae* are deficient for virulence and growth under low iron conditions. *Mol. Plant Microbe. Interact. 15*, 463-471.

[140] Zhao, Y., Qian, G., Yin, F., Fan, J., Zhai, Z., Liu, C., Hu, B. & Liuao, F. (2011). Proteomic analysis of the regulatory function of DSF-dependent quorum sensing in *Xanthomonas oryzae* pv. *oryzicola*. *Microb. Pathog. 50*, 48-55.

[141] Schugegger, R., Ihring, A., Gantner, S., Bhanweg, G., Knappe, C., Vogg, G., Hutzler, P., Schmid, M., Van Breusegem, F., Eberl, L., Hartmann, A. & Langebartels, C. (2006). Induction of systemic resistance in tomato by N-acyl-homoserine lactone producing rhizosphere bacteria. *Plant Cell Environ. 29*, 909-918.

[142] von Rad, U., Klein, I., Dobrev, P. I., Kotova, J., Zazimalove, E., Fekete, A., Hartmann, A., Schmitt-Koppli, P. & Durner, J. (2008). Response of *Arabidopsis thaliana* to N-hexanoyl-DL-homoserine lactone, a bacterial quorum sensing molecule produced in the rhizosphere. *Planta*, *229*, 73-85.

[143] Schwessinger, B. & Zipfel, C. (2008). News from the frontline: recent insights into PAMP-triggered immunity in plants. *Curr. Opin. Plant Biol. 11*, 1–17.

[144] Zipfel, C. (2008). Pattern-recognition receptors in plant innate immunity. *Curr. Opin. Immunol. 20*, 10–16.

[145] Dong, Y. H., Wang, L. Y. & Zhang, L. H. (2007). Quorum-quenching microbial infections: mechanisms and implications. Philos. *Trans. R. Soc. Lond. B. Biol. Sci. 362*, 1201–1211.

[146] Bauer, W. D. & Teplitski, M. (2001). Can plants manipulate bacterial quorum sensing? *Aust. J. Plant Physiol. 28*, 913-921.

[147] McDougald, D., Rice, S. A. & Kjelleberg, S. (2007). Bacterial quorum sensing and interference by naturally occurring biomimics. *Anal. Bioanal. Chem. 387*, 445–453.

[148] Manefield, M., de Nys, R., Kumar, N., Read, R., Givskov, M., Steinberg, P. & Kjelleberg, S. (1999). Evidence that halogenated furanones from *Delisea pulchra* inhibit acylated homoserine lactone (AHL)-mediated gene expression by displacing the AHL signal from its receptor protein. *Microbiol. 145*, 283-291.

[149] Dong, Y. H., Wang, L. H., Xu, J. L., Zhang, H. B., Zhang, X. F. & Zhang, L. H. (2001). Quenching quorum sensing dependent bacterial infection by an N-acyl homoserine lactonase. *Nature*, *411*, 813-817.

[150] Dong, Y. H., Gusti, A. R., Zhang, Q., Xu, J. l. & Zhang, L. H. (2002). Identification of quorum quenching N-acylhomoserine lactonases from *Bacillus* species. *Appl. Environ. Microbiol. 68*, 1754-1759.

[151] Givskov, M., de Nys, R., Manefield, M., Gram, L., Maximilien, R., Eberl, L., Molin, S., Steinberg, P. D. & Kjelleberg, S. (1996). Eukaryotic interference with homoserine lactone-mediated prokaryotic signaling. *J. Bacteriol. 178*, 6618–6622.

[152] Dong, Y. H., Xu, J. L., Li, X. Z. & Zhang, L. H. (2000). AiiA, an enzyme that inactivates the acylhomoserine lactone quorum-sensing signal and attenuates the virulence of *Erwinia carotovora*. *Proc. Nat. Acad. Sci. USA*, *97*, 3526–3231.

[153] Uroz, S., Dangelo, C., Carlier, A., Faure, D., Petit, A., Oger, P., Sicot, C. & Dessaux, Y. (2003). Novel bacteria degrading N-acylhomoserine lactones and their use as quenchers of quorum sensing regulated functions of plant pathogenic bacteria. *Microbiol. 149*, 1981-1989.

[154] Zhang, L. H. & Dong, Y. H. (2004). Quorum sensing and signal interference: diverse implications. *Mol. Microbiol.* 53, 1563-1571.

[155] Chan, K.-G., Wong, C.-S., Yin, W.-F., Sam, C.-K. & Koh, C.-L. (2010). Rapid degradation of N-3-oxoacylhomoserine lactones by a *Bacillus cereus* isolate from Malaysian rainforest soil. *Anton. Van Leeuwen. 98*, 299–305.

[156] Molina, L., Constantinescu, F., Michel, L., Reimmann, C., Duffy, B. & Defago, G. (2003). Degradation of pathogen quorum-sensing molecules by soil bacteria: a preventive and curative biological control mechanism. *FEMS Microbiol. Ecol. 45*, 71-81.

[157] Pierson, E. A., Wood, D. W., Cannon, J. A., Blachere, F. M. & Pierson III, L. S. (1998). Interpopulation signaling via N-acyl-homoserine lactones among bacteria in the wheat rhizosphere. *Mol. Plant Microbe. Interact. 11*, 1078-1084.

[158] Chun, C. K., Ozer, E. A., Welsh, M. J., Zabner, J. & Greenberg, E. P. (2004). Inactivation of a *Pseudomonas aeruginosa* quorum-sensing signal by human airway epithelia. *Proc. Natl. Acad. Sci. USA*, *101*, 3587–3590.

[159] Mayville, P., Ji, G., Beavis, R., Yang, H., Goger, M., Novick, R. P. & Muir, T. W. (1999). Structure–activity analysis of synthetic autoinducing thiolactone peptides from *Staphylococcus aureus* responsible for virulence. *Proc. Natl. Acad. Sci. USA*, *96*, 1218–1223.

[160] Wu, H., Song, Z., Hentzer, M., Andersen, J. B., Molin, S., Givskov, M. & Hoiby, N. (2004). Synthetic furanones inhibit quorum-sensing and enhance bacterial clearance in *Pseudomonas aeruginosa* lung infection in mice. *J. Antimicrob. Chemother. 53*, 1054–1061.

[161] Fedi, S., Tola, E., Moenne-Loccoz, Y., Dowling, D. N., Smith, L. M. & O'Gara, F. (1997). Evidence for signaling between the pathogenic fungus *Pythium ultimum* and *Pseudomonas fluorescens* F113: *P.ultimum* represses the expression of genes in *P.fluorescens* F113, resulting in altered ecological fitness. *Appl. Environ. Microbiol. 63*, 4261-4266.

In: Microbes in Soil and Their Agricultural Prospects
Editors: K. Kishore Choudhary and D. Wattal Dhar
ISBN: 978-1-63482-824-6

Chapter 2

RHIZOBIUM-LEGUME SYMBIOSIS AND EFFECTS OF ENVIRONMENTAL STRESSES ON THE SYMBIOSIS

***Akhilesh Yadav*[*1], *Asha Lata Singh*[1], *Ajay Kumar*[1], *Amit Kishore Singh*[2], *Kapil Deo Pandey*[1] and *Major Singh*[†2]**

[1]Department of Botany, Faculty of Science,
Banaras Hindu University, Varanasi, India
[2]Division of Crop Improvement,
Indian Institute of Vegetable Research, Varanasi, India

ABSTRACT

Biological nitrogen fixation represents the major source of nitrogen input in agricultural soils. The symbiotic nitrogen fixation play a significant role in improving the fertility and productivity of low-nitrogen soils. Several environmental problems affect the host-microbe interaction and influence the crop yield. The rhizobium-legume symbiosis has received most attention as they are widely deployed in agricultural practices for nitrogen recovery. Major environmental stresses such as, desiccation and salinity not only suppress the growth and symbiotic characteristics of rhizobia but also change the structural patterns of the associated proteins involved in nodulation processes. During environmental stress condition molecular chaperones enable denatured proteins to acquire their native folding state in the cell. However, several strains distributed among various species of rhizobia are tolerant to desiccation and salt stress effects. Reclamation and improvement in the fertility of arid lands by application of organic and inorganic fertilizers are expensive and can be a source of pollution. The rhizobium-legume symbiosis is suggested to be the ideal solution to the improvement of soil fertility and the rehabilitation of arid lands and is an important direction for future research.

* akhileshgzp@gmail.com
† Corresponding Author address: Division of Crop Improvement, Indian Institute of Vegetable Research, Shahanshahpur, Varanasi-221305, India. Email: singhvns@gmail.com.

Keywords: stress, chickpea, legumes, *Mesorhizobium*, molecular chaperones, nitrogen fixation, symbiosis

INTRODUCTION

Legumes are important crops because its food grain is a relevant source of protein for both human and animal. Rhizobia are able to establish nitrogen fixing symbiosis with legumes for biological nitrogen fixation to fulfil the demand of nitrogen and improve crop productivity. Therefore, the use of environmental stress tolerant, *Rhizobium* spp. could enhance the production of legumes in arid and semiarid regions of the world. Desiccation and salinity may negatively affect the several processes of rhizobia-plant symbiosis such as: growth and survival of rhizobia in soil, root colonization, infection, nodule development and functioning in the rhizosphere zone. The rhizobia isolated from chickpea nodules and cultured *in vitro* are usually much more tolerant to stress than their host [1].

Proteins are of permanent risk for unfolding during stress condition. Molecular chaperones enable denatured proteins to acquire their native folding state [2]. The *DnaK* machinery comprises the co-chaperone *DnaJ* and the nucleotide exchange factor *GrpE*, whereas the *GroEL* system includes the co-chaperone *GroES* [3].

However, in chickpea rhizobia, little is known about the expression of chaperone genes under stress conditions.

LEGUMES

Legumes are members of the bean family Fabaceae, which includes all types of beans and peas. The legumes are divided into three subfamilies, Mimosoideae, Caesalpinoideae and Papilionoideae. Most cultivated legumes, such as common bean (*Phaseolus vulgaris*), soybean (*Glycine max*) and chickpea (*Cicer arietinum*), belongs to the Papilionoideae subfamily. Mostly legumes have five-petaled flower, with bilateral symmetry. The fruit is a pod with one row of seeds; the seeds contain two prominent food-storing cotyledons. The seeds of many legumes are important as food because they are rich in oil and protein. The high protein content of legumes are correlated with the presence of root nodules, which contain nitrogen-fixing bacteria. These bacteria are species of the genus *Rhizobium* able to convert free atmospheric nitrogen into ammonia. That can be used by plants for the synthesis of protein and other nitrogen-containing compounds [4].

HOST LEGUME - CHICKPEA (*CICER ARIETINUM* L.)

Chickpea ranks third in world production among food legumes followed by beans (*Phaseolus* spp.) and field pea (*Pisum sativum*) [5]. The major chickpea producing countries in terms of annual production include India (65%), Pakistan (10%), Turkey (7%), Iran (3%), Myanmar (2%), Mexico (1.5%) and Australia (1.5%). Chickpea seed is a protein-rich supplement to cereal-based diets, especially critical in developing countries where people cannot afford animal protein. Two rhizobia species were described as chickpea symbionts, namely *Mesorhizobium ciceri* and *Mesorhizobium mediterraneum* [6, 7, 8]. The phylogenetic analysis of two symbiosis genes (*nifH* and *nodC*) of chickpea rhizobia in *Mesorhizobium ciceri* and *Mesorhizobium mediterraneum* type strains, suggesting that the symbiosis genes were horizontally transferred [9].

RHIZOBIA

Rhizobia are gram negative, aerobic or facultative anaerobic, rod shaped, and do not produce endospores. Rhizobia cells are motile with one polar or sub-polar flagellum with two to six peripheral flagella. Rhizobia can exist as legume-host-specific nitrogen fixing symbionts or as free living [10]. At the end of the growing season, nodule senescence leads to the release of a large number of rhizobia into soil.

Once the symbiosis is established, the rhizobia convert atmospheric nitrogen into ammonia to their legume host plant. These bacteria may be considered as plant growth promoting bacteria (PGPB) because they directly affect plant growth and development [11, 12]. PGPB may use to fix atmospheric nitrogen, synthesize siderophores and phytohormones, which can act to enhance various stages of plant growth; solubilize minerals such as phosphorus and synthesize the enzyme 1-aminocyclopropane-1-carboxylate (ACC) deaminase, which can lower plant ethylene levels [13, 14].

Most of these bacterial species belongs to the *α-Proteobacteria* class including species of *Rhizobium, Bradyrhizobium, Azorhizobium, Ensifer* (formely *Sinorhizobium*) and *Mesorhizobium* [15, 16, 17]. β-Proteobacteria comprises three genera containing rhizobia species, namely *Burkholderia*, *Cupriavidus* and *Herbaspirillum* [18, 19, 20]. Nevertheless, both classes comprise non-nodulating legume bacteria, such as *Rhizobium larrymoorei* [21].

Rhizobial genomes are composed of core and accessory elements. Core genome consist of housekeeping genes, which are essential for proper functioning of the cell, as well as other genes necessary for the maintenance and basic metabolism while accessory genome is responsible for special features other than symbiosis or other kinds of ecological niche adaptation [22, 23]. *Rhizobium* species possess symbiosis genes on plasmids, usually designated by p*Sym* [24].

THE *MESORHIZOBIUM GENUS*

The *Mesorhizobium* genus is the most recently described rhizobial genus within the order *Rhizobiales*. *Mesorhizobium* type strains were isolated from the root nodules of legume, with the exception of *M. thiogangeticum*, which was obtained from a legume rhizosphere [25]. Within the *Mesorhizobium* genus, the *16S rRNA* gene used as phylogenetic marker to show discrimination between species. However, some *Mesorhizobium* species have 100% identical *16S rRNA* gene sequence, such as *M. metallidurans* and *M. gobiense*. Some phylogenetic marker such as *dnaK*, *dnaJ*, *atpD* and *recA* can also be used for identification of mesorhizobia isolates for better resolution compared to the *16S rRNA* gene [26, 27, 28].

THE LEGUME-RHIZOBIA NODULATION PROCESS

The legume-rhizobia symbiosis results in the formation of root nodules that provide suitable environment for nitrogen fixation by rhizobia. Nodules developmental require two processes: bacterial infection and nodule organogenesis [4].

Generally, leguminous plants excrete specific chemical substances such as; phenolic flavonoid via roots that promote rhizobial proliferation in rhizosphere (Fig 1) [29]. Flavonoid perception attracts the bacteria towards the root and activates the nodulation process through the *nod* genes expression, via the *nod* gene activator *NodD* [30]. *NodD* activates transcription of *nod* boxes promoters, and represents the first level of host-specific recognition [31]. The *nod* genes expression leads to the production of strain-specific lipo-chito-oligosaccharides, also called as Nod factors (NFs). NFs are considered as the second level of host-specific recognition [32, 33]. For instance, *nodC* is involved in the first step of the synthesis of *Nod* factors and is also important in determining the length of the chitin oligosaccharide chain, which is one of the host determinant factors [34, 35].

The presence of compatible rhizobia species and their corresponding NF are enough to trigger nodule development. Normally, the tip of the emerging root hair is the primary target for infection by rhizobia. Attachment of rhizobia to root hairs stimulates root hair deformation and also promotes cortical cell divisions (Figure 1).

Rhizobial infection can occur through root hairs and formation of infection threads (IT) in growing root hairs (Figure 1) [32]. Bacteria released near the growing tip of the IT as infection droplet in the host cell cytoplasm though a process of endocytosis. Bacteria are surrounded by a plant-derived membrane, called peribacteroid membrane, latter it forms the symbiosome (Figure 2) [36].

The membrane-enveloped bacteria continue to divide within the host cells before they differentiate into bacteroids and start nitrogen fixation (Figure 2) [37, 38]. Atmospheric nitrogen is converted into ammonia by bacteroids and is subsequently assimilated into the plant.

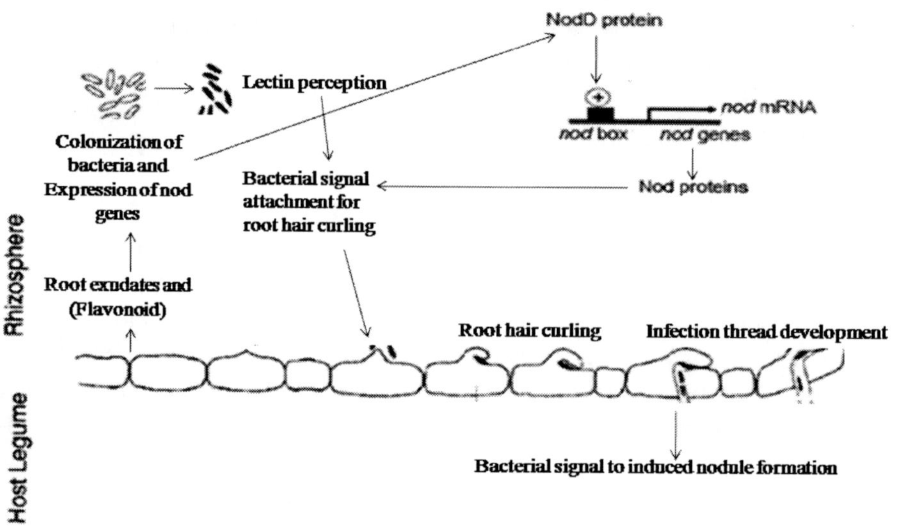

Figure 1. Infection process of the legume-rhizobia symbiosis.

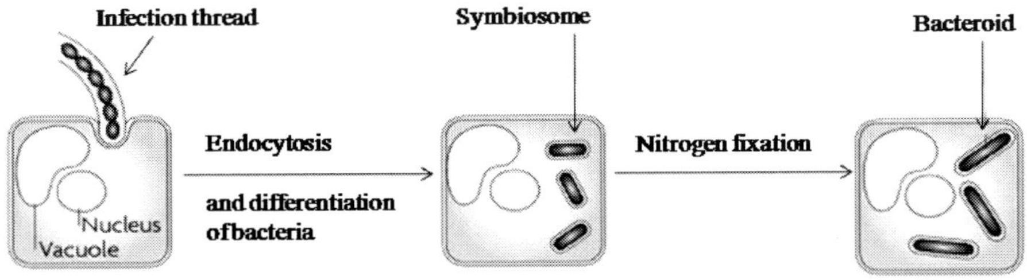

Figure 2. Process of endocytosis of bacteria.

THE IMPORTANCE OF NITROGEN FIXATION

Climate changes and anthropogenic activities lead to eutrophification of soils and fresh water resources, soil degradation, loss of soil fertility, and desertification [39, 40]. Nitrogen is an essential nutrient for plant growth and it can be provided using chemical nitrogen fertilizers. However, the use of nitrogen fertilizers accelerates the depletion of large amounts of fossil fuels, non-renewable energy resources; substantially culminate to environmental pollution through atmospheric emission and leaching of nitrogenous compounds to ground or surface water [41, 42].

Therefore, the biological nitrogen fixation (BNF) is one of the most important contributors to the agricultural sustainability as it is less susceptible to volatilization, denitrification and leaching, avoiding soil and water pollution [43]. Thus, BNF in agrosystems reduces the need for chemical nitrogen fertilizers and consequently reduces global warming and water contamination [41].

MECHANISMS OF BIOLOGICAL NITROGEN FIXATION

The conversion of dinitrogen into ammonia is catalyzed by the nitrogenase enzyme complex by ATP-dependent manner in all diazotrophs. This enzyme complex is composed by two components that are; dinitrogenase reductase (Fe protein) and the dinitrogenase (Mo-Fe protein) [44, 45]. The nitrogenase complex is encoded by the *nif* genes.

The biological nitrogen fixation can be represented by the following equation:

$$N_2 + 8H^+ + 8e^- + 16ATP = 2NH_3 + H_2 + 16ADP + 16Pi$$

The nitrogenase enzyme complex is highly sensitive to oxygen, due to the fact that oxygen reacts with the iron component of the proteins. The free-living aerobic bacteria, such as *Azotobacter* species, developed several mechanisms to overcome such limitation in soils either by maintaining a very low level of oxygen in their cells or by producing extracellular polysaccharides. In the symbiotic nitrogen-fixing organisms, the plant leghemoglobin regulate the supply of oxygen to the nodule tissues. A complete and efficient nitrogen fixation in legume rhizobia symbiosis requires the interaction of several genes present in rhizobia: the *nif* genes and *fix* genes for atmospheric nitrogen fixation, and the *nod*, *nol* and *noe* genes for nodulation [46].

Associations of nitrogen-fixing microorganisms and non-legumes are also important in agriculture. For example, symbiosis of *Frankia* bacteria with non-legume angiosperm families is important to control soil erosion and as well as wood production for economic importance [47]. On the other hand, the photosynthetic cyanobacteria can also fix nitrogen with waterfern *Azolla* and cycads. Nitrogen-fixing cyanobacteria are an important symbiont of coral reefs [48].

EFFECTS OF ENVIRONMENTAL STRESSES ON THE SYMBIOSIS

The biological nitrogen fixation process is known to be influenced by numerous environmental factors such as desiccation and salinity [1]. These factors can directly influence the plant growth during post-nodulation events and consequently the efficient functioning of the nitrogenase enzyme complex, or indirectly affect the nodulation process itself and thus, affect nitrogen fixation.

DESICCATION

Desiccation negatively affects the several important steps in the process of symbiosis like nodulation process and nitrogen fixation [49]. Prokaryotes including bacterium have capacity to tolerate desiccation stress by the synthesis or accumulation of the osmolytes [50]. Trehalose is one of such osmolyte that can adjust osmosis and protect macromolecules of the cell [51]. A number of genes involved in trehalose metabolism are known and trehalose-6-phosphate synthetase (*TPS)* is one of them [52]. Trehalose concentration in rhizobia reported to increase in desiccation stress; stabilizes membranes and proteins by glass formation [53].

Desiccation stress is encountered by terrestrial bacteria but the genetic mechanisms to understand the resistance to this stress is still infancy. Bacterial systems impart functional roles such as energy storage as glycogen [54] and glycogen debranching enzyme (*glgX*) undergo metabolism of glycogen to increases the viability of cell under stress condition [55].

SALT STRESS

Salinity in soil affects about 800 mha of total land area that is more severe in arid and semiarid regions [56, 57]. Salt stress may inhibit the steps of the symbiosis and nitrogen fixation [1]. Rhizobia subjected to salt stress may lead to changes in cell morphology and also modified the pattern of extracellular polysaccharides (EPS) [58, 59]. Many microbes, including rhizobia, use distinct mechanisms for osmotic adaptation under salt stress [1]. Intracellular accumulation of osmolytes, including amino acids, sugars, and polyamines have been observed in some species of rhizobia under salt stress [60]. Some compatible solutes can be used either as nitrogen or carbon sources for growth purpose during salt stress such as, glutamate and proline [60, 61]. The *glgA2*, *glgB2*, and *glgX* genes involved in glycogen metabolism are expressed at higher levels when exposure to salt stress, indicating that glycogen accumulates during salt stress [61].

The transport systems of ions may also be involved in rhizobia in response to salt stress. Due to the intracellular accumulation of potassium and some polyamines identified by *kup* gene, to confer salt tolerance in *Rhizobium tropici* and BetS (betaine/proline transporter) gene in *E. meliloti* [[1, 62, 63]. Several genes such as *omp*10 (outer membrane lipoprotein), *ntrY* (nitrogen regulator), *alaS* (alanyl-tRNA synthase), *greA* (transcription elongation factor), *nuoL* (NADH dehydronase I chain L protein) and *dnaJ* (heat-shock chaperone) were also involved in salt stress response in rhizobia [62, 64].

But the tolerance mechanisms of rhizobia to overcome desiccation and salt stress remains unknown because it is a complex phenomenon involving many physiological and biochemical processes that reflect the gene expression pattern.

Molecular Chaperones

Proteins are of permanent risk of unfolding under environmental stress conditions [65]. Bacteria have evolved several mechanisms that ensure protein folding and promote homeostasis under stress conditions. One of them is the activation of proteins such as molecular chaperones, proteases, and regulatory factors under stress condition [2, 66]. The molecular chaperones enable denatured proteins to acquire their native structure.

Protein Folding

In the bacterial cytosol, the folding of new proteins is assisted by three major molecular chaperone complexes such as *DnaKJ-GrpE* and *GroELS* complexes [67]. Chaperone first interacts with nascent polypeptide chains coming from the bacterial ribosome.

DnaKJ-GrpE and *GroELS* are multi-component molecular machines that promote folding through ATP and cofactor-regulated binding. The two systems act sequentially, where by *DnaKJ* system interacts upstream with nascent and newly synthesized polypeptides and the *GroESL* function downstream in the final folding of those proteins that fail to reach native state by cycling on *DnaK* system alone [68].

DnaK binds to solvent-exposed hydrophobic regions of unfolded polypeptide chains, assisting the folding in an ATP-driven process that is regulated by the co-chaperone *DnaJ* and the nucleotide-release cofactor *GrpE* [67]. Hydrolysis of ATP to ADP is strongly accelerated by *DnaJ* leading to stable peptide binding and can recruit *DnaK* to protein substrates (Fig. 3) [69].

The folding activity of *GroEL* requires the cooperation of the co-chaperone *GroES* to form the *GroESL* complex in the presence of ATP [3]. The apical domains of *GroEL* present hydrophobic amino-acid residues for substrate binding in the ring centre and substrate encapsulation by *GroES* (Fig. 3). The complex *GroESL* can assist the protein folding by two different ways: a *cys-* folding action or a *trans-* folding mechanism.

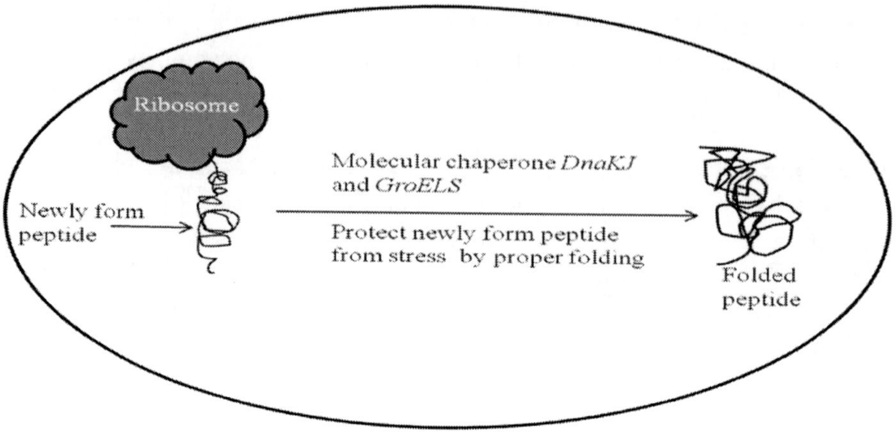

Figure 3. Process of protein folding.

Major Molecular Chaperones in Rhizobia and Their Role in Symbiosis

The molecular chaperones systems are less studied in rhizobia. Several copies of the *groEL* gene are found in rhizobia genomes. *Bradyrhizobium japonicum*, *Ensifer meliloti*, *Mesorhizobium sp.* MAFF303099, *Rhizobium etli* and *Rhizobium leguminosarum,* all have three or more *groEL* homologues and these are differentially induced [70]. For example, in *B. japonicum*, *groESL1,4,5* are heat inducible and *groESL3* is induced by low oxygen conditions [71]. In *R. Leguminosarum* only one of the three *groEL* homologues is needed for normal growth [72]. In *E. meliloti*, two of the five *groESL* copies are heat inducible and each one is regulated by a distinct mechanism [73, 74]. The *DnaKJ* system is less studied compared to the *GroESL* system in rhizobia. As in the most bacteria, including rhizobia the *dnaK* gene is found as a single copy gene. While, its co-chaperone *dnaJ* has several copies. *Rizobium tropici dnaJ* mutants showed higher sensitivity to salt stress condition while *B. japonicum dnaJ* mutants displayed slower growth at high temperatures [62, 75].

The involvement of molecular chaperones in the nitrogen-fixing symbiosis seems to be controversial, probably due to their involvement in different stages in the symbiosis process of rhizobia. For example, *dnaJ* is required for effective symbiosis of *R. leguminosarum* bv. *Phaseoli*, however in *B. japonicum* the symbiotic performance of *dnaJ* mutants was not altered [75, 76, 77]. Nogales et al. (2002) found that a *dnaJ* mutant of *Rhizobium tropici* was able to form nodules in *Phaseolus vulgaris*, however this mutant showed low nitrogenase activity that account to reduced plant growth as well as less nitrogen content in the plant shoots. In *E. meliloti,* a *groESL*5 mutant strain has the ability of normal symbiotic nitrogen fixation [73]. From the five *groESL* operons in the *E. meliloti* genome, only one operon (*groEL1*) was found to be involved in symbiosis [70]. All single mutants in *E. meliloti were* viable but double mutants are depleted in their symbiotic phenotype [78, 79]. Similarly, *B. japonicum* mutants that individually lack one *groEL* gene do not change the symbiotic phenotype while double mutation on *groEL*3 and *groEL*4 genes affects the symbiotic performance [78]. These two copies are most abundant in *GroEL* pool in bacteroid and are required for the formation of a functional nitrogenase in *B. japonicum* [78].

CONCLUSION

Impact of adverse environmental conditions plays essential role in the control of legume-rhizobia interactions. They can arrest the growth, multiplication and survival of rhizobia in soil rhizosphere. The adverse environmental conditions may also have depressive effect on the steps involved in legume-rhizobium symbiosis such as molecular signalling, infection process, nodule development and function, resulting in low nitrogen fixation and crop yield. Selection of hosts and their nitrogen-fixing endosymbionts that are tolerant to a broad range of environmental stresses is important for agriculture system.

The various chaperone factors protect non-native protein chains from misfolding and aggregation in adverse condition. Evidence accumulated over the last decade indicates that many newly synthesized proteins require a complex cellular machinery of molecular chaperones and the input of metabolic energy to reach their native states efficiently. Recent years have seen major advances in understanding of the basic mechanisms of chaperone-assisted protein folding. Future efforts will define more comprehensively the rules for the thousands of different proteins in a cell undergoing the chaperone machinery. Global analyses of chaperone usage and folding properties in eukaryotes and prokaryotes will address these questions through a combination of proteomics and high-throughput protein expression.

REFERENCES

[1] Zahran, H.H., Moharram, A.M., Mohammad, H.A., (1992). Some ecological and physiological-studies on bacteria isolated from salt-affected soils of Egypt. *J. Basic Microbiol.* 32, 405-13.

[2] Hartl, F.U., Bracher, A., Hayer-Hartl, M., (2011). Molecular chaperones in protein folding and proteostasis. *Nature* 475, 324-332.

[3] Chaudhuri, T.K., Verrna, V.K., Maheshwari, A., (2009). GroEL assisted folding of large polypeptide substrates in *Escherichia coli*: Present scenario and assignments for the future. *Prog. Biophys. Mol. Biol.* 99, 42-50.

[4] Oldroyd, G.E.D., Murray, J.D., Poole, P.S., Downie, J.A., (2011). The rules of engagement in the legume-rhizobial symbiosis. *Annu. Rev. Genet.* 45, 119-144.

[5] FAO., (2008). FAOSTAT. Food and agriculture organization of the United Nations.

[6] Nour, S.M., Fernandez, M.P., Normand, P., Cleyet-Marel, J.C., (1994). *Rhizobium ciceri* sp. nov., consisting of strains that nodulate chickpeas (*Cicer arietinum* L.). *Int. J. Syst. Bacteriol.* 44, 511-522.

[7] Nour, S.M., Cleyet-Marel, J.C., Normand, P., Fernandez, M.P., (1995). Genomic heterogeneity of strains nodulating chickpeas (*Cicer arietinum* L.) and description of *Rhizobium mediterraneum* sp. nov. *Int. J. Syst. Bacteriol.* 45, 640-648.

[8] Romdhane, S., Aouani, M.E., Mhamdi, R., (2007). Inefficient nodulation of chickpea (*Cicer arietinum* L.) in the arid and Saharan climates in Tunisia by *Sinorhizobium meliloti* biovar *medicaginis*. *Ann. Microbiol.* 57, 15-19.

[9] Laranjo, M., Alexandre, A., Rivas, R., Velázquez, E., Young, J.P.W., Oliveira, S., (2008). Chickpea rhizobia symbiosis genes are highly conserved across multiple *Mesorhizobium* species. *FEMS Microbiol. Ecol.* 66, 391-400.

[10] McDermott, T.R., Graham, P.H., Brandwein, D.H., (1987). Viability of *Bradyrhizobium japonicum* bacteroids. *Arch. Microbiol.* 148, 100-106.

[11] Park, M., Kim, C., Yang, J., Lee, H., Shin, W., Kim, S., Sa, T., (2005). Isolation and characterization of diazotrophic growth promoting bacteria from rhizosphere of agricultural crops of Korea. *Microbiol. Res.* 160, 127-133.

[12] Vassilev, N., Vassileva, M., Nikolaeva, I., (2006). Simultaneous P solubilizing and biocontrol activity of microorganisms: potentials and future trends. *Appl. Microbiol. Biotechnol.* 71, 137-144.

[13] Spaepen, S., Vanderleyden, J., Okon, Y., (2009). Plant growth-promoting actions of rhizobacteria. In: VanLoon LC. (ed). *Plant Innate Immunity,* p283-320.

[14] Vilchez, M.,(2011). Biotechnological uses of desiccation-tolerant microorganisms for the rhizoremediation of soils subjected to seasonal drought. *Appl. Microbiol. Biotechnol.* 91, 1297-1304.

[15] Beijerinck, M.W., (1888). Die Bacterien der Papilionaceenknölchen. Botanische Zeitung 46, 797-804.

[16] Dreyfus, B., Garcia, J.L., Gillis, M., (1988). Characterization of *Azorhizobium caulinodans* gen nov, sp nov, a stem-nodulating nitrogen-fixing *bacterium* isolated from *Sesbania rostrata*. *Int. J. Syst. Bacteriol.* 38, 89-98.

[17] Jarvis, B.D.W., van Berkum, P., Chen, W.X., Nour, S.M., Fernandez, M.P., Cleyet-Marel, J.C., Gillis, M., (1997). Transfer of *Rhizobium loti*, *Rhizobium huakuii*, *Rhizobium ciceri*, *Rhizobium mediterraneum*, and *Rhizobium tianshanense* to *Mesorhizobium* gen. nov. *Int. J. Syst. Bacteriol.* 47, 895-898.

[18] Moulin, L., Munive, A., Dreyfus, B., Boivin-Masson, C., (2001). Nodulation of legumes by members of the beta-subclass of *Proteobacteria*. *Nature* 411, 948-950.

[19] Chen, W.M., Laevens, S., Lee, T.M., Coenye, T., De Vos, P., Mergeay, M., Vandamme, P., (2001). *Ralstonia taiwanensis* sp nov., isolated from root nodules of *Mimosa* species and sputum of a cystic fibrosis patient. *Int. J. Syst. Evol. Microbiol.* 51, 1729-1735.

[20] Valverde, A., Velázquez, E., Gutiérrez, C., Cervantes, E., Ventosa, A., Igual, J.M., (2003). *Herbaspirillum lusitanum* sp. nov., a novel nitrogen-fixing bacterium associated with root nodules of *Phaseolus vulgaris*. *Int. J. Syst. Evol. Microbiol.* 53, 1979-1983.

[21] Young, J.M., Kuykendall, L.D., Martínez-Romero, E., Kerr, A., Sawada, H., (2001). A revision of *Rhizobium* Frank 1889, with an emended description of the genus, and the inclusion of all species of *Agrobacterium* Conn 1942 and *Allorhizobium undicola* de Lajudie et al. 1998 as new combinations: *Rhizobium radiobacter*, *R. rhizogenes*, *R. rubi*, *R. undicola* and *R. Vitis*. *Int. J. Syst. Evol. Microbiol.* 51, 89-103.

[22] Vinuesa, P., Rojas-Jimenez, K., Contreras-Moreira, B., Mahna, S.K., Prasad, B.N., Moe, H., Selvaraju, S.B., Thierfelder, H., Werner, D., (2008). Multilocus Sequence analysis for assessment of the biogeography and evolutionary genetics of four *Bradyrhizobium* species that nodulate soybeans on the Asiatic Continent. *Appl. Environ. Microbiol.* 74, 6987-6996.

[23] MacLean, A.M., Finan, T.M., Sadowsky, M.J., (2007). Genomes of the symbiotic nitrogen-fixing bacteria of legumes. *Plant Physiol.* 144, 615-622.

[24] Sullivan, J.T., Trzebiatowski, J.R., Cruickshank, R.W., Gouzy, J., Brown, S.D., Elliot, R.M., Fleetwood, D.J., McCallum, N.G., Rossbach, U., Stuart, G.S., Weaver, J.E.,

Webby, R.J., de Bruijn, F.J., Ronson, C.W., (2002). Comparative sequence analysis of the symbiosis island of *Mesorhizobium loti* strain R7A. *J. Bacteriol.* 184, 3086-3095.

[25] Ghosh, W., Roy, P., (2006). *Mesorhizobium thiogangeticum* sp. nov., a novel sulfuroxidizing chemolithoautotroph from rhizosphere soil of an Indian tropical leguminous plant. *Int. J. Syst. Evol. Microbiol.* 560, 91-97.

[26] Stepkowski, T., Czaplinska, M., Miedzinska, K., Moulin, L., (2003). The variable part of the *dnaK* gene as an alternative marker for phylogenetic studies of rhizobia and related alpha *Proteobacteria*. *Syst. Appl. Microbiol.* 26, 483-494.

[27] Vinuesa, P., Silva, C., Lorite, M.J., Izaguirre-Mayoral, M.L., Bedmar, E.J., Martínez-Romero, E., (2005). Molecular systematic of rhizobia based on maximum likelihood and Bayesian phylogenies inferred from rrs, *atpD*, *recA* and *nifH* sequences, and their use in the classification of *Sesbania* microsymbionts from Venezuelan wetlands. *Syst. Appl. Microbiol.* 28, 702-716.

[28] Alexandre, A., Laranjo, M., Young, J.P.W., Oliveira, S., (2008). *dnaJ* is a useful phylogenetic marker for alpha-proteobacteria. *Int. J. Syst. Evol. Microbiol.* 58, 2839-2849.

[29] Redmond, J.W., Batley, M., Djordjevic, M.A., Innes, R.W., Kuempel, P.L., Rolfe, B.G., (1986). Flavones induce expression of nodulation genes in *Rhizobium*. *Nature* 323, 632-635.

[30] Lindström, K., Terefewrok, Z., Suominen, L., Lortet, G., (2002). Signalling and development of *Rhizobium*-legume symbioses. Biology and Environment. *Proc. Royal Irish Acad.* 1, 61-64.

[31] Schultze, M., Kondorosi, A., (2008). Regulation of symbiotic root nodule development. *Ann. Rev. Gene* 32, 33-57.

[32] Oldroyd, G.E.D., Downie, J.M., (2008). Coordinating nodule morphogenesis with rhizobial infection in legumes. *Annu. Rev. Plant. Biol.* 519-546.

[33] Perret, X., Staehelin, C., Broughton, W., (2000). Molecular basis of symbiotic promiscuity. *Microbiol Mol Biol Rev.* 64, 180-201.

[34] Kamst, E., Pilling, J., Raamsdonk, L.M., Lugtenberg, B.J.J., Spaink, H.P., (1997). *Rhizobium* nodulation protein NodC is an important determinant of chitin oligosaccharide chain length in nod factor biosynthesis. *J. Bacteriol.* 179, 2103-2108.

[35] Djordjevic, M.A., Weinman, J.J., (1991). Factors determining host recognition in the clover-*Rhizobium* symbiosis. *Aust. J. Plant Physiol.* 18, 543-557.

[36] Jones, K.M., Kobayashi, H., Davies, B.W., Taga, M.E., Walker, G.C., (2007). How rhizobial symbionts invade plants: the *Sinorhizobium–Medicago* model. *Nature Rev. Microbiol.* 5, 619-633.

[37] Roth, L.E., Stacey, G., (1989a). Bacterium release into host cells of nitrogen-fixing soybean nodules: the symbiosome membrane comes from three sources. *Eur. J. Cell Biol.* 49, 13-23.

[38] Roth, L.E., Stacey, G., (1989b). Cytoplasmic membrane systems involved in bacterium release into soybean nodule cells as studied with two *Bradyrhizobium japonicum* mutant strains. *Eur. J. Cell Biol.* 49, 24-32.

[39] Spiertz, J.H.J., (2010). Nitrogen, sustainable agriculture and food security. A review *Agrom. Sustain. Dev.,* 30, 43-55.

[40] Gordon, L.J., Finlayson, C.M., Falkenmark, M., (2010). Managing water in agriculture for food production and other ecosystem services. *Agric. Water Manage.* 97, 512-519.

[41] Bohlool, B.B., Ladha, J.K., Garrity, D.P., George, T., (1992). Biological nitrogen fixation for sustainable agriculture: a perspective. *Plant Soil* 141, 1-11.

[42] Velthof, G.L., Oudendag, D., Witzke, H.P., Asman, W.A.H., Klimont, Z., Oenema, O., (2009). Integrated assessment of nitrogen emissions form agriculture in EU-27 using miterra europe. *J. Environ. Qual.* 38, 402-417.

[43] Garg, N., Geetanjali, (2007). Symbiotic nitrogen fixation in legume nodules: process and signaling. A review. *Agron. Sustain. Dev.* 27, 59-68.

[44] Peters, J.W., Fisher, K., Dean, D.R., (1995). Nitrogenase structure and function: a biochemical-genetic perspective. *Annu. Rev. Microbiol.* 49, 335-366.

[45] Hageman, R.V., Burris, R.H., (1978). Dinitrogenase and nitrogenase reductase associate and dissociate with each catalytic cycle. *Proc. Natl. Acad. Sci. USA* 75, 2699-2702.

[46] Kaminski, P.A., Batut, J., Boistard, P., (1998). A survey of symbiotic nitrogen fixation by rhizobia. In: (Spaink, H.P., Kondorosi, A., Hooykaas, P.J.J., Eds.), The Rhizobiaceae, Kluwer Academic Publishers, 431-460.

[47] Benson, D.R., Dawson, J.O., (2007). Recent advances in the biogeography and genecology of symbiotic Frankia and its host plants. *Physiol. Plant.* 130, 318-330.

[48] Lesser, M.P., Mazel, C.H., Gorbunov, M.Y., Falkowski, P.G., (2004). Discovery of symbiotic nitrogen-fixing cyanobacteria in corals. *Science* 305, 997-1000.

[49] Zahran, H.H., (1999). Rhizobium-Legume Symbiosis and Nitrogen Fixation under Severe Conditions and in an Arid Climate. *Microbiol. Mol. Biol. Rev.* 63, 968-989.

[50] Potts, M., (1994). Desiccation tolerance of prokaryotes. *Microbiol. Rev.* 58, 755-805.

[51] Halverson, L.J., Firestone, M.K. (2000). Differential effects of permeating and nonpermeating solutes on the fatty acid composition of *Pseudomonas putida*. *Appl. Environ. Microbiol.* 66, 2414-2421.

[52] Harris, R. F. 1981. Effect of water potential on microbial growth and activity, p. 23–95. *In* J. F. Parr, W. R. Gardner, and L. F. Elliott (ed.), Water potential relations in soil microbiology. Soil Science Society of America special publication no. 9. Soil Science Society of America, Madison, Wis.

[53] Crowe, J.H., Carpenter, J.F., Crowe, L.M., (1998). The role of vitrification in anhydrobiosis. *Annu. Rev. Physiol.* 60, 73-103.

[54] Preiss, J., Romeo, T., (1994). Molecular biology and regulatory aspects of glycogen biosynthesis in bacteria. *Prog. Nucleic Acid Res. Mol. Biol.* 47, 299-329.

[55] Romeo, T., Kumar, A., Preiss, J., (1988). Analysis of the *E. coli* glycogen gene cluster suggests that catabolic enzymes are encoded among the biosynthetic genes. *Gene* 70, 363-376.

[56] Munns, R., Tester, M., (2008). Mechanisms of salinity tolerance. *Annu. Rev. Plant Biol.* 651-681.

[57] Tester, M., Davenport, R., (2003). Na+ tolerance and Na+ transport in higher plants. *Ann. Bot.* 91, 503-527.

[58] Vanderlinde, E.M., Harrison, J.J., Muszynski, A., Carlson, R.W., Turner, R.J., Yost, C.K., (2010). Identification of a novel ABC transporter required for desiccation tolerance, and biofilm formation in *Rhizobium leguminosarum* bv. Viciae 3841. *FEMS Microbiol. Ecol.* 71, 327-340.

[59] Vanderlinde, E.M., Muszynski, A., Harrison, J.J., Koval, S.F., Foreman, D.L., Ceri, H., Kannenberg, E.L., Carlson, R.W., Yost, C.K., (2009). *Rhizobium leguminosarum*

biovar viciae 3841, deficient en 27- hydroxyoctacosanoate-modified lipopolysaccharide, is impaired in dessication tolerance, biofilm formation and motility. *Microbiol. SGM* 155, 3055-3069.

[60] Rüberg, S., Tian, Z.X., Krol, E., Linke, B., Meyer, F., Wang, Y., Pühler, A., Weidner, S., Becker, A., (2003). Construction and validation of a *Sinorhizobium meliloti* whole genome DNA microarray: genome-wide profiling of osmoadaptive gene expression. *J. Biotechnol.* 106, 255-268.

[61] Domínguez-Ferreras, A., Perez-Arnedo, R., Becker, A., Olivares, J., Soto, M.J., Sanjuan, J., (2006). Transcriptome profiling reveals the importance of plasmid pSymB for osmoadaptation of *Sinorhizobium meliloti*. *J. Bacteriol.* 188, 7617-7625.

[62] Nogales, J., Campos, R., BenAbdelkhalek, H., Olivares, J., Lluch, C., Sanjuan, J., (2002). *Rhizobium tropici* genes involved in free-living salt tolerance are required for the establishment of efficient nitrogen-fixing symbiosis with *Phaseolus vulgaris*. *Mol. Plant Micr. Int.* 15, 225-232.

[63] Boscari, A., Mandon, K., Dupont, L., Poggi, M.C., Le Rudulier, D., (2002). BetS is a major glycine betaine/proline betaine transporter required for early osmotic adjustment in *Sinorhizobium meliloti*. *J. Bacteriol.* 184, 2654-2663.

[64] Wei, W., Jiang, J., Li, X., Wang, L., Yang, S.S., (2004). Isolation of salt-sensitive mutants from *Sinorhizobium meliloti* and characterization of genes involved in salt tolerance. *Lett. Appl. Microbiol.* 39, 278-283.

[65] Jahn, T.R., Radford, S.E., (2005). The Yin and Yang of protein folding. *FEBS J.* 272, 5962-5970.

[66] Wickner, S., Maurizi, M.R., Gottesman, S., (1999). Post-translational quality control: folding, refolding, and degrading proteins. *Science* 286, 1888-1893.

[67] Genevaux, P., Keppel, F., Schwager, F., Langendijk-Genevaux, P.S., Hartl, F.U., Georgopoulos, C., (2004). In vivo analysis of the overlapping functions of DnaK and trigger factor. *EMBO Rep.* 5, 195-200.

[68] Mogk, A., Mayer, M.P., Deuerling, E., (2002). Mechanisms of Protein Folding: Molecular Chaperones and Their Application in Biotechnology. *Chem. Bio. Chem.* 9, 807-814.

[69] Kampinga, H.H., Craig, E.A., (2010).The HSP70 chaperone machinery: J proteins as drivers of functional specificity. *Nature Rev. Mol. Cell Biol.* 11, 579-592.

[70] Ogawa, J., Long, S.R., (1995). The *Rhizobium meliloti groELc* locus is required for regulation of early nod genes by the transcription activator *nodD*. *Genes Dev.* 9, 714-729.

[71] Babst, M., Hennecke, H., Fischer, H.M., (1996). Two different mechanisms are involved in the heat-shock regulation of chaperonin gene expression in *Bradyrhizobium japonicum*. *Mol. Microbiol.* 19, 827-839.

[72] Rodriguez-Quinones, F., Maguire, M., Wallington, E.J., Gould, P.S., Yerko, V., Downie, J.A., Lund, P.A., (2005). Two of the three *groEL* homologues in *Rhizobium leguminosarum* are dispensable for normal growth. *Arch. Microbiol.* 183, 253-265.

[73] Mitsui, H., Sato, T., Sato, Y., Ito, N., Minamisawa, K., (2004). *Sinorhizobium meliloti* RpoH1 is required for effective nitrogen-fixing symbiosis with alfalfa. *Mol. Genet. Genomics* 271, 416-425.

[74] Bittner, A.N., Oke, V., (2006). Multiple *groESL* operons are not key targets of RpoH1 and RpoH2 in *Sinorhizobium meliloti*. *J. Bacteriol.* 188, 3507-3515.

[75] Minder, A.C., Narberhaus, F., Babst, M., Hennecke, H., Fischer, H.M., (1997). The *dnaKJ* operon belongs to the sigma (32)-dependent class of heat shock genes in *Bradyrhizobium japonicum*. *Mol. General Genetics* 254, 195-206.

[76] Karunakaran, R., Ramachandran, V.K., Seaman, J.C., East, A.K., Mouhsine, B., Mauchline, T.H., Prell, J., Skeffington, A., Poole, P.S., (2009). Transcriptomic analysis of *Rhizobium leguminosarum* biovar *viciae* in symbiosis with host plants *Pisum sativum* and *Vicia cracca*. *J. Bacteriol.* 191, 4002-4014.

[77] Labidi, M., Laberge, S., Vezina, L.P., Antoun, H., (2000). The *dnaJ* (hsp40) locus in *Rhizobium leguminosarum* bv. *phaseoli* is required for the establishment of an effective symbiosis with *Phaseolus vulgaris*. *Mol. Plant Micr. Int.* 13, 1271-1274.

[78] Fischer, H.M., Schneider, K., Babst, M., Hennecke, H., (1999). GroEL chaperonins are required for the formation of a functional nitrogenase in *Bradyrhizobium japonicum*. *Arch. Microbiol.* 171, 279-289.

[79] Bittner, A.N., Foltz, A., Oke, V., (2007). Only one of five *groEL* genes is required for viability and successful symbiosis in *Sinorhizobium meliloti*. *J. Bacteriol.* 189, 1884-1889.

In: Microbes in Soil and Their Agricultural Prospects
Editors: K. Kishore Choudhary and D. Wattal Dhar
ISBN: 978-1-63482-824-6

Chapter 3

METHYLOTROPHIC BACTERIA IN RELATION TO SOIL AND PLANT HEALTH

D. D. Giri [1,3,*], ***Ajay Kumar*** [3], ***Lata Kumari*** [2], ***Pardeep Singh*** [2], ***D. B. Pal*** [1], ***K. D. Pandey*** [3] ***and P. K. Mishra*** [1]

[1]Department of Chemical Engineering and Technology, Indian institute of Technology, Banaras Hindu University, Varanasi, India
[2]Department of Chemistry, Indian institute of Technology, Banaras Hindu University, Varanasi, India
[3]Department of Botany, Banaras Hindu University, Varanasi, India

ABSTRACT

Methylotrophs are a diverse group of microorganisms that can use reduced one-carbon compounds, such as methanol or methane, as the carbon source for their growth; and multi-carbon compounds that contain no carbon bonds, such as dimethyl ether and dimethylamine. Utilization of various types of organic compounds present in the coal mine spoil as carbon substrate is the the reason for better survival of facultative methylotrophic bacteria in stressful and nutrient poor coal mine spoils. Methylotrophic bacteria significantly modify the physicochemical properties of degraded soil. The nutrients released from death and decay of methylotrophic bacteria facilitate growth of heterotrophic bacteria and increases microbial diversity of the degraded land. This chapter gives brief description of methylotrophic bacteria and their role in improvement of soil and plant health have given special attension.

Keywords: Methylotrophic bacteria, N transformation, nitrification, phytohormone, root nodulation

* Correspondence address: Department of Chemical Engineering & Technology, IIT (Banaras Hindu University), Varanasi-221005, India. E-mail –ddgiri1@gmail.com.

INTRODUCTION

Methylotrophs are the group of organisms that utilize reduced one carbon compounds and multicarbon compounds lacking carbon–carbon bonds [1]. Some examples of C_1 compounds utilized by methylotrophs are methanol and methylamines. Majority of methylotroph utilize other non-methylated one-carbon compounds such as formaldehyde, formic acid, thiocyanate, carbon disulfide and carbonyl sulfide. Some facultative methylotrophs are capable to use a range of substituted halomethanes, methanethiol and the chlorofluorocarbons. The multicarbon substrates for facultatative methylotrophs are ethanol, propenol and butenol [2]. These substrates are very commonly available in the environment and many of them are part of in plant metabolism. Methanol originates from decomposition of lignin and demethylation of pectin of cell wall in the plants. The amount of methanol emitted from the plants is more than 10^5 ton carbon that account for about half of the total volatile organic carbon of the atmosphere [3]. Some higher plants and algae produce halomethanes (< 5 million ton carbon per year), methylsulfurous compounds, and methylated amines that contribute significantly to global carbon turnover [3]. The substrates for the methylotrophs growth are abundant in the soil as well as in plants and animal tissues [4]. Many methylotrophic bacteria are facultative heterotrophs that grow on multicarbon substrates. Actually, some heterotrophic bacteria exhibit methylotrophic nature [4], and some well-studied methylotrophs grow chemolithoautotrophically e.g., *Xanthobacter* sp and *Paracoccus* sp, [5]. *Methylobacterium* sp. grow mixotrophically with thiosulfate and methanol [5]. Synergistic metabolism of a broad range of C_1 compounds is common in the marine methylotrophic bacterium HTCC2181. This aquatic methylotrophsurvive in the marine habitat having low methanol concentration (~100nM). Growth rate and cell density of HTCC2181 increased with increasing methanol concentration. The bacteria was unable to grow on methylated (C_1) compounds, methyl chloride, trimethylamine-oxide (TMAO) or di-methylsulfoniopropionate (DMSP) when they were the sole substrates but bacterial growth significantly enhanced when any of these methylated compounds were provided in the presence of a limiting concentration of methanol [6].

Plant Tissues

Methylotrophs have been isolated from leaf and seed surfaces of monocotyledonous and dicotyledonous plants [7-8]. The cell density of the methylotrophs was very high on the plant surfaces. They were very densely colonized in seeds of 140 angiospermic plants species. The number of methylotrophs on the leaf surface differed with the plants species. The most widely distributed methylotrophs on the plant surface are pink pigmented facultative methylotroph (PPFMs) which occurs on majority of the plants species. In commercial green perilla plant, number of methylotrophic bacteria ranged between 2.0 - 4.1 x 10^7 CFU/g of the leaf weight. The number corresponds near 15% of the total bacteria present on leaf surface.

Rhizosphere

In addition to the plant surfaces, methylotrophs were densely colonized in rhizosphere of 140 angiospermic plants species [7, 9, 10]. A facultative methylotroph *Flavobacterium glycines* sp. nov., a isolated from the rhizosphere of soybean (9) and *Methylophilus rhizosphaerae* sp. nov., a restricted facultative methylotroph isolated from rice rhizosphere soil [11].

Bulk Soil

Methylotrophic bacterial has been isolated from the bulk soil as well as from the degraded coalmine spoils [10]. Population of methylotrophs was lower in the bulk soil and it was in the range of half to one tenth to that of the rhizospheric population.The reason could be the abundant availability of the nutrient in the rhizosphere.

Mechanism of C_1 compound Utilization

Formaldehyde is produced from methanol in the periplasm of the cell and it is transferred into the cytoplasm. Farmaldehyde is common substrate for catabolic and anabolic pathway. In the catabolic pathway, formaldehyde is oxidized to CO_2, whereas its assimilation via the serine cycle results in the cell biomass synthesis. The basic steps have been presented in the following figure

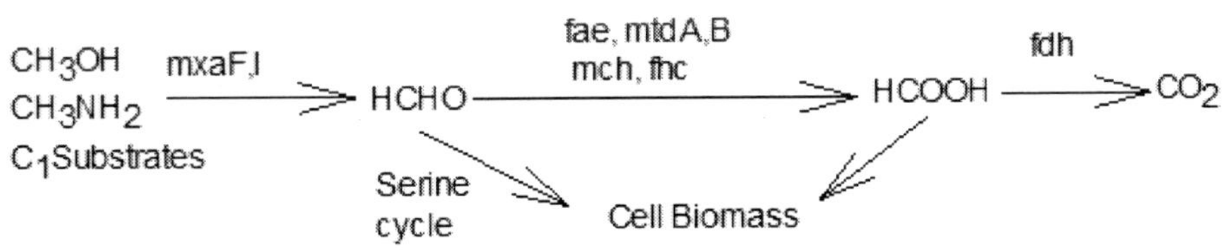

Genes involved in the process are-
mxaF,I - methanol dehydrogenase,
fae, tetrahydromethanopterin (H4MPT)-dependent formaldehyde-activating enzyme
mtdA,B NADP-dependent methylene tetrahydromethanopterin/tetrahydrofolate dehydrogenase
mch, methenyl-tetrahydromethanopterin cyclohydrolase
fhc, formyltransferase/hydrolase complex
fdh, formate dehydrogenase

Figure 1. Pathway for C substrates utilization by methylotrophs.

Plant Species Specificity of Methylotrophs

Seeds were collected from the perilla plant grown in the Queto and shown at different geographical locations of the Japan. The leaves and seeds of the plants grown at different geographical locations of Japan were investigated for methylotrophic population. The PPFMs isolated from leaves and seeds the plants at different geographical localities were very much similar. The result clearly indicated that PPMF directly transmitted from seeds to leaves and not from the environmental sources. The result also emphasized on host plant species-level specificity of methylotrophs [12]. However, there is also a strong possibility of methylotrophs transfer on the leaf surface via air transfer of soil particles.

Plant Methylotrophs Interaction

There are numerous indications for interaction of methylotrophs with plants, however their biochemical and genetic details are lacking. In the genome sequence of *Methylobacterium extorquens* AM1 contains a number of open reading frames related to plant-microbes interaction are similar to Rhizobia and *Agrobacterium*. Molecular basis of *Methylobacterium*-plant interactions come with the work of Koenig et al. [13] who stated that *Methylobacterium* excrete *trans*-zeatin derived from tRNA at low levels in the pure culture. Actually tRNA-derived zeatin in the bacteria and the plants are mainly of *cis*-isomer. The low-level of *trans*-zeatin affect seed germination in the host plants. The *Methylobacterium*-specific stimulation of seed germination was absent in a mutant incapable of producing *trans*-zeatin[14-15]. This work provided mechanistic evidence for cytokinin production in *Methylobacterium*.

PPFMs actively participate in plant growth promotion [15-16]. They aggregate on the aerial parts by quorum-sensing (QS) and affect the multicellular behavior of bacterial community. The occurrence of QS systems in *Methylobacterium* is widespread [18]. The two component QS systems, with production of a novel long chained C14:2 (N-tetra decenoyl) AHL molecules has been described in *Methylobacterium extorquens* AM1 [19].

Methylotrophs in Improvement of Soil health and Fertility

Soil Reclamation

The methylotrophs play an important role in soil C and N transformation process, so they affect soil fertility. The bare coal mine spoil population of methylotrophs increased continuously in the revegetated soil with time (3-12 years) and reached comparable to the natural forest soils. It indicated that the methylotrophs could be useful in reclaiming the coalmine dump. The growth of methylotrophs in the coal mine soil facilitated growth of other group of microorganisms. The methylotrophs enhance soil nutrient availability by mobilizing the nutrients available in the organic residues of the coal. However, the process of reclamation of dump could be further enhanced by applying some nutrients like urea because majority of methylotrophs present in the studied soil were utilizing it as carbon and or nitrogen source [10].

Methylotrophic Nitrification

Methanotrophic nitrifying consortium of humisol showed CH_4-dependent nitrification due to the presence of an obligately methylotrophic *Methylobacillus* sp. growing on CH_3OH and but it was unable to nitrify. Some *Pseudomonas* growing on the culture filtrate of *Methylobacillus* produced NO_2^-, NO_3^- and N_2O from NH_2OH, and NO_3^- from NO_2^- whereas other *Pseudomonas* isolates produced NO_3^- from NH_4^+ or NO_2^-, and N_2O from NH_2OH [20]. So, it can be concluded that methylotrophs modify the environment for better nitrification in the soil by other bacterial species present.

Methylotrophic Denitrification

The Methylophaga sp. strain JAM1 isolated from a denitrification system was well adapted to denitrifying conditions. Its complete genome sequences of this bacteria predicted presence of gene clusters involved in denitrification process [21]. The strain grew under denitrifying conditions by reducing NO_3^- into NO_2^- due to presence of two nitrate reductase (narG) genes. There was two nar operons and two nor operons (nitric oxide reductase) and one nos operon (nitrous oxide reductase) in the genome sequence. The presence of an nirK sequence encoding an 82-amino-acid truncated nitrate reductase explained the bacterial capability to reduce only NO_3^- into NO_2^- [21]. This methylotrophic denitrifying bacteria have the specific electron transfer chains for 'methanol oxidation' and 'denitrification', in the periplasm. A unique blue copper protein (HdBCP, 14.5 kDa) play important role in methylotrophic denitrification by Hyphomicrobium denitrificans [22]

Methylotrophic Degradation of Insecticide

Methylotrophic bacterial population increased in the soils sprayed with the organophosphorus insecticide fenitrothion [O,O-dimethyl O-(3-methyl-p-nitrophenyl) phosphorothioate, abbreviated as MEP] in deep sequencing analysis. Methylotrophic genera *Methylobacillus*, *Methylobacterium*, and *Methylophilus* increased dramatically in soils after MEP-spray. Actually, methanol is one of the by product of MEP degradation. The presence of methanol promote growth of methylotrophic bacteria. Thus methylotrophs growth of methylotrophs remove toxic methanol from the soil environment and indirectly facilitate better growth of MEP-degrading bacteria [23].

Methylotrophic Association and Benefits to Host Plant

The association of the methylotrophs on the plant tissues is well established. The methylotrophic association is very much useful for the various stages of the plant growth and development. Some important aspects of methylotrophs are described below.

Phytohormone Synthesis

Cytokinins are adenine derived phytoharmones that stimulate plant cell division. The phyllospheric methylotrophs produce cytokinin in the phyllosphere [14-15]. Cytokinin production by *Methylobacterium* sp. in the liquid medium has been proved by several biochemical and cytological procedures [7]. *The Methylobacterium* sp. have been isolated from several crop plants and their cell free extracts have been effectively used in seed germination of the respective crops.

Root Nodulation and N_2 Fixation

Plant- bacteria symbiosis are very specific [23]. Jounard et al. (2005) reported non-pigmented bacterial strain *Methylobacterium nodulans* inducing nitrogen-fixing root nodules in the legume *Crotalaria sp.* The lacZ fusion to the mxaF gene indicated methylotroph genes expression in the root nodules during symbiosis. Loss of the bacterial methylotrophic function significantly affected plant development. Inoculation of mutant *M. nodulans* in host plant decreased the total root nodule and plant nitrogen fixation as well as total dry plant biomass compared with the wild-type bacterial strain. In contrast, inoculation of the host with non methylotrophic mutants complemented with functional mxa genes and restored the symbiotic wild phenotype [25]. Madhaiyan et al. [23] isolated methylotrophic rhizobia from legumes that possessed nitrogenase activity. High nitrogenase activity was observed in methylotrophic strains CMCJ317 from root nodule of *C. juncea* (162.72 nmol C_2H_4 h^{-1} mg^{-1} protein) and *Sesbania aculeate*(120.48 nmol C_2H_4 h^{-1} mg^{-1} protein).

In *Methylobacterium* spp. utilization of methanol is an advantage for plant colonization and specific for *C. podocarpa / M. nodulans* symbiosis [25, 27] In contrast, Yates et al. [28] have shown that the symbiosis between *Lotononis angolensis* and their nodulating methylobacteria is highly effective and inability to utilize methanol was not deleterious for effective colonization or symbiosis. They stated that factors other than methylotrophy was responsible for specificity of the symbiosis.

Flavour Biosynthesis

Methylobacterium strains have been suggested to contribute to the flavor of strawberries [1]. This endosymbiotic methylotrophic bacteria localized within bud cells of scotch pine dominated just before its development and differentiation of buds. The bacteria disappeared in differentiated buds indicated their role in flavor biosynthesis. The increased endophytic metabolic activity have also been observed only during the growing seasons in *Pinus sylvestris* [29].

Cellulase Production

Several methylotrophic bacteria like *Methylobacterium extorquens, M. organophilum, M. gregans*, and *M. komagatae* were responsible for cellulase production. Among various PPMF isolated from the rivers of the South India, *M. gregans* shown cellulose production in the wide range of temperature (35-65 °C) and pH(5-8) [30]

Ecotine Production

Some methylotrophs produce ectoine which is a universal bioprotectant of proteins, nucleic acids, and whole cells. Ectoine is used in medicine, cosmetic and scientific practices [31]. Similar to other osmoprotactants like betains, trehalose, and citrulline, it inhibits formation of amyloid plaques, and is a potential drug for treatment of diabetes and Alzheimer's disease [31]. Ectoin also protects skin by scavenging singlet oxygen produced by UV light.

Exopolysaccharides

The methylotrophs synthesize exopolysaccharides matrix for their effective colonization on the surface of the host plant [31]. After achieving high population density they interact with the host plant [33]. The capsules and extra-capsular slime merge to form a biopolymeric matrix (acidic polysaccharides and glycosylphosphate biopolymers). This matrix represents a system of microtubules intended for the movement of cells and the transport of various substances. It also performs several protective functions like prevention of cells from dehydration, extreme temperatures, hydrolytic enzymes and UV radiation. The matrix transports microbial exo-metabolites and the products of cell autolysis including signal substances produced in undetected amount in the culture liquid.

Temperature Tolerance and Disease Resistance

PPFMs have been shown to affects agronomic traits like branching, vigor, rooting, and heat/cold tolerance in the host plant [28]. The PPMF Methylobacterium decrease susceptibility of rice crops to fungus Rhizoctonia solani by increasing pathogenesis related proteins and phenolic contents in the host plant [34].

Conclusion

Methylotrophic bacteria are important for the health of soil and plant. They play important role in soil nutrient cycling and nutrients level improvement, degradation of toxic chemicals [10, 20, 23].They modify the soil physicochemical properties leading to increased microbial diversity of the soil [34]. Their association with plant are very specific and vital for growth and development of plant as well as their protection from extreme environmental variations.

REFERENCES

[1] Chistoserdova L, Chen SW, Lapidus A, Lidstrom ME, 2003. Methylotrophy in *Methylobacterium extorquens* AM1 from a genomic point of view. *J Bacteriol* 185: 2980-2987.

[2] Dahl JS, Hoare DS, Mehta RJ, 1972. New Obligate Methylotroph. J Bacteriol 109, 916.

[3] Trotsenko YA, Ivanova EG, Doronina NV 2001. Aerobic methylotrophic bacteria as phytosymbionts. *Microbiology* 70 (6): 623–632.

[4] Borodina E, Kelly DP, Schumann P, Rainey FA, Ward-Rainey NL, 2002 Enzymes of dimethylsulfone metabolism and the phylogenetic characterization of the facultative methylotrophs *Arthrobacter sulfonivorans* sp. nov and *Hyphomicrobium sulfonivorans* sp. nov. *Wood Arch Microbiol.* 177(2):173-83.

[5] Kelly DP, Wood AP 2010. Isolation and Characterization of Methanotrophs and Methylotrophs: Diversity of Methylotrophic Organisms and of One-Carbon Substrates. 3827-3845.

[6] Halsey KH, Carter AE, Giovannoni SJ 2012. Synergistic metabolism of a broad range of C1 compounds in the marine methylotrophic bacterium HTCC2181. *Environ Microbiol* 14: 630-640.

[7] Doronina NV, Trotsenko YA, Kuznetsov BB, Tourova TP, Salkinoja-Salonen MS 2002. *Methylobacterium suomiense* sp. nov. and *Methylobacterium lusitanum* sp. nov., aerobic, pink pigmented, facultatively methylotrophic bacteria. *Int J Syst. Evol. Microbiol* 52: 773–776.

[8] Doronina NV, Trotsenko YA, Tourova TP, Kuznetsov BB, Leisinger T. 2000. *Methylopila helvetica* sp. nov. and *Methylobacterium dichloromethanicum* sp. nov. – novel aerobic facultatively methylotrophic bacteria utilizing dichloromethane. *Syst Appl Microbiol* 23: 210–218.

[9] Madhaiyan M, Poonguzhali S, Lee JS, Lee KC, Sundaram S, 2010. *Flavobacterium glycines* sp. nov., a facultative methylotroph isolated from the rhizosphere of soybean. *Int J Syst Evol Micr* 60: 2187-2192.

[10] Giri DD, Shukla PN, Kumar A, Singh R, Pandey KD 2013. Salt stress tolerance of *Methylophilus* and *Methylobacterium* sp. isolated from coal mine. *Polish Journal of Microbiology* 63(3):273-280.

[11] Madhaiyan, M., S. Poonguzhali, S. W. Kwon and T. M. Sa (2009). "Methylophilus rhizosphaerae sp nov., a restricted facultative methylotroph isolated from rice rhizosphere soil." *International Journal of Systematic and Evolutionary Microbiology* ***59***: 2904-2908.

[12] Mizuno M, Yurimoto H, Yoshida N, Iguchi H, Sakai Y, 2012. Distribution of pink-pigmented facultative methylotrophs on leaves of vegetables. *Biosci Biotech Bioch* 76: 578-580.

[13] Koenig RL, Morris RO, Polacco JC, 2002. tRNA is the source of low-level trans-zeatin production in *Methylobacterium* spp. *J. Bacteriol.* 184:1832-1842.

[14] Holland MA, Polacco. JC 1994. PPFMs and other contaminants: Is there more to plant physiology than just plant? *Annu. Rev. Plant Physiol. Plant Mol. Biol.* 45: 197- 209.

[15] Lee HS, Madhaiyan M, Kim CW, Choi SJ, Chung KY, Sa TM 2005. Physiological enhancement of early growth of rice seedlings *Oryza sativa*) by production of phytohormone of N_2-fixing methylotrophic isolates. *Biol Fertil Soils* 2:402–408.

[16] Idris R, Kuffner M, Bodrossy L, Puschenreiter M, Monchy S, Wenzel WW, Sessitsch A. 2006. Characterization of Ni-tolerant methylobacteria associated with the hyper accumulating plant *Thlaspi goesingense* and description of *Methylobacterium goesingense* sp. nov. *Syst Appl Microbiol* 29: 634–644.

[17] Ryu JH, Madhaiyan M, Poonguzhali S, Yim WJ, Indiragandhi P, Kim KA, Anandham R, Yun JC, Sa TM 2006. Plant growth substances produced by *Methylobacterium* spp. and their effect on the growth of tomato (*Lycopersicon esculentum* L.) and red pepper (*Capsicum annuum* L.). *J. Microbiol Biotechnol.* 16:1622–1628.

[18] Poonguzhali S, Madhaiyan M Sa TM. 2007. Quorum-sensing signals produced by plant-growth promoting *Burkholderia* strains under *in vitro* and *in planta* conditions. *Res. Microbiol.* 158:287–294.

[19] Penalver CGN, Morin D, Cantet F, Saurel O, Milon A, Vorholt JA 2006. *Methylobacterium extorquens* AM1 produces a novel type of acyl-homoserine lactone with a double unsaturated side chain under methylotrophic growth conditions. *FEBS Lett,* 580:561–567.

[20] Megraw SR, Knowles R, 1989. Isolation, characterization, and nitrification potential of a methylotroph and two heterotrophic bacteria from a consortium showing methane-dependent nitrification. *Fems Microbiol. Lett.* 62: 367-374.

[21] Villeneuve C, Martineau C, Mauffrey F, Villemur R, 2012. Complete Genome Sequences of *Methylophaga* sp Strain JAM1 and *Methylophaga* sp Strain JAM7. *J. Bacteriol.* 194: 4126-4127.

[22] Hira D, Nojiri M, Yamaguchi K, Suzuki S, 2007. Identification of a blue copper protein from *Hyphomicrobium denitrificans* and its functions in the periplasm. *J. Biochem.* 142: 335-341.

[23] Madhaiyan M, Poonguzhali S, Senthilkumar M, Sundaram SP, Sa T, 2009b. Nodulation and plant-growth promotion by methylotrophic bacteria isolated from tropical legumes. *Microbiol. Res.* 164: 114-120.

[24] Itoh, H., R. Navarro, K. Takeshita, K. Tago, M. Hayatsu, T. Hori and Y. Kikuchi (2014). "Bacterial population succession and adaptation affected by insecticide application and soil spraying history." *Front Microbiol.* 5: 457.

[25] Jourand P, Renier A, Rapior S, de Faria SM, Prin Y, Galiana A, Giraud E, Dreyfus B, 2005. Role of methylotrophy during symbiosis between Methylobacterium nodulans and Crotalaria podocarpa. *Mol. Plant. Microbe* 18: 1061-1068.

[26] Madhaiyan M, Poonguzhali S, Kwon SW, Sa TM., 2009a. *Methylophilus rhizosphaerae* sp nov., a restricted facultative methylotroph isolated from rice rhizosphere soil. *Int .J. Syst. Evol. Micr.* 59: 2904-2908.

[27] Renier A, De Faria SM, Jourand P, Giraud E, Dreyfus B, Rapior S, Prin Y, 2011. Nodulation of Crotalaria podocarpa DC. by *Methylobacterium nodulans* displays very unusual features. *J. Exp. Bot.* 62: 3693-3697.

[28] Yates RJ, Howieson JG, Reeve WG, Nandasena K, Law IJ, Brau L, Ardley JK, Nistelberger H, Real DO. Hara GW 2007. *Lotononis angolensis* forms nitrogen fixing,

lupinoid nodules with phylogenetically unique fast-growing, pink-pigmented bacteria which do not nodulate L. bainesii or L. listii. *Soil Biol. Biochem.* 39: 1680–1688.

[29] Jaftha JB, Strijdom BW, Steyn PL, 2002. Characterization of Pigmented Methylotrophic Bacteria which Nodulate *Lotononis bainesii*. *Syst Appl Microbiol* 25: 440-449.

[30] Jayashree, S., R. Lalitha, P. Vadivukkarasi, Y. Kato and S. Seshadri (2011). "Cellulase production by pink pigmented facultative methylotrophic strains (PPFMs)." *Appl Biochem Biotechnol.* 164(5): 666-680.

[31] Trotsenko Y A, Doronina NV, Li TsD, Reshet nikov AS, *Mikrobiologiya,* (2007), 76 : 293–305.

[32] Kanapathipillai M, Lentzen G, Sierks M, Park CB, *FEBS Lett.,* (2005) 579: 4775–4780. Oleskin AV, Botvinko IV, Tsavkelova EA 2000. *Colonial Organization and intercellular Communication in Microorganisms, Mikrobiologiya*, 69(3), 309–327.

[33] Madhaiyan M, Poonguzhali S, Sundaram SP, Sa T, 2006. A new insight into foliar applied methanol influencing phylloplane methylotrophic dynamics and growth promotion of cotton (*Gossypium hirsutum* L.) and sugarcane (*Saccharum officinarum* L.). *Environmental and Experimental Botany* 57, 168-176.

[34] Giri, D. D., P. N. Shukla, S. Ri*tu, A. Kumar and K.* D. Pandey (2013). *"Substrate utilization* of stress tolerant methylotrophs isolated from revegetated heavy metal polluted coalmine spoil." *World J. Microbiol. Biotechnol.* 29(4): 635-643.

In: Microbes in Soil and Their Agricultural Prospects
Editors: K. Kishore Choudhary and D. Wattal Dhar
ISBN: 978-1-63482-824-6

Chapter 4

MICROBIAL-INDUCED ABIOTIC STRESS TOLERANCE IN PLANTS

Dilfuza Egamberdieva*
Institute for Landscape Biogeochemistry, Leibniz Centre for Agricultural Landscape Research (ZALF), Müncheberg, Germany

ABSTRACT

The biotic and abiotic factors including drought, salinity, extreme temperature and phytopathogens limit the crop productivity. The use of stress-tolerant microbial strain associated with roots of agronomic crops which stimulate plant growth, protect plants from soil borne disease, improve stress tolerance of plants and stimulate soil microbial activity, can lead to improved fertility of salt affected soils. The traits involved in mitigating plants stress and plant growth stimulation by PGPR include production of phytohormones, decreasing ethylene levels by the enzyme ACC deaminase, and production of osmoprotectants. This process may be a good mean of protection against salt stress, promotion of plant growth and symbiotic performance of legumes under harsh soil conditions. The objectives of this paper are to emphasize salinity stress effects on plants and discuss recent developments and advances in our understanding the interactions between the plant and beneficial rhizobacteria and their role in improvement of plant resistance to abiotic stresses.

Keywords: abiotic stress, salinity, drought, rhizobacteria, nutrient uptake, phytohormones

1. INTRODUCTION

The abiotic stresses such as drought, salinity, and extreme temperature has main impact on the crop productivity worldwide and remains as a big challenge while addressing the problem of food insecurity [1, 2]. Salinity and drought exert negative effects not only on plant

* Corresponding author: Dilfuza Egamberdieva. Institute for Landscape Biogeochemistry, Leibniz Centre for Agricultural Landscape Research (ZALF), Eberswalder str. 86, 15374 Müncheberg, Germany.

growth but also soil biological activity, which play an important role in maintaining soil quality [3]. It has been also revealed that an abiotic stress causes a disturbance of plant-microbe interaction which is a critical ecological factor to help further plant growth in degraded ecosystems [4]. The use of beneficial microbes associated with plant roots could be helpful to improve the crop yield under hostile environmental conditions [5-6]. Evidently, plant growth promoting rhizobacteria (PGPR) holds enormous prospects in enhanced plant tolerance to stress, better plant nutrient uptake, and reduced use of chemical inputs [7]. The activity of root associated beneficial microbs is very important for ensuring sufficient nutrient supply to the plant and plays a significant role in regulating the dynamics of organic matter decomposition and the availability of plant nutrients such as N, P and S in the soil [8-9]. The roles of PGPR in stress management are emerging areas in agriculture that is not yet well understood; consequently, the benefits are yet to be maximized anywhere in the world [7].

The effect of plant growth promoting bacteria on plant growth was well addressed by Lugtenberg et al. [10], and Egamberdieva [11-12]. There is now increasing evidence that the use of beneficial microbes can enhance plants' resistance to adverse environmental stresses. In addition, mixed inoculation with PGPR and *Rhizobium* strain creates synergistic interactions that may result in a significant increase in growth, symbiotic performance, and uptake of mineral nutrients by plants [13]. Understanding the interactions of microbes in the rhizosphere of plants and their ameliorative effects on plants under stress is important for crop improvement under hostile conditions.

In this review, we attempt to discuss the impact of salt and drought stresses on plant growth and the ameliorative effects of root associated microbes on plant health under hostile environment. In addition this paper describes the mechanisms involved in plant growth stimulation, improved symbiotic performance and alleviation of salt stress.

2. Plant Growth and Abiotic Stress

The most cultivated crops in human/animal nutrition are susceptible to abiotic stresses and their productivity is considerably reduced due to improper nutrition of the plant [14-15]. Climate change will result in alter towards again a more arid climate, which is conducive to salt accumulation [1]. Plants exposed to stress factors in early seedling growth stage, because the seedling root is in direct contact with soil [16-17]. Many studies have demonstrated that salinity inhibits seed germination rate, seedling root and shoot length and dry weight of various crops like chickpea [18], wheat [19], faba bean [20], rice [21], maize [22], soybean [23], cabbage, sugar beet, and pak-choi [17]. Several explanations for these effects have been proposed, such as preventing water uptake on germinating seed [24], disturbance of the hormonal balance [25], and inhibition of the activity of enzymes involved in nucleic acid metabolism [26].

Dolatabadian et al. [27] and Golezani and Yengabad [28] reported that the toxic effect of salt stress on root growth may be caused by a reduction of water uptake and an unbalanced nutrient uptake by the seedling. The higher concentration of Na and Cl in plant leaves was observed with increased level of salinity, whereas mineral nutrients such as N, P, Ca, K and Mg were decreased [29]. The leaf chlorosis, leaf bleaching and necrosis were also observed by salt stress effect on soybean [30]. Salinity significantly reduced the growth of shoot and root as well as the essential oil content of *Ammolei majus*, *Hyoscyamus niger* and *Matricaria chamomile* [31- 33].

The plant growth and symbiotic performance of leguminous plants are strongly influenced by abiotic stress factors such as drought, salinity and nutrient deficiencies [34-35,18]. Previous studies have shown that salinity and drought stresses led to a significant decline in plant biomass accumulation (root and shoot), rhizobial colonisation, nodule development, and nitrogenase activity [36]. There are several reports indicating negative effect of salinity and drought on nodule function by inhibiting of its nitrogenase activity and leghemoglobin accumulation through restricting the supply of carbon to the nodules [37, 20]. Salt stress also decreased the number of *Rhizobium* cells able of colonize *G. officinalis* root tips [13], through disturbance of infection and nodulation process. For example, according to Bouhmouch et al. [38], salt inhibits the absorption of Ca_2+ ions, which causes reduction of the growth of roots, root tips and root hairs, thereby decreasing sites for potential rhizobial infection and further nodule development. Singleton and Bohlool [39] reported that salinity inhibited survival and proliferation of *Rhizobium* spp. in the soil and rhizosphere, and infection process. In other study Slattery et al. [40] observed negative effect of low (4.5) and high (8.0) soil pH and temperature on rhizobial population in soil. Drought has also a pronounced effect on the colonization of rhizobia in the rhizosphere, N_2 fixation rates and nodulation [34].

The phytohormones such as indole acetic acid (IAA), gibberellic acid (GA), and cytokinins (CK) play an important role in plant physiology and regulate responses to stress or coordinate plant growth under stress conditions [41]. Exposure of plants to drought and salinity results in a decrease in the level of cytokinins and auxins [42]. For example Figueiredo et al. [43] observed that drought stress causes a change in the balance of cytokinins, indole acetic acid and gibberellic acid in common bean. The decrease of hormone levels in root system of plants resulted in inhibition of root growth of soybean and disturbance of nutrient uptake from soil [44]. Similar results were observed by Dunlap and Binzel [45] where salinity caused 75% reduction in IAA levels of tomato. Drought stress also cause a change in hormonal balance, including a decrease in IAA and GA3 and a sharp fall in zeatin content in the leaves of the common bean [43]. There are many reports on the mitigation of sat and drought stress and stimulation of root and shoot growth of plants by exogenous application of phytohormones such as gibberellins [46- 47], auxins [19], and cytokinins [48]. For example pretreatment of wheat seeds with plant growth regulators like IAA and GA alleviated the growth inhibiting effect of salt stress [49, 19]. GA3 has also been shown to alleviate the effects of salt stress on pigment content, and water-use efficiency [50] and promote plant length and plant fresh/dry biomass under salt stress conditions [51]. In other study the plant growth, stress tolerance and nitrogen fixation were improved after treatment of seeds with abscisic acid (ABA) [52].

3. Plant Beneficial Microbes

The rhizosphere is colonized more intensively by microorganisms than the other regions of the soil [10, 12). These microbes can be beneficial, neutral or pathogenic and compete for nutrients and niches under strongly varying conditions [53- 54]. Beneficial rhizosphere bacteria are of two general types, those forming a symbiotic relationship with the plant and those that are free-living in the soil and root [55- 57]. Plant growth promoting rhizobacteria (PGPR) play important role in crop production as they: (i) stimulate root growth, (ii) make soil nutrients available to the plant root, (iii) fix nitrogen from air and improve soil fertility, (iv) suppresses soil borne pathogens and protect plants from various disease, (v) enhance plant tolerance to various environmental stresses including drought, salinity, high temperature and heavy metals [7, 44].

Frequent root associated plant beneficial bacteria include species such as *Azospirillum*, *Arthrobacter*, *Azotobacter*, *Bacillus*, *Burkholderia*, *Cellulomonas*, *Enterobacter*, *Flavobacterium*, *Klebsiella*, *Microbacterium*, *Pseudomonas*, *Serratia*, *Stenotrophomonas*, and *Rhizobium* may contribute to the beneficial effects on the growth and yield of various agricultural important plants [58, 54, 59- 61]. Among those species *Bacillus* considered as ecologically significant group of bacteria, which are also well adapted to the arid and salt affected environment [62- 63]. They form endospores, thus can tolerate high temperature, pH, and osmotic conditions [64]. This enables these bacteria to be effective as plant growth stimulator and biocontrol agent in many environmental conditions (saline, drought, heavy metal contaminated etc.) [11].

The plant growth promotion activity of rhizobacteria is primarily related to its impact on root growth and morphology and has proven able to increase nutrient availability in the rhizosphere of plant [65, 55]. The positive effect of plant growth promoting bacteria on the growth and nutrient uptake of cotton [57], chickpea [66], bean [67], wheat [9], soybean [44], and corn [68], were reported. Creus et al. [69] reported that bacterial inoculation stimulated the production of lengthy root hairs, lateral roots, and improved the root diameter and area. In soybean, the application of *Bradyrhizobium japonicum* enhanced the number of nodules, dry weight of plant, grain yield and protein content in soybean grown in salinated soils of Uzbekistan [70]. Several PGPR strains such as *Pseudomonas putida*, *P. extremorientalis*, *P. chlororaphis*, and *P. aureafaciens* significantly increased shoot length, root length and dry matter of wheat under saline arid soil [6]. Further studies also confirmed the enhanced growth, nodulation and yield of chickpea by *Mezorhizobium* under saline arid soil condition [18]. Karnwal and Kumar [71] reported that the *P. aeruginosa* increased seed germination (62%), shoot, root length (92% and 35%), and dry matter of chickpea (43%). Inoculation of chickpea seeds with *Mezorhizobium ciceri* significantly induced higher numbers of nodules on the roots and the nodule number showed a high correlation with shoot and root weights, which is an indication of the connection between the nodules and plant growth [18].

Plant growth requires significant quantities of nitrate, phosphate, and other minerals which are often not available in free form in the soil, or available in limited quantities under stress soil conditions [72]. Root associated microbes may supply of available nutrients to plants, through solubilization of minerals and increase concentration of P, N, K, and Mg in plant root and shoots [73]. For example, the inoculation of cotton seeds with salt tolerant phosphate solubilizing bacteria *Rhizobium meliloti* URM1combined with phosphate had a significant stimulatory effect on total dry matter, shoot and root dry weight, yield and P content under saline soil [74]. Sheng [75] observed that *Bacillus edaphicus* NBT strain increased K content of cotton and rape plants by 30% when the soil was treated with insoluble K sources. *Pseudomonas fluorescens* PsIA12 and *Pantoea agglomerans* increased the plant growth and N, P, K and Mg uptake of maize under field condition [76]. Plant growth promoting *B. megaterium* and *B. mucilaginous* strains improved nutritional assimilation of plant total NPK [77]. The radish seeds inoculated with *B. subtilis* and *P. fluorescens* caused significantly increase in fresh and dry masses of roots and leaves, photosynthetic pigments, proline, total free amino acids, crude protein and N, P, K+, Ca_2+, and Mg_2+ uptake compared to uninoculated control plants under saline condition [78]. This rationale is consistent with the observation that plants inoculated with PGPR take up N, P, K and microelements more efficiently from the soil [58].

Phosphate solubilizing bacteria (PSB) help plants to acquire more phosphorus from soil, thus stimulate P uptake by plants [79]. El-Azouni [80] observed significant increase of dry matter, N, P uptake and yield of soybean inoculated with phosphate -solubilizing fungi *A. niger* and *P. italicum*. In other study S-oxidizing bacteria *Thiobacillus* sp. stimulated nodule number, plant biomass, yield of groundnut and increased soil available S [81]. The PGPR strains *Pseudomonas fluorescens* PsIA12, and *Pantoea agglomerans* significantly increased root development, shoot growth and K uptake of maize [82]. In the rhizosphere, a synergism between various bacterial genera such as *Azotobacter, Azospirillum, Alcaligenes, Bacillus, Enterobacter, Pseudomonas, Rahnella, Stenotrophomonas*, and *Rhizobium* has been shown to promote plant growth of various plants and protect them from soil borne pathogens [83-85]. For example co-inoculation of some *Trichoderma* strains with effective *Rhizobium* spp. stimulated chickpea nodulation and nitrogen fixation and provide more nitrogen is offered to the crop [86]. In other study the combined inoculation of *Azotobacter, Azospirillum, Pseudomonas*, and *Mezorhizobium* resulted in promotion of grain yield and biomass in chickpea [87]. Similar results were observed by Khurana and Sharma [88] where combined inoculation of *Rhizobium* and phosphate-solubilizing bacteria (PSB) *Pseudomonas striata* and *Bacillus polymyxa* increased in nodulation, growth, and yield of chickpea under greenhouse conditions. Galal et al. [89] demonstrated the beneficial influence of co-inoculation of *Azospirillum lipoferum* and *Bacillus megaterium* for providing balanced nitrogen and phosphorus nutrition of wheat plants. Ordookhani et al. [90] found an increased shoot, root dry weight, N,P,K content and essential oils in *Ocimum basilicum* inoculated with (PGPR) *Pseudomonas putida* and *Azotobacter chroococcum*. A significant increase in N content of root and shoot of fodder galega (*Galega orientalis*) was also observed after co-inoculation of *Pseudomonas trivialis* strain 3Re27 strains with *R. galegae* HAMBI 540 compared to *R. galegae* HAMBI 540 alone [72]. Karthikeyan et al. [91] reported that PGPR strains *Pseudomonas fluorescens*, and *Bacillus megaterium* significantly increased plant height, root length, root girth, alkaloid content and N, P, K, Ca and Mg uptake in *Madagascar periwinkle* (*Catharanthus roseus*) in comparison to the control. In other studies combined inoculation of *Begonia malabarica* and *Calamus thwaitessii* with *G. mosseae*, *B. coagulans* and *T. viride* enhanced the growth, biomass, nutrients, and production of secondary metabolites [92]. The increase of secondary metabolites such as total phenols, orthodihydroxy phenols, flavanoids, alkaloids, saponins and tannins in the roots or shoots of medicinal plants were explained due to positive interactions of PGPR with mycorrhizal fungi [93-94].

4. Microbial Effect on Stressed Plants

Promising measures for improving plant productivity under hostile environment are the use of microbial inoculants, which opens a new advanced technology for combating stress factors [95- 97]. The improvement of plant growth and yield by PGPR inoculation under stress conditions has been reviewed by Nadeem et al. [98], and Egamberdieva and Lugtenberg [83] and more details are given in Table 1. Inoculation of wheat with halotolerant *A. brasilense* strain NH [99], *P. fluorescens* 153 and *P. putida* 108 [100], *P. extremorientalis*, and *P. chlororaphis* improved germination, and plant growth under saline soil conditions [6]. Cotton growth and development under saline soil condition were improved by PGPR strains *P. alcaligenes PsA15*, *P. chlororaphis* TSAU13, *P. extremorientalis* TSAU20, and *B. amyloliquefaciens* BcA12 in comparison with the uninoculated control plants [101]. In other study *Klebsiella oxytoca* increased plant tolerance to salt stress and promote the plant height and dry weight of cotton by 14.9 and 26.9%, respectively [102]. Seed inoculation of cucumber by *Serratia plymuthica* RR-2-5-10, *Stenotrophomonas rhizophila* e-p10, *Pseudomonas fluorescens* SPB2145, *P. extremorientalis* TSAU20, and *P. fluorescens* PCL1751 resulted in improved root and shoot biomass in salinated arid soil [103]. Those strains were reduced *Fusarium* root rot of cucumber to as low as 10 % and showed a significant stimulatory effect on plant growth, increasing the dry weight of whole cucumber plants up to 62% and fruit yield up to 32 % in comparison to the non-bacterized control [103].

Table 1. Summary of reported effects of PGPR on plant growth under stress condition

Bacteria	Plant benefits	Stress	References
Pseudomonas spp.	Plant growth, nutrient uptake and yield of pea (*Pisum sativum* L.)	Drought	Arshad et al. [115]
Rhizobium sp. *Azotobacter* sp.	Plant growth, and nodulation of faba bean (*Vicia faba*)	Drought	Dashadi et al. [118]
Rhizobium tropici *Paenibacillus polymyxa*	Shoot dry matter, nodule number and dry matter and N uptake of common bean (*Phaseolus vulgaris* L.)	Drought	Figueiredo et al. [43]
Pseudomonas sp.	Shoot, root growth and yield of madagascar periwinkle (*Catharanthus roseus*)	Drought	Jaleel et al. [110]
Bacillus sp.	Seed germination, shoot length and biomass, and pod yield of cowpea (*Vigna unguiculata)*	Drought	Minaxi et al. [161]
Rhizobium spp. *Glomus intraradices*	Root, shoot growth, nodulation and nitrogen fixation, grain yield of chickpea (*Cicer arietinum*)	Drought	Rasaei et al. [114]
Bacillus sp.	Root, shoot growth, N,P, K uptake of lettuce (*Lactuca sativa* L.)	Drought	Vivas et al. [116]
Pseudomonas mendocina *Glomus intraradices*	Root, shoot growth, and catalase activity in lettuce (*Lactuca sativa* L.)	Drought	Yang et al. [113]
P. fluorescens *P. putida*	Root, shoot length, dry weight and yield biomass of wheat (*Triticum aestivum*)	Salt	Abbaspoor et al. [100]
Bacillus pumilus *Exiguobacterium oxidotolerans*	Plant growth and bacoside-A content of brahmi (*Bacopa monnieri*)	Salt	Bharti et al. [119]

Bacteria	Plant benefits	Stress	References
Mycobacterium phlei *Mycoplana bullata*	Root, shoot length, m dry weight and N, P, K and Mg uptake of wheat (*Triticum aestivum*)	Salt	Egamberdiyeva and Hoflich [59]
Bradyrhizobium japonicum	Root, shoot growth, the number of nodules, dry weight, grain yield and protein content in soybean (*Glycine max* L.)	Salt	Egamberdiyeva et al. [70]
P. alcaligenes, *B. polymyxa*	Shoot and root growth and nitrogen (N), phosphorus (P) and potassium (K) uptake of maize (*Zea mays*)	Salt	Egamberdiyeva [82]
Pseudomonas extremorientalis, *P. chlororaphis*	Root, shoot length, dry weight and yield biomass of wheat (*Triticum aestivum*)	Salt	Egamberdieva and Kucharova [6]
Serratia plymuthica *Stenotrophomonas rhizophila* *P. fluorescens* *P. extremorientalis*	Root and shoot length, dry weight, plant biomass and yield of cucumber (*Cucumis sativum*)	Salt	Egamberdieva et al. [11]
P. chlororaphis	Root, shoot length, biomass, and fruit yield of cucumber (*Cucumis sativus*), and tomato (*Solanum lycopersicum*)	Salt	Egamberdieva [113]
P. extremorientalis	Root, shoot length and dry weight of milk thistle (*Silybum marianum*)	Salt	Egamberdieva et al. [105]
P. alcaligenes, *P. chlororaphis*, *P. extremorientalis* *B. amyloliquefaciens*	Root, shoot length, dry weight of cotton (*Gossypium hirsutum*)	Salt	Egamberdieva and Jabborova [57]
Rhizobium galegae P. extremorientalis	Root and shoot growth and nodulation of goats rue (*Galega officinalis* L.)	Salt	Egamberdieva et al. [13]
Rhizobium tropici *Chryseobacterium balustinum*	Root, shoot growth and nodulation of soybeans (*Glycine max* L.)	Salt	Estevez et al. [117]
Pseudomonades sp. *Bacillus lentus*	Plant growth, photosynthesis, mineral content and antioxidant enzymes of basil (*Ocimum basilicum*)	Salt	Golpayegani and Tilebeni [120]
Serratia sp. *Rhizobium* sp.	Root, shoot growth, N, P and K uptake of lettuce (*Lactuca sativa*)	Salt	Han and Lee [107]
P. putida, *B. japonicum*	Plant growth, N, P uptake, and protein content of soybean (*Glycine max* L.)	Salt	Jabborova et al. [44]
B. subtilis *P. fluorescens*	Fresh and dry weight of roots and leaves, photosynthetic pigments, proline, total free amino acids and crude protein contents of radish (*Raphanus sativus*)	Salt	Mohamed and Gomaa [78]
Pseudomonas sp.	Root/shoot length, biomass, and the level of cellular metabolites of wheat (*T. aestivum*)	Salt	Mishra et al. [162]

Table 1. (Continued)

Bacteria	Plant benefits	Stress	References
A. brasilense	Root, shoot growth, dry weight, yield of wheat (*Triticum aestivum*)	Salt	Nabti et al. [99]
Azospirillum brasilens *Glomus clarum*	Root, shoot growth, P, N uptake, nodule number, protein content of faba bean (*Vicia faba*)	Salt	Rabie and Almadini [20]
P. fluorescens	Plant growth, plant biomass of groundnut (*Arachis hypogea*)	Salt	Saravanakumar and Samiyappan [145]
Serratia proteamaculans *Mesorhizobium ciceri*	Plant growth, nodule number and grain yield of chickpea (*Cicer arietinum* L.)	Salt	Shahzad et al. [141]
Staphylococcus kloosii *Kocuria erythromyxa*	Shoot, root dry weight, leaf number, relative water content of the leaf and chlorophyll content of radish (*Raphanus sativus* L.)	Salt	Yildirim et al. [104]
Klebsiella oxytoca	Plant height, and dry weight of cotton (*Gossypium hirsutum*)	Salt	Yue et al. [102]
P. putida	Germination rate, plant height, and dry weight of cotton (*Gossypium hirsutum* L.)	Salt	Yao et al. (2010)
Rhizobium *Pseudomonas*	Shoots and root dry weight of maize (*Zea mays* L.)	Salt	Bano and Fatima [112]
Azospirillum brasilense	Plant growth, root branching, secretion of flavonoids and lipochitooligosaccharides in bean (*Phaseolus vulgaris*)	Salt	Dardanelli et al. [121]

Staphylococcus kloosii and *Kocuria erythromyxa* showed positive effect on plant growth and development in radishes under salt affected soil condition [104], stimulating shoot/root dry weight, leaf number per plant, relative water content of the leaf and chlorophyll content of fruit. These results agree with Egamberdieva et al. [105], who reported that root and shoot length and dry weight of milk thistle (*Silybum marianum*) and bean (*Phaseolus vulgaris*) inoculated by *P. extremorientalis* TSAU20 were increased compared to the control [106]. The application of *Bradyrhizobium japonicum* enhanced the number of nodules, dry weight, grain yield and protein content in soybean grown in salinated soils [70].

Inoculation of lettuce with *Serratia* sp. and *Rhizobium* sp. alleviated the negative effects of salinity on the plant and stimulated N, P and K uptake [107]. The nutrient (N, P, K and Mg) uptake of wheat was increased by *Mycobacterium phlei* MbP18 and *Mycoplana bullata* MpB46 [59]. The pepper and cucumber with a higher potassium content showed more tolerance to salt stress than other plants with low P uptake [108]. Yildirim et al. [109] reported that plants treated with PGPR have lower Na^+ and Cl^- contents and higher NPK contents compared with noninoculated plants under salt stress. Similar results observed by Jaleel et al. [110], where inoculation with plant growth promoting *Pseudomonas* strains stimulated the shoot, root growth and yield of medicinal plant *Catharanthus roseus* under drought stress.

There are also several studies which indicate increased plant growth and symbiotic performance of leguminous plants by PGPR [67, 13]. It has been also reported that mixed inoculation with PGPR and *Rhizobium* or arbuscular-mycorrhizal fungi creates synergistic interactions that may result in a significant increase in growth, symbiotic performance, an enhancement in the uptake of mineral nutrients such as phosphorus, nitrogen, potassium and increase stress resistance of plants [111, 56, 72]. For example Rabie and Almadini [20] observed positive effect of the tri-partite interactions among bacteria (*Azospirillum brasilens*), mycorrhiza (*Glomus clarum*), and legume (*Vicia faba*) under saline conditions. The combined inoculation increased salinity tolerance of bean, phosphorus and nitrogen uptake, nodule number, protein content, phosphatase and nitrogenase enzymes. Similar observation reported by Bano and Fatima [112] where co-inoculation with *Rhizobium* and *Pseudomonas* increased the dry mass of shoots and roots of maize in both unstressed and salt-stressed conditions. In other study co-inoculation of lettuce (*Lactuca sativa* L.) with PGPR *P. mendocina* and arbuscular mycorrhizal fungi (*Glomus intraradices* or *G. mosseae*) improved catalase activity under drought conditions, suggesting that they can alleviate the oxidative damage elicited by drought [113]. In comparison with plants inoculated with rhizobia alone, co-inoculation of salt-stressed goat's rue with *Rhizobium galegae* HAMBI 1141 and *P. extremorientalis* TSAU20 significantly improved root and shoot growth as well as nodulation of the plants [13]. Co-inoculation of common bean (*Phaseolus vulgaris* L.) with *Rhizobium tropici* and *Paenibacillus polymyxa* showed improved plant growth, shoot dry matter, nodule dry matter and N uptake as well as higher nodule numbers than those inoculated with *R. tropici* alone under conditions of drought stress [43]. Rokhzadi et al. [87] showed that the combined inoculation of *Azotobacter, Azospirillum, Pseudomonas*, and *Mezorhizobium* resulted in promotion of the grain yield and biomass in chickpea. The co-inoculation of chickpea with *Rhizobium* spp. and mycorrhizal fungi (*Glomus intraradices*), had enhanced nodulation and nitrogen fixation, plant biomass and grain yield under drought condition [114]. Similar results were also obtained by Arshad et al. [115] where *Pseudomonas* spp. resulted in enhanced plant growth, nutrient uptake and yield of pea under drought condition. *Bacillus* sp. increased root, shoot growth of lettuce and N,P, K uptake (up to 50%) in lettuce under drought stress conditions compared to the control [116]. For example Estevez et al. [117] observed that co-inoculation of *Rhizobium tropici* CIAT899 with *Chryseobacterium balustinum* Aur9 improved growth and symbiotic performance of salt-stressed soybeans compared with the single inoculation (CIAT899). Dual inoculation of *Rhizobium* with *Azotobacter* showed increased nodulation, plant growth of faba bean under drought condition [118].

PGPR strains are able to stimulate production of biological active compounds of medicinal plants [33]. For example Bharti et al. [119] observed that salt tolerant *Bacillus pumilus* and *Exiguobacterium oxidotolerans* stimulated plant growth and bacoside-A content of brahmi (*Bacopa monnieri*). These results were somewhat similar to that got by Golpayegani and Tilebeni [120] where inoculation of basil (*Ocimum basilicum*) with *Pseudomonas* sp. and *Bacillus lentus*, alleviated the salinity effects on the growth photosynthesis, mineral content and antioxidant enzymes. In other study, the increased root branching in been seedling and secretion of flavonoids and lipochitooligosaccharides were observed after inoculation of bean with *Azospirillum brasilense* [121].

5. Biomechanisms to Enhance Plant Stress Tolerance

PGPR may use several mechanisms for plant growth promotion and increased plant stress tolerance such as synthesis of phytohormones like indole acetic acid (IAA), gibberellic acid, cytokinins [122- 123], production of 1-aminocyclopropane-1-carboxylate (ACC) deaminase to reduce the level of ethylene in the roots of developing plants [84], solubilization of minerals such as phosphorus, and potassium [55], and production of exopolysaccharides (EPS) [124].

5.1. Bacterial Phytohormones

Rhizobacteria synthesize and release phytohormones as secondary metabolites because of the rich supplies of substrates exuded from the roots [10, 123]. Phytohormones such as gibberellic acid (GA) and indole-3-acetic acid (IAA) are known to have a major role in plant growth development and in stress responses [125]. According Bianco and Defez [126] they may enhance different cellular defense systems for the protection of plants from external stress conditions. The ability of plant species to adapt to stress conditions often appears to depend on their association with certain microbes, which produce phytohormones under stress conditions [53]. The most of root associated microorganisms are able to synthesize and release phytohormones like gibberellins, auxin, and cytokinins as secondary metabolites [127, 96], which subsequently re-taken up by plants for growth and development. Some root associated beneficial bacteria may produce IAA under saline conditions as well. For example *P. alcaligenes*, *P. aurantiaca*, *P. aureofaciens*, *P. denitrificans*, *P. mendocina*, *P. rathonis*, and *P. stutzeri* isolated from saline arid soil showed the ability to growth in media which contain 6% NaCl and produce phytohormone auxin (IAA) [9]. Other strains such as *Acinetobacter* sp., *P. aeruginosa*, *S. saprophyticus*, *B. cereus, Enterobacter hormaechei*, *Pantoae agglomerans*, and *Alcaligenes faecalis* isolated from the rhizosphere of wheat grown in saline soil also showed the ability to produce IAA (up to 34 μg/ ml) [96]. Nabti and others [128] isolated the halotolerant A. brasilense strain NH which is able to produce auxin at a concentration of 200 mM NaCl.

The positive effect of cytokinin and giberellin producing PGPR strains such as *Pseudomonas*, *Bacillus*, *Azospirillum*, and *Stenotrophomonas* on root, shoot growth and dry weight of various plants have been reported by several authors [129, 122]. According Egamberdiyeva and Hoflich [65] the major mechanism of plant growth stimulation of cotton and pea under saline arid soil conditions by PGPR *P. alcaligenes* PsA15*, P. denitrificans* PsD6, and *Bacillus polymyxa* BcP26 related to the production of indole-3-acetic acid. Phytohormones produced by PGPR will be taken up by plant cells, and can stimulate plant cell proliferation; this mechanism might be responsible for the enlarged root system under stressed conditions and thereby provides the plant greater access to soil nutrients [130, 47]. The auxin levels in the host legume plants are necessary for nodule formation [127], which are involved in cell division, differentiation and vascular bundle formation, which are considered essential for nodule formation [131].

Camerini et al. [132] reported that *Vicia hirsute* inoculated with IAA producing *Rhizobium leguminosarum* bv. *viciae* produced potential nitrogen fixing root nodules containing more IAA than nodules formed by the wild-type counterpart. Jabborova et al. [44] observed that *Bradyrhizobium japonicum* and *P. putida* which produce IAA) in culture media stimulated root shoot growthand nodule number of chickpea under salt stress condition.

The phytohormone producing PGPR strains are also known to stimulate the level of phytohormones in plant tissue. For example Arkhipova et al. [133] observed increased cytokinin concentrations in plants by treatment with a cytokinin-producing PGPR strain. *Azospirillum* strains increased levels of GA3 in the roots after inoculation of maize seedling [134]. Increased production of IAA in inoculated plants by PGPR may be a good mean of protection against salt stress and promotion of plant growth in the harsh soil conditions [135]. *Pseudomonas* strain which produce significant amount of indole acetic acid (IAA) stimulated the root and shoot growth of plant, nodule number and nodule biomass by *Mesorhizobium* sp. cicer compared to untreated control [136]. The production of phytohormones by PGPR has been correlated with stimulation of root growth and development, through which enhanced uptake of nutrients by the associated plants [122].

Application of the IAA producing bacteria *P. denitrificans, P. rathonis, B. laevolacticus, B. amyloliquefaciens*, and *A. simplex* resulted in enhanced plant growth and nitrogen (N), phosphorus (P) and potassium (K) uptake of wheat and maize under saline soil condition [9,82]. Inoculation of *Pinus pinea* plants with gibberellins producing bacterial strains *B. licheniformis* and *B. pumilus* increased growth and development [137]. According Khan and Doty [138] IAA production by endophytic bacteria *Enterobacter*, *Pseudomonas* and *Stenotrophomonas* involved in root growth regulation and protecting the cells against adverse conditions in potato. The production of IAA by *P. putida* was closely linked to the stimulation of the root growth and development in canola and mung bean [10].

The beneficial effect of plant growth promoting rhizobacteria on growth of many plants can be partly explained by their ability to produce phytohormones. They can have multiple impacts on the phytohormone status, modifying root-to-shoot signalling and shoot hormone concentrations, which may improve growth, development and physiological processes of plants under salt stress [139].

5.2. ACC Deaminase Enzyme

It has been reported that PGPR releasing the enzyme ACC deaminase, that may decrease the ethylene level in the root and enhance salt tolerance of plants [11]. The enzyme ACC deaminase is present in many rhizosphere bacteria, which cleaves the ethylene precursor ACC to α-ketobutyrate and ammonium and thereby lowers the level of ethylene in developing or stressed plants [140]. For example inoculation of chickpea with ACC-deaminase producing *P. fluorescens* stimulated root elongation (46%) and root dry weight (94%) over the respective uninoculated drought stressed control [115]. Shahzad et al. [141] reported that *Serratia proteamaculans* (J119) and *Mesorhizobium ciceri* (S14) which were positive for ACC-deaminase enzyme activity, increased the grain yield (up to 76%), nodule number (up to 94%), nodule dry weight (up to 96.6%) compared to uninoculated control. It has been reported that PGPR strain *P. trivialis* 3Re27 was able to utilize ACC as N source indicating the presence of ACC deaminase and increased salt tolerance of goats'rue, stimulating shoot, root growth under salinated soil condition [13]. Similar results were observed by Shaharoona et al. [142] where co-inoculation of *Bradyrhizobium* with PGPR isolates strains possessing

ACC-deaminase activity enhanced the nodulation in mung bean compared with inoculation with *Bradyrhizobium* alone. In following studies PGPR strains which produce ACC deaminase showed positive effect on plant growth of chickpea [143], and lentil [144], resulting increased number of nodules, root, shoot growth, and yield of plant under stress conditions. The ACC deaminase producing strain *P. fluorescens*, improved the plant growth parameters and the salt stress resistance of groundnut seedlings under saline condition as compared to plants inoculated with *Pseudomonas* strains lacking ACC deaminase activity [145]. Ma et al. [146] observed that ACC deaminase producing *R. leguminosarum* could lower ethylene production in pea roots and improved nodulation.

5.3. Osmoprotectants

Plants may protect themselves from drought and salt stress by accumulating compatible solutes such as sugars and amino acids to osmotically adjust themselves [147- 148]. Glycine betaine (GB) and proline are two major organic osmolytes that accumulate in a variety of plant species in response to environmental stresses such as drought, salinity, extreme temperatures, UV radiation and heavy metals [149].

Glycine betaine (N,N,N-trimethylglycine) is a preferred compatible solute for the majority of prokaryotes [150]. It has been reported that exogenous application of glycine betaine to low-accumulating or non-accumulating plants may help reduce adverse effects of environmental stresses [151]. Verbruggen and Hermans [152] reported that the accumulation of proline is one of the best known alterations induced by water and salt stress in plants. In addition production of exopolysaccharides (EPS) by the bacteria protects them against unfavorable conditions and enhances their survival [124]. In addition, the exopolysaccharides protect the plant from desiccation, through formation of protective layer around soil aggregates [153]. Jha et al. [154] reported that paddy rice (*Oryza sativa* L.) inoculated with *P. pseudoalcaligenes* showed a significantly higher concentration of glycine betaine-like quaternary compounds and a higher shoot biomass under salinity conditions. *Azospirillum* inoculation leads to an increased content of proline [155], and free amino acids in maize under drought stress conditions [156]. Several PGPR strains, such as *Burkholderia* [157], *Arthrobacter* and *Bacillus* [158], enhance proline synthesis in stressed plants, which helps in maintaining the cell water status, thereby helping the plant to cope with the salinity stress.

Inoculation of plant roots with an EPS-producing *Rhizobium* strain was reported to improve soil structure [159]. The rhizobial population having enhanced EPS production colonises roots and rhizosphere, and can bind cations including Na+ and may decrease the content of Na+ available for plant uptake, and thus help alleviating salt stress in plants growing in saline environments [31]. Similar observations were done in drought-affected soybean roots [160]. Co-inoculation of *Phaseolus vulgaris* L. with *R. tropici* and the PGPR *Paenibacillus polymyxa* (which produces trehalose) increased plant growth, N content, and nodulation under drought stress [43].

Conclusion and Future Prospects

As discussed in this review, abiotic stresses such as drought and salinity affects physiological processes of plants, including decline of endogenous levels of plant growth regulators. PGPR are able to stimulate plant growth, development, nutrient uptake, alleviate salt stress and improve symbiotic performance of leguminous plants with its rhizobia. They may increase levels of phytohormones in the roots after inoculation of plant seedling and may affect the metabolism of endogenous phytohormones in the plant under stress. This process may be a good mean of protection against salt stress and promotion of plant growth.

The beneficial microbes should therefore be considered as a seed dressing in field trials to improve crop yield in salt affected soils. The traits involved in mitigating plants stress and plant growth stimulation by PGPR include production of phytohormones, decreasing ethylene levels by the enzyme ACC deaminase, and production of osmoprotectants.

However, there is still a lot that is not understood regarding the functioning of these organisms under stressed soil conditions and also their interactions with the host plant.

More detailed studies are needed on the role of abiotic factors in altering the activity of rhizobacteria and managing plant–microbe interactions, with respect to their adaptability to extreme environments.

Future research should explore how root exudation could affect the activity of root associated bacteria and their interactions under hostile environmental condition. Hopefully, new research will provide farmers with novel control strategies for development of more effective and longer shelf-lived microbial inoculants that replace chemical fertilizers in agriculture.

Acknowledgments

The research activity of Dilfuza Egamberdieva was supported by a Georg Forster Research Fellowship for experienced Researchers (HERMES), Alexander Von Humboldt Foundation.

References

[1] Othman, Y., G. Al-Karaki, A.R. Al-Tawaha and A. Al- Horani. (2006). Variation in germination and jon uptake in barley genotypes under salinity conditions. *World J Agric Scie.* 2 (1): 11-15.

[2] Parvaiz, A., Satyawati, S. (2008). Salt stress and phyto-biochemical responses of plants-a review. *Plant Soil Env.* 54(3): 89.

[3] Egamberdieva, D., Berg, G., Lindstrom, K., Rasanen, L. (2010). Root colonising Pseudomonas spp. improve growth and symbiosis performance of fodder galega (Galega orientalis LAM) grown in potting soil. *Eur J Soil Biol.* 46(3-4): 269-272.

[4] Paul, D., Nair, S. (2008). Stress adaptations in a Plant Growth Promoting Rhizobacterium (PGPR) with increasing salinity in the coastal agricultural soils. *J. Basic Microbiol.* 48 (5): 378-384.

[5] Mayak, S., Tirosh, T., Glick, B.R. (2004). Plant growth-promoting bacteria that confer resistance to water stress in tomatoes and peppers. *Plant Science* 166: 525–530.

[6] Egamberdieva, D., Kucharova, Z. (2009). Selection for root colonising bacteria stimulating wheat growth in saline soils. *Biol Fert Soils.* 45: 561-573.

[7] Adesemoye, A.O., Egamberdieva, D. (2013). Beneficial effects of plant growth promoting rhizobacteria on improved crop production: the prospects for developing economies. In: Maheshwari, D.K., Saraf, M., Aeron, A. (Eds.) Bacteria in agrobiology: crop productivity. *Springer-Verlag Berlin Heidelberg,* pp 45-63.

[8] Khalid, A., Arshad, M., Zahir, Z. A. (2004). Screening plant growth-promoting rhizobacteria for improving growth and yield of wheat. *J. Appl. Microbiol.* 96 (3): 473-480.

[9] Egamberdiyeva, D. (2005). Plant growth promoting rhizobacteria isolated from calcisol soil in a semiarid region of Uzbekistan: biochemical characterisation and effectiveness. *Plant Nut Soil Sci.* 168:94-99

[10] Lugtenberg, B.J.J., Dekkers, L. and Bloemberg, G.V. (2001). Molecular determinants of rhizosphere colonization by Pseudomanas. *Annu Rev Phytopath* 39: 461-490.

[11] Egamberdieva, D. (2012). Pseudomonas chlororaphis: A salt tolerant bacterial inoculant for plant growth stimulation under saline soil conditions. *Acta Physiol Plant.* 34: 751-756.

[12] Egamberdieva, D. (2012). The management of soil quality and plant productivity in stressed environment with rhizobacteria. In: Maheshwari, D.K. (Ed), Bacteria in agrobiology: stress management, *Springer-Verlag Berlin Heidelberg.* pp 27-40.

[13] Egamberdieva, D., Berg, G., Lindström, K., Räsänen, L.A. (2013). Alleviation of salt stress of symbiotic Galega officinalis L. (goat's rue) by co-inoculation of Rhizobium with root colonising Pseudomonas. *Plant Soil.* 369(1): 453-465.

[14] Chinnusamy, V., Jagendorf, A., Zhu, J.K. (2005). Understanding and improving salt tolerance in plants. *Crop Sci.* 45: 437-448.

[15] Mantri, N., Patade, V., Penna, S., Ford, R., Pang, E. (2012). Abiotic stress responses in plants: present and future. In: Ahmad, P., Prasad, M.N.V. (Eds) Abiotic stress responses in plants: metabolism, productivity and sustainability. *Springer,* New York, pp 1–19.

[16] Rahman, M.S., Matsumuro, T., Miyake, H., Takeoka, Y. (2000). Salinity-induced ultrastructural alternations in leaf cells of rice (Oryza sativa L.). *Plant Prod Sci.* 3: 422-429.

[17] Jamil, M., Lee, D.B., Jung, K.Y., Ashraf, M., Lee, S.C., Rha1, E.S. (2006). Effect of salt (NaCl) stress on germination and early seedling growth of four vegetables species. *Central European Agr.* 7: 273-282.

[18] Egamberdieva, D., Shurigin, V., Gopalakrishnan, S., Sharma, R. (2014). Growth and symbiotic performance of chickpea (Cicer arietinum) cultivars under saline soil conditions. *J Biol Chem Res.* 31(1): 333-341.

[19] Egamberdieva, D. (2009). Alleviation of salt stress by plant growth regulators and IAA producing bacteria in wheat. *Acta Physiol Plant.* 31: 861-864.

[20] Rabie, G.H., Almadini, A.M. (2005). Role of bioinoculants in development of salt-tolerance of Vicia faba plants under salinity stress. *Afr J Biotech.* 4 (3): 210-222.

[21] Xu, G.Y., Rocha, P.S., Wang, M.L., Xu, M.L., Cui, Y.C., Li, L.Y., Zhu, Y.X., Xia, X. (2011). A novel rice calmodulin-like gene, OsMSR2, enhances drought and salt tolerance and increases ABA sensitivity in Arabidopsis. *Planta.* 234: 47-59.

[22] Khodarahmpour, Z., Ifar, M., Motamedi, M. (2012). Effects of NaCl salinity on maize (Zea mays L.) at germination and early seedling stage. *Afr J Biotechnol.* 11: 298–304.

[23] Essa, T.A. (2002). Effect of salinity stress on growth and nutrient composition of three soybean (Glycine max (L.) Merrill) cultivars. *J Agron Crop Scie.* 188(2): 86-93.

[24] Atak, M., Kaya, M.D., Kaya, G., Cikili, Y., Ciftci, C.Y. (2006). Effects of NaCl on the germination, seedling growth and water uptake of triticale. *Turk J Agric.* 30: 39-47.

[25] Prakash, L., Parthapasenan, G. (1990). Interactive effect of NaCl salinity and gibberelic acid on shoot growth, content of absisic acid and gibberelin like substances and yield of rice (Oruza sativa). *Plant Science.* 100: 173-181.

[26] Arbona, V., Marco, A.J., Iglesias, D.J., Lopez-Climent, M.F., Talon, M., Gomez-Cadenas, A. (2005). Carbohydrate depletion in roots and leaves of salt-stressed potted Citrus clementina L. *Plant Growth Reg.* 46: 153–160.

[27] Dolatabadian, A., Modarres Sanavy, S.A.M., Ghanati, F. (2011). Effect of salinity on growth, xylem structure and anatomical characteristics of soybean. *Not Sci Biol.* 3:41-45.

[28] Golezani, K.G., Yengabad, F.M. (2012). Physiological responses of lentil (Lens culinaris Medik.) to salinity. *Int J Agr Crop Sci.* 4(20): 1531-1535.

[29] Heidari, M., Jamshid, P. (2010). Interaction between salinity and potassium on grain yield, carbohydrate content and nutrient uptake in pearl millet. *J Agric Biol Sci.* 5: 39-46.

[30] Parker, M.B., Gaines, T.P., Hook, J.E., Gascho, G.J., Maw, B.W. (1987). Chloride and water stress effects on soybean in pot culture. *J Plant Nut.* 10: 517-538.

[31] Ashraf, M. (2004). Photosynthetic capacity and ion accumulation in a medicinal plant henbane (Hyoscyamus niger L.) under salt stress. *J Appl Bot.* 78: 91–96.

[32] Razmjoo, K., P. Heydarizadeh and M.R. Sabzalian, 2008. Effect of salinity and drought stresses on growth parameters and essential oil content of Matricaria chamomila. *Int J Agri Biol.* 10: 451–454.

[33] Teixeira da Silva, J.A., Egamberdieva, D. (2013). Plant-growth promoting rhizobacteria and medicinal plants. In: Recent progress in medicinal plants, 38, Essential Oils III and Phytopharmacology, *Studium Press LLC, TX, USA,* pp 26-42.

[34] Giller, K.E. (2001). Nitrogen fixation in tropical cropping systems: Cabi, Wallingford.

[35] Egamberdieva, D., Jabborova, D., Wirth, S. (2013). Alleviation of salt stress in legumes by co-inoculation with Pseudomonas and Rhizobium In: Arora NK (Ed) Plant Microbe Symbiosis- Fundamentals and Advances, *Springer India*, pp 291 – 303.

[36] Garg, N., Baher, N. (2013). Role of arbuscular mycorrhizal symbiosis in proline biosynthesis and metabolism of Cicer arietinum L. (chickpea) genotypes under salt stress. *J Plant Growth Regul.* 32: 767–778.

[37] Garg, N., Singla, R. (2004). Growth, photosynthesis, nodule nitrogen and carbon fixation in the chickpea cultivars under salt stress. *Braz J Plant Physiol.* 416: 137–146.

[38] Bouhmouch, I., Souad-Mouhsine, B., Brhada, F., Aurag, J. (2005). Influence of host cultivars and Rhizobium species on the growth and symbiotic performance of Phaseolus vulgaris under salt stress. *J Plant Physiol.* 162: 1103–1113.

[39] Singleton, P.W., Bohlool, B. (1984). Effect of salinity on the nodule formation by soybean. *Plant Physiol.* 74: 72-6.

[40] Slattery, J., Coventry, D.R., Slattery, W. (2001). Rhizobial ecology as affected by the soil environment. *Animal Production Science.* 41 (3): 289-298.

[41] Tuteja, N., Sopory, S.K. (2008). Chemical signaling under abiotic stress environment in plants. *Plant Signal Beh.* 3: 525–536.

[42] Sakhabutdinova, A.R., Fatkhutdinova, D.R., Bezrukova, M.V., Shakirova, F.M. (2003). Salicylic acid prevents the damaging action of stress factors on wheat plants. *Bulg J Plant Physiol.* 314–319.

[43] Figueiredo, M.V., Burity, H.A., Martinez, C.R., Chanway, C. (2008). Alleviation of drought stress in the common bean (Phaseolus vulgaris L.) by co-inoculation with Paenibacillus polymyxa and Rhizobium tropici. *Appl Soil Ecol.* 4: 182 – 188.

[44] Jabborova, D., Qodirova, D., Egamberdieva, D. (2013). Improvement of seedling establishment of soybean using IAA and IAA producing bacteria under saline conditions. *Soil Water Journal.* 2(2): 531-539.

[45] Dunlap, J.R,. Binzel, M.L. (1996). NaCl reduces indole-3-acetic acid levels in the roots of tomato plants independent of stress induced abscisic acid. *Plant Physiol.* 112: 379-384.

[46] Afzal, I, Basra, Sh, Iqbal, A. (2005). The effect pf seed soaking with plant growth regulators on seedling vigor of wheat under salinity stress. *J. Stress Physiol. Biochem.* 1(1): 6-14.

[47] Jabborova, D., Egambertdieva, D., Davranov, K., Jabborov, M. (2013). Restoration of growth of soybean under saline conditions due to inoculation with Bradyrhizobium japonicum and Pseudomonas putida. *Uzbek Biol J.* 5: 23-26.

[48] Maggio, A., Barbieri, G., Raimondi, G., De Pascale, S. (2010). Contrasting effects of GA3 treatments on tomato plants exposed to increasing salinity. *J Plant Growth Regul.* 29: 63–72.

[49] Sastry, E.V.D., Shekhawa, K.S. (2001). Alleviatory effect of GA3 on the effect of salt at seedling stage in wheat (Triticum aestivum). *Indian J. Agric. Res.* 35: 226-231.

[50] Shah, S.H. (2007). Effects of salt stress on mustard as affected by gibberellic acid application. *Gen Appl Plant Physiol.* 33: 97–106.

[51] Hamayun, M., Khan, S.A., Shinwari, Z.K., Khan, A.K., Ahmad, N., Lee, I.J. (2010). Effect of polyethylene glycol induced drought stress on physio-hormonal attributes of soybean. *Pak J Bot.* 42: 977-986.

[52] Khadri, M., Tejera, N.A., Lluch, K. (2006). Alleviation of salt stress in common bean (Phaseolus vulgaris) by exogenous abscisic acid supply. *J Plant Growth Regul.* 25: 110–119.

[53] Berg, G., Alavi, M., Schmidt, C.S., Zachow, C., Egamberdieva, D., Kamilova, F., Lugtenberg, B. (2013). Biocontrol and osmoprotection for plants under saline conditions. In: Frans J. de Bruijn (Ed), Molecular microbial ecology of the rhizosphere, *Wiley -Blackwell,* USA

[54] Barassi, C.A., Ayrault, G., Creus, C.M., Sueldo, R.J., Sobrero, M.T. (2006). Seed inoculation with Azospirillum mitigates NaCl effects on lettuce. *Scientia Horticult.* 109(1): 8-14.

[55] Lugtenberg, B., Kamilova, F. (2009). Plant growth-promoting rhizobacteria. *Ann. Rev. Microbiol.* 63: 541-556.

[56] Hameed, A., Egamberdieva, D., Abd-Allah, E.F., Hashem, A., Kumar, A., Ahmad, P. (2014). Salinity stress and arbuscular mycorrhizal symbiosis in plants. In: Miransari, M. (Ed) Use of microbes for the alleviation of soil stresses, *Springer New York* , 1: 139-159.

[57] Egamberdieva, D., Jabborova, D. (2013). Improvement of cotton production in arid saline soils by beneficial microbes. In: Huang, L., Zhao, Q. (Eds), Crop yields: production, management practices and impact of climate change, *Nova Publishers* USA. pp 109-122.

[58] Cakmakci, R., Donmez, D., Aydin, A., Sahin, F. (2005). Growth promotion of plants by plant growth-promoting rhizobacteria under greenhouse and two different field soil conditions. *Soil Biol Bioch.* 38: 1482-1487.

[59] Egamberdiyeva, D., Hoflich, G. (2003). The effect of associative bacteria from different climates on plant growth of pea at different soils and temperatures. *Arch Agron Soil Sci.* 49(2): 203-213.

[60] Sharma, N., Sharma, S. (2008). Control of foliar diseases of mustard by Bacillus from reclaimed soil. *Microb Res*. 163(4): 408-413.

[61] Rajasekar, S., Elango, R. (2011). Effect of microbial consortium on plant growth and improvement of alkaloid content in Withania somnifera (Ashwagandha), *Current Botany.* 2(8): 27-30.

[62] Logan, N.A., Lebbe, L., Verhelst, A., Goris, J., Forsyth, G., Rodriguez-Diaz, M., Heyndrickx, M., De Vos, P. (2002). Bacillus luciferensis sp. nov., from volcanic soil on Candlemas Island, South Sandwich archipelago. *Int J Syst Evol Microbiol.* 52: 1985–1989.

[63] Egamberdiyeva, D., Hoflich, G. (2004). Importance of plant growth promoting bacteria on growth and nutrient uptake of cotton and pea in semi-arid region Uzbekistan. *J Arid Env.* 56: 293-301.

[64] Ashwini, N., Srividya, S. (2014). Potentiality of Bacillus subtilis as biocontrol agent for management of anthracnose disease of chilli caused by Colletotrichum gloeosporioides OGC1. 3 *Biotech* 4: 127-136.

[65] Egamberdiyeva, D., Hoflich, G. (2003). Influence of grwth-promoting bacteria on the growth of wheat in different soils temperatures. *Soil Biol Biochem.* 35: 973-978.

[66] Akhtar, M.S., Siddiqui, Z.A. (2009). Use of plant growth-promoting rhizobacteria for the biocontrol of root-rot disease complex of chickpea. *Aust Plant Path.* 38: 44–50.

[67] Yadegari, M., Rahmani, A. (2010). Evaluation of bean (Phaseolus vulgaris) seeds inoculation with Rhizobium phaseoli and plant growth promoting Rhizobacteria (PGPR) on yield and yield components. *Afr J Agric Res.* 5: 792-799.

[68] Mehnaz, S., Lazarovits, G. (2006). Inoculation effects of Pseudomonas putida, Gluconacetobacter azotocaptans, and Azospirillum lipoferum on corn plant growth under greenhouse conditions. *Microb Ecol.* 51: 326–335.

[69] Creus, C.M., Sueldo, R.J., Barassi, C.A. (2004). Water relations and yield in Azospirillum inoculated wheat exposed to drought in the field. *Can J Bot.* 82: 273-281.

[70] Egamberdiyeva, D., Qarshieva, D., Davranov, K. (2004). Growth and yield of soybean inoculated with Bradyrhizobium spp in calcareous soil. *Biol Fert Soils.* 4: 144-146

[71] Karnwal, A., Kumar, V. (2012). Influence of plant growth promoting rhizobacteria (PGPR) on the growth of chickpea (Cicer arietinum l.). *Ann Food Science Tech.* 43-48.

[72] Egamberdieva D, Berg G, Lindstrom K,Rasanen L (2010) Root colonising Pseudomonas spp. improve growth and symbiosis performance of fodder galega (Galega orientalis LAM) grown in potting soil. *Eur J Soil Biol* 46(3-4): 269-272.

[73] Vazquez, P., Holguin, G., Puente, M.E., Lopez-Cortes, A., Bashan, Y. (2000). Phosphate-solubilizing microorganisms associated with the rhizosphere of mangroves in a semiarid coastal lagoon. *Biol Fertil Soils.* 30: 46–468.

[74] Egamberdiyeva D,Juraeva D, Poberejskaya S, Myachina O, Teryuhova P, Seydalieva L, Aliev A (2004) Improvement of wheat and cotton growth and nutrient uptake by phospate solubilising bacteria. In *proceeding book of 26th Annual Conservation Tillage Conference for Sustainable Agriculture, Auburn,* USA, 58-65.

[75] Sheng, X.F. (2005). Growth promotion and increased potassium uptake of cotton and rape by a potassium releasing strain of Bacillus edaphicus. *Soil Biol Biochem.* 37: 1918–1922.

[76] Egamberdiyeva D, Hoflich G (2005) Increase in forage maize production by bacterial fertilizers. In O' Mara FP, Wilkins RJ, Mannetje L, Lovett DK, Rogers PAM, Boland TM (eds), offered papers of XX International Grassland Congress, 734

[77] Wu, S. C., Caob, Z. H., Lib, Z. G., Cheunga, K. C., Wong,a M. H. (2005). Effects of biofertilizer containing N-fixer, P and K solubilizers and AM fungi on maize growth: a greenhouse trial. *Geoderma.* 125: 155-166.

[78] Mohamed, H.I., Gomaa, E.Z. (2012). Effect of plant growth promoting Bacillus subtilis and Pseudomonas fluorescens on growth and pigment composition of radish plants (Raphanus sativus) under NaCl stress. *Photosynthetica.* 50 (2): 263-272.

[79] Elkoca, E., Turan, M., Donmez, M.F. (2010). Effects of single, dual and triple inoculations with Bacillus subtilis, Bacillus megaterium and Rhizobium leguminosarum bv. phaseoli on nodulation, nutrient uptake, yield and yield parameters of common bean (Phaseolus vulgaris l. cv.'elkoca-05'). *J Plant Nutrition* 33(14): 2104-2119.

[80] El-Azouni, I. (2008). Effect of phosphate solubilizing fungi on growth and nutrition uptake of soybean (Glycine max L.) *Plants J Appl Sci Res.* 4(6): 592-598.

[81] Anandham, R., Sridar, R., Nalayini, P., Poonguzhali, S., Madhaiyan, M., Tongminsa. (2007) Potential for plant growth promotion in groundnut (Arachis hypogaea L.) cv. ALR-2 by co-inoculationof sulfur-oxidizing bacteria and Rhizobium. *Microb Res.* 162: 139-153.

[82] Egamberdiyeva, D. (2007). The growth and nutrient uptake of maize inoculated with plant growth promoting bacteria affected by different soil types. *Appl Soil Ecol.* 36: 184-189.

[83] Egamberdieva, D., Lugtenberg, B. (2014). PGPR to alleviate salinity stress on plant growth. In: Miransari, M. (Ed) Use of microbes for the alleviation of soil stresses. *Springer* New York. 1: 73-96.

[84] Dey, R., Pal, K.K., Bhatt, D.M., Chauhan, S.M. (2004). Growth promotion and yield enhancement of peanut (Arachis hypogaea L.) by application of plant growth-promoting rhizobacteria. *Microb Res.* 159 (4): 371-394.

[85] Ratti, N., Kumar, S., Verma, H.N., Gautam, S.P. (2001). Improvement in bioavailability of tricalcium phosphate to Cymbopogon martinii var. motia by rhizobacteria, AMF and Azospirillum inoculation. *Microb Res.* 156(2): 145-149.

[86] Rudresh, D.L., Shivaprakash, M.K., Prasad, R.D. (2005). Effect ofcombined application of rhizobium, phosphate solubilizing bacteriumand Trichoderma spp. on growth, nutrient uptake and yield of chickpea (Cicer aritenium L.). *Appl Soil Ecol.* 28: 139-146.

[87] Rokhzadi, A., Asgharzadeh, A., Darvish, F., Nour-Muhammadi, G., Majidi, E. (2008). Influence of plant growth promoting rhizobacteria on dry matter accumulation and yield of chickpea (Cicer arietinum L.) under field conditions. *Amer Eur J Agr Env Sci.* 3(2): 253-257.

[88] Khurana, A.S., Sharma, P. (2000). Effect of dual inoculation of phosphate solubilizing bacteria, Bradyrhizobium sp. and phosphorus on nitrogen fixation and yield of chickpea. *Indian J Pulses Res.* 13: 66–67.

[89] Galal, Y.G., EI-Gandaour, J.A., EI-Akel, F.A. (2001). Stimulation of wheat growth and N fixation through Azospirillum and Rhizobium inoculation: In: Horsts, W.J. (ed) Plant nutrition – food security and sustainability of agro ecosystems. *A field trust with* 15 N technique, pp 66-67.

[90] Ordookhani, K., Sharafzadeh, S., Zare, M. (2011). Influence of PGPR on growth, essential oil and nutrients uptake of sweet basil. *Adv Environ Biology.* 5(4): 672-677.

[91] Karthikeyan, B., Joe, M.M., Jaleel, C.A., Deiveekasundaram, M. (2010). Effect of root inoculation with plant growth promoting rhizobacteria (pgpr) on plant growth, alkaloid content and nutrient control of Catharanthus roseus (L.) G. Don. *Natura Croatica.* 1: 205-212.

[92] Lakshmipathy, R., Chandrika, K., Gowda, B., Balakrishna, A.N., Bagyaraj, D.J. (2002). Response of Calamus thwaitessii var. canaranus Wilde to inoculation with Glomus mosseae, Bacillus coagulans and Trichoderma harzianum. *J Soil Biology Ecology.* 22: 16-21.

[93] Elango, K.V. (2004). Studies on the effect of native AM fungi and PGPR's on growth and productivity of Gloriosa superba L., PhD. Thesis, Bharathidasan University, India.

[94] Mani, N. (2004). Phytochemical and antimicrobial studies on Alpinia galanga and Coleus amboinicus as influenced by native AM fungi, PhD. Thesis, Bharathidasan Univiersity, India.

[95] Lugtenberg, B., Malfanova, N., Kamilova, F., Berg, G. (2013). Plant growth promotion by microbes. In: de Bruijn F J (ed) Molecular microbial ecology of the rhizosphere. *Wiley-Blackwell,* Hoboken, N.J., USA, pp 561-73.

[96] Egamberdieva, D., Kamilova, F., Validov, S., Gafurova, L., Kucharova, Z., Lugtenberg, B. (2008). High incidence of plant growth-stimulating bacteria associated

with the rhizosphere of wheat grown in salinated soil in Uzbekistan. *Env Microb*. 19: 1-19.

[97] Pliego, C., Kamilova, F., Lugtenberg, B. (2011). Plant Growth-promoting bacteria: Fundamentals and exploitation. In: Maheshwari, D.K. (Ed) Bacteria in Agrobiology: Crop Ecosystems, *Springer,* Germany, pp 295-343.

[98] Nadeem, S.M., Zahir, Z.A., Naveed, M., Asghar, H.N., Arshad, M. (2010). Rhizobacteria capable of producing ACC-deaminase may mitigate the salt stress in wheat. *Soil Sci Soc Am J.* 74: 533–542.

[99] Nabti, E., Sahnoune, M., Ghoul, M., Fischer, D., Hofmann, A., Rothballer, M., Schmid, M., Hartmann, A. (2010). Restoration of growth of durum wheat (Triticum durum var. waha) under saline conditions due to inoculation with the rhizosphere bacterium Azospirillum brasilense NH and extracts of the marine alga Ulva lactuca. *J Plant Growth Regul.* 29: 6–22.

[100] Abbaspoor, A.S. Movafegh, Akbari Asl, M.H. (2009). The efficiency of plant growth promoting rhizobacteria (PGPR) on yield and yield components of two varieties of wheat in salinity condition. *Amer Euras J Sust Agric.* 3(4): 824-828.

[101] Egamberdieva D, Jabborova D (2013) Improvement of cotton production in arid saline soils by beneficial microbes. In: Lijuan Huang and Qiao Zhao (eds), Crop yields: production, management practices and impact of climate change, *Nova Publishers* USA, 109-122, ISBN: 978-1-62618-176-2

[102] Yue, H.T., Mo, W.P., Li, C., Zheng, Y.Y., Li, H. (2007). The salt stress relief and growth promotion effect of Rs-5 on cotton. *Plant Soil.* 297: 139–145.

[103] Egamberdieva, D., Kucharova, Z., Davranov, K., Berg, G., Makarova, N., Azarova, T., Chebotar, V., Tikhonovich, I., Kamilova, F., Validov, S., Lugtenberg, B. (2011). Bacteria able to control foot and root rot and to promote growth of cucumber in salinatedsoils. *Biol Fert Soils.* 47: 197-205

[104] Yildirim, E., Turan, M., Donmez, M.F. (2008). Mitigation of salt stress in radish (Raphanus sativus l.) by plant growth promoting rhizobacteria. *Roum Biotech Let.* 13(5): 3933-3943.

[105] Egamberdieva D, Jabborova D, Mamadalieva N (2013) Salt tolerant Pseudomonas extremorientalis able to stimulate growth of Silybum marianum undersalt stress condition. *Med Aromatic Plant Science Biotech.* 7(1): 7-10.

[106] Egamberdieva, D. (2011). Survival of Pseudomonas extremorientalis TSAU20 and P. chlororaphis TSAU13 in the rhizosphere of common bean (Phaseolus vulgaris) under saline conditions. *Plant Soil Env.* 57(3): 122-127.

[107] Han, H.S., Lee, K.D. (2005). Plant growth promoting rhizobacteria effect on antioxidant status, photosynthesis, mineral uptake and growth of lettuce under soil salinity. *Research J Agric Biol Sciences.* 1(3): 210-215.

[108] Kaya, C., Ak, B.E., Higgs, D. (2003). Response of salt-stressed strawberry plants to supplementary calcium nitrate and/or potassium nitrate. *J Plant Nutr.* 26(3): 543-560.

[109] Yildirim, E., Turan, M., Ekinci, M., Dursun, A., Cakmakci, R. (2011). Plant growth promoting rhizobacteria ameliorate deleterious effect of salt stress on lettuce. *Scie Res Essays* 6(20): 4389-4396.

[110] Jaleel, C.A., Manivavannan, P., Sankar, P., Krishnakumar, B., Gopi, A.R., Somasundaram, R., Pannerselvam (2007). Psudomonas fluorescens enhances biomass yield and Ajmalicine production in Catharanthus roseus under water deficit stress. *Colloid and Surface B: Biointerfaces.* 60(1): 7–11.

[111] Adesemoye, A.O., Torbert, H.A., Kloepper, J.W. (2009). Plant growth-promoting rhizobacteria allow reduced application rates of chemical fertilizers. *Microbial Ecology.* 58 (4): 921-929.

[112] Bano, A., Fatima, M. (2009). Salt tolerance in Zea mays (L). following inoculation with Rhizobium and Pseudomonas. *Biol Fert Soils.* 45(4): 405-413.

[113] Yang, J., Kloepper, J. and Ryu, C.N. (2008). Rhizosphere bacteria help plants tolerate abiotic stress. *Trends Plant Science,* 14: (1).

[114] Rasaei, B., Ghobadi, M.E., Ghobadi, M., Nadjaphy, A., Rasaei, A. (2012). The study effects of some biological agents on chickpea (Cicer arietinum L.) under semi-dry conditions in Kermanshah. *Eur J Exp Biol.* 2(4): 1113-1118.

[115] Arshad, M., Shaharoona, B., Mahmood, T. (2008). Inoculation with Pseudomonas spp. containing acc-deaminase partially eliminates the effects of drought stress on growth, yield, and ripening of pea (Pisum sativum L.). *Pedosphere* 18(5): 611-620.

[116] Vivas, A., Marulanda, A., Ruiz-Lozano, J.M., Barea, J.M., Azcon, R. (2003). Influence of a Bacillus sp. on physiological activities of two arbuscular mycorrhizal fungi and on plant responses to PEG - induced drought stress. *Mycorrhiza.* 13: 249-256.

[117] Estevez J, Dardanelli MS, Megias M, Rodriguez-Navarro DN (2009). Symbiotic performance of common bean and soybean co-inoculated with rhizobia and Chryseobacterium balustinum Aur9 under moderate saline conditions. *Symbiosis* 49: 29-36.

[118] Dashadi, M., Khosravi, H., Moezzi, A., Nadian, H., Heidari, M., Radjabi, R. (2011). Co-Inoculation of Rhizobium and Azotobacter on growth of faba bean under water deficit conditions. *Amer-Eur J Agr Env Sci.* 11(3): 314-319.

[119] Bharti, N., Yadav, D., Barnawal, D., Maji, D., Kalra, A. (2013). Exiguobacterium oxidotolerans, a halotolerant plant growth promoting rhizobacteria, improves yield and content of secondary metabolites in Bacopa monnieri (L.) pennell under primary and secondary salt stress. *World J Microbiol Biotech.* 29(2): 379-87.

[120] Golpayegani, A., Tilebeni, H.G. (2011). Effect of biological fertilizers on biochemical and physiological parameters of basil (Ociumum basilicm L.) medicine plant. *Amer-Eurasian J Agric Environ Scence.* 11 (3): 411-416.

[121] Dardanelli, M.S., De Cordoba, F.J.F., Espuny, M.R., Carvajal, M.A.R., Diaz, M.E.S., Serrano, A.M.G., Okon, Y., Megias, M. (2008). Effect of Azospirillum brasilense coinoculated with Rhizobium on Phaseolus vulgaris flavonoids and Nod factor production under salt stress. *Soil Biol Bioch.* 40: 2713-2721.

[122] Spaepen, S., Vanderleyden, J., Okon, Y. (2009). Plant Growth-promoting actions of rhizobacteria. In: van Loon, L.C., Ed Kader, J.C., Delseny, M. (Eds) *Adv Botanical Res.* 51: 283-320.

[123] Egamberdieva, D. (2013). The role of phytohormone producing bacteria in alleviating salt stress in crop plants. In: Miransari, M. (Ed) Biotechnological techniques of stress tolerance in plants, *Stadium Press USA,* 21-39.

[124] Upadhyay, S.K., Singh, J.S., Singh, D.P. (2011). Exopolysaccharide-producing plant growthpromoting rhizobacteria under salinity condition. *Pedosphere*. 2: 214-222.

[125] Shaterian, J., Waterer, D., De-Jong, H., Tanino, K.K. (2005). Differential stress response to NaCl salt application in early and late maturing diploid potato (Solanum sp.) clones. *Envir Exper Bot.* 54: 202-212.

[126] Bianco, C., Defez, R. (2009). Medicago truncatula improves salt tolerance when nodulated by an indole-3-acetic acid-overproducing Sinorhizobium meliloti strain. *J Exp Bot.* 60: 3097–3107.

[127] Spaepen, S., Vanderleyden, J., Remans, R. (2007). Indole-3-acetic acid in microbial and microorganism-plant signalling. *FEMS Microb Rev.* 31: 425-448.

[128] Nabti, E., Sahnoune, M., Adjrad, S., Dommelen, A.V., Ghoul, M., Schmid, M., Hartmann, A. (2007). A halophilic and osmotolerant Azospirillum brasilense strain from Algerian soil restores wheat growth under saline conditions. *Eng Life Scien.* 7(4): 354-360.

[129] Gamalero, E., Glick, B.R. (2011). Mechanisms used by plant growth-promoting bacteria. In: Maheshwari, D.K. (ed) Bacteria in agrobiology: plant nutrient management. *Springer,* Berlin Heidelberg, pp 17–46.

[130] Tanimoto, E. (2005). Regulation of root growth by plant hormones: Roles for auxin and gibberellin. Critical Reviews Plant Sciences. 24: 249–265.

[131] Ahemad, M., Khan, M.S. (2011). Effect of tebuconazole-tolerant and plant growth promoting Rhizobium isolate MRP1 on pea-Rhizobium symbiosis. *Sci Hortic.* 129: 266–272.

[132] Camerini, S., Senatore, B., Lonardo, E., Imperlini, E., Bianco, C., Moschetti, G., Rotino, G.L., Campion, B., Defez, R. (2008). Introduction of a novel pathway for IAA biosynthesis to rhizobia alters vetch root nodule development. *Arch Microbiol.* 190: 67–77.

[133] Arkhipova, T.N., Prinsen, E.A., Veselov, S.U., Martinenko, E.V., Melentiev, A.I., Kudoyarova, G.R. (2007). Cytokinin producing bacteria enhance plant growth in drying soil. *Plant Soil* 292: 305–315.

[134] Fulchieri, M., Lucangeli, C., Bottini, R. (1993). Inoculation with Azospirillum lipoferum affects growth and gibberellin status of corn seedling roots. *Plant Cell Physiol.* 34:1305–1309.

[135] Egamberdiyeva, D., Islam, K.R. (2008). Salt tolerant rhizobacteria: Plant growth promoting traits and physiological characterization within ecologically stressed environment. In: Plant-Bacteria Interactions: Strategies and techniques to promote plant growth. Ahmad, I., Pichtel, J., Hayat, S. (Eds) Wiley-VCH Verlag GmbH & Co., KGaA, *Weinheim,* Germany, pp 257-281.

[136] Malik, D.K., Sindhu, S.S. (2011). Production of indole acetic acid by Pseudomonas sp.: effect of coinoculation with Mesorhizobium sp. cicer on nodulation and plant growth of chickpea (Cicer arietinum). *Physiol Mol Biol Plants.* 17(1): 25–32.

[137] Probanza, A., García, J.A.L., Palomino, M.R., Ramos, B., Manero, F.J.G. (2002). Pinus pinea L. seedling growth and bacterial rhizosphere structure after inoculation with PGPR Bacillus (B. licheniformis CECT 5106 and B. pumilus CECT 5105). *Appl Soil Ecol.* 20: 75–84.

[138] Khan, Z., Doty, S.L. (2009). Characterization of bacterial endophytes of sweet potato plants. *Plant Soil.* 322: 197–207.

[139] Dodd, I.C., Zinovkina, N.Y., Safronova, V.I., Belimov, A. (2010). Rhizobacterial mediation of plant hormone status. *Ann Appl Biol.* 157: 361–379.

[140] Glick, B.R. (2010). Using soil bacteria to facilitate phytoremediation. *Biotechnology Advances.* 28(3): 367-374

[141] Shahzad, S.M., Khalid, A., Arshad, M., Rehman, K. (2010). Improving nodulation, growth and yield of Cicer arietinum L. through bacterial ACC-deaminase induced changes in root architecture. *Eur J Soil Biol.* 46(5): 342-347.

[142] Shaharoona, B., Arshad, M., Zahir, Z.A. (2006). Effect of plant growth promoting rhizobacteria containing ACC-deaminase on maize (Zea mays L.) growth under axenic conditions and on nodulation in mung bean (Vigna radiata L.). *Lett Appl Microbiol.* 42(2): 155-159.

[143] Roopa, B., Maya, C., Makari, H.K. (2012). Effect of different PGPR strain along with rhizobium on nodulation and chick pea productivity. *Asian J Exp Biol Sci.* 3: 424-426.

[144] Zafar-ul-Hye, M., Ahmad, M., Shahzad, S.M. (2013). Synergistic effect of rhizobia and plant growth promoting rhizobacteria on the growthand nodulation of lentil seedlings under axenic conditions. *Soil Env.* 32(1): 79-86.

[145] Saravanakumar, D., Samiyappan, R. (2007). ACC deaminase from Pseudomonas fluorescens mediated saline resistance in groundnut (Arachis hypogea) plants. *J Appl Microbiol.* 102(5): 1283-1292.

[146] Ma, J.H., Yao, J.L., Cohen, D., Morris, B. (1998). Ethylene inhibitors enhance in vitro root formation from apple shoot cultures. *Plant Cell Rep.* 17: 211–214.

[147] Serraj, R.H., Vasquez- Diaz, G., Hernandez, Drevon, J.J. (2001). Genotypic difference in response of nitrogenase activity (C2H2 Reduction) to salinity and oxygen in common bean. *Agronomie* 21: 645-650.

[148] Evelin, H., Kapoor, R., Giri, B. (2009). Arbuscular mycorrhizal fungi in alleviation of salt stress: a review. *Ann Bot.* 104: 1263–1280.

[149] Ashraf, M., Foolad, M.R. (2007). Roles of glycine betaine and proline in improving plant abiotic stress resistance. *Env Exp Bot.* 59: 206–216.

[150] Sleator, R.D., Hill, C. (2001). Bacterial osmoadaptation: The role of osmolytes in bacterial stress and virulence. *FEMS Microbiol Rev.* 26: 49–71.

[151] Yang X, Lu C (2005) Photosynthesis is improved by exogenous glycine betaine in salt-stressed maize plants. Physiologia Plant. 124 (3): 343-352.

[152] Verbruggen, N., Hermans, C. (2008). Proline accumulation in plants: a review. *Amino Acids.* 35: 753-759.

[153] Tisdall, J.M., Oades, J.M. (1982). Organic matter and water stable aggregates in soils. *J Soil Sc*i. 33: 141–163.

[154] Jha, Y., Subramanian, R.B., Patel, S. (2010). Combination of endophytic and rhizospheric plant growth promoting rhizobacteria in Oryza sativa shows higher accumulation of osmoprotectant agaist saline stress. *Acta Phys Plant.* 33(3): 797-802.

[155] Kandowangko, N.Y., Suryatmana, G., Nurlaeny, N., Simanungkalit, R.D.M. (2009). Proline and abscisic acid content in droughted corn plant inoculated with Azospirillum sp. and arbuscular mycorrhizae fungi. *Hayati J Biosci.* 16(1): 15-20.

[156] Sandhya, V., Ali, Sk.Z., Grover, M., Reddy, G., Venkateswarlu, B. (2010). Effect of plant growth promoting Pseudomonas spp. on compatible solutes, antioxidant status and plant growth of maize under drought stress. *Plant Growth Reg.* 62: 21-30.

[157] Barka, E.A., Nowak, J., Clément, C. (2006). Enhancement of chilling resistance of inoculated grapevine plantlets with a plant growthpromoting rhizobacterium, Burkholderia phytofirmans strain PsJN. *Appl Env Microb.* 70: 7246–7252.

[158] Sziderics, A.H., Rasche, F., Trognitz, F., Wilhelm, E., Sessitsch, A. (2007). Bacterial endophytes contribute to abiotic stress adaptation in pepper plants (Capsicum annuum L.). *Can J Microb.* 53: 1195-1202.

[159] Alami, Y., Achouak, W., Marol, C., Heulin, T. (2000). Rhizosphere soil aggregation and plant growth promotion of sunflowers by an exopolysaccharide-producing Rhizobium sp. strain isolated from sunflower roots. *Appl Environ Microbiol.* 66: 3393–3398.

[160] Williams, P.M., Sicardi de Mallorca, M.S. (1984). Effect of osmotically induced leaf moisture stress on nodulation and nitrogenase activity of Glycine max. *Plant Soil.* 80: 267–283.

[161] Minaxi, Nain, L.,Yadav, R.C., Saxena, J. (2012). Characterization of multifaceted Bacillus sp. RM-2 for its use as plant growth promoting bioinoculant for crops grown in semi arid deserts. *App Soil Ecol.* 59: 124–135.

[162] Mishra, R.K., Prakash, O., Alam, M., Dikshit, A. (2010). Influence of plant growth promoting rhizobacteria (PGPR) on the productivity of Pelargonium Graveolens l. herit. *Recent Res Sci Tech.* 2(5):53–57.

[163] Yao, L., Wu, Z., Zheng, Y., Kaleem, I., Chun, Li. (2010). Growth promotion and protection against salt stress by Pseudomonas putida Rs-198 on cotton. *Eur J Soil Biol.* 46(1): 49–54.

In: Microbes in Soil and Their Agricultural Prospects
Editors: K. Kishore Choudhary and D. Wattal Dhar
ISBN: 978-1-63482-824-6

Chapter 5

RHIZOSPHERE AND THEIR ROLE IN PLANT-MICROBE INTERACTION

Ajay Kumar, Vandana, Akhilesh Yadav, Deen Dayal Giri, P. K. Singh and Kapil D. Pandey*

Centre of Advance study in Botany, Banaras Hindu University,
Varanasi, India

ABSTRACT

Plant growth promoting rhizobacteria (PGPR) plays an important role in sustainable agriculture. These microbes directly or indirectly associated with the plants for growth promotion, disease management and yield enhancement. Genotypes and diversity of individual plants influences the composition of the associated communities. Microbial population of PGPR present in the rhizosphere depends upon the amount and composition of root exudates. The molecules present in the root exudates act as signaling molecules and helps in microbe interaction. This chapter describes the role of root exudates and mode of plant –microbe interaction in the rhizosphere.

Keywords: PGPR, rhizosphere, microbial interaction, plant growth

INTRODUCTION

Plants are members of complex communities, which function as a link between above and below ground organisms that consist of microbes, insects, vertebrates and invertebrates [1]. Theseare directly or indirectly influenced by the biotic like microbes, insects, humans and abiotic factors like soil, temperature, pH etc. To survive, plants need to allocate optimal resources for the growth and defences. Plants form associations with non-pathogenic root

* Corresponding Author address: CAS in Botany, Banaras Hindu University, Varanasi -221005, India, E-mail- kdpandey2005@gmail.com

associated microbes such as rhizobacteria, mycorrhizae, rhizobia that can promote plant growth by increasing their access to soil minerals [2, 3].

The number of prokaryotic cells on earth is approximately more than 10^{30} of which largest fraction occurs in the soil [4]. The soil matrix is a favourable niche for bacteria since both temperature and humidity are relatively stable [5]. The community structure of indigenous bacterial microflora in soil is determined by many variables including geographic location and soil structure [6], soil particle size [7], and mineral composition [8].

Soil serves as natural medium for growth of the plants. Most of the plants depend on soil for nutrients, but plants and their associated microorganisms play a crucial role in the formation and restoration of soil [9, 10]. Soils are the product of activities of plants which supply organic matter, organic acid and play pivotal role in weathering of rocks and minerals. Lambers et al. [11] described formation of soil influenced by combined effect of various factors –

$$\boldsymbol{S} \int \boldsymbol{b, p, c, t1, t2}$$

where as S= soil; b= biological activity; p= parental rock; c= climate; t_1= time; t_2 = topography.

Plants require essential mineral nutrientsfor their growth, which also obtained from the soil as well as interaction between microbes present within the rhizosphere. The complex interaction between plants and rhizospheric microorganisms meet most of the mineral nutrition and essential elements. Soil microorganisms have to compete for nutrients and other resources that are sparsely available in soil. Because of these limiting circumstances, bacterial proliferation in soil is greatly influenced by plant roots [12]. Plants release an array of chemical signals through the root surface which are carbon containing metabolites in soil matrix, which result in rhizodepostion. Rhizodeposition includes shedding of root cells, exudation secretion, leakage of sugars, organic acids and amino acids in soil [12, 13, 14]. Microorganisms use these compounds as substrates, resulting in an increased microbial biomass and activities around the roots. During seed germination and seedling growth, the developing plant interacts with a range of microorganisms present in the surrounding soil. As seeds germinate and roots grow through the soil, the release of organic material provides development of active microbial populations in the zone that includes plant root and surrounding soil in a few mm of thickness. This phenomenon is referred as the rhizospheric effect [15]. The term rhizosphere, means the soil compartment influenced by plant roots was first defined by Lorentz Hiltner [16, 17].

The population of microorganisms in the root zone are influenced by many factors: nature and age (or stage of growth) of plant; type, moisture content, reaction, and treatment of the soil. Soil treatment may stimulate plant growth which in turn, may cause an increase in number of organisms on the root surface. An increase in number of organisms may be aspected following certain soil treatments, but a "rhizospheric effect" need not necessarily results unless the plant itself is stimulated in some manner. The extent to which the roots influence soil microflora may then be expected by the rhizosphere: soil (R: S) ratio, that is the number of organisms in the rhizosphere soil divided by the number in the soil at a distance from the root (calculated on a dry weight basis).The interactions of plant roots and rhizospheric microorganisms are based largely on interactive modification of the soil environment by processes such as water uptake by the plant and release of organic chemicals by the roots. The release of Glucose, sucrose and mucilaginous exudates from plant roots stimulate growth of microorganisms rapidly within the plant parts or root surface and within the mucilaginous sheath (rhizoplane). After the plant matures, autolysis of some of the root materials (i.e., root cap senescent cells, root border cells) takes place and simple sugars and amino acids are released into the soil. This further stimulates the growth of bacteria like Pseudomonas and Azotobacter with high intrinsic growth rates.

The rhizosphere is colonized predominantly by Gram negative microbial community. Microflora consists of higher proportion of Gram- negative rods and a lower proportion of Gram- positive rods, cocci and pleomorphic forms. A relatively higher proportion of motile, rapidly growing bacteria are also seen. The release of rhizodeposition or carbon compounds from plants into the soil results in interaction of greater Gram negative microbial populations in the rhizosphere relative to the bulk soil where as Gram positive bacteria are reported to be inhibited [18].

Root exudates interact number of microorganisms in the plant rhizosphere. The microorganisms and their products, also interact with plant roots in a variety of positive, negative and neutral ways, such interaction can influence plant growth, development and nutrient dynamics. These interactions help to alter the plant's susceptibility to diseases and abiotic stresses [19, 20]. The microbial population present in the rhizosphere include various group of microorganisms such as bacteria, actinomycetes, fungi, protozoa, nematodes and micro arthropods in decreasing order. These rhizosphere inhabiting microorganisms compete with each other for water, nutrients, space and sometimes improve their competitiveness by developing an intimate association with plant. The rhizosphere supports diverse bacteria that can stimulate growth of plants. Such plant growth promoting rhizobacteria (PGPR) operate a wide variety of mechanisms, including N_2 fixation, enhanced P solubilisation, and phytohormones production [21]. Plant associated microbes sometimes give physiological and environmental advantages to their host plants. Although, rhizosphere-plant and microbe interactions are very complex, but, four main types of interactions occur when two or more organisms live in the zone: commensalism or symbiosis, competition or antibiosis, mutualism and rarely neutralism.

Rhizosphere

The rhizosphere includes plant root and the surrounding soil. The term and definition were given by Hiltner [16] to describe a thin layer of soil adhering on the plant root and extend 2mm outwards on the root surface. The size of rhizosphere depends on particular root system, root structure where contact area with soil is very large. It is necessary to understand the basic principles of rhizosphere microbial ecology, including the function and diversity of microorganisms that reside there, before soil microbial technology can be applied in the rhizosphere. Rhizosphere is a zone where root activity significantly influences biological properties of microorganism. The rhizosphere has three distinct components that are rhizospheric soil, rhizoplane and the root itself. The rhizosphere soil is thought to be quite different from the bulk soil and highly influenced by root exudates which affect microbial activity. The rhizoplane is the root surface having strongly adhering soil particles. The root itself is a part of the system, because endophytic microorganisms reside in the inner root tissues.

Plant Root

Plant roots lie below the surface of soil, have four major functions; absorption of water and inorganic nutrients, anchoring of the plant body to the ground, storage of food and nutrients and to prevent soil erosion. Besides these four functions, roots provide most suitable habitat for the growth of many microorganisms.

Interaction between the plant root and soil microorganisms satisfies important nutritional requirement for both, the plant and the associated microorganisms [22]. Plant roots divided into tap roots, fibrous and adventitious roots.

The roots possess root caps, root border cells and unicellular root hairs which play the most significant role in the interaction between the roots and associated microorganisms.

The root cap is a complex and dynamic plant organ which sense and transmit the environmental signal, synthesize and secrete the biochemicals and shed metabolically active cells.

The root cap contains essential components for signalling system and its ablation altered root architecture [23]. The root hairs are found just behind the root tip where they are continuously formed. They are the tubular outgrowths of trichoblasts, the hair forming cells and are the lateral extension of single cells which are rarely branched. Root hairs are generally 5-17 μm in diameter, 80 to 1500 μm in length and can survive for three to four weeks before they die off [24].

The Root Border Cells

The replacement, turnover or programmed cell separation of the root cap results in the delivery of detached dead cells of roots in the rhizosphere termed as root border cells or root border like cells depending upon their release and organization [25].

Root boarder cells consist of tissues which differ from other tissues of plants due to production of specific chemicals, which dramatically alter the behaviour of root associated microflora population. The root border cells remain viable in the rhizospheric environment for several days, thus they control the direction of root growth and have the ability to engineer the chemical and physical properties of the root external environment [24].

Root Secretion

Roots secrete a large numbers of organic compounds in surrounding soil where plant microbe soil interaction occurs. The substrate released from the root has many origins and may be classified as Exudates: (compound of low molecular weight that leak non-metabolically from the intact plant cell), Compound metabolically released from active plant cells secrete, Lysates (compound released by autolysis of older cells) Plant Mucilage's (polysaccharides from the root cap, root cap cells, primary cell wall and other cells)and Mucigel (gelatinous material of plant and microbial origin)

Root exudates are the main compounds which help with the interaction of plant microbe. They are mainly composed of water soluble sugars, organic acids, amino acids and also contain hormones, vitamins, amino compound, phenolic and sugar phosphate esters [26, 27]. Release of these low molecular weight compounds is a passive process along the steep concentration gradient which usually exists between the cytoplasm of intact root cells (milli molar range) and the external (soil) solution (micro molar range). Mucilage is secreted by plant roots as well as microbes in the rhizosphere. On contact with the soil, mucilage forms the gel which has a number of beneficial properties. The gel binds soil particles and microbes together with the root to form a rhizosheath. The beneficial effects of rhizosheath are maintenance of hydraulic conductivity of plant, aiding lubrication as roots move through the soil, absorption of ions, including $Fe^{2+/3+}$, Ca^{2+} and $H_2PO_4^-$ and prevent the incursion of pathogens and herbivores [20].

Root Exudates and Soil Microbial Community

Soil has limited nutrients and resources, so organisms have to compete for nutrients and other resources. Microbial activity in soil is greatly influenced by plant roots [12]. Plant roots secrete rhizodeposition which includes shedding of root cells in the form of loss of carbon-containing metabolites from the roots into the soil matrix and the exudation, secretion and leakage of sugars, organic acids, and amino acids into the soil [12, 19], Microorganisms can use these compounds as substrate, increased microbial biomass and activity around the roots. Plants have special cell organelles in root cells containing mitochondria, Golgi bodies and special secretary vesicles which involve in active secretion of metabolites [28]. In this secretion ATP-binding cassette transporters are involved in the translocation of phytochemicals into the rhizosphere [29, 30].

The composition of root exudates depends on plant species, cultivar, developmental stage, plant growth substrate, and stress factors [26]. Composition is also influenced by the rhizosphere microflora. Application of the bacterial consortia on the rhizosphere of plants influence the amount of organic acids and their amount may be increased or decreased. The presence of plant biocontrol strain *Pseudomonas fluorescens* WCS365 (WCS365) on tomato roots resulted in increased levels of total organic acids, whereas the amount of succinic acid decreased [31].

Regulation of PGPR by Root Exudates

Root exudates are important factors that structure the rhizosphere bacterial community. It is established that root exudates of plants can be processed as nutrients to enhanced growth and interaction of microorganism. Rhizosphere microflora can interact with root agglutinins present in the root exudates to form the rhizoplane community.

Root exudates are responsible for the changes in physiological active micro-flora of the roots in the early stage of plant development because they have the ability to secrete both low and high molecular weight molecules into the rhizosphere in response to biotic and abiotic stresses [32]. The amount of production of root exudates varies both qualitatively and quantitatively due to different environmental conditions. The organic compounds exuded by roots in the rhizosphere attract more bacterial population in the rhizosphere as compared to bulk soil. It has been found that approx 30 to 40% more bacterial population resides in the rhizospheric soil as compared to bulk soil [18].

The release of root exudates affects the bacterial gene expression, especially genes encoding plant-beneficial traits. To analyze the effect of root exudates on bacterial gene expression study was carried out on phlA, which involved in DAPG biosynthesis in *Pseudomonas* protegens [33]. The expression of phl increased four- fold in the rhizosphere of monocots (maize and wheat) compared to the rhizosphere of dicots (bean and cucumber). Many plants component and product like phenolics, phytohormones, sugars, flavinoids play a significant role in plant-microbe interactions and modulate the expression of phlA and pltA in *Pseudomonas* protegens CHA0 [34].

Efficient root colonization by plant beneficial rhizobacteria is assumed to be essential for the biocontrol of root pathogens [35]. Plant root exudates contain various organic compounds likes aminoacids, lipids, polysaccharides etc. and its composition depends upon the plant physiology, seasons and weather. These exudates effectively colonized the roots and roots interiors to check the phytopathogens and effectively interacts the beneficals microorganisms. In a case study of tomato, the major components of amino acids in root exudates are glutamic acid, aspartic acid, leucine, isoleucine, and lysine [36], organic acids especially citric acid, malic acid and succinic acid [31], and sugars glucose and xylose as major components [37].

Previous studies demonstrated that *P. fluorescens* strains exhibit chemotactic responses toward 20 L aminoacids and weak towards aminoacids Cys and proline [14, 38]. *Azotobacter* shows strong chemotactic behavior towards (glutamic acid, arginine threonine) and organic acids like citric acid, succinic acid, maleic acid and malonic acid [39]. Therefore, it is proposed that chemotaxis to components of plant root exudates is involved in effective root colonization. Methyl-accepting chemotaxis proteins (MCPs) are chemotaxis sensory proteins responsible for the detection of chemotactic ligands [40]. There are two classes of MCPs for amino acids in bacteria. One class includes Tar and Tsr . Tsr is an MCP for the attractants Ser, Ala and Gly, while Tar is an MCP for attractants Asp and Glu [41].

These MCPs possess short periplasmic domains (ca. 150 amino acid residues), and their ligand specificity is relatively narrow. which detects 18 commonly-occurring L-amino acids, shows broader ligand specificity, and its periplasmic domain (ca. 240 amino acid residues) is longer than those of Tsr and Tar (Kuroda et al.1995). CtaA and CtaB, the main MCPs for amino acids in *P. fluorescens* detect 16 amino acids and possess long periplasmic domains, suggesting that CtaA and CtaB belong to the class of PctA-type MCPs. Thus, amino acids are supposed to be major chemoattractants for *Pseudomonas P. fluorescens* Pf0-1 wild-type and mutant strains were tested for chemotaxis to tomato root exudate to assess the involvement of MCPs for amino acids.

The wild-type strain was strongly attracted by tomato root exudate, while ctaA ctaB ctaC triple mutant showed much decreased responses .The double mutants showed stronger responses to root exudate than the triple mutant, but weaker responses than the wild-type strain. This result suggests that amino acids are the major chemo-attractants of *P. fluorescens* Pf0- and CtaA, CtaB, and CtaC are responsible for chemotaxis to tomato root exudates [14]

Regulation of Plant Growth Function by Plant Hormones

Plant hormones function as signal molecules to regulate plant growth, development and responses to biotic and abiotic factors. The phytohormone Jasmonic acid (JA) is a lipid-derived compound playing a prominent role in regulating plant growth and defense against various microbes [42, 43, 44].

JA regulates various aspects of plant growth and development such as seed germination, root growth and flower development [43]. Induction of JA-signaling mainly occurs after attack by necrotrophic pathogens, tissue-chewing insects such as caterpillars, and cell-content feeding insects such as thrips [45].

Role of Ethylene in Plant Microbe Interaction

The Plant hormone ethylene is essential for the growth and development of plants and found in all higher plants having different effects on plant growth depending on its concentration in root tissues. At high concentrations, it can be harmful, as it inhibits the root elongation and auxin transport, promotes senescence and abscission of various organs, and leads to fruit ripening [46]. It also play important role in rhizobia nodulation of legumes and in the rooting of cutting plant parts [47]. Under different types of environmental streses, such as cold, draught, flooding, infections with pathogens, presence of heavy metals, plants respond by synthesizing 1-aminocyclopropane-1-carboxylate (ACC), which is the precursor for ethylene [46].

Mechanism of Ethylene Action

Some of the ACC secreted into the rhizosphere re-adsorbed by the roots, where it is converted into ethylene. This accumulation of ethylene diminished ability to acquire water and nutrients resulting in poor root growth. PGPR with the ability to degrade ACC in the rhizosphere can help to break this downward cycle and re-establish a healthy root system that is needed to cope with environmental stress. The primary mechanism, that is used by PGPR is to degrade ethylene via the enzyme ACC deaminase. This enzyme can diminish or prevent some of the harmful effects of the high ethylene levels [48]. The ACC deaminase acts on ACC, an immediate ethylene precursor in higher plants, and degrades this chemical to alpha-ketobutyrate and ammonium, [46, 48, 49].

Another mechanism of plant growth stimulation by PGPR is the production of ACC deaminase [46]. ACC is the immediate precursor of the plant hormone ethylene (ET), which is involved in stress signaling and in the negative regulation of root elongation. The bacterial enzyme ACC deaminase hydrolyzes ACC to ammonia and α-ketobutyrate. Glick et al. [48] postulate that plants releases ACC into the rhizosphere, and ACC is hydrolyzed by the bacterial ACC deaminase, which reduced ethylene-mediated suppression of root growth. This interaction is also beneficial for the bacteria, as ammonia and α-ketobutyrate are respectively sources of N and C.

Regulation of PGPR Functions by Microbial Signals

Many Plant growth-promoting rhizobacteria (PGPR) exchange cell-to-cell communication signals between each other and regulate their gene expression in response to changes in population density with other rhizosphere inhabiting bacteria and fungi in a process quorum-sensing (QS) signals which mediated through small diffusible signal molecules [49, 50] Quorum-sensing relies on the synthesis and perception of small diffusible molecules, such as N-acyl-homoserine lactones (AHLs). In fluorescent *pseudomonads*, colonization properties and biosynthesis of phenazines (antimicrobial metabolites) is AHL-based QS regulation [51]. Production of pyrrolnitrin in *Serratia plymuthica* is able to protect crops against *Verticillium wilt* [52]. Virulance of *Erwinia cardovora* is attenuated in the prescence of lactones enzyme produced by *Bacillus*. Moreover, a proteomic approach indicates that AHL-based QS regulation in *Azospirillum* controls functions linked to rhizosphere competence and adaptation to plant roots [53].

Plants being able to produce molecules which interfere in the QS system and their amount depend upon the inhabiting microbes which enhance or inhibit the phenotype regulated by QS [50]. Several studies have shown that bacterial communication of a specific bacterial population could be checked by other microbes, Some soil bacteria can inactivate AHL (notably members of the genus Bacillus),whereas others can intercept AHL or can act as a physical barrier preventing their diffusion [53]. Consequently, other members of the bacterial rhizosphere community can compromise expression of biocontrol traits in PGPR. Plant compounds designated as AHL-mimics can also interfere with bacterial QS and may influence the expression of plant-beneficial functions [54]. Some *Pseudomonas fluorescens* strains unable to synthetize AHLs but possessing the cognate receptor may even recognize a plant compound to trigger expression of genes involved in biocontrol properties [55].

Rhizospheric Effect

Rhizosphere effect occurs at the time of seed germination and seedling growth, when plant interacts with a range of microorganisms present in the surrounding soil.

The rhizospheric effect is caused by the release of organic and inorganic materials from the plant roots which provide the driving force for the development of active microbial population in a rhizosphere.

These substances are released from the root border cell of the root. Rhizospheric effect can be represented by the ratio of microorganisms in the rhizosphere soil (R) to the number of corresponding microorganisms remote from roots (S) (the R/S ratio). Generally R/S ratios range from 5 to 10 but it is common to find R/S ratio of 100 or even more. The rhizospheric effect can thus be viewed as the creation of a dynamic environment where microbes can develop and interact.

Rhizospheric Environment

The rhizospheric environment interfaces between the root and soil microorganisms, a habitat where the root exudates are released as rhizodeposition.

In that rhizospheric environment, important and intensive interactions take place between the plant, soil, and microorganisms.

The rhizosphere inhabiting microorganisms compete for water, nutrients, space and sometimes improve their competitiveness by developing an intimate association with plant [17].

Plant microbes interaction play important role in the growth and ecological fitness of that particular plant from which microbes associate. Rhizospheric environment promotes two types of microorganisms, the one, deleterious rhizospheric microorganism (DRMOs) and the others have been shown and thought to improve the plant growth by colonizing the root system (Plant Growth Promoting Rhizobacteria; PGPR) and preventing the colonization of DRMOs [24].

Plant Growth Promoting Rhizobacteria (PGPR)

Plant growth in agricultural soils is influenced by many abiotic and biotic factors. A large number of microorganisms such as bacteria, fungi, protozoa and algae coexist in the rhizosphere. Bacteria are the most abundant among them. Plants select those bacteria contributing most to their fitness by releasing organic compounds through exudates [56] creating a very selective environment where diversity is low [57]

Since bacteria are the most abundant microorganisms in the rhizosphere, it is highly probable that they influence the plants physiology to a greater extent, especially considering their competitiveness in root colonization [13] Microorganisms that colonize the rhizosphere can be classified according to their effects on plants and the way they interact with roots, some being pathogens whereas other trigger beneficial effects. Rhizobacteria inhabit plant roots and exert a positive effect ranging from direct influence mechanisms to an indirect effect. The bacteria inhabiting the rhizosphere and beneficial to plants are termed Plant growth promoting rhizobacteria (PGPR) [58]. In the last few years, the number of PGPR that have been identified has seen a great increase, mainly because the role of the rhizosphere as an ecosystem has gained importance in the functioning of the biosphere. Various species of bacteria like *Pseudomonas, Azospirillum, Azotobacter, Klebsiella, Enterobacter, Alcaligenes, Arthrobacter, Burkholderia, Bacillus and Serratia* have been reported to enhance the plant growth [59].

Table 1. Plant growth promoting rhizobacteria

Organisms	Growth hormones	References
Pseudomonas sp.	IAA, Siderophore	[60]
Penibacillus polymyxa	IAA, Siderophore	[61]
Bacillus sp.	IAA, Siderophore,NH_3,HCN	[62]
Rhizobium	IAA	[63]

Table 2. (Continued)

Organisms	Growth hormones	References
Mesorhizobium sp.	IAA, Siderophore,NH_3,HCN	[62]
Klebsiella oxytoca	IAA, Phosphate solubilization,	[64]
Pseudomonas aeruginosa	ACC deaminse, IAA, Siderophore,	[65]
Azospirillum amazonense	IAA, nitrogenase activity	[66]
Enterobacter sp.	ACC deaminse, IAA, Siderophore,	[67]

There are several PGPR inoculants currently commercialized that seem to promote growth through suppression of plant disease improved nutrient acquisition (Biofertilizers), or phytohormone production (Biostimulants). Inoculant development has been most successful to deliver biological control agents of plant disease i.e., organisms capable of killing other organisms pathogenic or disease causing to crops. Various bacteria which are predominantly studied and increasingly marketed as plant growth promoting agents and biological control agents includes the genera *Bacillus, Streptomyces, Pseudomonas, Burkholderia* and *Agrobacterium* some of them described in Table 1.

CONCLUSION

Microbes in the rhizosphere are greatly influenced by the rhizodeposition or the root exudates released by the plant. The rhizospheric microbial population depends upon biotic and abiotic factors. Plant species can also influence the microbial population in the rhizosphere at the species level. The amount and composition (flavanoids, amino acids, strigolacttans and sesquiterpenes) of the plants act as signaling molecules and regulate the functions of plant microbe interaction. PGPR plays an important role in sustainable agriculture. These microbes directly or indirectly associated with the plants for growth promotion, disease management and yield enhancement.

REFERENCES

[1] Dicke, M., Badwin, I. T., (2010). The evolutionary context for herbivore-induced plant volatiles: Beyonds the "cry for help". *Trends Plant Sci.* 15, 167-175.

[2] Berendsen, R. L., Pieterse, C.M.J., Bakker, P.A.H.M., (2012). The rhizosphere microbiome and plant health. *Trends Plant Sci,* 17, 478-86.

[3] Bulgarelli, D., Schlaeppi, K., Spaepen, S., Loren, V., Themaat, E., Schulze-Lefert, P., (2013). Structure and functions of the bacterial microbiota of plants. *Annu. Rev. Plant Biol.* 64, 807-38.

[4] Whitman, W.B., Coleman, D.C., Wiebe, W.J. (1998) Prokaryotes, The unseen majority. *Proc. Natl Acad. Sci. USA*. 95, 6578-6583.

[5] Lavelle, P., Spain, A. V., (2001). Soil ecology. Kluwer, Dordrecht.

[6] Gelsomino, A., Keijzer-Wolters, A. C., Cacco, G., Van Elsas, J.D., (1999). Assessment of bacterial community structure in soil by polymerase chain reaction and denaturing gradient gel electrophoresis. *J. Microbiol. Meth.* 38, 1-15.

[7] Postma, J., Van Veen, J. A., (1990). Habitable pore space and survival of Rhizobium leguminosarum biovar trifolii introduced into soil. *Microb. Ecol.* 19, 149-161.

[8] Carson, J. K., Campbell, L., Rooney, D., Clipson, N., Gleeson, D.B., (2009). Minerals in soil select distinct bacterial communities in their microhabitats. *FEMS Microbiol. Ecol.* 67,381-388.

[9] Pate, J. S., Verboom, W. H., (2009). Contemporary biogenic formation of clay pavements by Eucalyptus further support for the phytotarium concept. *Ann. Bot.* 103, 673-685.

[10] Taylor, L. L., Leake, J. R., Quirk, J., Hardy, K., Banwarts, S.A., Beerling, D. J., (2009). Biological weathering and the long term carbon cycle: integrating mycorrhizal evolution and function into the paradigm. *Geobiology* 7, 171-191.

[11] Lambers, H., Mougel, C., Jaillard, B. and Hinsinger, P., (2009). Plant- microbe- soil interactions in the rhizosphere: an evolutionary perspective. *Plant Soil* 321, 83-115.

[12] Bais, H. P., Weir, T. L., Perry, L. G., Gilroy, S., Vivanco, J. M., (2006). The role of root exudates in rhizosphere interactions with plants and other organisms. *Annu. Rev. Plant Biol.* 57, 233–266.

[13] Barriuso, J., Solano, B. R., Lucas, J. A., Lobo, A. P., Villaraco, A.G., Mañero, F.J.G., (2008). Ecology, Genetic Diversity and Screening Strategies of Plant Growth Promoting Rhizobacteria (PGPR). Plant-Bacteria Interactions: Strategies and Techniques to Promote Plant Growth (Ahmad, I., Pichtel, J., Hayat, S., Eds.), Wiley-VCH Verlag GmbH & Co. KGaA, Weinheim, pp. 1-17.

[14] Oku, S., Komastu, A., Tajima, T., Nakashimada, Y. and Kato, J., (2012). Identification of chemotaxis sensory proteins for aminoacids in *Pseudomonas fluorescens* Pf0-1 and their involvement in chemo taxis to tomato root exudates and root colonization. *Microbes Environ.* 27(4), 462-469.

[15] Morgan, J. A. W., Whipps, J. M., (2001). Methodological approaches to the study of rhizosphere carbon flow and microbial population dynamics. In: *The rhizosphere. Biochemistry and organic substances at the soil-plant interface,* (Pinton, A., Varanini, Z., Nannipieri, P., Eds.) New York, USA, Marcel Dekker, pp. 373-409.

[16] Hiltner, L., (1904). Uber neuere Erfarungen und problem auf dem Gebiete der Bodenbakteriologie unter besonderer Berucksichtigung der Grundungung und Brache. Arb. Dtsch. 98,59-78.

[17] Hartmann, A., Schmid, M., van Tuinen, D., Berg, G., (2009). Plant- driven selection of microbes. *Plant Soil* 321, 235-257.

[18] Ambardar, S., Vakhlu, J., (2013). Plant growth promoting bacteria from *Crocus sativus* rhizosphere. *World J. Microbiol. Biotechnol.* 29, 2271-2279.

[19] Glick, B. R., (2014). Bacteria with ACC deaminase can promote plant growth and help tofeed the world. *Microb. Res.* 169, 30-39.

[20] Bakker, P.A.H.M., Berendsen, R. L., Doornbos, R.F., Wintermans, P.C.A., Pieterse, C.M.J., (2013). The rhizosphere revisited: root microbiomics. *Front. Plant Sci.* 4, 165.

[21] Lucy, M., Reed, E., Glick, B. R., (2004). Applications of free living plant growth-promoting rhizobacteria. Rev. Antonie Van Leeuwen. 86, 1-25.

[22] Bowen, G. D., (1980) "Misconceptions, concepts and approaches in Rhizosphere Biology". In: *Contemporary Microbial Ecology*, (Elwood, D.C., Hedger, J. N., Latham, M. J., Lynch, J. M., Slater, J. M., Eds.), Acad. Press, London, pp. 283-304.

[23] Tsugeki, R., Fedoroff, N. V., (1999). Genetic ablation of root cap cells in *Arabdiopsis*. *Proc. Natl Acad. Sci*. USA 96, 12941-12946.

[24] Pepper, I. L., (2000). Beneficial and Pathogenic microbes in Agriculture. In: *Environmental Microbiology*, (Maier, R. M., Pepper, I. L., Gerba, C. P., Eds.), Academic Press London. pp. 425-446.

[25] Driouich, A., Durand, C., Cannesan, M. A., Percoco, G., Vicré-Gibouin, M., (2010). Border cells versus border-like cells: are they alike? *J. Exp. Bot*. 61, 3827-3831.

[26] Uren, N. C., (2000) Types, amount, and possible functions of compounds released into the rhizosphere by soil-grown plants. In: The *rhizosphere: biochemistry and organic substances at the soil-plant interface*, (Pinton, R., Varanini, Z., Nannipieri, P., Eds.). Marcel Dekker, New York, pp 19-40.

[27] Doornbos, R. F., van Loon, L.C., Peter, A.H.M., (2012). Impact of root exudates and plant defense signaling on bacterial communities in the rhizosphere. A review. *Agron. Sustain. Dev*. 32(1), 227-243.

[28] Vicré ,M., Santaella, C., Blanchet, S., Gateau, A., Driouich, A., (2005). Root border-like cells of Arabidopsis. Microscopical characterization and role in the interaction with rhizobacteria. *Plant Physiol*. 138, 998–1008.

[29] Loyola-Vargas, V. M., Broeckling, C. D., Badri, D., Vivanco, J.M., (2007). Effect of transporters on the secretion of phytochemicals by the roots of Arabidopsis thaliana. *Planta* 225, 301–310.

[30] Badri, D. V., Loyola-Vargas, V. M., Broeckling, C.D., De-la-Pena, C., Jasinsk, M., Santelia, D., Martinoia, E., Sumner, L.W., Banta, L.M., Stermitz, F., Vivanco, J. M., (2008). Altered profile of secondary metabolites in th root exudates of Arabidopsis ATP-binding cassette transporte mutants. *Plant Physiol*. 146, 762–771.

[31] Kamilova, F., Kravchenko, L. V., Shaposhnikov, A. I., Makarova, N., Lugtenberg, B., (2006). Effects of the tomato pathogen *Fusarium oxysporum* f. sp *radicis lycopersici* and of the biocontrol bacterium *Pseudomonas fluorescens* WCS365 on the composition of organic acids and sugars in tomato root exudate. Mol Plant- *Microb. Interact*. 19, 1121–1126.

[32] Bertin, C., Yang, X., Weston, L. A., (2003). The role of root exudates and allelochemicals in the rhizosphere. *Plant Soil* 256, 67-83.

[33] Notz, R., Maurhofer, M., Dubach, H., Haas, D., Défago, G., (2002). Fusaric Acid-producing strains of *Fusarium oxysporum* alter 2,4-diacetylphloroglucinol biosynthetic gene expressions in *Pseudomonas fluorescens* CHA0 in vitro and in the rhizosphere of Wheat. *Appl. Environ. Microbiol*. 68, 2229-2235.

[34] de Werra, P., Huser, A., Maurhofer, M., (2011). Plant- and Microbe-Derived Compounds Affect the Expression of Genes Encoding Antifungal Compounds in a Pseudomonad with Biocontrol Activity. *Appl. Environ. Microbiol*. 77(8), 2807-2812.

[35] Suslow, T. V., (1982). Role of root-colonizing bacteria in plant growth. In: *Phytopathogenic Prokaryotes,* (Mount, M.S., Lacy, G.H., Ed.), Vol 1, Academic Press, London. pp. 187-223.

[36] Simons, M., Permentier, H. P., de Weger, L. A., Wijffelman, C. A., Lugtenberg, B.J.J., (1997). Amino acid sysnthesis is necessary for tomato root colonization by Pseudomonas fluorescens strain WCS365. *Mol. Plant Microbe Interact.* 10, 102-106.

[37] Lugtenberg, B.J.J., Kravchenko L. V., and Simons, M., (1999). Tomato seed and root exudate sugars: composition, utilization by Pseudomonas biocontrol strains and role in rhizosphere colonization. *Environ. Microbiol.* 1, 439-446.

[38] Singh, T., and Arora, D. K., (2001). Motility and chemotactic response of Pseudomonas fluorescens toward chemoattractants present in the exudate of Macrophomina phaseolina. *Microbiol. Res.* 156, 343-351.

[39] Kumar, R., Bhatia, R., Kukreja, K., Behl, R. K., Dudeja, S. S.,Narula, N., (2007). Establishment of *Azotobacter* on plant roots: chemotactic response, development and analysis of root exudates of cotton (*Gossypium hirsutum* L) and wheat (*Triticum aestivum* L.). *J. Basic Microbiol.* 47 (5), 436-439.

[40] Kato, J., (2008). Pseudomonas motility and chemotaxis, In: *Pseudomonas* (Rem, B.H.A., Ed.), Wiley-VCH, Weinheim, pp. 109-128.

[41] Stewart, R. C. and Dahlquist, F. W., (1987). Molecular components of bacterial chemotaxis. *Chem. Rev.* 87, 997–1025.

[42] Browse, J., (2005). Jasmonate:an oxylipinsignalwithmanyrolesin plants. *Plant Hormones* 72, 431-456.

[43] Wasternack, C., (2007). Jasmonates: an updateonbiosynthesis,signal transduction and action in plant Stress response, growth and development. *Ann. Bot.* 100, 681-697.

[44] Pieterse, C, M,, Van Der Does, D., Zamioudis, C., Leon-Reyes, A., VanWees, S.C., (2012). Hormonal modulation of plant immunity. *Annu. Rev. Cell Dev. Biol.* 28, 489-521.

[45] De, Vos, Van Oosten M, Van V. R., Poecke, R.M.P., Van Pelt, J. A., Pozo, M. J., Mueller, M.J., (2005).Signal signature and transcriptome changes of Arabidopsis during pathogen and insect attack. *Mol. Plant Microbe Interact.* 18, 923-937.

[46] Glick, B. R., Cheng, Z., Czarny, J., Duan, J., (2007). Promotion of plant growth by ACC deaminase-containing soil bacteria. *Eur. J. Plant Pathol.* 119, 329-39.

[47] Montero-Calasanz, M. C., Santamaría, C., Albareda, M., Daza, A., Duan, J., Glick, B. R., (2013). Rooting induction of semi-hardwood olive cuttings by several auxin-producingbacteria. *Span J. Agric. Res.* 11, 146–54.

[48] Glick, B. R., Penrose, D. M., Li, J., (1998). A model for the lowering of plant ethylene concentrationsby plant growth promoting bacteria. *J. Theor. Biol.* 190, 63-68.

[49] Fuqua, W. C., Winans, S. C. and Greenberg, E. P. (1994). Quorum sensing in bacteria: the LuxR LuxI family cell density-responsive transcriptional regulators. *J. Bacteriol.* 176,269-275.

[50] Pérez-Montaño, F., Alías-Villegas, C., Bellogín, R. A., del Cerro, P., Espuny, M. R., Jiménez-Guerrero, I., López-Baena, F. J., Ollero , F.J., Cubo, T., (2014). Plant growth promotion in cereal and leguminous agricultural important plants: from microorganism capacities to crop production. *Microbiol. Res.* 169(5-6), 325-36.

[51] De, Maeyer., KD, Aes, J., Hua, G.K., Perneel, M., Vanhaecke, L., Noppe, H., (2011). N-Acyl homoserine lactone quorum sensing signalingin antagonistic phenazine producing Pseudomonas isolates from the red cocoyam rhizosphere. *Microbiology* 157, 459-472.

[52] Liu, X., Bimerew, M., Ma, Y., Muller, H,, Ovadis, M., Eberl, L., (2007). Quorum-sensing signalling is required for production of the antibiotic pyrrolnitrinina rhizospheric biocontro lstrain of Serratia plymuthica. *FEMS Microbiol. Lett.* 270, 299-305.

[53] Boyer, M. and Wisniewski-Dyé, F., (2009). Cell–cell signalling in bacteria: not simply a matter of quorum. *FEMS Microbiol. Ecol.* 70, 1-19.

[54] Vandeputte, O.M., Kiendrebeogo, M., Rajaonson, S., Diallo, B., Mol, A., ElJaziri, M., (2010). Identification of catechinasone of the flavonoids from Combretum albiflorum bark extract that Reduces the production of quorum-sensing-controlled virulence factors in Pseudomonas aeruginosa PAO1. *Appl. Environ.Microbiol.* 76, 243-253.

[55] Subramoni, S., Gonzalez, J. F., Johnson, A., Pechy-Tarr, M., Rochat, L., Paulsen, I., (2011). Bacterial subfamily of Lux R regulators that respond to plant compounds. *Appl. Environ. Microbiol.* 77, 4579-4588.

[56] Lynch, J. M., (1990). The Rhizosphere. John Wiley & Sons Ltd, Chichester, pp. 458.

[57] García, J. L., Probanza, A., Ramos, B., Mañero, F.J.G., (2001). Ecology, genetic diversity and screening strategies of plant growth promoting rhizobacteria. *J. Plant Nutr. Soil Sci.*, 164, 1-7.

[58] Kloepper, J. W., Schroth, M. N., (1978). Plant growth promoting rhizobacteria on radish. In: Proceedings of the 4th International Conference on Plant Pathogenic Bacteria, vol 2. Station de Pathologie Vegetale et Phytobacteriologie, INRA, Angers, France, pp. 879-882.

[59] Joseph, B., Patra, R. R., Lawrence, R., (2007). Characterization of plant growth promoting Rhizobacteria associated with chickpea (Cicer arietinum L). *Internat. J. Plant Prod*, 1 (Suppl 2), 141-152.

[60] Ma, J. F., (2005). Plant root responses to three abundant soil minerals: silicon, aluminum and iron." *Criti. Rev. Plant Sci.* 24(4), 267-281.

[61] Phi, Quyet.-Tien , Yu-Mi , P., Keyung-Jo, S. , Choong-Min, R., Seung-Hwan, P., Jong-Guk, K., Sa-Youl, G., (2010). Assessment of root-associated *Paenibacillus polymyxa* groups on growth promotion and induced systemic resistance in pepper. *J. Microbiol. Biotechnol.* 20, 1605-1613.

[62] Wani, P. A., Khan, M. S., Zaidi, A., (2008). Chromium-reducing and plant growth-promoting Mesorhizobium improves chickpea growth in chromium-amended soil. *Biotechnol. Lett.* 30, 159-163.

[63] Zahir, Z. A., Shah, M. K., Naveed, M., Akhter, M. J., (2010). Substrate dependent auxin production by *Rhizobium phaseoli* improves the growth and yield of *Vigna radiata* L. under salt stress conditions. *J. Microbiol. Biotechnol.* 20, 1288-1294.

[64] Jha, P. N. and Kumar, A., (2007). Endophytic colonization of *Typha australis* by a plant growth-promoting bacterium *Klebsiella oxytoca* strain GR-3. *J. Appl. Microbiol.* 103, 1311-1320.

[65] Ganesan, V., (2008). Rhizoremediation of cadmium soil using a cadmium-resistant plant growth-promoting rhizopseudomonad. *Curr. Microbiol.* 56, 403-407.

[66] Rodrigues, E. P., Rodrigues, L. S., de., Oliveira, A.L.M., Baldani, V.L.D., Teixeira, K.R.S., Urquiaga, S., Reis, V.M., (2008). *Azospirillum amazonense* inoculation: effects on growth, yield and N2 fixation of rice (*Oryza sativa* L.). *Plant Soil* 302, 249-261.

[67] Kumar, K. V., Singh, N., Behl, H. M., Srivastava, S., (2008). Influence of plant growth promoting bacteria and its mutant on heavy metal toxicity in *Brassica juncea* grown in fly ash amended soil. *Chemosphere* 72, 678-683.

In: Microbes in Soil and Their Agricultural Prospects
Editors: K. Kishore Choudhary and D. Wattal Dhar
ISBN: 978-1-63482-824-6

Chapter 6

CYANOBACTERIA OR BLUE-GREEN ALGAE: SUSTAINABLE SOURCE OF SOIL FERTILITY AND CROP PRODUCTIVITY

***Kaushal Kishore Choudhary*[1,†] *and Dolly Wattal Dhar*[2,‡]**

[1]Department of Botany, Dr Jagannath Mishra College, Muzaffarpur, Bihar, India

[2]Centre for Conservation and Utilisation of Blue Green Algae, Division of Microbiology, Indian Agricultural Research Institute, New Delhi, India

ABSTRACT

Cyanobacteria or blue-green algae constitute a group of prokaryotic photosynthetic microorganisms that exhibit diversity in their forms and functions. These are ubiquitous in their distribution and have been reported to contribute to the agricultural productivity.

These play an important role towards biological nitrogen fixation (symbiotic and asymbiotic), phosphate mobilization, and production of several primary and secondary metabolites that affect the soil fertility and crop productivity either directly or indirectly.

These organisms convert atmospheric nitrogen into ammonia with the help of the enzyme nitrogenase. They can also transform complex organic phosphate into soluble and easily available inorganic forms of phosphate with the help of enzyme 'phosphatase'. They synthesize and/or release growth promoting hormones (Auxins, Gibberellins, cytokinins and ethylene), amino acids, carbohydrates, enzymes, polysaccharides, proteins, vitamins into the soil.

These attributes enhance the physiological and biochemical processes and support protection against biotic and abiotic stresses. This chapter discusses the contribution of cyanobacteria towards soil fertility with particular reference to augmentation of soil with nitrogen, phosphorus and other beneficial growth promoting substances.

† Corresponding author: Department of Botany, Dr Jagannath Mishra College, Muzaffarpur – 842001 (Bihar), India. E-mail: kkc1970@gmail.com.

‡ Dolly Wattal Dhar: Centre for Conservation and Utilisation of Blue Green Algae, Division of Microbiology, Indian Agricultural Research Institute, New Delhi – 110012, India.

Keywords: cyanobacteria; nitrogen fixation; phosphate solubilization; phytohormones; polysaccharides; stress tolerance

INTRODUCTION

High fertilizer doses which are a non-renewable source of energy are required to meet the ever increasing demand for food of the huge population of the Indian subcontinent and to exploit the high yield potential of crop plants. The long term fertilizer experiments have revealed that continuous application of sub-optimal doses of chemical fertilizers has resulted in the deterioration of soil health, environmental pollution and decrease in crop productivity. Further, such a practice has aggravated micronutrient deficiency causing significant decline in crop productivity [1]. Current agriculture practices are governed and manipulated by using limited sources of inorganic chemical fertilizers and pesticides (synthetic abiotic natural resources). These in turn deteriorate the soil fertility and crop productivity and subsequently influence the human health [2]. Therefore, alternate or supplementary sources of fertilizer are required that can replace or reduce the dependency on chemical fertilizer (abiotic fertilizer) for healthy agricultural practices. Some beneficial organisms have been recognized as important inputs in sustainable agriculture. Biofertilizer can be defined as "a product that contains living microorganisms, which exert direct or indirect beneficial effects on plant growth and crop yield through different mechanisms" [3]. These are apparently eco-friendly, low cost and non-bulky agricultural inputs, renewable and pollution free. They also play a significant role in plant nutrition acting as supplementary and complementary sources of plant nutrients and have no deleterious effect on the environment. Studies have shown that plant growth promoting rhizobacteria, mycorrhizal fungi and symbiotic bacteria are the important biological components of soil and have been widely used in agricultural field for better crop productivity [4, 5, 6, 7, 8]. Cyanobacteria or commonly known as blue green algae received attention for their agricultural prospects only after De [9] who reported their role in nitrogen economy. These organisms are not only known to add N and P to soil but also help in enhancing the soil aggregation property as well as water holding capacity by building up of extracellular polysaccharides [10]. These also produce growth promoting hormones [11] and biologically active metabolites which can protect the plants from stress [12, 13, 14].

Cyanobacteria, commonly known as blue green algae have been widely studied for their distribution and other attributes pertaining to soil fertility and crop productivity particularly in rice [15, 16]. These are the most potent candidates of present day sustainable agricultural practices due to their ability to colonize wide range of habitats with minimal nutritional requirement. This chapter describes the cyanobacterial distribution and their agronomic potential with particular respect to augmentation of soil with N, P and growth promoting hormones in enhancement of soil fertility and crop productivity.

CYANOBACTERIA

Cyanobacteria or blue green algae [17] are extremely diverse and most important ancient group of oxygenic photosynthetic prokaryotic organisms on the earth [18, 19].

These show metabolic functioning as those of eukaryotes and have cellular organization like prokaryotes [20]. They originated in 'Proterozoic era' which is also known as 'Age of Cyanobacteria' as unicellular fresh water forms [21, 22, 23]. Cyanobacteria have received attention in recent years for their potential attributes with a great ecological adaptability to various ecological niches [24]. These organisms are unique as they possess a specialized structure called 'heterocyst' that enables them to convert unavailable form of atmospheric nitrogen (N_2) into easily available form of ammonium compounds (NH_4^+) deriving required energy directly from the sun [25]. Additionally, their ability to release inorganic phosphate (Pi) from combined complex organic phosphates with the help of enzyme 'phosphatase' make them more compatible under stressed environmental conditions.

Habitats and Cellular Organization

Cyanobacteria show a large variation in their forms and habitats. Originating as free living unicells, these diversified from simple unicells to multiseriate, colonial and filamentous to filamentous branched forms [26]. They occupy all possible biomes ranging from fresh water to marine [26, 27, 28], arctic to antarctic [29], psychrophilic to thermophilic [30, 31, 32] and acidophilic to alkaliphilic environments [33, 34]. They also form symbiotic associations with various groups of organisms ranging from algae to angiospermic plants and can grow epiphytically on the bark of trees, leaves, roots and on submerged roots and stems of flooded fields [35].

They possess four kinds of cellular structures namely vegetative cells, heterocyst, akinetes and hormogonia. Vegetative cells of unicellular cyanobacteria range from 0.4 μm to 40 μm and that of filamentous forms of Oscillatoriaceae measure around 100 μm. Variation in cell size is attributed to short-lived prokaryotic mRNA [36]. These may possess multiple copies of genome e.g., six chromosomes are present in simple morphological forms of *Synechococcus* sp. and *Synechocystis* sp. [37, 38, 39]. Vegetative cells are the sites of oxygenic photosynthesis and atmospheric dinitrogen fixation occurs in heterocysts. Akinetes are specialized structures formed amongst heterocystous cyanobacteria under adverse conditions and are much larger than vegetative cells [40]. Hormogones are the chains of uniform cells and are considered as means of multiplication and dispersal of organisms in filamentous forms of cyanobacteria [41]. These are rich in nitrogen, phosphorus and other nutrients [42] and play significant role in a number of physiological processes [43]. Growth rate of cyanobacteria varies largely from thousand years for isolates of Antarctica [44] to 2.1 h for *Anacystis nidulans* (*Synechococcus* PCC 6803) [45].

Symbiotic Association

Cyanobacteria form symbiotic associations with algae, fungi, bryophytes, pteridophytes, gymnosperms, angiosperms and animals [35]. Symbiotic associations with algae are *Richellia intracellularis* in the cells of marine diatom *Rhizosolenis* sp., *Nostoc symbioticum* in a siphonous green alga *Geosipho pyriformis* sp. and *Cyanothece* sp. in the freshwater centric diatom *Rhopalodia gibba* as inclusion bodies [46].

Other genera such as *Calothrix*, *Cyanodictyon*, *Lyngbya* and *Phormidium* have also been reported to be present in the mucilaginous sheath of other algae [26]. Likewise, cyanobacteria form symbiotic associations with fungi *i.e.,* in lichens and the common cyanobacteria forming associations with fungi are *Calothrix*, *Dichothrix*, *Nostoc*, *Scytonema* and *Stigonema* [47]. The association of different genera of cyanobacteria like *Calothrix* with *Lichina*, *Dichothrix* with *Placynthium*, *Nostoc* with *Collema*, *Leptogium*, *Lobaria* and *Peltigera, Scytonema* with *Dendriscocaulon*, *Massalongia*, and *Polychidium*, and *Stigonema* with *Ephebe* has been reported.

Meeks [48] reported association of cyanobacteria with non-vascular plants of bryophytes like liverworts and hornworts. He proposed that out of 340 genera of liverworts only four viz. *Merchantia Porella* form epiphytic association while *Blasia* and *Cavicularia* form endophytic association. On the other hand, four genera (out of six) of hornworts namely *Anthoceros*, *Phaeoceros*, *Notothylus* and *Dendroceros* form association with cyanobacteria [48]. The most common cyanobacterium forming symbiotic associations is *Nostoc*, however, *Calothrix* and *Chlorogloeopsis* have been reported to form association with liverworts *viz. Blasia* and *Cavicularia* [49, 50]. Association of cyanobacteria with mosses has also been reported under natural systems. The mosses such as *Sphagnum* sp., *Bryum* sp., *Weisia controvera*, *Funaria hygrometrica*, *Polytrichum junperinum*, *Ceratodon purpureus*, *Campylopus introflexus* form epiphytic or endophytic association with cyanobacteria like *Chroococcus*, *Phormidium*, *Oscillatoria*, *Lyngbya*, *Anabaena*, *Nostoc*, and *Scytonema* [51, 52].

Symbiotic association of cyanobacterium with aquatic fern *Azolla* has been well studied and is extensively used in agricultural fields particularly paddy fields as biofertilizer due to nitrogen-fixation and biomass production [53]. The most studied and common cyanobacterium forming symbiotic association in the leaf cavity of *Azolla* is filamentous heterocystous species of *Anabaena i.e., A. azollae* [54]. *Anabaena* and *Nostoc* infect coralloid roots of the members of Cycadaceae family (9 genera and 150 species) of gymnosperms and form nodule like structure by establishing themselves intercellularly or intracellularly. *Anabaena* and *Nostoc* form symbiotic association with others and the most common genera are *Ceratozamia*, *Cycas*, *Dioon*, *Macrozamia* and *Zamia* [55, 56]. Symbiotic associations of cyanobacteria with angiosperms have been reported only with monogeneric family of Gunneraceae and *Nostoc punctiforme* is a microsymbiont with genus *Gunnera* [57]. Endosymbiotic association of cyanobacteria has also been reported with coral-reef sponges [58]. Wilkinson and Fay [59] experimentally showed that sponges with cyanobacteria demonstrated nitrogenase activity whereas without cyanobacteria no nitrogenase activity was observed.

Distribution of Cyanobacteria in Crop Plants

Agronomic potential of cyanobacteria witnessed a sharp increase in enumeration and evaluation of cyanobacterial community in various kinds of ecological habitats worldwide. Distribution of these under different environmental situations is the function of climatic factors such as moisture, pH, mineral nutrient, nitrogen and phosphorus sources available in that particular habitat [60, 61].

It is well established that the cyanobacteria are in abundance in paddy fields as these provide optimum growth conditions of light, temperature, pH, moisture, nutrient etc. for their multiplication [62]. Distribution of cyanobacterial species composition has been studied around world [61, 63, 64, 65]. These constitute more than half the number of total algae in north and south India [66 and 0-76 percent in acidic soils of Kerala state [67]. Out of the total algae reported from south-East Iraq, 86 percent belonged to cyanobacteria [68]. A study on cyanobacterial composition of rice fields of 24 Parganas showed dominance of heterocystous forms [64]. Prasanna et al. [69] showed that *Anabaena* was the most dominant genera in the rice rhizosphere constituting around 80 % of total population followed by the *Hapalosiphon*, *Westiellopsis* and *Calothrix*. Similar studies of genera wise distribution of cyanobacteria have been studied by different groups around the world [63, 70, 71, 72, 73].

Cyanobacterial distribution in rice fields varies with respect to time and space of the cultivation cycle [63, 70, 74, 75, 76]. Genera which are widespread in Indian rice field soils and are known to contribute significantly to their fertility in India include mainly species of *Nostoc, Anabaena, Tolypothrix, Aulosira, Cylindrospermum, Scytonema, Westiellopsis* [77] whereas reports show that up to 50% of the total algae was cyanobacteria in some of the southern and eastern states. Common genera found in Indian rice soils include *Anabaena, Nostoc, Aulosira, Calothrix* and *Tolypothrix* etc. [78, 79].

Choudhary [70] showed that cyanobacterial diversity varies during cultivation cycle of rice. It has been observed that these may dominate the rice fields just after the plantation with availability of more light and temperatures whereas in some cases cyanobacteria may dominate after two weeks [74] or after one month [75]. He enumerated members of Chroococcaceae during rice cultivation cycle for 60 days and showed 21 species around 30th day, 8 species around 10th day and 13 species around 60th day of paddy cultivation.

Further enumeration of cyanobacterial diversity carried out in paddy field with members of microchaetaceae, rivulariaceae and scytonemataceae showed maximum dominance of 26 species (8 genera), 12 species (7 genera) and 6 species (4 genera) around 60th day, 40th day and 20th day of rice crop under field conditions [63]. The distribution of cyanobacteria is also greatly affected due to fertilizer application as there was a marked difference in the distribution and composition in fertilized and unfertilized paddy field [80]. Thirty-two species of nitrogen-fixers were recorded from unfertilized paddy fields whereas only twenty-five species of nitrogen-fixers were recorded from fertilized paddy fields during sixty days of crop growth indicating that the application of nitrogenous fertilizer has a negative effect on growth and emergence of nitrogen-fixing cyanobacteria.

Algal succession in rice fields is manifested in terms of differing populations of unicellular/colonial, non-heterocystous, followed by heterocystous - initially unbranched and by late harvest stage, branched heterocystous forms [81]. Nitrogen fixers do not dominate either in the beginning or late stages of crop growth and the phase between tillering and panicle development leads to production of maximum biomass [82].

Various microorganisms such as protozoans, fungi and grazers such as copepods, ostracods, snails and larvae have a significant effect in reducing cyanobacterial populations [83]. Management practices such as tillage, transplantation and weeding or water management also influence the proliferation of these organisms in the field. The application of insecticides or fungicides at recommended doses does not significantly influence the growth of cyanobacteria [77], which are known to exhibit a wide range of tolerance.

Analysis of soil samples of 11 districts of Dhaka depicted 84 cyanobacterial strains of which about 50% were heterocystous in nature and they predominantly belonged to *Fischerella*, *Nostoc* and *Calothrix* [84].

Agricultural Significance of Cyanobacteria

Like other microbes, cyanobacteria promote plant growth by augmenting the soil nutrient status with nitrogen, phosphorus and growth promoting hormones. Details of cyanobacterial contribution with respect to augmentation of soil with nitrogen, phosphorus and growth promoting hormones have been discussed below:

Nitrogen Metabolism and Agricultural Significance

Nitrogen is the most important vital nutrients for all living organisms required for synthesis of nucleic acids, enzymes, proteins and chlorophyll [85]. Atmospheric molecular dinitrogen gas exists in abundance (78%), but, this inert gas is unavailable for direct assimilation by living systems as this needs to be converted into available form by a process called as nitrogen fixation. Nitrogen fixation is carried out by atmospheric lightening, synthetic process (Haber-Bosch process) and biological nitrogen fixation.

Out of theses, biological nitrogen fixation mediated through microbes is the most efficient and sustainable process. Amongst microbes, cyanobacteria are the most efficient, reliable and ecofriendly nitrogen-fixers for their simple nutritional requirement and fix atmospheric nitrogen with the help of the enzyme nitrogenase [86].

Heterocyst and Nitrogen Fixation

Cyanobacteria are endowed with specialized structure called 'heterocyst' which contains enzyme nitrogenase that enables these organisms to convert unusable form of atmospheric nitrogen into available forms [25]. Heterocyst develops in filamentous forms of cyanobacteria under deprived condition of dissolved inorganic nitrogen. These consist of thick multilayered wall and photosystem I which supports the energy requirement (ATP production) for nitrogen fixation. The PS II is absent in heterocysts which protects the highly oxygen sensitive nitrogenase enzyme from oxygen.

The outer layer of heterocyst consists of polysaccharides and inner layer is composed of glycolipids. Heterocysts get fixed carbon from vegetative cells and transfer glutamine outside into the vegetative cells. Glycolytic substrates such as phosphoenolpyruvate (PEP) and pyruvate remain inactive. The reductants are generated through the oxidative pentose phosphate pathway (PPP) and possibly by isocitrate dehydrogenase.

Thus, NADPH generated transfers electron *via* ferrodoxin: NADP reductase to heterocyst specific ferredoxin (fdxH) and finally to two components of nitrogenase complex *viz*. Fe protein and Mo-Fe protein.

NAD(P)H and hydrogen also donate electrons to electron transport system to generate sufficient amount of energy to complete nitrogen fixation by nitrogenase complex.

Nitrogenase Complex

The enzyme 'Nitrogenase complex' consists of two distinct components: Fe protein (dinitrogenase reductase) and Mo-Fe protein (dinitrogenase). Fe protein (Mol. wt. 60 000 Daltons) consists of two identical subunits together with cluster of 4 iron and 4 labile sulphur atoms [87, 88]. Mo-Fe protein (Mol. wt. 200 000 Daltons) consists of 1-2 Mo, 12-32 Fe and 24 labile sulphur atoms per molecule [87]. Fe protein functions as an electron carrier to the tetrameric Mo-Fe protein where reduction of molecular nitrogen into ammonia takes place. This is the reason why Mo-Fe protein is called as nitrogenase and Fe protein as nitrogenase reductase [89] and both components of nitrogenase complex are highly oxygen sensitive. Filamentous heterocystous cyanobacteria are unique as they develop heterocyst either in between or at the end of vegetative cells where oxygen is evolved. Nitrogenase enzyme is oxygen sensitive and it is protected in heterocysts as these lack oxygen producing PS II and allow nitrogen fixation by creating an anaerobic environment. The thick walls of the heterocysts create a diffusion barrier of gases and allow conversion of atmospheric nitrogen into ammonium compounds [90].

The nitrogen fixation precedes with transfer of low redox potential reductant to nitrogenase *via* ferredoxin from vegetative cell and generation of ATP by PS I or oxidative phosphorylation in the heterocyst [91, 92]. Reductant molecule donates an electron to nitrogenase reductase that enables it to react with Mg-ATP [87].

In the meantime, nitrogen molecule combines with Mo Fe nitrogenase to get reduced. The two subunits of the enzyme complex unite together to complete the metabolic process of nitrogen fixation.

Flows of electrons occur singly from nitrogenase reductase to nitrogenase with concomitant hydrolysis of two ATP molecules [91]. The transfer of electrons is followed by dissociation of nitrogenase reductase: nitrogenase complex. The electrons are then transferred to nitrogen molecule that is reduced to ammonia through three two-electron steps [87].

Therefore, one nitrogen molecule requires 6 dinitrogenase reductase-Mg-ATP complexes. The heterocystous forms of cyanobacteria predominantly fix nitrogen in light than dark [91, 93] and the low rate of nitrogen fixation in dark might be attributed to poor generation of reductant [94]. The overall reaction summary of nitrogen fixation may be expressed as

$$N_2 + 8H^+ + 8e^- + 16\,ATP \rightarrow 2NH_3 + H_2 + 16\,ADP + 16\,P_i$$

The non-heterocystous forms of cyanobacteria also fix nitrogen under anaerobic conditions and it was first reported by Wyatt and Silvey [95]. They proposed that all vegetative cells of non-heterocystous cyanobacteria consist of nitrogenase complex and the process of nitrogen fixation is stimulated by light. Later studies showed that a number of unicellular (*Gloeocapsa*, *Aphanothece*, *Gloeothece* etc.) and non-heterocystous filamentous (*Oscillatoria*, *Plectonema, Trichodesmium etc.*) cyanobacteria fix nitrogen under field conditions [91, 96, 97] and nitrogen fixation can occur in dark.

Augmentation of Soil with Nitrogen

The distribution pattern of cyanobacteria and their contribution to the nitrogen economy of the agricultural fields has been extensively studied [9, 78, 79]. The global application of around 180 million tones of industrial N fertilizer annually accounts only for 1/3 of total nitrogen required for production of agricultural crops and the remaining 2/3 is contributed by the biological sources [16]. The cyanobacteria occupy a position next to legume-*Rhizobium* symbiosis in terms of total nitrogen contribution to the soil [16]. Nitrogen fixed by cyanobacteria is released into soil by autolysis or exudation or microbial decomposition on cell death [98]. Partial release of fixed nitrogen occurs extracellularly during growth and some soil nitrogen can be contributed by decomposition of dead biomass after harvesting of crop. Cyanobacteria have mostly been studied for their role in nitrogen economy of the paddy fields as these provide an optimum growth condition of light, water, humidity, temperature and available nutrients for their growth [62, 82, 83, 99].

Besides increasing the yield and augmenting N and P, regular build up of cyanobacteria population improves the physico-chemical properties of the soil in terms of an increase in residual N and C, soil pH and electrical conductivity [16]. These organisms account for 33% of total algal population in Indian rice fields [16, 77] and can contribute, in general, 10-30 kg N/ha with increase in productivity of rice grains up to 10-15 percent [82, 100]. In some studies, nitrogen fixation by cyanobacteria has also been reported to be 90 Kg N/ha/Year [101]. Prasad [102] collected the cyanobacterial crust from various soils after crop harvest and estimated total nitrogen contribution to soil by cyanobacteria can range between 12 to 16 Kg N/ha. De and Mandal [103] showed that the nitrogen fixation is high in cropped soils than uncropped soils. The nitrogen fixation rate ranges from 13.8 to 44.4 Kg N/ ha in six rice growing soils [103].

Watanabe et al. [104] reported augmentation of 20 kg N/ha by *Tolypothrix tenuis* from Japan. MacRae and Castro [105] demonstrated that cyanobacteria can contribute 10-15 kg N/ha in rice fields. Different values of nitrogen contribution viz. 40-80 kg N/ha [106], 15-51 kg N/ha [107], 18-38 kg/ha [108], 0-30 kg/ha [109], 50-80 kg/ha [110] and 11-23 kg/ha [65] have been reported. Understanding of nitrogen contribution through cyanobacterial population has been studied and enumerated in terms of acetylene reduction assay (ARA). Kolte and Goyal [111] reported the nitrogen-fixing capacity (mg N 50 mL^{-1}) for *Anabaena* (0.84-2.45), *Nostoc* (0.44-4.04), *Calothrix* (0.77-5.83), *Scytonema* (0.30-3.41), *Tolypothrix* (2.55-5.67), *Westiellopsis* (0.60-2.34) and *Hapalosiphon* (0.31-1.15) in twenty days old cultures. Cyanobiont, *Anabaena azollae* fixes more nitrogen than free-living cyanobacteria andit can fix more than 3.6 kg N/ha/day. This symbiotic system has been studied in detail and has been reported to be beneficial in paddy cultivation [112, 113, 114]. Such a system has various beneficial effects such as residual effects on other crop yield [115, 116], suppression of weed growth [117] and methane efflux in flooded rice fields [118].

Cyanobacterial nitrogen fixation is influenced by fertilizer application. The nitrogenous fertilizer application has a negative effect whereas P application exhibits a positive effect on establishment and growth of nitrogen fixing cyanobacteria [16]. The application of 40 ppm NH_4-N [119] and 30 ppm of urea [120] didn't reduce the nitrogenase activity in paddy field. However, Stewart [121] reported that application of 75 ppm combined nitrogen is toxic to growth and establishment as well as nitrogen fixation ability. On the other hand, 100 ppm nitrate-N may not have any negative effect on cyanobacterial nitrogen fixation [16].

PHOSPHATE METABOLISM AND AGRICULTURAL SIGNIFICANCE

Phosphorus

Phosphorus (P) is the most vital element, next to nitrogen, for plant growth and development. It is the key element of several biomolecules such as nucleic acids, phospholipids, co-enzymes, ATP and NADPH [122]. Phosphorus exists in both inorganic (bound, fixed or labile) and organic (bound) forms and is unavailable for direct uptake by plants [85, 123]. The major contributors of organic phosphates are agricultural, municipal and industrial effluents [124]. Although complex phosphates molecules present in the soil are sufficient [125] to supply P for the plant growth, their dissociation or mobilization to release P to support plant growth is essential. The availability of P to plants is regulated in a coordinated manner and is influenced by several factors such as pH, compaction, aeration, moisture, temperature, texture and organic matter of soils, crop residues, extent of plant root systems and root exudates and available soil microbes [85]. Studies suggested that microbes are capable of releasing consumable forms of phosphate such as monobasic ($H_2PO_4^-$) and dibasic ($H_2PO_4^{2-}$) forms for plants [5]. The application of phosphatic fertilizer is sufficient to supply P to plants, but, excessive application may reduce the soil fertility and productivity. Maintenance of consumable P pool from complex phosphates mediated through biological sources is the urgent need of current agricultural practices. Amongst different microbial components of the soil, cyanobacteria are most efficient systems endowed with enzyme 'phosphatase' to release available P to plants [126, 127].

Phosphatase

Boavida [128] has defined 'Phosphatase' as a 'whole bunch of enzymes' with different saturation constants, temperature, pH optima and substrate specificity [129]. This enzyme has been classified differently such as cellular [130], cell-bound [127], surface or surface-bound extracellular [131, 132] and intracellular [133] depending on their locations.

All enzymatic activity outside cytoplasmic membrane is extracellular [134, 135] and activity inside cytoplasmic membrane is intracellular [133]. The organic phosphates hydrolyzed by extra or intracellular enzymes are released external to the cytoplasmic membrane [136, 137, 138]. Ammerman [139] used the term ecto-phosphohydrolase and defined it as hydrolysis of organic or other complex phosphorus compounds, soluble or particulate, in which the hydrolysed phosphate is released outside the cell. In cell surface of cyanobacteria, phosphomonoesterase is located in the periplasm [140].

Cyanobacteria have evolved number of strategies to cope with phosphate deprivation [141]. Phosphorus metabolism in cyanobacteria involves three main components viz. uptake of inorganic phosphate (Pi), hydrolysis of dissolved organic phosphorus (DOP) with biosynthesis and catabolism of polyphosphate (polyp) [142].

These derive P from accumulated polyphosphate bodies for cellular metabolism during short period of P starvation [143] and can show 3-4 cell divisions in completely depleted condition of dissolved phosphate [144].

They synthesize phosphatases under long period of P starvation and scavenge P from complex organic phosphate molecules converting them into consumable inorganic P and organic moiety [145, 146]. Phosphatases (phosphomonoester hydrolases - PME) are a group of inducible catabolic phosphohydrolases enzymes present in cyanobacteria and are responsible for release of available P from complex organic phosphates in aquatic ecosystems [146, 147, 148]. Expression and synthesis of phosphatase enzymes is largely determined by the concentration of inorganic phosphate present externally and also N:P ratio in the cell [129]. Primary accumulation of enzyme 'phosphatase' occurs in periplasmic space and is transported outside at later stage [141]. Biosynthesis of phosphates is influenced by light, temperature, pH, heavy metals and salinity [149]. Based on their action in acidic and basic environments, phosphatases have been classified broadly into two type's viz. alkaline phosphatase and acidic phosphatase. Alkaline phosphates are group of inducible isoenzymes [150] that optimally function in the pH range of 7.6 to 9.6 [148], whereas acid phosphatases react in the range of 4.0 to 5.5 [148]. Chróst and Suida [148] reported that biosynthesis of alkaline phosphatase is regulated under repression-depression mechanism and competitive inhibition in the presence of P whereas synthesis of acid phosphatase is independent of repression in presence of inorganic phosphorus. Solubilization of complex phosphates by microorganisms involve: a) release of organic acids to dissolve compounds (dissolution-precipitation), b) liberation of extracellular enzymes (sorption-desorption), and c) release of soluble forms of P after degradation (mineralization-immobilization) [151, 152]. Liberation of available forms of inorganic phosphates by cyanobacteria is brought about by acidic and alkaline phosphatase depending on their external environments. It has been demonstrated that cyanobacteria can add 30 kg of P_2O_5 in soil and can save 188 kg SSP/ha with an additional yield in paddy and wheat [16].

Growth Promoting Hormones from Cyanobacteria

Cyanobacteria also synthesize and release a wide range of biologically active substances such as proteins, vitamins, carbohydrates, aminoacids, polysaccharides and phytohormones into soil that function as signaling molecule and may indirectly promote plant growth [153, 154, 155]. Therefore, their contribution is not only restricted to their N and P augmentation of soil, they may also enhance soil microbial biomass, carbon, growth promoting hormones which increase seed germination, root and shoot growth and grain yield [155]. These organisms not only contribute to the growth promoting effect under flooded environment, they can also support growth of the plants such as wheat, maize, cotton etc. under non-flooded growing conditions [155].

Presence of more numbers of non-nitrogen-fixing cyanobacteria over nitrogen-fixing cyanobacteria may indicate that the growth promoting effect on crops is not related to nitrogen fixation [73]. Their contribution towards growth promotion is not restricted to rice plant, but also wheat [11, 73, 156, 157, 158, 159], cotton [160], sorghum, maize and lentil [161], soyabean and clover crop [162] as well as several vegetables such as *Solanum lycopersicum*, *Cucurbita maxima*, *Cucumis sativus*, *Mentha spicata* and *Satureia horiensis* [72, 155, 163]. It is now established fact that cyanobacteria synthesize and release a number of growth promoting hormones that accelerate the plant growth [164, 165, 166].

These compounds are the key components of soil that play crucial role in plant growth and development [167], plant pathogenesis [168], and plant-microbe interactions [169]. These are synthesized and secreted by rhizospheric, epiphytic and symbiotic microorganisms [170, 171, 172]. Phytohormones govern not only the physiological processes from seed germination to fruit ripening [173, 174, 175, 176], they also provide resistivity against environmental stresses [177], and also induce or suppress the gene expression and synthesis of enzymes, pigments and metabolites [155, 178, 179]. Like other microorganisms, cyanobacteria can also release phytohormones and other metabolites which enable them to fight against environmental stresses in soil.

Cyanobacteria produce phytohormones such as auxins, gibberellins, cytokinins, ethylene etc. which are responsible for promoting plant growth under natural conditions. These phytohormones have been detected and quantified using advanced techniques such as HPLC, ELISA and GC-MS. Mishra and Kaushik [180] identified auxin like substances in *Nostoc* and *Hapalosiphon* having estimates of around 3.76 and 4.48 μg/g, respectively. Sergeeva et al. [181] studied various morphotypes of cyanobacteria, ranging from unicellular to branched filamentous forms, for the presence of IAA using ELISA and GC-MS. They confirmed the presence of IAA in *Gloeothece*, *Synechocystis*, *Nostoc*, *Anabaena*, *Cylindrospermum*, *Anabaenopsis*, *Calothrix*, *Plectonema*, *Chlorogloeopsis* and *Hapalosiphon*. Karthikeyan et al. [73] carried out experiments on estimation of IAA in selected cyanobacterial strains viz. *Calothrix* sp., *C. ghosei*, *C. membranacea*, *Hapalosiphon intricatus*, *Nostoc* sp., *N. muscorum*, *Westiellopsis prolifica* grown in dark and light conditions and showed maximum IAA for the culture grown in light (0.6-3.37 μg/g) compared to dark (0.5-1.39 μg/g) condition. *In-vitro* studies on accumulation of IAA showed that addition of tryptophan in nutrient medium increased the IAA content in cell extract [11, 182, 183]. Beside IAA, IPA (indole-3-propionic acid) and IBA (indole-3-butyric acid) have also been reported from *Anabaena vaginicola* and *Nostoc calcicola* [163] and IBA was more than IAA. David et al. [184] reported that IBA is the stored form of auxin and can convert into IAA during stress. Extract from IAA producing strains of mangrove root-associated cyanobacterium, *Phormidium* sp., was screened for their effect on tobacco seeds and exhibited a positive effect in terms of seed germination and callus differentiation [185].

Besides IAA, cyanobacteria such as *Anabaena*, *Oscillatoria*, *Phormidium*, *Chroococcidiopsis* and *Synechocystis*, produce five different types of cytokinins *i.e.,* trans zeatin, cis zeatin, zeatin riboside, dihydrozeatin riboside and zeatin-o-glucoside [159, 186]. They reported that *Chroococcidiopsis* sp. produce 22.7 pmol mg^{-1} chl-*a* cytokinins and 38 pmol mg^{-1} chl-*a* IAA. Cytokinin production has been also reported from *Nostoc entophytum*, *Lyngbya*, *Chlorogloeopsis* and *Calothrix* [155, 187, 188].

Ördög and Pulz [189] demonstrated that cytokinin production in a cyanobacterium *Arthronema africanum* is regulated differently under light-dark cycle of the callus. Moreover, abscisic acid (ABA) and gibberellin like substances have also been reported [187, 188, 190]. Gibberellin production has been reported from *Phormidium* sp. [190], *Aulosira fertilissima* [191], *Oscillatoria angustissima*, *Cylindrospermum* sp. and *Anabaenopsis* sp. [155, 187, 188]. Like higher plants, abscisic acid production in these is regulated under physiological stresses. Maršálek et al. [192] demonstrated the production of abscisic acid by *Nostoc muscorum*, *Trichomorus variabilis* and *Synechococcus leopoliensis* under salt stress of 0.05 and 0.1 M NaCl supplemented nutrient medium.

Abscisic acid production has been reported for *Cylindrospermum musicola* and *Anabaena oryzae* [193] and these organisms also produce ethylene [187] and jasmonic acid [187, 194].

Effect of the cyanobacterial growth promoting hormones on seed germination, crop yield, root and shoot development, branching tillering and fruit ripening has been studied in various crop plants. Gupta [76] demonstrated that rice seeds treated with extract of *Scytonema hofmanni*, *Fischerella muscicola* and *Nostoc* sp showed earlier germination with improved protein content than non-treated rice seeds. Comparative study with rice seeds treated with water and cyanobacterial extract (ether extract) of non-nitrogen-fixing strains of *Phormidium* sp. (*P. foveolarum*, *P. corium* and *P. autumnale*) clearly demonstrated the presence of growth promoting hormones in cyanobacterial extract of *P. foveolarum* which showed increased germination and growth of rice seedlings [195]. Gupta and Shukla [196] reported the presence of growth promoting hormones in another species of *Phormidium i.e., P. tenue* which showed enhanced root and shoot growth of rice seedlings.

Osman et al. [188] demonstrated that inoculation of soil with suspension of *Nostoc entophytum* and *Oscillatoria angustissima (auxin and cytokinin producers) increased both germination percentage and* other growth parameters along with photosynthetic pigment fractions of pea. Increase in rice grain yield has been reported for the soil inoculated with *Anabaena torulosa*, *Nostoc carneum*, *Nostoc piscinale* and *Anabaena doliolum* [197].

Study carried out on plant growth promoting effect of certain cyanobacterial species such as *Nostoc*, *Anabaena*, *Calothrix*, *Hapalosiphon*, *Oscillatoria*, *Lyngbya* and *Phormidium* isolated from rhizosphere of rice and wheat showed positive response to growth attributes [72, 198].

Conclusion

Cyanobacteria or blue green algae have been reported to offer an economically feasible and supplementary alternative to chemical fertilizers for enhancing soil fertility and crop productivity. Application of these in the Integrated Nutrient Management strategies can reduce the environmental hazards generated by the use of synthetic fertilizers. Cyanobacteria can augment the soil with nitrogen, phosphorus and growth promoting hormones essential for healthy and sustainable agricultural practices. Concerted efforts are required with respect to quality inoculum preparation with enhanced shelf life and maintenance of optimal moisture content in the soil to have the best benefits of cyanobacterial inoculation under field conditions with maximum agricultural potential.

References

[1] Singh, G. B., Swarup, A., (2000). Lessons from long-term fertility experiments. *Fert. News* 45,13-24.

[2] Pimentel, D., (1996). Green revolution agriculture and chemical hazards. *Sci. Total Environ.* 188 (supplement), S86-S98.

[3] Fuentes-Ramirez, L. E., Caballero-Mellado, J., (2005). Bacterial biofertilizers. In: *PGPR: biocontrol and biofertilization*, (Siddiqui, Z. A., Ed.), Springer, The Netherland, pp. 143-172.

[4] Lugtenberg, B., Kamilova, F., (2009). Plant growth promoting rhizobacteria. *Ann. Rev. Microbiol.* 63, 541-556.

[5] Bhattacharya, P. N., Jha, D. K., (2012). Plant growth-promoting rhizobacteria (PGPR): emergence in agriculture. *World J. Microbiol. Biotechnol.* 28, 1327-1350.

[6] Vessey, K. J., (2003). Plant growth promoting rhizobacteria as biofertilizers. *Plant Soil* 255,571-586.

[7] Uren, N. C., (2007). Types, amounts, and possible functions of compounds released into the rhizosphere by soil-grown plants. In: *The rhizosphere: biochemistry and organic substances at the soil–plant interface*, (Pinton, R., Varanini, Z., Nannipieri, P., Eds.), CRC Press, Boca Raton, pp. 1-22.

[8] Sharma, C., Gupta, R. K., Pathak, R. K., Choudhary, K. K., (2014). Impact of fluoride and VAM fungi on growth and foliar F accumulation of tea plants. *Fluoride* 47, 15-22.

[9] De, P. K., (1939). The role of blue-green algae in nitrogen fixation in rice fields. *Proc. R. Soc. Lond.* B 127, 121-139.

[10] Belnap, J., (2003). Biological soil crusts in deserts: a short review of their role in soil fertility, stabilization and water relations. *Algol. Stud.* 109, 113-126.

[11] Mazhar, S., Cohen, J. D., Hasnain, S., (2012). Auxin producing non-heterocystous cyanobacteria and their impact on the growth and endogenous auxin homeostasis of wheat. *J. Basic Microbiol.* 9, 1-8.

[12] Nikkinen, H., Hakkila, K., Gunnelius, L., Huokko, T., Pollari, M., Tyystja¨rvi, T., (2012). The SigB σ Factor Regulates Multiple Salt Acclimation Responses of the Cyanobacterium Synechocystis sp. PCC 6803. *Plant Physiol.* 158, 514-523.

[13] de Caire, G., Zaccaro, M. C., de Cano, S. M. M., (1979). Productos extracelulares de *Nostoc muscorum* Ag. (cepa 79a) obtenidos en medios cony sin nitr geno combinado. I:Sus efectos sobre plὑntulas de arroz. *Phyton* 37, 1-13.

[14] Chaudhary, V., Prasanna, R., Nain, L., Dubey, S. C., Gupta, V., Singh, R., Jaggi, S., Bhatnagar, A. K., (2012). Bioefficacy of novel cyanobacteria-amended formulations in suppressing damping off disease in tomato seedlings. *World J. Microbiol. Biotechnol.* 28, 3301-3310.

[15] Mishra, U., Choudhary, K. K., Pabbi, S., Dhar, D. W., Singh, P. K., (2005). Influence of Blue-Green Algae and Azolla inoculation on specific soil enzymes under paddy cultivation. *Asian Jr. Microbiol. Biotechnol. Env. Sc.* 7, 9-12.

[16] Kaushik, B. D., (2014). Developments in cyanobacterial biofertilizer. *Proc. Indian Natn. Sci. Acad.* 80, 379-388.

[17] Haselkorn, R., (1978). Heterocyst. *Annual Rev. Plant Physiol.* 29, 319-344.

[18] Berry, J. P., Gantar, M., Perez, M. H., Berry, G., Noriega, F. G., (2008). Cyanobacterial toxins as allelochemicals with potential applications as algaecides, herbicides and insecticides. *Marine Drugs* 6, 117-146.

[19] Waterbury, J. B., (2006). The Cyanobacteria- Isolation, purification and denitrification. *Prokaryotes* 4, 1053-1073.

[20] Atzenhofer, W., Regelsberger, G., Jacob, U., Peschek, G., Furtmuller, P., Huber, R., Obinger, C., (2002). The 2.0A resolution structure of the catalytic portion of a

cyanobacterial membrane-bound manganese superoxide dismutase. *J. Mol. Biol.* 321, 479-489.

[21] Schopf, J. W., Walter, M. R., (1982). Origin and early evolution of cyanobacteria. The Geological evidence. In: *The biology of cyanobacteria*, (Carr, N. G., Whitton, B. A., Eds.), Blackwell, Oxford and Univ. of California Press, Berkeley, pp. 543-564.

[22] Schopf, J. W., (2000). The fossil record: tracing the roots of the cyanobacterial lineage. In: *The ecology of cyanobacteria: their diversity in time and space*, (Whitton, B. A., Potts, M., Eds.), Kluwer Academic, Boston, pp. 13-35.

[23] Sanchez-Baracaldo, P., Haynes, P. K., Blank, C. E., (2005). Morphological and habitat evolution in the cyanobacteria using a compartmentalization approach. *Geobiology* 3, 145-165.

[24] Oren, A., (2000). Salts and Brines. In: *The Ecology of Cyanobacteria: Their Diversity in Time and Space*, (Whitton, B. A., Potts, M., Eds.), Kluwer Academic Publishers, Dordrecht, The Netherlands, pp. 281-306.

[25] Adams, D. G., Duggan, P. S., (1999). Heterocyst and akinete differentiation in cyanobacteria. *New Phytol.* 144, 3-33.

[26] Thajuddin, N., Subramanian, G., (2005). Cyanobacterial biodiversity and potential applications in biotechnology. *Curr. Sci.* 89, 47-57.

[27] Thajuddin, N., Subramanian, G., (1992). Survey of cyanobacterial flora of the southern east coast of India. *Bot. Mar.* 35, 305-311.

[28] Hoffman, L., (1999). Marine cyanobacteria in tropical regions: diversity and ecology. *Eur. J. Phycol.* 34, 371-379.

[29] Taton, A., Grubisic, S., Balthasart, P., Hodgson, D. A., Laybourn-Parry, J., Wilmotte, A., (2006). Biogeographical distribution and ecological ranges of benthic cyanobacteria in east Antarctic lakes. *FEMS Microbiol. Ecol.* 57, 272-289.

[30] Vincent, W. F., (2000). Cyanobacterial dominance in the polar regions. In: *Ecology of the Cyanobacteria: Their Diversity in Space and Time*, (Whitton, B., Potts, M., Eds.), Kluwer Academic Press, Dordrecht, The Netherlands, pp. 321-340.

[31] Sompong, U., Hawkins, P., Besley, C., Peerapornpisal, Y., (2005). The distribution of cyanobacteria across physical and chemical gradient in hot springs in northern Thailand. *FEMS Microbiol. Ecol.* 52, 365-376.

[32] Mohamed, Z. A., (2008). Toxic cyanobacteria and cyanotoxins in public hot springs in Saudi Arabia. *Toxicon* 51, 17-27.

[33] Gopalaswamy, G., Karthikeyan, C. V., Raghu, R., Udaysuryan, V., Apte, S. K., (2007). Identification of acid-stress-tolerant proteins from promising cyanobacterial isolates. *J. Appl. Phycol.* 19, 631-639.

[34] Gerasimenko, L. M., Dubinin, A. V., Zavarzin, G. A., (1996). Alakaliphilic cyanobacteria from soda lakes of Tuva and their ecophysiology. *Microbiol.* 65, 736-740.

[35] Rai, A. N., Soderback, E., Bergman, B., (2000). Tansley Review No. 116, Cyanobacterium plant symbioses. *New Phytol.* 147, 449-481.

[36] Demoulin, V., Janssen, M. P., (1981). Relationship between diameter of the filament and cell shape in blue-green algae. *Br. Phycol. J.* 16, 55-58.

[37] Mann, N. H., Carr, N. G., (1974). Control of macromolecular composition and cell division in the blue-green alga, *Anacystis nidulans*. *J. Gen. Microbbiol.* 83, 399-405.

[38] Binder, B. J., Chisholm, S. W., (1990). Relationship between DNA cycle and growth rate in *Synechococcus* sp. Strain 6304. *J. Bacteriol.* 172, 235-241.

[39] Catenholz, R. W., (1992). Species usage, concept and evolution in the cyanobacteria (blue-green algae). *J. Phycol.* 28, 737-745.

[40] Whitton, B. A., (1992). Diversity, ecology and taxonomy of the cyanobacteria. In: *Photosynthetic prokaryotes*, (Mann, N. H., Carr, N. G., Eds.), Plenum Press, New York, pp. 1-51.

[41] Desikachary, T. V., (1959). *Cyanophyta*. Indian Council of Agricultural Research, New Delhi, India. pp. 1-686.

[42] Rippka, R., Josette, D., Waterbury, J. B., Herdman, M., Stanier, R. Y., (1979). Generic assignments, strain histories and properties of pure cultures of cyanobacteria. *J. Gen. Microbiol.* 111(1), 1-61.

[43] Tandeau de Marsac, N., (1994). Differentiation of hormogonia and relationships with other biological surfaces. In: *The Molecular Biology of Cyanobacteria*, (Bryant, D. A., Ed.), Kluwer Academic Publisher, Dordrecht, The Netherland, pp. 825-842.

[44] Nienow, J. A., Friedlmann, E. I., (1993). Terrestrial lithopytic (rock) communities. In: *Antarctic microbiology*, (Friedlmann, E. I., Ed.), Wille-Liss, New York, pp. 353-412.

[45] Kratz, W. A., Myers, J., (1955). Nutrition and growth of several blue-green algae. *Am. J. Bot.* 42, 275-280.

[46] Prechtl, J., Kneip, C., Lockhart, P., Wenderoth, K., Maier, U. G., (2004). Intracellular spheroid bodies of *Rhopalodia gibba* have nitrogen-fixing apparatus of cyanobacterial origin. *Mol. Biol. Evol.* 21, 1477-1481.

[47] Hitch, C. J. B., Millbank, J. W., (1975). Nitrogenase metabolism in lichens. VII; Nitrogenase activity and heterocyst frequency in lichens with blue–green phycobionts. *New Phytol.* 73, 239-244.

[48] Meeks, J. C., (1990). Cyanobacterial-bryophyte associations. In: *Handbook of Cyanobacteria*, (Rai, A. N., Ed.). CRC Press, Boca Raton, Florida, pp. 43-63.

[49] Adams, D. G., Bergman, B., Nierzwicki-Bauer, S. A., Rai, A. N., and Schubler, A., (2006). Cyanobacterial-plant symbiosis. In: *Prokaryotes* (Dworkin, M., Falkow, S., Rosenberg, E., Karl-Heinz, S., Stackebrandt, E., Eds.), 1, 331-363.

[50] West, N., Adams, D. G., (1997). Phenotypic and genotypic comparison of symbiotic and free-living cyanobacteria from a single field site. *Appl. Environ. Microbiol.* 63, 4479-4484.

[51] Madhusoodan, P. V., Dominik, T. K., (1996). Epiphytic cyanobacteria on mosses from Western Ghats of Kerala. *J. Econ. Tax Bot.* 20, 355-359.

[52] Solheim, B., Zielke, M., (2002). Associations between cyanobacteria and mosses. In: *Cyanobacteria in Symbiosis*, (Rai, A. N., Bergman, B., Rasmussen, U., Eds.). Kluwer Academic Publishers, Netherlands, pp. 137-152.

[53] Nierzwicki-Bauer, S. A., (1990). *Azolla-Anabaena* symbioses: use in agriculture. In: *Handbook of symbiotic cyanobacteria*, (Haselkorn, R., Ed.), CRC Press, Florida, pp. 119-136.

[54] Braun-Howland, E. B., Nierzwicki-Bauer, S. A., (1990). Biochemistry, physiology, ultrastructure and molecular biology of the *Azolla-Anabaena* symbiosis. In: *Handbook of Symbiotic Cyanobacteria*, (Rai, A. N., Ed.), CRC Press, Boca Raton, FL, pp. 65-118.

[55] Lindblad, P., Bergman, B., (1990). The cyacad-cyanobacterial symbiosis. In: *Handbook of Symbiotic Cyanobacteria*, (Rai, A. N., Ed.), CRC Press, Boca Raton, Florida, pp. 137-159.

[56] Ow, M. C., Gantar, M., Elhai, J., (1999). Reconstitution of cycad-cyanobacterial association. *Symbiosis* 27, 125-134.

[57] Bergman, B., Johansson, C., Söderbäck, E., (1992). The *Nostoc-Gunnera* symbiosis. *New Phytol.* 122, 379-400.

[58] Wilkinson, C. R., (1978). Microbial associations in sponges. III. Ultrastructure of the *in situ* associations in coral reef sponges. *Mar. Biol.* 49, 177-185.

[59] Wilkinson, C. R., Fay, P., (1979). Nitrogen fixation in coral reef sponges with symbiotic cyanobacteria. *Nature* 279, 527-529.

[60] Granhall, U., (1975). Nitrogen fixation by blue-green algae in temperate soils. In: *Nitrogen fixation by free-living microorganisms*, (Stewart, W. D. P., Ed.), Cambridge Univ. Press, pp. 189-198.

[61] Choudhary, K. K., Singh, R. K., (2013). Culturable cyanobacterial diversity along altitudinal gradient in Eastern Himalayas. *J. Algal Biomass Utln.*, 4(2), 53-58.

[62] Watanabe, A., Yamamoto, Y., (1971). Algal nitrogen fixation in the tropics. *Plant Soil spl.* vol., 403-413.

[63] Choudhary, K. K., (2011). Occurrence of nitrogen-fixing cyanobacteria during different stages of paddy cultivation. *Bangladesh J. Plant Taxon.* 18(1), 73-76.

[64] Singh, B. V., Choudhary, K. K., Dhar, D. W., Singh, P. K., (2001). Occurrence of some *Nostocales* from 24 Parganas of West Bengal. *Phykos.* 40, 83-87.

[65] Okuda, A., Yamaguchi, M., (1955). Nitrogen-fixing microorganisms in paddy soils. I. Characteristics of the nitrogen-fixation in paddy soils. *Soil Plant Food* 1, 102-104.

[66] Mitra, A. K., (1951). The algal flora of certain Indian soil. *Indian J. Agric. Sci.* 2, 357.

[67] Aiyer, R. S., (1965). Comparative algological studies in rice-fields in Kerala state. *Agric. Res. J.* Kerala 3, 100-104.

[68] Al-Kaisi, K. A., (1976). Contributions to the algal flora of the rice fields of south-eastern Iraq. *Nova Hedwigia* 27, 813-827.

[69] Prasanna, R., Jaiswal, P., Nayak, S., Sood, A., Kaushik, B. D., (2009a). Cyanobacterial diversity in the rhizosphere of rice and its ecological significance. *Ind. J. Microbiol.* 49, 89-97.

[70] Choudhary, K. K., (2009). Occurrence of Chroococcaceae during rice cultivation in northern Bihar, India. *Bangladesh J. Plant Taxon.* 16(1), 57-63.

[71] Song, T., Mårtensson, L., Eriksson, T., Zheng, W., Rasmussen, U., (2005). Biodiversity and seasonal variation of the cyanobacterial assemblage in a rice paddy field in Fujian, China. *FEMS Microbiol. Ecol.* 54, 131-40.

[72] Shariatmadari, Z., Riahi, H., Hastroudi, M. S., Ghassempour, A., Aghashariatmadary, Z., (2013). Plant growth promoting cyanobacteria and their distribution in terrestrial habitats of Iran. *Soil Sci. Plant Nutr.* 59, 535-547.

[73] Karthikeyan, N., Prasanna, R., Sood, A., Jaiswal, P., Nayak, S., Kaushik, B. D., (2009). Physiological characterization and electron microscopic investigations of cyanobacteria associated with wheat rhizosphere. *Folia Microbiol.* 54, 43-51.

[74] Kurusawa, H., (1956). The weekly succession in the standing crop of plankton and zoobenthos in the paddy field. Part I and II. *Bull. Res. Sci. Jpn.* 41-42, 86-98.

[75] Ichimura, S., (1954). Ecological studies on the phytoplankton in paddy fields. I. Seasonal fluctuations in the standing crop and productivity of plankton. *Jpn. J. Bot.* 14, 269-279.

[76] Gupta, A. B., (1966). Algal flora and its importance in the economy of rice fields. *Hydrobiologia* 28, 213-222.

[77] Venkataraman, G. S., (1975). Nitrogen fixation by free living micro-organisms, Cambridge Uni Press, pp. 207-268.

[78] Gupta, R. D., Khajuria, M. R., (1996). Prospects and constraints of biofertilizers for rice with special reference to Jammu region of Jammu and Kashmir state. *Farmers and Parliament* 31, 21-22.

[79] Swarnalakshmi, K., Dhar, D. W., Singh, P. K., (2006). Blue-green algae: a potential biofertilizer for sustainable rice cultivation. *Proc. Indian Natl. Sci. Acad.* 72, 135-143.

[80] Choudhary, K. K., Bimal, R., (2010). Distribution of nitrogen-fixing cyanobacteria (Nostocaceae) during rice cultivation in fertilized and unfertilized paddy fields. *Nord. J. Bot.* 28(1), 100-103.

[81] Singh, P. K., Bisoyi, R. N., (1989). Blue-green algae in rice fields. *Phykos*, 28, 181-195.

[82] Roger, P. A., Kulasooriya, S. A., (1980). *Blue-green algae and rice*, IRRI, Los Banos.

[83] Venkataraman, G. S., (1972). *Algal biofertilizer and rice cultivation.* Today and Tomorrow's, Publ. New Delhi.

[84] Khan, Z. U. M., Begum, Z. N. T., Mandal, R., Hossain, M. Z., (1994). Cyanobacteria in rice soils. *World J. Microbiol. Biotech.* 10, 296-298.

[85] Gopalakrishnan, S., Sathya, A., Vijayabharathi, R., Varshney, R. K., Laxmipathi Gowda, C. L., Krishnamurthy, L., (2014). Plant growth promoting rhizobia: challenges and opportunities. *3 Biotech* DOI 10. 1007/s13205-014-0241-x.

[86] Stacey, G. H., Burris, R. H., Evans, H. J., (1992). *Biological nitrogen fixation.* Chapman and Hall, New York.

[87] Sprent, J. I., (1979). The metabolic problems of biological nitrogen fixation. In: *The biology of nitrogen-fixing organisms*, (Sprent, J. I., Ed.), McGraw-Hill Book Company Ltd., UK, pp. 51-74.

[88] Orne-Johnson, W. H., (1977). Biochemistry of nitrogenase. In: *Genetic engineering for nitrogen fixation*, (Hollaender, A., Ed.), Plenum press, New York, pp. 317-332.

[89] Hagemann, R. V., Burris, R. H., (1978). Nitrogenase and nitrogenase reductase associate and dissociate with each catalytic cycle. *Proc. Natl. Acad. Sci.* US 75, 2699-2702.

[90] Scherer, S., Almon, H., Boger, P., (1988). Interaction of photosynthesis, respiration and nitrogen fixation in cyanobacteria. *Photosynth. Res.* 15, 95-114.

[91] Issa, A. A., Abd-Alla, M. H., Ohyama, T., (2014). Nitrogen-fixing cyanobacteria: future prospect. In: *Advances in Biology and Ecology of Notrogen Fixation*, (Ohyama, T., Ed.), INTECH Science, pp. 23-48.

[92] Stewart, W. D. P., Rowell, P., (1986). Biochemistry and physiology of nitrogen fixation with particular emphasis on nitrogen fixing phototrophs. *Plant Soil* 90, 167-191.

[93] Horne, A. J., (1979). Nitrogen fixation in Clear Lake, California. 4. Diel studies on *Aphanizomenon* and *Anabaena* blooms. *Limnol. Oceanogr.* 24, 329-341.

[94] Ernst, A., Rohme, H., (1984). Control of hydrogen-dependent nitrogenase activity by adenylates and electron flow in heterocysts of *Anabaena variabilis*. *Biochim. Biophys. Acta* 767, 362-268.

[95] Wyatt, J. T., Silvey, J. K. G., (1969). Nitrogen fixation by *Gloeocapsa*. *Sci.* 165, 908-909.

[96] Ohki, K., Fujita, Y., (1988). Aerobic nitrogenase activity measured as acetylene reduction in the marine non-heterocystous cyanobacterium *Trichodesmium* spp. grown under artificial conditions. *Marine Biol.* 98, 111-114.

[97] Fredriksson, C., Bergman, B., (1995). Nitrogenase quantity varies diurnally in a subset of cells within colonies of the non-heterocystous cyanobacteria *Trichodesmium* spp. *Microbiol.* UK, 141, 2471-2478.

[98] Roger, P. A., Reynaud, P. A., (1979). *Nitrogen and Rice*, International Rice Res Institute Los Banos Philippines, pp. 289-309.

[99] Singh, R. N., (1961). *Role of Blue Green Algae in Nitrogen Economy of Indian Agriculture, Indian Council of Agricultural Research*, New Delhi.

[100] Kaushik, B. D., (1998). Use of cyanobacteria biofertilizers in rice cultivation: a technology improvement. In: *Cyanoobacterial Biotechnology*, (Subramanian, G., Kaushik, B. D., Venkataraman, G. S., Eds.), Science Publishers, Inc., Enfield US, pp. 211-222.

[101] Metting, B., (1981). The systematic and ecology of soil algae. *Bot. Rev.* 147, 195-312.

[102] Prasad, S., Mehrotra, R. K., Singh, Y., (1949). Nitrogen recuperation by blue-green algae in soils of Bihar and their growth on different types. *J. Proc. Inst. Chem. India* 21, 135-140.

[103] De, P. K., Mandal, L. N., (1956). Fixation of nitrogen by algae in rice soils. *Soil Sci.* 81, 453-458.

[104] Watanabe, A., Nishigzaki, A., Konishi, C., (1951). Effect of nitrogen fixing blue green algae on the growth of rice plant. *Nature* 168, 748-749.

[105] MacRae, I. C., Castro, T. F., (1969). Nitrogen fixation in some tropical rice soils. *Soil Sci.* 103, 277-280.

[106] Yoshida, T., Ancazans, P. R., (1973). Nitrogen fixing activity in upland and flooded rice fields. *Soil Sci. Soc. Amer. Proc.* 37, 42-46.

[107] Henriksson, E., (1971). Biological Nitrogen Fixation in Natural and Agricultural Habitats, *Plant Soil Spl.* Vol., p. 415.

[108] Alimango, B. V., Yoshida, T., (1977). In-situ acetylene-ethylene assay of biological nitrogen-fixation in lowland rice soils. *Plant Soil* 47, 239-244.

[109] Reynaud, P. A., Roger, P. A., (1978). N_2-fixing algal biomass in Senegal rice fields. *Ecol. Bull. Stockholm* 26, 148-157.

[110] Traore, T. M., Roger, P. A., Renaud, P. A., et al. (1978). Nitrogen-fixation by blue-green algae in paddy field in Mali. *Cab ORSTOM Ser. Biol.* 13, 181-185.

[111] Kolte, S. O., Goyal, S. K., (1989). Natural variation in nitrogen potential of cyanobacteria. *Phykos* 25, 166-170.

[112] Watanabe, I., (1987). Summary report of the *Azolla* programme of the international network on soil fertility and fertilizer evaluation for rice. In: *Azolla Utilization*, (Watanabe, I., Ed.), International Rice Research Institute, Manila, pp. 197-201.

[113] Singh, A. L., Singh, P. K., (1990). Phosphorus fertilization on growth and nitrogen fixation of *Azolla* and BGA in rice fields. *Ind. J. Plant Physiol.* 33, 21-26.

[114] Singh, P. K., Singh, D. P., (1992). *Azolla*-a potential way towards organic farming in rice cultivation in Indian conditions. In: *Proc. Natl. Seminar Organic Farming*, (Rai, M. M., Verma, L. N., Eds.), Ravi Printing Press, Jabalpur, pp. 233-240.

[115] Mahapatra, B. S., Sharma, G. L., (1989). Integrated management of *Sesbania*, *Azolla* and urea nitrogen in low land rice under rice wheat cropping system. *J. Agric. Sci.* (Cambridge) 113, 203-206.

[116] Kannaiyan, S., (1990). *Biotechnology of biofertilizers for rice crop.* Tamil Nadu Agricultural University, Tamil Nadu, p. 1-225.

[117] Satpathy, K. B., Singh, P. K., (1985). Control of weeds by *Azolla pinnata* in rice fields. *J. Aquat. Plant Manag.* 23, 40-42.

[118] Bharati, K., Mohanty, S. R., Singh, D. P., Rao, V. R., Adhya, T. K., (2000). Influence of incorporation or dual cropping of Azolla on methane emission from a flooded alluvial soil planted to rice in eastern India. *Agric. Ecosyst. Environ.* 79, 73-83.

[119] Venkataraman, G. S., (1979). *Nitrogen and Rice*, International Rice Res Institute Los Banos Philippines, pp. 311-321.

[120] Mekonen, P. K., (2002). Response of *Anabaena sp.* to different nitrogen sources. *Acta Bot. Hung.* 53, 367-380.

[121] Stewart, W. D. P., (1964). Nitrogen fixation by Myxophyceae from marine environments. *J. Gen. Microbiol.* 36, 415-422.

[122] Vance, C. P., Uhde-Stone, C., Allan, D. L., (2003). Phosphorus acquisition and use: Critical adaptation by plants securing a non-renewable resource. *New Phytol.* 157, 423-447.

[123] Wagner, K. U., Masepohl, B., Pistorius, E. K., (1995). The cyanobacterium *Synechococcus* sp. Strain PCC 7942 contains a second alkaline phosphatase encoded by Pho V. *Microbiology* 141, 3049-3058.

[124] Singh, S. K., Tiwari, D. N., (2000). Control of alkaline phosphatase activity in *Anabaena oryzae* Fritsch. *J. Plant Physiol.* 157, 467-472.

[125] Buol, S., Eswaran, H., (2000). Oxisols. *Adv. Agron.* 68, 151-195.

[126] Whitton, B. A., Potts, M., Simon, J. W., Grainger, S. L. J., (1990). Phosphatase activity of the blue-green alga (cyanobacterium) *Nostoc commune* UTEX 584. *Phycologia* 29, 139-145.

[127] Whitton, B. A., Grainger, S. L. J., Hawley, G. R. W., Simon, J. W., (1991). Cell-bound and extracellular phosphatase activity of cyanobacterial isolates. *Microb. Ecol.* 21, 85-98.

[128] Boavida, M. J., (1990). Natural plankton phosphatases and the recycling of phosphorus. *Verhandlung internationale Vereinigung Limnologie* 24, 258-259.

[129] Hoppe, H.-G., (2003). Phosphatase activity in the sea. *Hydrobiologia* 493, 187-200.

[130] Grainger, S. L. J., Peat, A., Tiwari, D. N., Whitton, B. A., (1989). Phosphomonoesterase activity of the cyanobacterium (blue-green alga) *Calothrix parietina*. *Microbios* 59, 7-17.

[131] Whitton, B. A., Yelloly, J. M., Christmas, M., Hernández, I., (1998). Surface phosphatase activity of benthic algae in a stream with highly variable ambient phosphate concentrations. *Verhand. Internat. Verein. Limnol.* 26, 967-972.

[132] Nedoma, J., Štrojsova, A., Vrba, J., Komárková, J., Šimek, K., (2003). Extracellular phosphatase activity of natural plankton studied with ELF97 phosphate: fluorescence quantification and labeling kinetics. *Environ. Microbiol.* 5, 462-472.

[133] Whitton, B. A., Al-Shehri, A. M., Ellwood, N. T. W., Turner, B. L., (2005). Ecological aspects of phosphatase activity in cyanobacteria, eukaryotic algae and bryophytes. In:

Organic Phosphorus in the environment, (Turner, B. L., Frossard, E., Baldwin, D. S., Eds.), CAB International, pp. 205-241.

[134] Lee, T.-M., (2000). Phosphate starvation induction of acid phosphatase in *Ulva lactuca* L. (Ulvales, Chlorophyta). *Bot. Bull. Acad.* Sin. 41, 19-23.

[135] Priest, F. G., (1984). *Extracellular Enzymes: Aspects of Microbiology*, 9. Van Nostrand Reinhold, Wokingham, UK, pp. 79.

[136] Heath, R. T., Edinger, A. C., (1990). Uptake of ^{32}P-phosphoryl from glucose-6-phosphate by plankton in an acid bog lake. *Verhand. Internat. Verein. Limnol.* 24, 210-213.

[137] Bjoerkman, K., Karl, D. M., (1994). Bioavailability of inorganic and organic phosphorus compounds to natural assemblages of microorganisms in Hawaiian coastal waters. *Mar. Ecol. Prog. Ser.* 111, 265-275.

[138] Štrojsova, A., Vrba, J., Nedoma, J., Komárková, J., Znachor, P., (2003). Seasonal study of extracellular phosphatase expression in the phytoplankton of a eutrophic reservoir. *Eur. J. Phycol.* 38, 295-306.

[139] Ammerman, J. W., (1991). Role of eco-phosphohydrolases in phosphorus regeneration in estuarine and coastal ecosystems. In: *Microbial enzymes in aquatic environments*, (Chróst, R. J., Ed.), Springer, New York, pp. 165-186.

[140] Ihlenfeldt, M. J. A., Gibson, J., (1975). Phosphate utilization and alkaline phosphatase activity in *Anacystis nidulans* (*Synechococcus*). *Archives Microbiol.* 102, 23-28.

[141] Pandey, M., (2006). Nutrient modulated alkaline phosphatase and associated processes in diazotrophic cyanobacteria. *Polish J. Microbiol.* 55, 53-62.

[142] Duncan, O. E., (2010). Phosphorus physiology of the Marine Cyanobacterium *Trichodesmium*. Ph. D. Thesis. *MIT/WHOI*, 2010-06.

[143] Bhaya, D., Schwarz, R., Grossman, A. R., (2000). Molecular responses to environmental stress. In: *The Ecology of cyanobacteria: Their Diversity in Time and Space*, (Whitton, B. A., Potts, M., Eds.), Kluwer Academic Publishers, Dordrecht, The Netherlands, pp. 397-442.

[144] Chorus, I., Mur, L., (1999). Preventive measures. In: *Toxic cyanobacteria in water: A guide to their public health consequences, monitoring and management.* (Chorus, I., Bartram, J., Eds.), WHO E and FN Spon, London and New York, pp. 235-273.

[145] Stihl, A., Sommer, U., Post, A. F., (2001). Alkaline phosphatase activities among population of the colony-forming diazotrophic cyanobacterium *Trichodesmium* spp. (cyanobacteria) in the red sea. *J. Phycol.* 37, 310-317.

[146] Babić, O. B., Simeunović, J. B., Škrbić, N. Z., Kovač, D. J., Svirčev, Z. B., (2014). Detection of phosphatase activity in aquatic and terrestrial cyanobacterial strains. *Jour. Nat. Sci.* 125, 31-42.

[147] Matavulj, M., Flint, K. P., (1987). A model for acid and alkaline phosphatase activity in a small pond. *Microb. Ecol.* 13, 141-158.

[148] Chróst, R. J., Suida, W., (2002). Ecology of Microbial Enzymes in Lake Ecosystems. In: *Microbial Enzymes in the Environment. Activity, Ecology, and Applications*. (Burns, R. C., Dick, R. P. Eds.), New York: Marcel Dekker, Inc; pp. 35-72.

[149] Singh, S. K., Pandey, V., Pandey, K. D., (2007). Phosphate uptake kinetics and its regulation in N2-fixing cyanobacterium Anabaena oryzae Fritsch under salt stress. *Afr. J. Biotechnol.* 6, 2363-2368.

[150] Luo, M., Guo, Y. C., Deng, J. Y., Wei, H. P., Zhang, Z. P., Leng, Y., Men, D., Song, L. R., Zhang, X. E., Zhou, Y. F., (2010). Characterization of a monomeric heat- labile – classical alkaline phosphatase from *Anabaena* sp. PCC7120. *Biochemistry* (Moscow), 75, 655-664.

[151] Sharma, S. B., Sayyed, R. Z., Trivedi, M. H., Gobi, T. A., (2013). Phosphate solubilizing microbes: sustainable approach for managing phosphorus deficiency in agricultural soils. *Springer Plus* 2, 1-14.

[152] McGill, W. B., Cole, C. V., (1981). Comparative aspects of cycling of organic C, N, S and P through soil organic matter. *Geoderma* 26, 267-268.

[153] Zaccaro, M. C., Kato, A., Zulpa, G., Storni, M. M., Steyerthal, N., Lobasso, K., Stella, A. M., (2006). Bioactivity of *Scytonema hofmanni* (Cyanobacteria) in *Lilium alexandrae in vitro*propagation. *Electronic J. Biotechnol.* 9, 210-214.

[154] Mandal, B., Vlek, P. L. G., Mandal, L. N., (1998). Beneficial effect of blue-green algae and *Azolla* excluding supplying nitrogen, on wetland rice fields: a review. *Biol. Fert. Soils* 27, 329-342.

[155] Singh, S., (2014). A review on possible elicitor molecules of cyanobacteria: their role in improving plant growth and providing tolerance against biotic and abiotic stress. *J. Appl. Microbiol.* 117, 1221-1244.

[156] Obreht, Z., Kerby, N. W., Gantar, M., Rowell, P., (1993). Effects of root associated N2-fixing cyanobacteria on the growth and nitrogen content of wheat (*Triticum vulgare* L.) seedlings. *Biol. Fertil. Soils* 15, 68-72.

[157] Karthikeyan, N., Prasanna, R., Lata, Kaushik, B. D., (2007). Evaluating the potential of plant growth promoting cyanobacteria as inoculants for wheat. *Eur. J. Soil Biol.* 43, 23-30.

[158] Rana, A., Joshi, M., Prasanna, R., Shivay, Y. S., Nain, L., (2012). Biofortification of wheat through inoculation of plant growth promoting rhizobacteria and cyanobacteria. *Eur. J. Soil Biol.* 50, 118-126.

[159] Hussain, A., Hasnain, S., (2011). Phytostimulation and biofertilization in wheat by cyanobacteria. *J. Ind. Microbiol. Biotechnol.* 38, 85-92.

[160] Likhitkar, V. S., Tarar, J. L., (1995). Effect of presoaking seed treatment with *Nostoc muscorum* extracts on cotton. *Ann. Plant Physiol.* 9, 113-116.

[161] Adam, M. S., (1999). The promotive effect of the cyanobacterium *Nostoc muscorum* on the growth of some crop plants. *Acta Microbiol. Pol.* 48, 163-171.

[162] Tantawy, S. T. A., Musa, S. A. I., (2001). The influence of live Cyanobacteria filtrate of *Nostoc calcicola* and *Anabaena flos aquae* on seed germination and plant growth of some cultivars (wheat, soybean and clover). *Az. J. Microbiol.* 54, 260-272.

[163] Hashtroudi, M. S., Ghassempour, A., Riahi, H., Shariatmadari, Z., Khanjir, M., (2013). Endogenous auxin in plant growth-promoting Cyanobacteria-*Anabaena vaginicola* and *Nostoc calcicola*. *J. Appl. Phycol.* 25, 379-386.

[164] Muller, B., Sheen, J., (2008). Cytokinin and auxin interaction in root stem-cell specification during early embryo-genesis. *Nature* 453, 1094-1097.

[165] To, P. C., Kieber, J. J., (2008). Cytokinin signaling: two-components and more. *Trends Plant. Sci.* 13, 85-92.

[166] Venkataraman, G. S., Neelakantan, S., (1967). Effect of cellular constituents of the nitrogen fixing blue-green alga *Cylindrospermum muscicola* on the root growth of rice plants. *J. Gen. Appl. Microbiol.* 13, 53-61.

[167] Dobbelaere, S., Vanderleyden, J., Okon, Y., (2003). Plant growth promoting effects of diazotrophs in the rhizosphere. *Crit. Rev. Plant Sci.* 22, 107-149.

[168] Kado, C. I., (1984). Phytohormone-mediated tumorigenesis by plant pathogenic bacteria. In: *Genes Involved in Microbe-Plant Interactions, Plant Gene Res*earch (Verma, D. S., Hohn, T., Eds.), Vienna: Springer, pp. 311-336.

[169] Hirsch, A. M., Fang, Y., Asad, S., Kapulnik, Y., (1997). The role of phytohormones in plantmicrobe symbioses. *Plant Soil* 194, 171-184.

[170] Costacurta, A., Vanderleyden, J., (1995). Synthesis of phytohormones by plant-associated bacteria. *Crit. Rev. Microbiol.* 21, 1-18.

[171] Shi, Y., Lou, K., Li, C., (2009). Promotion of plant growth by phytohormone-producing endophytic microbes of sugar beet. *Biol. Fert. Soils* 45, 645-653.

[172] Ryan, F., Seipke1, Kaltenpoth, M., Matthew, I., (2012). Hutchings Streptomyces as symbionts: an emerging and widespread theme? *FEMS Microbiol. Rev.* 36, 862-876.

[173] Miyoshi, K., Mii, M., (1995). Phytohormone pre-treatment for the enhancement of seed germination and protocorm formation by the terrestrial orchid, *Calanthe discolor* (Orchidaceae), in asymbiotic culture. *Sci. Hortic. (Amsterdam)* 63, 263-267.

[174] Koshitaa, Y., Takaharaa, T., Ogataa, T., Goto, A., (1999). Involvement of endogenous plant hormones (IAA, ABA, GAs) in leaves and flower bud formation of satsuma mandarin (*Citrus unshiu* Marc.). *Sci. Hortic. (Amsterdam)* 79, 185-194.

[175] Saranga, J., Cameron, R., (2007). Adventitious root formation in *Anacardium occidentale* L. in response to phytohormones and removal of roots. *Sci. Hortic. (Amsterdam)* 111, 164-172.

[176] McAtee, P., Karim, S., Schaffer, R., David, K., (2013). A dynamic interplay between phytohormones is required for fruit development, maturation, and ripening. *Front Plant Sci.* 4, 1-7.

[177] Yang, J., Kloepper, J. W. and Ryu, C. M. (2009). Rhizosphere bacteria help plants tolerate abiotic stress. *Trends Plant Sci.* 14, 1-4.

[178] Bari, R., Jones, J. D., (2009). Role of plant hormones in plant defense responses. *Plant Mol. Biol.* 69, 473-88.

[179] Shan, X., Yan, J., Xie, D., (2012). Comparison of phytohormone signaling mechanisms. *Curr. Opin. Plant Biol.* 15, 84-91.

[180] Misra, S., Kaushik, B. D., (1989). Growth promoting substances of Cyanobacteria. I. Vitamin and their influence on rice plant. *Proc. Indian Nat. Sci. Acad. Part-B Biol. Sci.* 55, 295-300.

[181] Sergeeva, E., Liaimer, A., Bergman, B., (2002). Evidence for production of the phytohormones indole-3-acetic acid by cyanobacteria. *Planta* 215, 229-238.

[182] Ahmed, M., Stal, L. J., Hasnain, S., (2010). Association of non-heterocystous cyanobacteria with crop plants. *Plant Soil* 336, 363-375.

[183] Prasanna, R., Joshi, M., Rana, A., Nain, L., (2010). Modulation of IAA Production by Tryptophan and light. *Polish J. Microbiol.* 59, 99-105.

[184] David, A., Korasick, Tara, A., (2013). Enders and Lucia C. Strader Auxin biosynthesis and storage forms. *J. Exp. Bot.* doi:10.1093/jxb/ert080.

[185] Boopathi, T., Balamurugan, V., Gopinath, S., Sundararaman, M., (2013). Characterization of IAA Production by the mangrove cyanobacterium *Phormidium* sp. MI405019 and its influence on tobacco seed germination and organogenesis. *J. Plant Growth Regul.* doi:10.1007/s00344-013-9342-8.

[186] Hussain, A., Krischke, M., Roitsch, T., Hasnain, S., (2010). Rapid Determination of Cytokinins and Auxin in Cyanobacteria. *Curr. Microbiol.* 61, 361-369.

[187] Tsavkelova, E. A., Klimova, S. Y., Cherdyntseva, T. A., Netrusov, A. I., (2006). Hormones and hormone-like substances of microorganisms: a review. *Appl. Biochem. Microbiol.* 42, 229-235.

[188] Osman, M. E. H., El-Sheekh, M. M., El-Naggar, A. H., Gheda, S. F., (2010). Effect of two species of cyanobacteria as biofertilizers on some metabolic activities, growth, and yield of pea plant. *Biol. Fert. Soils* 46, 861-875.

[189] Ördög, V. L., Pulz, O., (1996). Diurnal changes of cytokinin-like activity in a strain of *Arthronema africanum* (Cyanobacteria), determined by bioassay. *Algol. Stud.* 82, 57-67.

[190] Gupta, A. B., Agarwal, P. R., (1973). Extraction, isolation and bioassay of a gibberellin-like substance from *Phormidium foveolarum*. *Ann. Bot.* 37, 737-741.

[191] Singh, V. P., Trehan, K., (1973). Effect of extracellular products of*Aulosira fertilissima* on the growth of rice seedlings. *Plant soil* 38, 457-464.

[192] Maršálek, B., Zahradníčková, H., Hronková, M., (1992). Extracellular Abscisic Acid Produced by Cyanobacteria under Salt Stress. *J. Plant Physiol.* 139, 506-508.

[193] Haroun, S. A., Hussein, M. H., (2003). The promotive effect of algal biofertilizers on growth, protein pattern and some metabolic activities of *Lupius termis* plants grown in siliceous soil. *Asian J. Plant Sci.* 2, 944-951.

[194] Rodgers, G. A., Bergman, B., Henriksson, U., Udris, M., (1979). Utilisation of blue green algae as biofertilisers. *Plant Soil* 52, 99-107.

[195] Gupta, A. B., Lata, K., (1964). Effect of algal growth hormones on the germination of paddy seeds. *Hydrobiologia* 24, 430-434.

[196] Gupta, A. B., Shukla, A. C., (1969). Effect of algal extracts of *Phormidium* species on growth and development of rice seedlings. *Hydrobiologia* 34, 77-84.

[197] Prasanna, R., Sharma, E., Sharma, P., Kumar, A., Kumar, R., Gupta, V., Pal, R. K., Shivay, Y. S., Nain, L., (2013). Soil fertility and establishment potential of inoculated cyanobacteria in rice crop grown under non-flooded conditions. *Paddy Water Environ.* 11, 175-183.

[198] Prasanna, R., Nain, L., Ancha, R., Srikrishna, J., Joshi, M., Kaushik, B. D., (2009b). Rhizosphere dynamics of inoculated cyanobacteria and their growth-promoting role in rice crop. *Egypt. J. Biol.* 11, 26-36.

In: Microbes in Soil and Their Agricultural Prospects
Editors: K. Kishore Choudhary and D. Wattal Dhar
ISBN: 978-1-63482-824-6

Chapter 7

METHODS FOR ANALYZING DIVERSITY OF MICROBIAL COMMUNITIES IN NATURAL ENVIRONMENTS

Md. Fakruddin*

Industrial Microbiology Laboratory, Institute of Food Science & Technology (IFST), Bangladesh Council of Scientific and Industrial Research (BCSIR), Dr. Kudrat I Khuda Road, Dhanmondi, Dhaka, Bangladesh

ABSTRACT

Difficulties in cultivating most of the microorganisms limit our ability to study natural microbial ecosystems. Molecular methods are valuable tools for investigating the diversity and structure of bacterial communities. These techniques can be used on cultivable as well as non-cultivable bacteria. Cultivation independent techniques based on nucleic acids extracted from the environment provide information on community structure and diversity. Analyzing of DNA can determine the numbers of different genomes. Ribosomal RNA (rRNA) or rDNA (genes coding for rRNA) fingerprinting, probing and sequencing can be used to detect and identify organisms. The combination of different methods that complemented each other is a useful strategy for monitoring changes in microbial communities and ecosystems.

Keywords: community, microbial diversity, biochemical methods, molecular methods, PLFA analysis, rRNA

INTRODUCTION

Our knowledge about bacteria in natural environments is limited, and studying microbial diversity in nature is not an easy task. There are probably several thousand species that have

---

* Corresponding author: Industrial Microbiology Laboratory, Institute of Food Science & Technology (IFST), Bangladesh. E mail: fakruddinmurad@gmail.com.

not yet been described. Furthermore the term species is also vague. One gram of soil or sediment may contain 10^{10} bacteria as counted in fluorescent microscopy after staining with a fluorescent dye. In pure sea water the number of bacteria is approximately 10^6 per milliliter [1]. There are bacteria that are adapted to almost all the different environments that exist on the earth, and also bacteria that are able to decompose all the chemical components made by living organisms. Important questions to be addressed when studying bacteria in their natural environment are how do the bacterial communities function, and how the qualitative variation on community composition due to environmental changes is [2]. To answer these questions, we need to put more effort into the basic knowledge about the community structure. The total bacterial community studied exhibited a tremendous amount of genetic information and therefore a very high genetic diversity [3].

Significance of Microbial Community Analysis for Sustainable Agriculture

It has been well established that soil microbial community has strong correlative relation with plant as well as agricultural practice. Microbial community of soil is believed to be involved in plant disease suppression & maintenance of soil health through their involvement in numerous soil functions [4]. A range of specific soil microorganisms are playing an important role in the suppression of soil-borne plant diseases as well as in plant growth promotion [5].

Soil microbial communities are among the most difficult to study and fully characterize, because of their immense phenotypic and genotypic diversity [6]. The bacterial population in soil top layers can rise up to more than 10^9 cells per gram soil and has been postulated to represents the world's greatest reservoir of biological diversity [2].

Manipulating soil microbial communities using soil and crop management practices is a basic strategy in developing sustainable agricultural systems [7]. There is growing concern that systems with reduced microbial diversity due to intensive soil management practices may suffer from impaired functioning in respect of the sustainability of microbial nutrient retention and plant disease suppression [8].

By directly accessing the microbial diversity of soil, the level of disease suppressiveness can be manipulated. Management of the resident soil microbial community holds promise as a means to diminish the activity of soil-borne plant pathogens and improve soil health [9]. Understanding of agro-ecosystems and the influence of management practices on soil microbial functional groups is key to determining effective farming system [4].

Advances in microbial research and ill effects of application of chemical inorganic fertilizer on agriculture product has challenged the current agricultural practices and are looking towards the biological sources particularly microbial resources as supplementary or complementary sources of fertilizer. The current article describes the different approaches of analysis/enumeration of microbial diversity.

THE CONCEPT OF MICROBIAL DIVERSITY

Biodiversity has been defined as the range of significantly different types of organisms and their relative abundance in an assemblage or community. The diversity has also been defined according to information theory, as the amount and distribution of information in an assemblage or community [3]. Microbial diversity refers unequivocally to biological diversity at three levels: within species (genetic), species number (species) and community (ecological) diversity [10]. The term species diversity consists of two components; the first component is the total number of species present which can be referred to as species richness. In other words it refers to the quantitative variation among species. The second component is the distribution of individuals among these species, which is referred to as evenness or equability (J). The diversity concept is hence most commonly used to describe both the information contents and how this information is distributed among a collection of microorganisms. One problem is that evenness often is unknown in bacterial systems because individual cells very seldom are identified to the species level. An attractive possibility for the measurement of biodiversity is to use divergence in molecular characters, especially the percentage of either nucleic acid homology or base sequence difference. In the past, diversity has been determined based on taxonomic species, which may limit the scope of information and relationship obtained. The diversity of Operational Taxonomic Unit (OTU) or even communities may give us a better estimation of the functioning of an ecosystem. Diversity studies can be used to retrieve ecological information about community structures. Species diversity is a community parameter that relates to the degree of stability of that community. Essentially, any diversity index must measure the heterogeneity of information stored within the community. Well-organized communities that contain a certain level of diversity are stable [11]. If some kind of stress is introduced to this community, the stability may collapse and the diversity will change. Diversity studies can therefore be used to monitor successions and effect of perturbations.

FUNDAMENTAL REASONS FOR DIVERSITY STUDIES

Within natural microbial populations a large amount of genetic information is "waiting" to be discovered. Culturable bacteria represent a minor fraction of the total bacterial population present [12]. Therefore it is important to continue the work both on the culturable as well as the non-culturable bacteria from different environments. Diversity studies are also important due to practical reasons, which means, that we can compare one sample to another.

The capability that one ecosystem has to resist extreme perturbations or stress conditions, can partly be dependent of the diversity within this system. Diversity analyses are therefore important in order to:

- Increase the knowledge of the diversity of genetic resources and understand the distribution of organisms
- Increase the knowledge of the functional role of diversity
- Identify differences in diversity associated with management disturbing

- Understand the regulation of biodiversity
- Understand the consequence of biodiversity. (To what extent does ecosystem functioning and sustainability, depend on maintaining a specific level of diversity)

One of the most compelling arguments for studying diversity is that so little is known about what exists and what is being lost. There is no consensus on how many species exist in the world, the potential usefulness of most of them, or the rate at which they are disappearing or emerging.

Factors Controlling Microbial Diversity

In a bacterial community many different organisms will perform the same processes and probably be found in the same niche [13]. Factors that affect microbial diversity can be classified in two groups- Abiotic factors and Biotic factors.

Abiotic factors include both physical and chemical factors such as water availability, salinity, oxic/anoxic conditions, temperature, pH, pressure, chemical pollution, heavy metals, pesticides, antibiotics etc. [14]. In general, all changes in the environmental variation affect different ways and to different degrees, resulting in a shift in the diversity profile.

Biotic factors include plasmids, phages, transposons that are types of accessory DNA that influences the genetic properties and in most cases, the phenotypes of their host and thus have great influence on microbial diversity [13]. Protozoa also reported to influence microbial diversity [15].

Methods for Describing Diversity

Since a minor fraction of the bacterial community is culturable, only a limited part of the bacteria has been fully characterized and given names. Prokaryotic organisms are difficult to classify, and the validity of the classification has been questioned very often. The morphological characteristics (cell shape, cell wall, movement, flagella, Gram staining, etc.) are so few that they alone are not fundamental for establishing a detailed classification. Advances in molecular and chemical ecology that provide an estimate of microbial diversity without having to isolate the organisms are promising [12]. Methods to measure microbial diversity in soil can be categorized into two groups: biochemical techniques (Table 1) and molecular techniques (Table 2) (Figure 1).

Biochemical Methods

In order to distinguish different types of microbes, microbiologists early turned to metabolic properties such as utilization of different carbon, nitrogen and energy sources in addition to requirement of growth factors. The diversity can be described using physiological diversity measures, which avoid the sometimes difficult grouping of similar bacteria into species or equivalents. These measures also include various indices (tolerance, nutrition etc.), but also multivariate data analyses have been used for extracting relevant information in the large data-sets frequently obtained in diversity studies [17]. The phylogenetic distributions of different types of carbon and energy metabolism among different organisms do not necessarily follow the evolutionary pattern of rRNA.

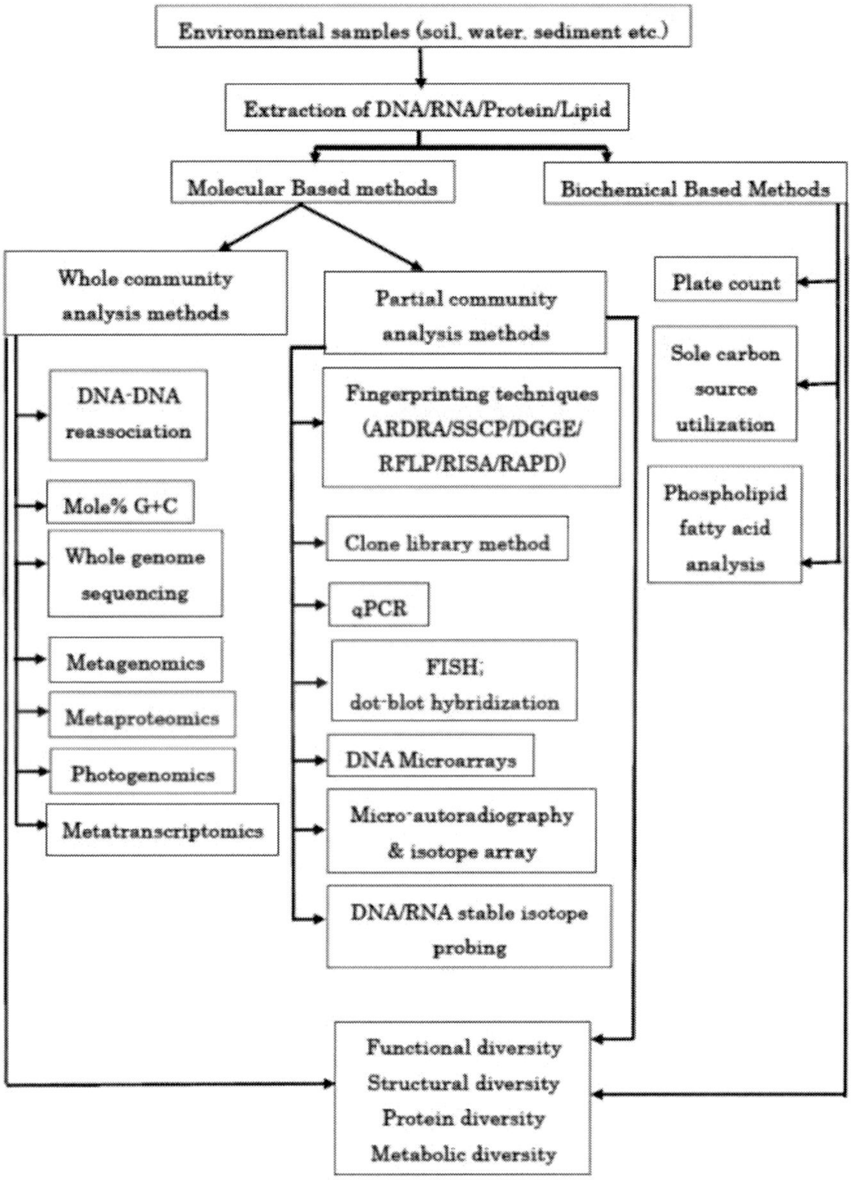

Figure 1. Summary of methods to characterize the structural and functional diversity of microorganisms in the environment.

Table 1. Advantages and disadvantages of biochemical methods to study microbial diversity [14]

Method	Advantages	Disadvantages
Plate counts	1. Fast 2. Inexpensive	1. Unculturable microorganisms not detected 2. Bias towards fast growing individuals 3. Bias towards fungal species that produce large quantities of spores
Community level physiological profiling (CLPP)/ Sole-carbon-source Utilization (SCSU) Pattern	1. Fast 2. Highly reproducible 3. Relatively inexpensive 4. Differentiate between microbial communities 5. Generates large amount of data 6. Option of using bacterial, fungal plates or site specific carbon sources (Biolog)	1. Only represents culturable fraction of community 2. Favours fast growing organisms 3. Only represents those organisms capable of utilizing available carbon sources 4. Potential metabolic diversity, not in situ diversity 5. Sensitive to inoculum density
Fatty acid methyl ester analysis (FAME)/ Phospholipid fatty acid (PLFA) analysis	1. No culturing of microorganisms, direct extraction from soil 2. Follow specific organisms or communities	1. If using fungal spores, a lot of material is needed 2. Can be influenced by external factors 3. Possibility results can be confounded by other microorganisms

Table 2. Advantages and disadvantages of some molecular-based methods to study soil microbial diversity [14]

Method	Advantages	Disadvantages
Guanine plus cytosine (G+C)	1. Not influenced by PCR biases 2. Includes all DNA extracted 3. Quantitative 4. Includes rare members of community	1. Requires large quantities of DNA 2. Dependent on lysing and extraction efficiency 3. Coarse level of resolution
Nucleic acid reassociation and hybridization	1. Total DNA extracted 2. Not influenced by PCR biases 3. Study DNA or RNA 4. Can be studied in situ	1. Lack of sensitivity 2. Sequences need to be in high copy number to be detected 3. Dependent on lysing and extraction efficiency
DNA microarrays and DNA hybridization	1. Same as nucleic acid hybridization 2. Thousands of genes can be analyzed 3. If using genes or DNA fragments, increased specificity	1. Only detect most abundant species 2. Need to be able to culture organisms 3. Only accurate in low diversity systems

Method	Advantages	Disadvantages
Denaturing and temperature gradient gel electrophoresis (DGGE and TGGE)	1. Large number of samples can be analyzed simultaneously 2. Reliable, reproducible and rapid	1. PCR biases 2. Dependent on lysing and extraction efficiency 3. Sample handling can influence community, i.e., if stored too long before extraction, community can change 4. One band can represent more than one species (co-migration) 5. Only detects dominant species
Single strand conformation polymorphism (SSCP)	1. Same as DGGE/TGGE 2. No GC clamp 3. No gradient	1. PCR biases 2. Some ssDNA can form more than one stable conformation
Restriction fragment length polymorphism (RFLP)	1. Detect structural changes in microbial community	1. PCR biases 2. Banding patterns often too complex
Terminal restriction fragment length polymorphism (T-RFLP)	1. Simpler banding patterns than RFLP 2. Can be automated; 3. large number of samples 4. Highly reproducible 5. Compare differences in microbial communities	1. Dependent on extraction and lysing efficiency 2. PCR biases 3. Type of Taq can increase variability 4. Choice of universal primers 5. Choice of restriction enzymes will influence community fingerprint
Ribosomal intergenic spacer analysis (RISA)/ automated ribosomal intergenic spacer analysis (ARISA)/ Amplified ribosomal DNA restriction analysis (ARDRA)	1. Highly reproducible community profiles	1. Requires large quantities of DNA (for RISA) 2. PCR biases

PLATE COUNTS

Traditionally, diversity was assessed using selective plating and direct viable counts. These methods are fast, inexpensive and can provide information on the active, heterotrophic component of the population. Limitations include the difficulty in dislodging bacteria or spores from soil particles or biofilms, growth medium selections [18], growth conditions (temperature, pH, light), the inability to culture a large number of bacterial and fungal species with current techniques and the potential for colony–colony inhibition or of colony spreading [19]. In addition, plate growth favours those microorganisms with fast growth rates and those fungi that produce large numbers of spores [20]. All of these limitations can influence the apparent diversity of the microbial community

Sole-Carbon-Source Utilization (SCSU)

Garland and Mills [21] developed a technique using a commercially available 96-well microtitre plate to assess the potential functional diversity of the bacterial population through SCSU patterns. It is also known as Community level physiological profiling (CLPP). The SCSU system (for example biochemical identification systems- API and Biolog) was initially developed as a tool for identifying pure cultures of bacteria to the species level, based upon a broad survey of their metabolic properties. SCSU examines the functional capabilities of the microbial population, and the resulting data can be analyzed using multivariate techniques to compare metabolic capabilities of communities [22]. A limitation of this application is that microbial communities often are composed of both fast and slow growing organisms. The slow growing organisms will thus not be included in the analysis. Growth on secondary metabolites may also occur during incubation. If progress is to be made in understanding patterns of microbial diversity we need to broaden our understanding of biodiversity and focus initially on approaches that have a reasonable change of deducing patterns that are biologically relevant. A multifaceted approach that includes both functional and taxonomic perspectives represents fertile grounds for future research. A limitation of this methodology is that many of the commercial available kits for measuring physiological diversity have been designed to cover the spectra of human pathogenic bacteria (API and Biolog). Few works are reported in order to optimize substrate combinations designed for environmental isolates [23]. This often leads to problems when identifying the isolates based on the available database. This method has been used successfully to assess potential metabolic diversity of microbial communities in contaminated sites [24], plant rhizospheres [25], arctic soils [26], soil treated with herbicides [27] or inoculation of microorganisms [28].

CLPPs can differentiate between microbial communities, are relatively easy to use, reproducible and produce a large amount of data reflecting metabolic characteristics of the communities [29]. Limitations of metabolic profiling are: the methods select for only culturable microorganisms capable of growing under the experimental conditions [21], favours fast growing microorganisms [30], is sensitive to inoculum density [31] and reflects the potential, and not the in situ, metabolic diversity [21]. In addition, the carbon sources may not be representative of those present in soil [30] and therefore the usefulness of the information can be questioned.

Phospholipid Fatty Acid (PLFA) Analysis

The fatty acid composition of microorganisms has been used extensively to aid microbial characterization. Taxonomically, fatty acids in the range C2 to C24 have provided the greatest information and are present across a diverse range of microorganisms [32]. The fatty acid composition is stable, and is independent of plasmids, mutations or damaged cells. The method is quantitative, cheap, robust and with high reproducibility. However it is important to notice the bacterial growth conditions are reflected in the fatty acid pattern. This method is also known as fatty acid methyl ester (FAME) analysis.

One way to examine the entire microbial community structure is to analyze the Phospholipid fatty acid (PLFA) compositions of the environment since different subsets of a community have different PLFA patterns [33]. It is usually not possible to detect individual strains or species of microorganisms with this method, but changes in the overall compositions of the community can be detected instead. Lipid analysis offers therefore an alternative method for the quantification of community structure that does not rely upon cultivation of microorganisms and is free of potential selections. It does not have the specificity to identify the members of microbial populations to species, rather the method produces descriptions of microbial communities based on functional group affinities [34]. Lipid has been the most often used signature components for determining the community composition of microorganisms in ecological studies [33]. Changes in such lipid profiles may be attributable to alterations in the physiological status of extant populations or to actual shifts in community structure. The estimation of such "signatures" may provide valuable insight to community structure, its nutritional status and activity.

Although FAME analysis is one method to study microbial diversity, if using total organisms, fatty acid analysis is a poor method fraught with limitations. This may obscure detection of minor species in the population. Cellular fatty acid composition can be influenced by factors such as temperature and nutrition, and the possibility exists that other organisms can confound the FAME profiles [35]. In addition, individual fatty acids cannot be used to represent specific species because individuals can have numerous fatty acids and the same fatty acids can occur in more than one species [4].

Molecular Methods to Study Microbial Diversity

Traditional methods for characterizing microbial communities have been based on analysis of the culturable portion of the bacteria. Due to the non-culturability of the major fraction of bacteria from natural microbial communities, the overall structure of the community has been difficult to interpret [36]. Recent studies have focused on the use of methods to characterize diversity that does not require cultivation, yet provide measures of diversity based on genetic diversity. The molecular-phylogenetic perspective is a reference framework within which to describe microbial diversity; the sequences of genes can be used to identify organisms [37]. A number of approaches have been developed to study molecular microbial diversity. These include DNA reassociation, DNA–DNA and mRNA-DNA hybridization, DNA cloning and sequencing, and other PCR-based methods such as denaturing gradient gel electrophoresis (DGGE), temperature gradient gel electrophoresis (TGGE), ribosomal intergenic spacer analysis (RISA) and automated ribosomal intergenic spacer analysis (ARISA).

Mole Percentage Guanine + Cytosine (Mol% G+C)

The first property of DNA used for taxonomical purpose was the base composition expressed as mole percentage guanine + cytosine (mol% G+C). Within bacteria this value is ranging from 25% up to 75%, though a value is constant for a certain organism. Closely related organisms have fairly similar GC profiles and taxonomically related groups only differ between 3% and 5% [38]. However, similar base composition is not a confirmation of relationship. On the other hand, if there is a difference in base composition this is a worthy evidence of missing relationship. Mol% G+C can be determined by thermal denaturation of DNA. Advantages of G+C analysis are that it is not influenced by PCR biases, it includes all DNA extracted, it is quantitative and it can uncover rare members in the microbial populations. It does, however, require large quantities of DNA (up to 50 μg) [38].

Nucleic Acid Hybridization

Nucleic acid hybridization using specific probes is an important qualitative and quantitative tool in molecular bacterial ecology [39]. These hybridization techniques can be done on extracted DNA or RNA, or *in situ*. Oligonucleotide or polynucleotide probes designed from known sequences ranging in specificity from domain to species can be tagged with markers at the 5'-end [40].

The sample is lysed to release all nucleic acids. rRNA sequences of interest are quantified relative to total rRNA by dot-blot hybridization with specific and universal oligonucleotide primers. The relative abundance may represent changes in the abundance in the population or changes in the activity and hence the amount of rRNA content [41]. Hybridization can also be conducted at the cellular level and can be done *in situ*. This provides valuable spatial distribution information on microorganisms in environmental samples.

The method, known as fluorescent in situ hybridization or FISH has been used successfully to study the spatial distribution of bacteria in biofilms [42]. One limitation of in situ hybridization or hybridization of nucleic acids extracted directly from environmental samples is the lack of sensitivity. Unless sequences are present in high copy number, i.e., from dominant species, they probably will not be detected.

DNA Reassociation

DNA reassociation is a measure of genetic complexity of the microbial community and has been used to estimate diversity [43]. Total DNA is extracted from environmental samples, purified, denatured and allowed to reanneal. The rate of hybridization or reassociation will depend on the similarity of sequences present. As the complexity or diversity of DNA sequences increases, the rate at which DNA reassociates will decrease [41]. Under specific conditions, the time needed for half of the DNA to reassociate (the half association value Cot1/2) can be used as a diversity index, as it takes into account both the amount and distribution of DNA reassociation [3]. Alternatively, the similarity between communities of two different samples can be studied by measuring the degree of similarity of DNA through hybridization kinetics [44].

Restriction Fragment Length Polymorphism (RFLP)

Restriction fragment length polymorphism (RFLP) is another tool used to study microbial diversity that relies on DNA polymorphisms. In this method, electrophoresed digests are blotted from agarose gels on to nitro-cellulose or nylon membranes and hybridized with appropriate probes prepared from cloned DNA segments of related organisms. RFLP has been found to be very useful particularly in combination with DNA-DNA hybridization and enzyme electrophoresis for the differentiation of closely related strains [45], and the approach seems to be useful for determination of intra species variation [46]. RFLPs may provide a simple and powerful tool for the identification of bacterial strains at and below species level. This method is useful for detecting structural changes in microbial communities but not as a measure of diversity or detection of specific phylogenetic groups [47]. Banding patterns in diverse communities become too complex to analyze using RFLP since a single species could have four to six restriction fragments [38].

However one should be aware that a similar banding pattern does not necessarily indicate a very close relationship between compared organisms. In the last couple of years RFLP applications have also been applied to estimate diversity and community structure in different microbial communities [48].

Terminal Restriction Fragment Length Polymorphism (T-RFLP)

Terminal restriction fragment length polymorphism (T-RFLP) is a technique that addresses some of the limitations of RFLP [49]. This technique is an extension of the RFLP/ ARDRA analysis, and provides an alternate method for rapid analysis of microbial community diversity in various environments. It follows the same principle as RFLP except that one PCR primer is labeled with a fluorescent dye, such as TET (4, 7, 2', 7'-tetrachloro-6-carboxyfluorescein) or 6-FAM (phosphoramidite fluorochrome 5-carboxyfluorescein). PCR is performed on sample DNA using universal l6S rDNA primers, one of which is fluorescently labeled. FLT-RFLP patterns can then be created by digestion of labeled amplicons using restriction enzymes. Fragments are then separated by gel electrophoresis using an automated sequence analyzer. Each unique fragment length can be counted as an Operational Taxonomic Unit (OTU), and the frequency of each OTU can be calculated. The banding pattern can be used to measure species richness and evenness as well as similarities between samples [47]. T-RFLP, like any PCR-based method, may underestimate true diversity because only numerically dominant species are detected because of the large quantity of available template DNA [47]. Incomplete digestion by restriction enzymes could also lead to an overestimation of diversity [50]. Despite these limitations, some researchers are of the opinion that once standardized, T-RFLP can be a useful tool to study microbial diversity in the environment [38], while others feel that it is inadequate [51].

T-RFLP is limited not only by DNA extraction and PCR biases, but also by the choice of universal primers. None of the presently available universal primers can amplify all sequences from eukaryote, bacterial and archaeal domains. Additionally, these primers are based on existing 16S rRNA, 18S rRNA or ITS databases, which until recently contained mainly sequences from culturable microorganisms, and therefore may not be representative of the true microbial diversity in a sample [52]. In addition, different enzymes will produce different community fingerprints [51].

T-RFLP has also been thought to be an excellent tool with which to compare the relationship between different samples [51]. T-RFLP has been used to measure spatial and temporal changes in bacterial communities [53], to study complex bacterial communities [54], to detect and monitor populations [38] and to assess the diversity of *arbuscular mycorrhizal fungi* (AMF) in the rhizosphere of *Viola calaminaria* in a metal-contaminated soil [55]. Tiedje et al. [38] reported five times greater success at detecting and tracking specific ribotypes using T-RFLP than DGGE.

Ribosomal Intergenic Spacer Analysis (RISA)/ Automated Ribosomal Intergenic Spacer Analysis (ARISA) /Amplified Ribosomal DNA Restriction Analysis (ARDRA)

Similar in principle to RFLP and T-RFLP, RISA, ARISA and ARDRA provide ribosomal-based fingerprinting of the microbial community. In RISA and ARISA, the intergenic spacer (IGS) region between the 16S and 23S ribosomal subunits is amplified by PCR, denatured and separated on a polyacrlyamide gel under denaturing conditions. This region may encode tRNAs and is useful for differentiating between bacterial strains and closely related species because of heterogeneity of the IGS length and sequence [56]. In RISA, the sequence polymorphisms are detected using silver stain while, in ARISA the forward primer is fluorescently labeled and is automatically detected [56]. Both methods provide highly reproducible bacterial community profiles but RISA, requires large quantities of DNA, is more time consuming, silver staining is somewhat insensitive and resolution tends to be low [56]. ARISA increases the sensitivity of the method and reduces the time but is still subject to the traditional limitations of PCR [56]. RISA has been used to compare microbial diversity in soil [57], in the rhizosphere of plants [57], in contaminated soil [58] and in response to inoculation [59].

DNA Microarrays

More recently, DNA–DNA hybridization has been used together with DNA microarrays to detect and identify bacterial species [60] or to assess microbial diversity [61]. This tool could be valuable in bacterial diversity studies since a single array can contain thousands of DNA sequences [62] with high specificity. The microarray can either contain specific target genes such as nitrate reductase, nitrogenase or naphthalene dioxygenase to provide functional diversity information or can contain a sample of environmental ''standards'' (DNA fragments with less than 70% hybridization) representing different species found in the environmental sample [61]. Reverse sample genome probing (RSGP) is a method used to analyze microbial community composition of the most dominant culturable species and uses genome microarrays. RSGP has four steps: (1) isolation of genomic DNA from pure cultures; (2) cross-hybridization testing to obtain DNA fragments with less than 70% cross-hybridization. DNA fragments with greater than 70% cross-hybridization are considered the same species. (3) Preparation of genome arrays onto a solid support; and (4) random labelling of a defined mixture of total community DNA and internal standard [61]. This method has been used to analyze microbial communities in oil fields, and in contaminated soils [63].

Like DNA–DNA hybridization, the use of RSGP and microarrays has the advantage that it is not confounded by PCR biases and microarrays can contain thousands of target gene sequences. However, it only detects the most abundant species. In general, the species need to be cultured, but in principle cloned DNA fragments of unculturables could be used. The diversity has to be minimal or enriched cultures used, otherwise cross-hybridization can become problematic. Using genes or DNA fragments instead of genomes on the microarray offers the advantages of eliminating the need to keep cultures of organisms growing as genes can be cloned into plasmids or PCR used to continually amplify the DNA fragments [64]. In addition, fragments would increase the specificity of hybridization over the use of genomes and functional genes in the community could be assessed [61].

Denaturant Gradient Gel Electrophoresis (DGGE) / Temperature Gradient Gel Electrophoresis (TGGE)

In denaturing gradient gel electrophoresis (DGGE) or temperature gradient gel electrophoresis (TGGE) DNA fragments of same length but with different base-pair sequences can be separated. DNA is extracted from natural samples and amplified using PCR with universal primers targeting part of the 16S or 18S rRNA sequences. The separation is based on the difference in mobility of partially melted DNA molecules in acrylamide gels containing a linear gradient of DNA denaturants (urea and formamide). Sequence variation within the DNA fragments causes a difference in melting behavior, and hence in a separation in denaturing gradient gels. The melting of the products occurs in different melting domains, which is stretches of nucleotides with identical melting temperatures [65].

Sequence variations in different fragments will therefore terminate migration at different positions in the gel according to the concentration of the denaturant [66]. Theoretically, DGGE can separate DNA with one base-pair difference [67]. TGGE uses the same principle as DGGE except the gradient is temperature rather than chemical denaturants.

DGGE/TGGE has the advantages of being reliable, reproducible, rapid and somewhat inexpensive. Multiple samples can also be analyzed concurrently, making it possible to follow changes in microbial populations [68]. Limitations of DGGE/ TGGE include PCR biases [69], laborious sample handling, as this could potentially influence the microbial community [66], and variable DNA extraction efficiency [41]. It is estimated that DGGE can only detect 1–2% of the microbial population representing dominant species present in an environmental sample [70]. In addition, DNA fragments of different sequences may have similar mobility characteristics in the polyacrylamide gel. Therefore, one band may not necessarily represent one species [71] and one bacterial species may also give rise to multiple bands because of multiple 16S rRNA genes with slightly different sequences [72].

DGGE profiles have been used successfully to determine the genetic diversity of microbial communities inhabiting different temperature regions in a microbial mat community [73], and to study the distribution of sulphate reducing bacteria in a stratified water column [74].

Single Strand Conformation Polymorphism (SSCP)

Another technique that relies on electrophoretic separation based on differences in DNA sequences is single strand conformation polymorphism (SSCP). Like DGGE/TGGE, this technique was originally developed to detect known or novel polymorphisms or point mutations in DNA [75]. Single-stranded DNA is separated on a polyacrylamide gel based on differences in mobility caused by their folded secondary structure [76]. When DNA fragments are of equal size and no denaturant is present, folding and hence mobility will be dependent on the DNA sequences. SSCP has all the same limitations of DGGE. Also, some single-stranded DNA can form more than one stable conformation. Therefore, one sequence may be represented by more than one band on the gel [38]. However, it does not require a GC clamp or the construction of gradient gels and has been used to study bacterial or fungal community diversity [77]. SSCP has been used to measure succession of bacterial communities [75], rhizosphere communities [78], bacterial population changes in an anaerobic bioreactor [79] and AMF species in roots [80].

Other Potential Molecular Methods

Other molecular methods that have the potential to be used as equally applicable as the above mentioned methods are Fluorescent In situ Hybridization (FISH) [36], Stable isotope probing [81], qPCR [82], Microautoradiography [83], DNA sequencing based community analysis such as Pyrosequencing based community analysis [84, 85], Illumina-based High throughput microbial community analysis [86, 87] etc. Though most of these methods are not as applicable as previously mentioned methods, they pose the potential to be method of choice in future.

Fluorescence in Situ Hybridization (FISH)

Fluorescent in situ hybridization is a powerful toll to detect and quantify specific microorganisms while maintaining their morphological integrity, without nucleic acid extraction and to follow the dynamics of bacterial populations [88]. FISH can be used to visualize microorganisms that have not yet been cultured, and, are useful in studying the ecological distribution of microorganisms throughout diverse habitats [89].

In FISH, whole cells are fixed, their 16S or 23S rRNA is hybridized with fluorescently-labeled taxon-specific oligonucleotide probes, and then the labeled cells are viewed by scanning confocal laser microscopy (SCLM). The FISH probes are generally 18–30 nucleotides long and contain a fluorescent dye at the 5' end that allows detection of probe bound to cellular rRNA by epifluorescence microscopy. FISH can be combined with flow cytometry for a high resolution automated analysis of mixed microbial populations.

FISH has several limitations such as low signal intensity, background fluorescence, target inaccessibility etc. FISH has also limited use in in situ quantification of cells with irregular shapes [37]. Advantages of FISH include ability to detect microorganism across all phylogenetic levels, more sensitivity than immunofluorescence, reduced non-specific probe binding, avoidance of artifacts arising from biases in DNA extraction, PCR amplification and cloning [90].

STABLE ISOTOPE PROBING

SIP is a combination of isotope labeling with the molecular biological approach and is used to identify microorganism in environmental samples and to examine microbial functions in various environmental systems. It can provide abundant information about microbial interactions and metabolic functions in complex communities. SIP has been applied in the functional analyses of methylotrphs, bacteria of organic pollutants degradation, rhizosphere-microorganisms ecology, syntrophic microorganism and metagenomics [81].

Environmental samples in situ or in microcosm are exposed to substrates labeled with stable isotopes (e.g., ^{13}C, ^{15}N) and microorganism can metabolize the stable isotope-enriched substrates as their carbon or nitrogen source for growth. The stable isotope assimilated by these microorganisms is then used to synthesize cellular component such as nucleic acids (DNA or RNA) and phospholipid fatty acids (PLFA) (Figure 2). By extracting and analyzing these stable isotope-labeled biomarkers, microbial identity and functions in a community can be determined [6].

Nucleic acid based SIP methods are subject to the various weaknesses of molecular methods, such as PCR bias and difficulties in recovering suitable quality nucleic acids from soil samples. In addition, timing of nucleic acid extraction relative to introduction of label must be experimentally determined [91]0.. The disadvantages of SIP include possible biases caused by the incubation with the isotope and the cycling of the stable isotope within the microbial community. Identification based on PFLA analysis is limited to previously cultivated organisms for which a PFLA database or authentic standard is available. PLFAs are much more amenable to strictly instrument analysis than nucleic acids, thus targeting this pool is not only more sensitive, but less labor intensive, and more compatible with laboratories outfitted primarily for process measurement rather than molecular ecology.

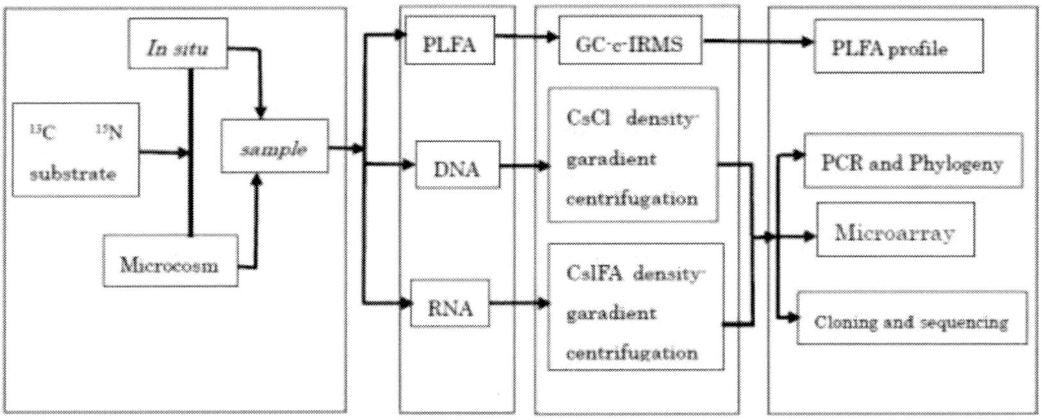

Figure 2. Overview of stable isotope probing [94].

In recent years, with advances in imaging and spectroscopic techniques, SIP has been combined with other techniques such as FISH and Raman microscopy to simultaneously investigate the taxonomic identities and activity of microbial communities at single-cell resolution [92]. The Raman–FISH provides much higher resolution and overcomes many of the limitations associated with conventional SIP/MAR–FISH techniques. One of the primary advantages of RNA-SIP over DNA-SIP is the potentially short incubation times possible (RNA synthesis occurs at a faster rate). However, for soil incubations, extremely short incubation times may result in problems with insufficient substrate distribution, particularly in unsaturated soils [93].

One fruitful area for application of SIP to soils research may be examining whether agricultural inputs (pesticides, fertilizers) impact which organisms are responsible for particular functions and if there is potential to influence these function through such changes in microbiology. SIP might also be used to help gain insight into the sole of microorganisms in soil fertility and how that role might be measured for predicting nutrient availability to plants.

QUANTITATIVE PCR

Quantitative PCR (Q-PCR), or real-time PCR, has been used in microbial investigations to measure the abundance and expression of taxonomic and functional gene markers [95]. q-PCR has been utilized to examine total microbial communities and the relative proportions of specific phylotypes within a number of unique environments. q-PCR can address metabolic potential of the microbial biomass by exploring specific biological functions utilizing probes targeting functional genes. Soil bacterial and fungal microbial communities can be determined quantitatively by qPCR with primers targeting 16S and 5.8S rRNA [96]. Q-PCR has also been successfully used in environmental samples for quantitative detection of important physiological groups of bacteria such as ammonia oxidizers, methane oxidizers, and sulfate reducers [97].

Unlike traditional PCR, which relies on end-point detection of amplified genes, Q-PCR uses either intercalating fluorescent dyes such as SYBR Green or fluorescent probes (TaqMan) to measure the accumulation of amplicons in real time during each cycle of the PCR. Specialized thermal cyclers equipped with fluorescence detection modules are used to monitor the fluorescence as amplification occurs. Software records the increase in amplicon concentration during the early exponential phase of amplification which enables the quantification of genes (or transcripts) when they are proportional to the starting template concentration. The cycle number at which enough amplified products accumulate to get a detectable signal is defined as the threshold cycle (Ct). Ct value of a reaction is dependent on the amount of the template present at the start of the reaction, hence ct value is inversely proportional to the amount of template [82].

Q-PCR is highly sensitive to starting template concentration and measures template abundance in a large dynamic range of around six orders of magnitude. q-PCR provides an extremely accurate quantification but is limited by the number of samples that can be directly tested as specific probes are designed for each bacteria of interest. q-PCR is limited in that it requires extremely accurate controls for inferring cell mass or gene copy. It is advantageous when highly quantitative results on specific microbes or particular functional potential of the community needs to be studied and prior knowledge of the sample is necessary for meaningful probes [98]. Careful attention needs to be developed in producing secondary microbial specific assays to allow for creation of standard curves required for quantification. Additionally, qPCR obviates the use of gel electrophoresis reducing time with increasing throughput. Post-amplification modification and chance of contamination also reduced [99].

Microautoradiography

Microbial assemblages are supplied with a radiolabeled substrate, then the cells are put in contact with an autoradiographic emulsion (e.g., radiation-sensitive silver halide emulsion). After exposure of the emulsion to their radioactive emissions, silver grains deposit around the cells that can be observed by transmission electron microscopy [100]. Commonly used radiolabeled substrates include glucose, acetate, and amino acids, which provide a general view of the overall metabolic diversity. More specific substrates along with selective growth (incubation) conditions can be used to identify important physiological processes in situ.

Radioisotopes are greatly preferable to other ways of labeling such as staining or use of fluorescence elements because their chemical and physical properties largely coincide with those of the natural isotopes of the same elements and molecules they compose. In addition, the rate of radioactive disintegration is a purely statistical process and does not depend on physical or chemical conditions such as temperature, pressure, or pH. Moreover, radioisotopes can be detected at quantities less than 10^{-10}-10^{-15} milligrams by counting radioactive disintegrations [101]. Drawbacks of this technique include the potentially hazardous work with radioactivity, and the limited range of available radiolabeled substrates. Moreover it is possible that the added substrates are degraded during the incubation, leading to labeling of organisms that did not take up the original molecule but metabolites derived from it.

When microautoradiography (MAR) is used in combination with FISH (MAR–FISH), it allows simultaneous phylogenetic identification of active cells that consume the radioactive substrate [102]. MAR–FISH has been modified to other methods such as STAR (*s*ubstrate *t*racking *a*uto*r*adiography)–FISH [103] and MICRO -FISH (Microautoradiography - fluorescence in situ hybridization [83]. MAR-FISH offers powerful insight into the microbial community structure and their functional activity. MAR-FISH were successfully used to quantify phenanthrene- and naphthalene-degrading Proteobacteria and Actinobacteria in a coal-tar impacted aquifer sediments [104] and to study the autotrophic nitrifying bacteria in biofilms [105].

Next Generation Sequencing Based Community Analysis

With the emergence of next-generation sequencing (NGS) technologies such as pyrosequencing, Illumina-based sequencing etc., the possibility of discovering new groups of microorganism in complex environmental systems without cultivated strains has been accrued and these real-time sequencing techniques are shedding light into the complexities of microbial populations [106]. Using NGS it is possible to resolve highly complex microbiota composition with greater accuracy as well as to link microbial community diversity with niche function. Next-generation sequencing strategies involve high throughput sequencing and can effectively provide deep insights in complex microbial communities in ecological niches [107]. Pyrosequencing, developed by Roche 454 Life Science, is one such example and being a high-throughput sequencing technique which can generate a huge amount of DNA reads [108]. Recently, it has been successfully applied in dissecting complex microbial environments such as the human gastrointestinal tract, soil, wastewater and marine sediments [109]. Pyrosequencing has provided a means to elucidate microbial members of the rare biosphere which occur in relatively low abundances. Besides eliminating the use of cloning vectors and library construction, and their associated biases, pyrosequencing can also read through secondary structures and produce vast amount of sequences of up to 100Mb per run [110]. In addition to the sequencing technology itself, various bioinformatics tools have emerged to process and analyze pyrosequenced raw data in silico to generate meaningful information. Software such as the Newbler Assembler and RDP Pyrosequencing Pipeline provides a systematic way of analyzing data to rapidly gain insights into the complex microbial composition and structure in environmental samples [111, 112].

Metagenomic Analysis of Microbial Communities

Metagenomics is defined as the functional and sequence-based analysis of the collective microbial genomes that are contained in an environmental sample [113]. In metagenomics, the collective genome (*metagenome* or *microbiome*) of coexisting microbes – called microbial *communities* [114] is randomly sampled from the environment and subsequently sequenced [115]. By directly accessing the collective genome of co-occurring microbes, metagenomics has the potential to give a comprehensive view of the genetic diversity, species composition, evolution, and interactions with the environment of natural microbial communities [116]. Community genomic datasets can also enable subsequent gene expression and proteomic studies to determine how resources are invested and functions are distributed among community members (figure 3). Ultimately, genomics can reveal how individual species and strains contribute to the net activity of the community [117].

Community Omics Analyzing Methods

Community genomics provides a platform to assess natural microbial phenomena that include biogeochemical activities, population ecology, evolutionary processes such as lateral gene transfer (LGT) events, and microbial interactions [117]. Applying community genomic data to DNA microarrays allows the analysis of global gene expression patterns and regulatory networks in a rapid, parallel format. Community microarray analyses can uncover apparent linkages between different genes and gene families and the distribution of metabolic functions in the community. Various genome assembly programmes such as ARACHNE, CAP, CELERA, EULER, JAZZ, PHRAP and TIGR assemblers are currently available to analysis community genomics data [118].

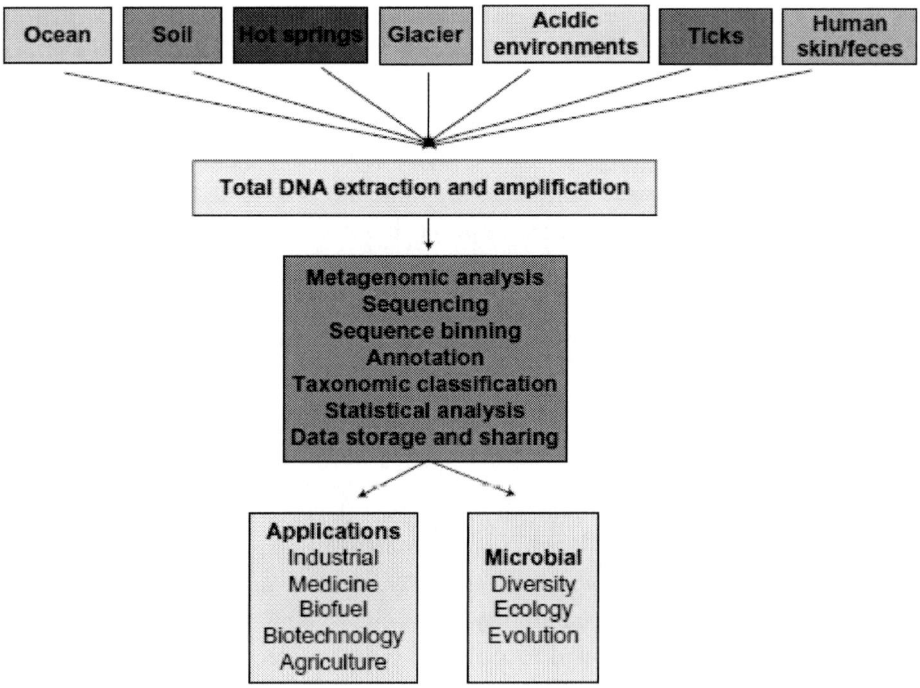

Figure 3. Metagenomic analysis of microbial communities.

Recently, sequencing and characterization of metatranscriptomes have been employed to identify RNA-based regulation and expressed biological signatures in complex ecosystems [113]. Technological challenges include the recovery of high-quality mRNA from environmental samples, short half-lives of mRNA species, and separation of mRNA from other RNA species. Metatranscriptomics had been limited to the microarray/high-density array technology or analysis of mRNA-derived cDNA clone libraries [116].

The proteomic analysis of mixed microbial communities is a new emerging research area which aims at assessing the immediate catalytic potential of a microbial community. Mass-spectroscopy-based proteomic methods are rapid and sensitive means to identify proteins in complex mixtures [115]. When applied to environmental samples, 'shotgun' proteomic analyses can produce surveys of prevalent protein species, which allows inferences of biological origin and metabolic function [114]. Challenges for metaproteomic analyses include uneven species distribution, the broad range of protein expression levels within microorganisms, and the large genetic heterogeneity within microbial communities [116]. Despite these hurdles, metaproteomics has a huge potential to link the genetic diversity and activities of microbial communities with their impact on ecosystem function.

STATISTICAL METHODS FOR ASSESSING FUNCTIONAL DIVERSITY OF MICROBIAL COMMUNITIES

Analyzing microbial diversity by metagenomics has limitations in processing the huge amount of data obtained from the community. To improve the efficiency of the analysis programs, statistical methods have been incorporated. Example of such methods in binning where the sequences derived from a mixture of different organisms is assigned to phylogenetic groups according to their taxonomic origins [118]. Depending on the quality of the metagenomic data set and the read length of the DNA fragments, the phylogenetic resolution can range from the kingdom to the genus level [117]. Examples of bioinformatic tools employing similarity-based binning are the Metagenome Analyzer (MEGAN), CARMA, or the sequence ortholog-based approach for binning and improved taxonomic estimation of metagenomic sequences (Sort-ITEMS) [116].

Abdo et al. [119] reported a statistical method named “R” (http:// www.ibest.uidah. edu /tools/ trflp_stats/index.php) for characterizing diversity of microbial communities by analysis of terminal restriction fragment length polymorphisms of 16S rRNA genes. R (R Development Core Team, 2003) functions can be implemented for identifying the ‘true’ peaks, binning the different fragment lengths, and for within cluster sampling.

Comparison of similarity indices can be used to statistically test for differences in community structure. One method frequently used for microbial communities is the analysis of similarity (ANOSIM) [120]. ANOSIM tests for statistically significant differences between community groups selected *a priori* (such as different sites or treatments). To create the ANOSIM statistic, a matrix of the similarity coefficients between each pair of samples is created, and this matrix is transformed to reflect the rank order of the similarities. The test statistic, *R*, is then calculated from the following equation:

$$R = \frac{r_b - r_w}{\frac{1}{4}(n(n-1))}$$

where *rb* is the mean rank of similarities between groups (average rank of similarities between community assemblages in different groups of samples), *rw* is the mean rank of similarities within groups (similarities between community assemblages in the same group), and *n* is the total number of samples. The statistic gives a value between -1 and 1, with 0 indicating that there are no differences in community similarity between groups, and 1 indicating that all communities within a group are more similar than they are to samples not in the group. Negative values are atypical, but indicate that members of the group are more different from each other than to samples not in the group.

‘quantitative insights into microbial ecology’ (QIIME; prounounced ‘chime’), an open-source software pipeline built using the PyCogent toolkit, to address the problem of taking sequencing data from raw sequences to interpretation and database deposition. QIIME is an open source software package for comparison and analysis of microbial communities, primarily based on high-throughput amplicon sequencing data (such as SSU rRNA) generated on a variety of platforms, but also supporting analysis of other types of data (such as shotgun metagenomic data). QIIME, available at http://qiime.sourceforge.net/, supports a wide range of microbial community analyses. QIIME takes users from their raw sequencing output through initial analyses such as OTU picking, taxonomic assignment, and construction of phylogenetic trees from representative sequences of OTUs, and through downstream statistical analysis, visualization, and production of publication-quality graphics [121].

Another tool for analyzing high throughput community data is EMPeror [122]. EMPeror is able to visualize gradients and categorical data, taxa as well as environmental samples, dynamically scale the axes according to the fraction of variance each explains, and display jackknifed-resampled data to assess statistical confidence in clustering. EMPeror, an open source and web browser enabled tool with a versatile command line interface allows researchers to perform rapid exploratory investigations of 3D visualizations of microbial community data. EMPeror increases the speed with which insight can be gained from large microbiome datasets.

Conclusion

The molecular perspective gives us more than just a glimpse of the evolutionary past; it also brings a new future to the discipline of microbial ecology. Because the molecular-phylogenetic identifications are based on sequences as opposed to metabolic properties, microbes can be identified without being cultivated. Consequently, all the sequence based techniques of molecular biology can be applied to the study of natural microbial ecosystems. Methods for studying diversity vary and diversity can be studied on several different levels: Global, community and population level. Biodiversity and function can be linked to global consequences and human activities. The actual contributors (or driving forces) to diversity to a system are unknown. It is therefore necessary for us to increase the knowledge of genetic and functional diversity to elucidate this problem. Microbial diversity in natural environments is extensive. In the future,, these techniques will be used to quantitatively analyze microbial diversity and expand our understanding of ecological processes. These methods characterize the process and suggest its cause, and should thereby be used in order to understand the extent of microbial diversity.

References

[1] Torsvik, V., Goksoyr, J. & Daae, F. L. (1990). High diversity in DNA of soil bacteria. *Appl. Environ. Microbiol.*, *56*, 782-787.

[2] Torsvik, V. & Øvreås, L. (2002). Microbial diversity and function in soil: from genes to ecosystems. *Curr. Opin. Microbiol.*, *5*, 240-245.

[3] Torsvik, V., Daae, F. L., Sandaa, R. A. & Ovreas, L. (1998). Review article: novel techniques for analysing microbial diversity in natural and perturbed environments. *J. Biotechnol.*, *64*, 53-62.

[4] Bossio, D. A., Scow, K. M., Gunapala, N. & Graham, K. J. (1998). Determinants of soil microbial communities: effects of agricultural management, season, and soil type on phospholipid fatty acid profiles. *Microb. Ecol.*, *36*, 1-12.

[5] Kennedy, A. C. & Smith, K. L. (1995). Soil microbial diversity and sustainability of agricultural soil. *Plant Soil.*, *170*, 75-86.

[6] Sims, G. K. (2007). Stable isotope probing to investigate microbial function in soil. *Rec. Res. Develop. Soil Sci.*, *2*.

[7] Van den Bogert, B., de Vos, W. M., Zoetendal, E. G. & Kleerebezem, M. (2011). Microarray Analysis and Barcoded Pyrosequencing Provide Consistent Microbial

Profiles Depending on the Source of Human Intestinal Samples. *Appl. Environ. Microbiol.*, *77*(6), 2071-2080.

[8] Read, D. J. (1998). Plants on the web. *Nat.*, *396*, 22-23.

[9] Abawi, G. S. & Widmer, T. L. (2000). Impact of soil health management practices on soilborne pathogens, nematodes and root diseases of vegetable crops. *Appl. Soil Ecol.*, *15*, 37-47.

[10] Harpole, W. (2010). Neutral theory of species diversity. *Nat Educ. Knowl.*, *1*, 31

[11] Yannarell, A. C. & Triplett, E. W. (2005). Geographic and environmental sources of variation in lake bacterial community composition. *Appl. Environ. Microbiol.*, *71*, 227-239.

[12] Giovannoni, S. J., Britschgi, T. B., Moyer, C. L. & Field, K. G. (1990). Genetic diversity in Saragasso Sea Bacterioplankton. *Nat.*, *345*, 60-62.

[13] Zhao, L., Ma, T., Gao, M., Gao, P., Cao, M., Zhu, X. & Li, G. (2012). Characterization of microbial diversity and community in water flooding oil reservoirs in China. *World J. Microbiol. Biotechnol.*, *28*(10), 3039-3052.

[14] Bååth, E., Frostegård, Å, Díaz-Raviña, M. & Tunlid, A. (1998). Microbial community-based measurements to estimate heavy metal effects in soil: the use of phospholipid fatty acid patterns and bacterial community tolerance. *Ambio.*, *27*, 58-61.

[15] Clarholm, M. (1994). The microbial loop in soil. In*: Beyond the biomass.* (Ritz, K., Dighton, J., Giller, K.E., Eds.), Chichester, John Wiley & Sons, 221-230.

[16] Kirk, J. L., Beaudette, L. A., Hart, M., Moutoglis, P., Klironomos, J. N., Lee, H. & Trevors, J. T. (2004). Methods of studying soil microbial diversity. *J. Microbiol. Meth.*, *58*, 169-188.

[17] Sørheim, R., Torsvik, V. L. & Goksøyr, J. (1989). Phenotypic divergences between populations of soil bacteria isolated on different media. *Microb. Ecol.*, *17*, 181-192.

[18] Tabacchioni, S., Chiarini, L., Bevivino, A., Cantale, C. & Dalmastri, C. (2000). Bias caused by using different isolation media for assessing the genetic diversity of a natural microbial population. *Microb. Ecol.*, *40*, 169-176.

[19] Trevors, J. T. (1998). Bacterial biodiversity in soil with an emphasis on chemically-contaminated soils. *Water Air Soil Poll.*, *101*, 45-67.

[20] Dix, N. J. & Webster J. (1995). Fungal Ecology. Chapman & Hall, London.

[21] Garland, J. L. & Mills, A. L. (1991). Classification and characterization of heterotrophic microbial communities on the basis of patterns of community-level-sole-carbon-source utilization. *Appl. Environ. Microbiol.*, *57*, 2351-2359.

[22] Preston-Mafham, J., Boddy, L. & Randerson, P. F. (2002). Analysis of microbial community functional diversity using sole-carbon-source utilisation profiles- a critique. *FEMS Microbiol. Ecol.*, *42*, 1-14.

[23] Derry, A. M., Staddon, W. J., Kevan, P. G. & Trevors, J. T. (1999). Functional diversity and community structure of micro-organisms in three arctic soils as determined by sole-carbon source-utilization. *Biodivers. Conserv.*, *8*, 205-221.

[24] Derry, A. M., Staddon, W. J. & Trevors, J. T. (1998). Functional diversity and community structure of microorganisms in uncontaminated and creosote-contaminated soils as determined by sole carbon- source-utilization. *World J. Microbiol. Biotechnol.*, *14*, 571-578.

[25] Grayston, S. J., Wang, S., Campbell, C. D. & Edwards, A. C. (1998). Selective influence of plant species on microbial diversity in the rhizosphere. *Soil Biol. Biochem.*, *30*, 369- 378.

[26] Konopka, A., Oliver, Jr. L. & Turco, R. F. (1998). The use of carbon source utilization patterns in environmental and ecological microbiology. *Microb. Ecol.*, *35*, 103-115.

[27] el Fantroussi, S., Verschuere, L., Verstraete, W. & Top, E. M. (1999). Effect of phenylurea herbicides on soil microbial communities estimated by analysis of 16S rRNA gene fingerprints and community- level physiological profiles. *Appl. Environ. Microbiol.*, *65*, 982- 988.

[28] Bej, A. K., Perlin, M. & Atlas, R. M. (1991). Effect of introducing genetically engineered microorganisms on soil microbial diversity. *FEMS Microbiol. Ecol.*, *86*, 169-175.

[29] Zak, J. C., Willig, M. R., Moorhead, D. L. & Wildman, H. G. (1994). Functional diversity of microbial communities: a quantitative approach. *Soil Biol. Biochem.*, *26*(9), 1101-1108.

[30] Yao, H., He, Z., Wilson, M. J. & Campbell, C. D. (2000). Microbial biomass and community structure in a sequence of soils with increasing fertility and changing land use. *Microb. Ecol.*, *40*, 223-237.

[31] Garland, J. L. (1996). Analytical approaches to the characterization of samples of microbial communities using patterns of potential C source utilization. *Soil Biol. Biochem.*, *28*, 213-221.

[32] Banowetz, G. M., Whittaker, G. W., Dierksen, K. P., Azevedo, M. D., Kennedy, A. C., Griffith, S. M. & Steiner, J. J. (2006). Fatty acid methyl ester analysis to identify sources of soil in surface water. *J. Environ. Qual.*, *3*, 133-140.

[33] Tunlid, A. & White, D. C. (1992). Biochemical analysis of Biomass, community structure, Nutritional Status and Metabolic Activity of Microbial Communities in soil. In: *Soil Biochemistry. (*Stotzky, G., Bollag, J.M., Eds.), New York, Marcel Dekker, Inc., *7*, 229-262.

[34] Findlay, R. H. (1996). The use of phospholipid fatty acids to determine microbial community structure. In: *Molecular Microbial Ecology manual* (Akkermans, A.D.L., van Elsas, J.D., DeBruijn F.J., Eds.) Dordrecht, Kluwer Academic Publishers, 215-247.

[35] Graham, J. H., Hodge, N. C. & Morton, J. B. (1995). Fatty acid methyl ester profiles for characterization of Glomalean fungi and their endomycorrhizae. *Appl. Environ. Microbiol.*, *61*, 58-64.

[36] Dokić, L., Savić M., Narančić, T. & Vasiljević, B. (2010). Metagenomic Analysis of Soil Microbial Communities. *Arch. Biol. Sci.-Belgrade.*, *62*(3), 559-564.

[37] Amann, R. I., Ludwig, W. & Schleifer, K. H. (1995). Phylogenetic identification and in situ detection of individual microbial cells without cultivation. *Microbiol. Rev.*, *59* (1), 143-169.

[38] Tiedje, J. M., Asuming-Brempong, S., Nusslein, K., Marsh, T. L. & Flynn, S. J. (1999). Opening the black box of soil microbial diversity. *Appl. Soil Ecol.*, *13*, 109-122.

[39] Clegg, C. D., Ritz, K. & Griffiths, B. S. (2000). %G+C profiling and cross hybridisation of microbial DNA reveals great variation in below-ground community structure in UK upland grasslands. *Appl. Soil Ecol.*, *14*, 125-134.

[40] Goris, J., Konstantinidis, K. T., Klappenbach, J. A., Coenye, T., Vandamme, P. & Tiedje, J. M. (2007). DNA–DNA hybridization values and their relationship to whole-genome sequence similarities. *Intl. J. System. Evolution. Microbiol.*, *57*, 81-91.

[41] Theron, J. & Cloete, T. E. (2000). Molecular techniques for determining microbial diversity and community structure in natural environments. *Crit. Rev. Microbiol.*, *26*, 37- 57.

[42] Schramm, A., Larsen, L. H., Revsbech, N. P., Ramsing, N. B., Amann, R. & Schleifer, K. H. (1996). Structure and function of a nitrifying biofilm as determined by in situ hybridization and the use of microelectrodes. *Appl. Environ. Microbiol.*, *62*, 4641.

[43] Torsvik, V., Sorheim, R. & Goksoyr, J. (1996). Total bacterial diversity in soil and sediment communities—a review. *J. Indust. Microbiol.*, *17*, 170-178.

[44] Griffiths, B. S., Ritz, K., Ebblewhite, N. & Dobson, G. (1999). Soil microbial community structure: effects of substrate loading rates. *Soil Biol. Biochem.*, *31*, 145-153.

[45] Hill, G. T., Mitkowskia, N. A., Aldrich-Wolfe, L., Emelea, L. R., Jurkonie, D. D., Ficke, A., Maldonado-Ramireza, S., Lyncha, S. T. & Nelsona, E. B. (2000). Methods for assessing the composition and diversity of soil microbial communities. *Appl. Soil Ecol.*, *15*, 25-36.

[46] Kauppinen, J., Pelkonen, J. & Katila, M. J. (1994). RFLP analysis of Mycobacterium malnroense strains using ribosomal RNA gene probes: an additional tool to examine intraspecies variation. *J. Microbiol. Meth.*, *19*, 261-267.

[47] Liu, W. T., Marsh, T. L., Cheng, H. & Forney, L. J. (1997). Characterization of microbial diversity by determining terminal restriction fragment length polymorphisms of genes encoding 16S rRNA. *Appl. Environ. Microbiol.*, *63*, 4516-4522.

[48] Moyer, C. L., Tiedje, J. M., Dobbs, F. C. & Karl, D. M. (1996). A computer-simulated restriction fragment length polymorphism analysis of Bacterial Small-subunit rRNA genes: Efficacy of selected tetraneric restriction enzymes for studies of microbial diversity in Nature. *Appl. Environ. Microbiol.*, *62*, 2501-2507.

[49] Thies, J. E. (2007). Soil microbial community analysis using terminal restriction fragment length polymorphisms. *Soil Sci. Soc. America J.*, *71*, 579-591.

[50] Osborn, A. M., Moore, E. R. B. & Timmis, K. N. (2000). An evaluation of terminal-restriction fragment length polymorphisms (TRFLP) analysis for the study of microbial community structure and dynamics. *Environ. Microbiol.*, *2*, 39-50.

[51] Dunbar, J., Ticknor, L. O. & Kuske, C. R. (2000). Assessment of microbial diversity in four southwestern United States soils by 16S rRNA gene terminal restriction fragment analysis. *Appl. Environ. Microbiol.*, *66*, 2943-2950.

[52] Rudi, K., Zimonja, M., Trosvik, P. & Næs, T. (2007). Use of multivariate statistics for 16S rRNA gene analysis of microbial communities. *Intl. J. Food Microbiol.*, *120*, 95-99.

[53] Lukow, T., Dunfield, P. F. & Liesack, W. (2000). Use of the T-RFLP technique to assess spatial and temporal changes in the bacterial community structure within an agricultural soil planted with transgenic and non-transgenic potato plants. *FEMS Microbiol. Ecol.*, *32*, 241-247.

[54] Moeseneder, M. M., Arrieta, J. M., Muyzer, G., Winter, C. & Herndl, G. J. (1999). Optimization of terminal-restriction fragment length polymorphism analysis for

complex marine bacterioplankton communities and comparison with denaturing gradient gel electrophoresis. *Appl. Environ. Microbiol.*, *65*, 3518-3525.

[55] Tonin, C., Vandenkoornhuyse, P., Joner, E. J., Straczek, J. & Leyval, C. (2001). Assessment of arbuscular mycorrhizal fungi diversity in the rhizosphere of Violoa calaminaria and effect of these fungi on heavy metal uptake by clover. *Mycorrhiza.*, *10*, 161-168.

[56] Fisher, M. M. & Triplett, E. W. (1999). Automated approach for ribosomal intergenic spacer analysis of microbial diversity and its application to freshwater bacterial communities. *Appl. Environ. Microbiol.*, *65*, 4630-4636.

[57] Borneman, J. & Triplett, E. C. (1997). Molecular microbial diversity in soils from eastern Amazonia: Evidence for unusual microorganisms and microbial population shifts associated with deforestation. *Appl. Environ. Microbiol.*, *63*(7), 2647-2653.

[58] Ranjard, L., Brothier, E. & Nazaret, S. (2000). Sequencing bands of ribosomal intergenic spacer analysis fingerprints for characterization and microscale distribution of soil bacterium populations responding to mercury spiking. *Appl. Environ. Microbiol.*, *66*, 5334-5339.

[59] Yu, Z. & Mohn, W. W. (2001). Bioaugmentation with resin-acid-degrading bacteria enhances resin acid removal in sequencing batch reactors treating pulp mill effluents. *Water Res.*, *35*, 883-890.

[60] Cho, J. C. & Tiedje, J. M. (2001). Bacterial species determination from DNA–DNA hybridization by using genome fragments and DNA microarrays. *Appl. Environ. Microbiol.*, *67*, 3677-3682.

[61] Greene, E. A. & Voordouw, G. (2003). Analysis of environmental microbial communities by reverse sample genome probing. *J. Microbiol. Meth.*, *53*, 211-219.

[62] DeSantis, T. Z., Brodie, E. L., Moberg, J. P., Zubieta, I. X., Piceno, Y. M. & Andersen, G. L. (2007). High-density universal 16S rRNA microarray analysis reveals broader diversity than typical clone library when sampling the environment. *Microb. Ecol..*, *53*, 371-383.

[63] Greene, E. A., Kay, J. G., Jaber, K., Stehmeier, L. G. & Voordouw, G. (2000). Composition of soil microbial communities enriched on a mixture of aromatic hydrocarbons. *Appl. Environ. Microbiol.*, *66*, 5282-5289.

[64] Gentry, T. J., Wickham, G. S., Schadt, C. W., He, Z. & Zhou, J. (2006). Microarray applications in microbial ecology research. *Microb. Ecol.*, *52*, 159-175.

[65] Mühling, M., Woolven-Allen, J., Murrell, J. C. & Joint, I. (2008). Improved group-specific PCR primers for denaturing gradient gel electrophoresis analysis of the genetic diversity of complex microbial communities. *The ISME J.*, *2*, 379-392.

[66] Muyzer, G., Hottentrfiger, S., Teske, A. & Wawer, C. (1996). Denaturant gradient gel electrophoresis of PCR amplified l6S rDNA - A new molecular approach to analyse the genetic diversity of mixed microbial communities. In: *Molecular Microbial Ecology Manual* (Akkermans, A.D.L., van Elsas, J.D., DeBruijn, F.J., Eds.), Dordrecht, kluwer Academic publishers, 45-83.

[67] Miller, K. M., Ming, T. J., Schulze, A. D. & Withler, R. E. (1999). Denaturing Gradient Gel Electrophoresis (DGGE): a rapid and sensitive technique to screen nucleotide sequence variation in populations. *BioTechniq.*, *27*, 1016-1030.

[68] Muyzer, G. (1999). DGGE/ TGGE a method for identifying genes from natural ecosystems. *Curr. Opin. Microbiol.*, *2*, 317-322.

[69] Wintzingerode, F. V., Gobel, U. B. & Stackebrandt, E. (1997). Determination of microbial diversity in environmental samples: pitfalls of PCR-based rRNA analysis. *FEMS Microbiol. Rev., 21*, 213-229.

[70] MacNaughton, S. J., Stephen, J. R., Venosa, A. D., Davis, G. A., Chang, Y. J. & White, D. C. (1999). Microbial population changes during bioremediation of an experimental oil spill. *Appl. Environ. Microbiol., 65*, 3566-3574.

[71] Gelsomino, A., Keijzer-Wolters, A. C., Cacco, G. & van Elsas, J. D. (1999). Assessment of bacterial community structure in soil by polymerase chain reaction and denaturing gradient gel electrophoresis. *J. Microbiol. Meth., 38*, 1-15.

[72] Maarit-Niemi, R., Heiskanen, I., Wallenius, K. & Lindstrom, K. (2001). Extraction and purification of DNA in rhizosphere soil samples for PCR-DGGE analysis of bacterial consortia. *J. Microbiol. Meth., 45*, 155–165.

[73] Ferris, M., Muyzer, G. & Ward, D. (1996). Denaturing gradient gel electrophoresis profiles of 16S rRNA-defined populations inhabiting a hot spring microbial mat community. *Appl. Environ. Microbiol., 62*(2), 340-346.

[74] Teske, A., Wawer, C., Muyzer, G. & Ramsing, N. B. (1996). Distribution of sulfate-reducing bacteria in a stratified flord (Mariager Fjord, Denmark) as evaluated by most-probable-number counts and denaturing gradient gel electrophoresis of PCR-amplified ribosomal DNA fragments. *Appl. Environ. Microbiol., 62*(4), 1405-1415.

[75] Peters, S., Koschinsky, S., Schwieger, F. & Tebbe, C. C. (2000). Succession of microbial communities during hot composting as detected by PCR-single-strand-conformation polymorphism based genetic profiles of small-subunit rRNA genes. *Appl. Environ. Microbiol., 66*, 930-936.

[76] Lee, D. H., Zo, Y. G. & Kim, S. J. (1996). Nonradioactive method to study genetic profiles of natural bacterial communities by PCR single strand conformation polymorphism. *Appl. Environ. Microbiol., 62*, 3112- 3120.

[77] Stach, J .E. M., Bathe, S., Clapp, J. P. & Burns, R. G. (2001). PCR-SSCP comparison of 16S rDNA sequence diversity in soil DNA obtained using different isolation and purification methods. *FEMS Microbiol. Ecol., 36*, 139-151.

[78] Schmalenberger, A., Schwieger, F. & Tebbe, C. C. (2001). Effect of primers hybridizing to different evolutionarily conserved regions of the small-subunit rRNA gene in PCR-based microbial community analyses and genetic profiling. *Appl. Environ. Microbiol., 67*, 3557-3563.

[79] Zumstein, E., Moletta, R. & Godon, J. J. (2000). Examination of two years of community dynamics in an anaerobic bioreactor using fluorescence polymerase chain reaction (PCR) single-strand conformation polymorphism analysis. *Environ. Microbiol., 2*, 69- 78.

[80] Kjoller, R. & Rosendahl, S. (2000). Detection of arbuscular mycorrhizal fungi (Glomales) in roots by nested PCR and SSCP (single stranded conformation polymorphism). *Plant Soil., 226*, 189-196.

[81] Neufeld, J. D., Dumont, M. G., Vohra, J. & Murrell, J. C. (2007). Methodological considerations for the use of stable isotope probing in microbial ecology. *Microb. Ecol., 53*(3),435-442.

[82] Asuming-Brempong S. (2014). Reverse-transcriptase-quantitative polymerase chain reaction (RT-qPCR) and its usefulness in soil microbial ecological studies-A review. *Afr. J. Biotechnol., 13*(6),723-728.

[83] Cottrell, M. T. & Kirchman, D. L. (2000). Natural assemblages of marine proteobacteria and members of the *Cytophaga-Flavobacter* cluster consuming lowand high-molecular-weight dissolved organic matter. *Appl. Environ. Microbiol.*, *66*, 1692-1697.

[84] Fakruddin, M., Mazumdar, R. M., Chowdhury, A., Hossain, M. N. & Mannan, K. S. B. (2012). Pyrosequencing- Prospects and Applications. *Intl. J. Life Sci. Pharma Res.*, *2*(2), 65-76.

[85] Lauber, C. L., Hamady, M., Knight, R. & Fierer, N. (2009). Pyrosequencing-based assessment of soil pH as a predictor of soil bacterial community structure at the continental scale. *Appl. Environ. Microbiol.*, *75*, 5111-5120.

[86] Caporaso, J. G., Lauber, C. L., Walters, W. A., Berg-Lyons, D., Huntley, J., Fierer, N., Owens, S. M., Betley, J., Fraser, L., Bauer, M., Gormley, N., Gilbert, J. A., Smith, G. & Knight, R. (2012). Ultra-high throughput microbial community analysis on the Illumina HiSeq and MiSeq platforms. *The ISME J.*, *6*, 1621-1624.

[87] Degnan, P. H. & Ochman, H. (2012). Illumina-based analysis of microbial community diversity. *The ISME J.*, *6*, 183-194.

[88] Caracciolo, A. B., Bottoni, P. & Grenni, P. (2010). Fluorescence in Situ Hybridization in soil and water ecosystems: a useful method for studying the effect of xenobiotics on bacterial community structure. *Toxicol. Environ. Chem.*, *92*, 567-579.

[89] Wullings, B. A., van Beuningen, A. R., Janse, J. D. & Akkermanns, A. D. L. (1998). Detection of *Ralstonia solanacearum* which causes brown rot of potato, by fluorescent in situ hybridization with 23S rRNA-targeted probes. *Appl. Environ. Microbiol.*, *64*, 4546-4554.

[90] Felske, A. & Akkermans, A. D. L. (1998). Prominent occurrence of ribosomes from an uncultured bacterium of the Verrucomicrobiales cluster in grassland soils. *Lett. Appl. Microbiol.*, *26*, 219-223.

[91] Manefield, M., Whiteley, A. S., Griffiths, R. I. & Bailey, M. J. (2002). RNA-stable isotope probing, a novel means of linking microbial community function to phylogeny. *Appl. Environ. Microbiol.*, *68*, 5367-5373.

[92] Huang, W. E., Stoecker, K., Griffiths, R., Newbold, L., Daims, H., Whiteley, A. S. & Wagner, M. (2007). Raman–FISH: combining stable-isotope Raman spectroscopy and fluorescence in situ hybridization for the single cell analysis of identity and function. *Environ. Microbiol.*, *9*, 1878-1889.

[93] Radajewski, S., Ineson, P., Parekh, N. R. & Murrell, J. C. (2000). Stable-isotope probing as a toll in microbial ecology. *Nat.*, *403*, 646-649.

[94] Ge, Y., He, J. Z., Zheng, Y. M., Zhang, L. M. & Zhu, Y. G. (2006). Stable isotope probing and its applications in microbial ecology. *Acta Ecolog. Sin.*, *26*, 1574-1582.

[95] Bustin, S. A., Benes, V., Nolan, T. & Pfaffl, M. W. (2005). Quantitative real-time RT-PCR – a perspective. *J. Molecular Endocrinol.*, *34*, 597-601.

[96] Fierer, N., Jackson, J. A., Vilgalys, R. & Jackson, R. B. (2005). Assessment of soil microbial community structure by use of taxon-specific quantitative PCR assays. *Appl. Environ. Microbiol.*, *71*,4117-4120.

[97] Foti, M., Sorokin, D. Y., Lomans, B., Mussman, M., Zacharova, E. E., Pimenov, N. V., Kuenen, J. G. & Muyzer, G. (2007). Diversity, activity, and abundance of sulfate-reducing bacteria in saline and hypersaline soda lakes. *Appl. Environ. Microbiol.*, *73*, 2093-3000.

[98] Smith, C. J. & Osborn, A. M. (2009). Advantages and limitations of quantitative PCR (Q-PCR)-based approaches in microbial ecology. *FEMS Microbiol. Ecol.*, *67*,6-20.

[99] Heid, C. A., Stevens, J., Livak, K. J. & Williams, P.M. (1996). Real time quantitative PCR. *Genome Res.*, *6*(10),986-994.

[100] Okabe, S., Kindaichi, T. & Tsukasa, I. (2004). MAR–FISH: an ecophysiological approach to link phylogenetic affiliation and in situ metabolic activity of microorganisms at a single-cell resolution. *Microbe. Environ.*, *19*,83-98.

[101] Nielsen, J. L., Christensen, D., Kloppenborg, M. & Nielsen, P. H. (2003). Quantification of cell-specific substrate uptake by probe- defined bacteria under in situ conditions by microautoradiography and fluorescence in situ hybridization. *Environ. Microbiol.*, *5*, 202-211.

[102] Rogers, S. W., Moorman, T. B. & Ong, S. K. (2007). Fluorescent in situ hybridization and Microautoradiography applied to ecophysiology in soil. *Soil Sci. Soc. America J.*, *71*,620-631.

[103] Ouverney, C. C. & Fuhrman, J. A. (1999). Combined microautoradiography-16S rRNA probe technique for determination of radioisotope uptake by specific microbial cell types in situ. *Appl. Environ. Microbiol.*, *65*,1746-1752.

[104] Rogers, S., Ong, S. & Moorman, T. B. (2005). Application of Fish and Microautoradiography to Determine Microbial Community Structure and Activity in Soil. Presented at Soil Science Society of America Annual Meeting, Salt Lake City, UT, November 06 – 10.

[105] Okabe, S., Kindaichi, T. & Ito, T. (2005). Fate of ^{14}C-labeled microbial products derived from nitrifying bacteria in autotrophic nitrifying biofilms. *Appl. Environ. Microbiol.*, *71*, 3987-3994.

[106] Bartram, A. K., Lynch, M. D. J., Stearns, J. C., Moreno-Hagelsieb, G. & Neufeld, J. D. (2011). Generation of Multimillion-Sequence 16S rRNA Gene Libraries from Complex Microbial Communities by Assembling Paired-End Illumina Reads. *Appl. Environ. Microbiol.*, *77*(11), 3846-3852.

[107] Fakruddin, M., 7 Mannan, K. S. B. (2012). Next Generation Sequencing- Prospects and Applications. *Res. Rev. Biosci.*, *6*(9), 240-247.

[108] Fakruddin, M. & Chowdhury, A. (2012). Pyrosequencing- An Alternative to Traditional Sanger Sequencing. *Am. J. Biochem. Biotechnol.*, *8*(1), 14-20.

[109] Claesson, M. J., Wang, Q., O'Sullivan, O., Greene-Diniz, R., Cole, J. R., Ross, R. P. & O'Toole, P. W. (2010). Comparison of two next-generation sequencing technologies for resolving highly complex microbiota composition using tandem variable 16S rRNA gene regions. *Nucl. Acids Res.*, *38*(22), e200.

[110] Royo, J. L., Hidalgo & M., Ruiz, A. (2007). Pyrosequencing protocol using a universal biotinylated primer for mutation detection and SNP genotyping. *Nat. Prot.*, *2*(7), 1734-1739.

[111] Fakruddin M., Mazumdar, R. M., Chowdhury, A., Hossain, M. N., Mahajan, S. & Islam, S. (2013). Pyrosequencing-A Next Generation Sequencing Technology. *World Appl. Sci. J.*, *24*(12), 1558-1571.

[112] Van Bruggen, A. H. C. (1995). Plant disease in high-input compared to reduced input and organic farming systems. *Plant Dis.*, *79*, 976-983.

[113] Zeyaullah, M., Kamli, M. R., Islam, B., Atif, M., Benkhayal, F. A., Nehal, M., Rizvi, M. A. & Ali, A. (2009). Metagenomics - An advanced approach for noncultivable micro-organisms. *Biotechnol. Mol. Biol. Rev.*, *4*(3), 49-54.

[114] Ghazanfar, S., Azim, A., Ghazanfar, M. A., Anjum, M. I. & Begum, I. (2010). Metagenomics and its application in soil microbial community studies: biotechnological prospects. *J. Anim. Plant Sci.*, *6*(2), 611-622.

[115] Schloss, P .D., Handelsman, J. (2003). Biotechnological prospects from metagenomics. *Curr. Opin. Biotechnol.*, *14*, 303-310.

[116] Simon, C. & Daniel, R. (2011). Metagenomic Analyses: Past and Future Trends. *Appl. Environ. Microbiol.*, *77*(4), 1153-1161.

[117] Allen, E. E. & Banfield, J. F. (2005). Community genomics in microbial ecology and evolution. *Nat. Rev. Microbiol.*, *3*, 489-498.

[118] Tyson, G. W., Chapman, J., Hugenholtz, P., Allen, E. E., Ram, R. J., Richardson, P. M., Solovyev, V. V., Rubin, E. M., Rokhsar, D. S. & Banfield, J. F. (2004). Community structure and metabolism through reconstruction of microbial genomes from the environment. *Nat.*, *428*, 37-43.

[119] Abdo, Z., Schüette, U. M. E., Bent, S. J., Williams, C. J., Forney, L. J. & Joyce, P. (2006). Statistical methods for characterizing diversity of microbial communities by analysis of terminal restriction fragment length polymorphisms of 16S rRNA genes. *Environ. Microbiol.*, *8*(5), 929-938.

[120] Clarke, K. R. (1993). Nonparametric multivariate analyses of changes in community structure. *Australian J. Ecol.*, *18*(1), 117-143.

[121] Caporaso, J. G., Kuczynski, J., Stombaugh, J., Bittinger, K., Bushman, F. D., Costello, E. K., Fierer, N., Pena, A. G., Goodrich, J. K., Gordon, J. I., Huttley, G. A., Kelley, S. T., Knights, D., Koenig, J. E., Ley, R. E., Lozupone, C. A., McDonald, D., Muegge, B. D., Pirrung, M., Reeder, J., Sevinsky, J. R., Turnbaugh, P. J., Walters, W. A., Widmann, J., Yatsunenko, T., Zaneveld, J. & Knight, R. (2010). QIIME allows analysis of high throughput community sequencing data. *Nature Methods.*, *7*(5), 335-336.

[122] Vázquez-Baeza, Y., Pirrung, M., Gonzalez, A. & Knight, R. (2013). EMPeror: a tool for visualizing high-throughput microbial community data. *GigaScience.*, *2*(1),16.

In: Microbes in Soil and Their Agricultural Prospects
Editors: K. Kishore Choudhary and D. Wattal Dhar
ISBN: 978-1-63482-824-6

Chapter 8

REVIEW ON UV-B RADIATION IMPACT ON CYANOBACTERIA AND POSSIBLE PROTECTION MECHANISMS

***B. Praveena*[1]*, P. Jyothsna*[1] *and S. D. S. Murthy*[1,•]**
[1]Department of Biochemistry, Sri Venkateswara University, Tirupathi, India

ABSTRACT

The depletion of ozone layer is due to industrially released atmospheric pollutants which results in increased solar ultraviolet radiation (UVR), especially UV-B (280-315 nm), on the earth's surface. Solar UV-B radiation has harmful effects to various sun-exposed organisms, including humans. They can penetrate deep into biologically significant depths in lakes, ponds, rivers, oceans etc. Cyanobacteria, one of the primitive O_2-evolving photosynthetic prokaryote are solely dependent upon the solar radiation for their growth and survival and thus they undergo UV-B stress. UVR affects cyanobacteria by inducing oxidative stress, causing loss of motility, low cell differentiation, damage to photosynthetic pigments, proteins, lipids, and nucleic acids, finally affecting their growth and survival. In response to these effects, these organisms have developed a number of defence mechanisms such as protection, repair and avoidance mechanisms. This chapter represents an overview not only on detrimental effects of UVR of cyanobacteria, but also various defence mechanisms employed by these organisms to withstand against UVR stress. Establishment of cyanobacteria under field condition may provide protective mechanism to other beneficial microorganisms from possible future threat of UV radiation to maintain the agricultural productivity sustainable.

Keywords: cyanobacteria, DNA damage, mycosporine-like amino acids (MAAs), protection mechanisms, scytonemin, ultraviolet radiation

• E-mail: sdsmurthy@rediffmail.com.

INTRODUCTION

UV radiation is one of the serious issues since past few decades due to industrialization. Increase in the industrialization results in the increase in anthropogenically important atmospheric pollutants such as chlorofluorocarbons (CFCs), halocarbons, chloroform (MCF) and dioxins (NOx). Considerable amounts of natural production of reactive nitrogen species (RNS) such as nitric oxide (NO^-), peroxynitrate ($ONOO^-$) and nitrous oxide (N_2O) from unpolluted aquatic and terrestrial ecosystems also contribute to the depletion of ozone layer [1]. These pollutants are being responsible for the depletion ozone layer in the stratosphere that helps in screening of UVR [2].

Due to increase of the pollutants depletion of ozone layer is occurring not only in the Antarctic region but also all over the earth surface resulting in the subsequent increase of UV radiation (UVR; 280-400 nm) entering to the earth surface [3]. The depletion of ozone layer in Antarctic region has been reported to be more than 70% [4].

These harmful UV rays reaching the earth surface effects the production of algae and photosynthetic macrophytes as they can easily absorb UV light by biomolecules such as nucleic acids and proteins [5, 6]. All this is because UV-B radiation can penetrate into the water upto a depth of 20-30 m [4].

Aquatic organisms like algae and photosynthetic macrophytes which grow in marine ecosystem are the support for entire life, because these aquatic organisms mainly produce food, for aquatic organisms like fishes sponges shelter, as O_2 supplement and as p^H regulators. This negative impact of UV-radiation is not only for the primary producers but also to various aquatic and terrestrial ecosystems ranging from prokaryotes to eukaryotes i.e., from lower to higher plants, animals and also human [7].

Cyanobacteria is a primitive group of gram negative, ubiquitous in nature, oxygenic photoautotrophic prokaryotes which have wide distribution ranging from hot springs to Arctic and Antarctic regions and are important biomass producers in both aquatic and terrestrial ecosystems [5, 8]. They are the valuable sources of natural products of medicinal and industrial importance [9] and also they are the ecologically important organisms which act as natural biofertilizers [10] which fix the atmospheric nitrogen as they consists of enzyme nitrogenase.

During this nitrogen fixation and photosynthetic processes they absorb light because of their light absorbing nature they can easily absorb the harmful UV radiation which leads to lethal effects [11]. Various factors like morphology, growth, survival, cell differentiation, pigmentation, motility, N_2 metabolism, phycobiliproteins, composition of protein, DNA, CO_2 uptake are severely affected by the harmful UV radiation when absorbed by cyanobacteria [12, 13].

The changes in these composition results in various primary UV-B mediated events such as loss of permeability membrane changes, pigment destruction, direct photosynthetic damage, protein and enzyme inactivation, reduction in protein and DNA synthesis, reduced uptake of nutrients, inactivation of hormones, effect in signal transduction [13, 14]. Cyanobacteria due to the adverse effects of harmful UV-rays have adopted many protection mechanisms.

The mechanisms evolved by cyanobacteria to cope with UV radiation are protection, repair and avoidance [15, 16].

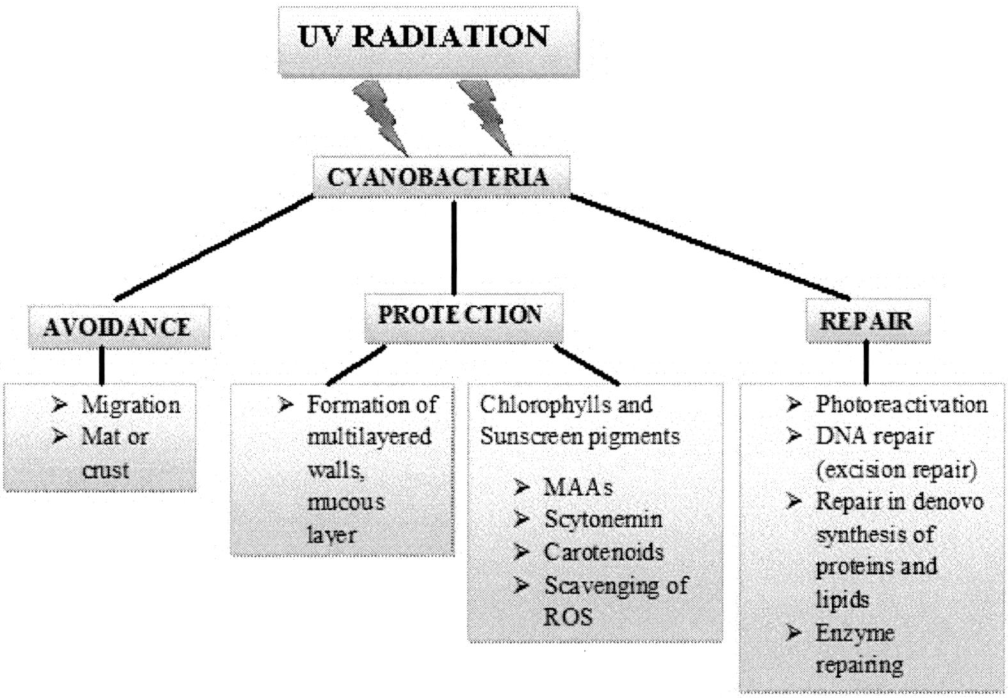

Figure 1. Mechanisms evolved by cyanobacteria to cope with UV-Radiation.

VARIOUS IMPACTS OF UVR ON CYANOBACTERIA

(a) *Effect on motility*

Motility of cyanobacteria is also adversely affected by UVR. It is the main behaviour change employed by cyanobacteria to cope up the harmful UVR. There is a significant decrease in the motile filaments of *Oscillatoria tenuis*, *Phormidium uncinatum* and *Anabaena variabilis* which show gliding motility was observed within 10-30 min of UV exposure [17]. UV-radiation results in the inhibiting of photophobic responses in cyanobacteria and hence reduces the ability of the organisms to adapt themselves in the photo environment leading to their death [18].

(b) *Effect on growth and survival*

Effect of UVR on growth and survival is severe and leads to complete killing of the organisms within 120-180 min exposure [19, 20]. The resistance capacity and sensitivity is different to the different species organisms. The growth of *Oscillatoria priestleyi,* Antarctica cyanobacterium was completely suppressed whereas *in Phormidium murrayi* it is only 62% when similar doses of UVR is exposed [21].

(c) *Effect on cell differentiation*

UV-radiation has detrimental effect on cell differentiation in certain cyanobacteria. UV exposure to cyanobacteria has been found to delay the differentiation of vegetative cells into heterocyst and akinete in *Anabaena aqualis* [22]. UV-radiation leads to the disruption of multi-layered heterocyst wall which is an essential component [23]. Recent studies have been reported that the spiral filaments are broken and compressed in *Anthrospira platensis* when exposed to UVR [24 and 25].

(d) *Effect on photosynthesis*

Photosynthesis is one of the most essential and sensitive metabolic process in cyanobacteria. As photosynthesis is directly linked to the biomass production and yield, it become essential to study response of photosynthesis to UV-B stress. Proteins, photosynthetic pigments, quinones such as plastoquinones and unsaturated fatty acids of galactolipids, all present in the bilayered structure of the thylakoid membrane, may be UV-B targets due to their electronic absorption in the UV-B region [26]. Major photosynthetic parameters like CO_2 uptake, O_2 evolution and ribulose-1, 5 bishosphate carboxylase (RUBISCO) activity in cyanobacteria are inhibited by UVR [27]. Photodegradation, fragmentation and denaturation of polypeptide chain, change in active site and increased solubility of membrane proteins are the various modifications occurs by RUBISCO [28]. Along with these modifications supply of ATP and $NADPH+H^+$ may also be impaired which may cause inhibition of CO_2 -fixing ability.

(e) *Effect on photosynthetic pigments*

Photosynthetic pigments like chlorophylls, phycobilins, carotenoids and xanthophylls which are present in the photosynthetic apparatus are also destroyed by UV-radiation [29]. The consequences of reduced synthesis of chlorophyll pigment complex which is encoded by the *cab* gene family decreases in the amount of pigment content due to UVR [30]. A special type of light harvesting complex is present in cyanobacteria called phycobiliproteins which consists of open chain tetra pyrole pigments can also be destroyed by UVR [31].

(f) *Effect on quinines*

Quinones are important components of various redox complexes in plant membranes and it consists of special role in photosynthesis as electron carries between the photosystems. There is a direct UV induced destruction on plastoquinones by UV-C radiation [32]. UV-B radiation on plastoquinones results in the decreased amount of these components in thylakoids [33]. By this the redox function of quinones in PS II complex is impaired [34, 35].

(g) *Effect on nucleic acids*

DNA is the most prominent target of UV-radiation in all living organisms including cyanobacteria. Both UV-A and UV-B radiation causes mutations during replication which is formed from DNA photoproducts [36]. UVR induces DNA lesions either directly by absorbing the radiation by native DNA or indirectly by oxidative stress [37]. Oxidative stress of DNA commonly results in the single or double-stranded breaks in the DNA molecule [38].The formation of UV-radiation induced thymine dimmers in three types of rice field cyanobacteria, *Anabaena*, *Nostoc* and *Scytonema* Sps. This has been demonstrated by Sinha group through blotting and chemiluminescence method [39].

(h) *Effect on protein and lipid contents*

Proteins and lipids are the main targets as the cell is majorly composed of proteins and lipids. Several studies have been conducted to show that UV-B radiation has major effects on protein profile of cyanobacteria [13,40]. As the proteins have the strong absorbance maxima at 280 nm, UV rays also absorbed by the organism as the wavelength of UV ranges between (200-400 nm). Majorly aromatic aminoacids like tyrosine, tryptophan, phenylalanine are the direct targets of UV-B radiation, hence UV induced destruction of these aminoacids are done either in free form or in proteins [41]. The main bonds responsible for the tertiary structure of proteins are disulphide bonds and they can also be split by UVR [42, 43]. Lipids are also the main targets of UVR, phospholipids and glycolipids are the main components present in cell membrane which contain unsaturated fatty acids are also destroyed by UV-B radiation in the presence of O_2 [44, 45].

PROTECTION MECHANISMS OF CYANOBACTERIA AGAINST UV-RADIATION

Protection mechanisms of cyanobacteria against UV radiation majorly include avoidance, protection and repair [46].

(i). Avoidance

Migration

To escape from the high solar radiation i.e., UV, motile cyanobacteria in mats migrate upward and downward depending on the spectral wave band [47]. In planktonic cyanobacteria sinking of floating regulated by gas vacuoles are also protective strategies against UVR [48].A downward movement of motile *Oscillatorials* from microbial mat surfaces has been reported by Ramsung et al. [49]. When the microbial mat surface of *O.priestleyi* exposed to UV-A and UV-B they migrated to the below surface of the mat and when it is kept in the shade for several hours they again migrated to the top surface [50 and 51].

Mat or Crust Formation

Cyanobacteria are the most successful crust forming organisms. They are closely associated with the substrate and produce mat or crust like structures, which can range from a thickness of few micrometers to decimetres. Mats are often composed of a varying number of different cyanobacterial taxa (10-40 Sps) [52]. They occur all over the world in rather extreme habitats, such as alkaline hot springs [53], arctic waters [54], marine and hypersaline environments like the interdial zone or evaporates [55], rice fields [56],hot arid areas [57] and terrestrial rock surfaces [58].

(ii). Protection

Protection by Physical Barriers (Multi Layered Cell Walls and Mucilaginous Sheath Layer)

Mandal et al., showed the presence of mucilaginous sheath layer among the adaptation mechanisms that allowed the intertidal cyanobacteria *Lyngbya majuscule* to withstand the

UVR [59]. Some dinoflagellates when exposed to artificial UVR may produce physical and chemical barriers to offset the deleterious effects of this radiation [60].

Protection by Antioxidant Compounds

Cyanobacteria perform oxygenic photosynthesis using water as an electron donor. Therefore they release molecular oxygen into the environment, which can be accumulated and converted into potentially harmful reactive oxygen species (ROS). The interaction between UVR, oxygen and certain organic compounds can produce toxic intermediates called ROS, such as superoxide anion (O_2), hydrogen peroxide (H_2O_2), Hydroperoxy radicals (HO_2^-) and Hydroxyl radicals (OH^-).

These ROS can cause extensive damage to proteins, nucleic acids and other biological structures [61]. To overcome these ROS cyanobacteria has developed various enzymatic and non-enzymatic antioxidants like carotenoids, tocopherols (vitamin-E), ascorbic acid (vitamin-C) and reduced glutathione (GSH). Enzymatic antioxidants are superoxide dismutase (SOD), catalase, and glutathione peroxidise, glutathione reductase (GR) [62 and 63]. Especially in plants, the generated O_2^- can be converted into H_2O_2 and O_2 by several SOD isoenzymes; they are mitochondrial manganese SOD (Mn- SOD), chloroplast iron SOD (Fe-SOD) and cytosolic copper and zinc SOD (Cu/Zn SOD). There is significant evidence that cyanobacteria when exposed to oxidative stress results in increase in the activities of ROS scavenging enzymes [64].

This indicates that higher and more stable antioxidant enzyme activities are associated with high stress tolerance in cyanobacteria. Thus the antioxidant defence mechanism against ROS is important for the survival of cyanobacteria under stress conditions. Reduced glutathione play an important role in protection against oxidative damage which is occurring from various stress conditions [65].

Carotenoids are organic pigments that are naturally occurring in chromoplasts of plants and some other photosynthetic organisms like algae, some types of fungus and some bacteria. There are over 600 known carotenoids that split into two classes; xanthophylls and carotenes which absorb blue light. Carotenoids serve two key roles in plant and algae: they absorb light energy used in photosynthesis and they protect chlorophyll from photodamage [66]. The common structure of carotenoids is long π-conjugated polyene structure with several methyl substituents.

These carotenoids have three absorption bands in the region between 400-500 nm. They are localized in biomembranes which are strongly hydrophobic due to the presence of long unsaturated alkene chain. The major carotenoids in cyanobacteria are β-carotene, zeaxanthin and nostoxanthin as hydroxyl derivatives and echinenone and canthaxanthin as ketoderivaties. Carotenoids are also well known for their antioxidant activity.

During long term exposure to highly natural or artificial radiation very high ratios of carotenoids to chlorophylls occur. These ratios are the prerequisite for a border tolerance against excessive irradiation, particularly at suboptimal growth temperature [67], and response to UVR.

Orange carotenoid protein (OCP) a water soluble 35 kDa protein was identified in three genera cyanobacteria. This OCP crystal structure analysis in *Anthrospira maxima* revealed that carotenoid, 3'- hydroxyechnenone binds to the protein non-covalently between N and C terminal domains. The light absorption by OCP induced structural changes in the carotenoid and protein which results in conversion of stable orange form into red which is unstable and active form. The red form accumulated under the condition in which photoprotection is required [68].

Protection by UV Absorbing Compounds

Apart from anatomical alterations in plants, the accumulation of UV-B absorbing compounds found in species of natural ecosystems, as well as UV-B reflecting waxes, may contribute to the protection of photosynthesis in nature [69, 70].

Low levels of UV-B stimulate the general phenylpropanoid pathway, resulting in accumulation of flavonoid and sinapic esters [71-73]. These compounds play a protective role by specific absorption in the wavelength region from 280- 340 nm. In barley the accumulation in epidermal mesophyll tissue of flavonoids reduces the UV-B induced DNA damage [74].

Flavonoids are the UV absorbing compounds. These are thought to protect photosynthetic tissues by acting as sunscreen pigments absorbing UV light. Flavonoids also possess free radicals scavenging activity, which offer additional protection to the cells accumulating these compounds [75].

Protective responses are also stimulated by UV-B radiation, including increased production of UV-B-absorbing compounds like flavonoids in higher plants Flavonoid production caused due to UV-radiation is generally occurred in the upper epidermis and has a maximal absorption around 0.300μm. Generally plants collected in naturally high UV-B radiation environments tend to be less affected by experimental exposure to UV-B radiation, this is due to adaptive accumulation of high levels of leaf flavonoids.

Sunscreen Pigments

UV radiation effect on cyanobacteria results in increase of two types of sunscreen pigments scytonemin and mycosporine like aminoacids (MAAs) [76].

These are also called as secondary metabolites. These secondary metabolites are thought to play multiple roles against several environmental stresses such as UV-radiation and desiccation [77].

Mycosporine Like Aminoacids (MAAs)

MAAs are ultraviolet absorbing molecules having absorption maxima between 320-360 nm. These are one of the pigment molecules produced in cyanobacteria and algae [78].

It is believed that these MAAs are derived from parasitic microorganisms or microorganisms taken in through ingestion. MAAs are small molecules which are <400 Da, these are colourless water-soluble compounds composed of cyclohexemone or cyclohexenimine chromophore conjugated with the nitrogen substituent of an amino acids or its imino alcohol.

Some MAAs also contains sulphate esters or glycosidic linkages through the imine substituents. MAAs are highly hydrophilic in nature this hydrophilic nature is due to their zwitter ionic form when compared to other UV absorbing antioxidants present in cyanobacteria like chlorophylls, carotenoids and scytonemin. The hydrophilic nature of MAAs is further increases due to modification by sulfonic acids and sugar molecules.

Sunscreen Properties of MAAs

The function that mycosporines play as natures sunscreen compounds has been the subject of a number of review articles. However in recent years evidences are accumulating that mycosporines play additional roles, for example they serve as antioxidant molecules scavenging toxic oxygen radicals, they also function as compatible solutes to protect the cells against salt stress.

They are also involved in protection against dessication or thermal stress in certain organisms and they may serve as an intercellular nitrogen reservoir [79]. By all these functions mycosporines are mentioned as multi response secondary metabolites. The finding that UV light is generally the strangest inducer for the biosynthesis of MAAs agrees well with their function as sunscreen components MAAs synthesis is promoted with UV light. MAAs synthesis in cyanobacteria enhanced by photosynthetically active radiation (PAR), UV-A (315-400 nm) UV-B (280-315 nm) light radiation [80]. UV-B light is the highest of three light regimes in enhancement of MAAs.

UV-B light can also enhances the ROS production because of their higher energy when compared to UV-A. The accumulated ROS can function as signal transfer molecules which effect antioxidant genes to turn on signal transduction for the biosynthesis of various antioxidants and antioxidant enzymes.

Therefore MAAs biosynthesis can be activated in an indirect manner via ROS accumulation which is am primary event caused by UV-B radiation. MAAs function as UV-A absorbing agent and can be release light energy as heat without ROS production [81, 82, 83]. This is because the range of absorbtion bands of MAAs overlaps well with the UV-A range and their molar extinction coefficients are very high.

Scytonemin

Scytonemin is a biological pigment synthesised by many strains of cyanobacteria. It is believed to be a bacterial sunscreen pigment which has the absorption from 325-425 nm and separate maxima at 250 nm. The biosynthesis of scytonemin is triggered by UV light. It is hydrophobic pigment whereas MAAs are hydrophilic pigments. The structure of this pigment is found by the dimerization of two polycyclic chromophores made by the condensation of cyclopentanone and indole rings [84].

These two chromophores are connected by a single bond in an oxidised state. Therefore these can be rotate freely and this can prevent steric repulsions. In living Cyanobacteria, scytonemin exists in an oxidised state, because of this; their presence is easily recognised by their brown colour which can be observed in a bright field optical microscope. Almost no scytonemin derivatives have been discovered unlike MAAs, except for methoxy substituted scytonemin derivatives.

These are isolated from *Scytonema* Sp. and their structures were determined by NMR and MS spectrometers [85].

Sunscreen Properties of Scytonemin

Scytonemin biosynthesis is activated during the exposure of light to the cyanobacteria to UVA/B (especially UV-A) [86, 87]. By this it is expected that Scytonemin functions as a photo protective compound. It is also known that repeated desiccation activities in some anhydrobiotic cyanobacteria activate the synthesis of scytonemin under UV-A exposure [88, 89]. It is clearly reported that molar extinction co-efficient of Scytonemin is large (250 l/g•cm) at wavelength 384 nm [90]. Therefore Scytonemin is an efficient photoprotective compound due to its large extinction co-efficient. The wide absorption range of Scytonemin also helps for photoprotection. This is because it can cover not only the UV-B range, but also UV-A range. Scytonemin can protect the cells by preventing 90% of UV-A light penetrating into cell [91].

DNA Repair Mechanisms in Cyanobacteria

Cyanobacteria have evolved three different repair mechanisms to minimize the UV-B induced damage to the genetic material. In addition DNA strand breaks, DNA- protein cross links, insertion or deletion of base pairs can also be induced by UV exposure [92]. UV radiation induced DNA damage can be repaired by Photoreactivation, excision or recombinational repair [93,94]. Photoreactivation and excision repair in dark has been reported in higher plants like Nicotiana, Petunia, Haplopappus mainly following UV-C exposure [95]. The process of excision repair can be divided into three steps: nicking of the damaged DNA near the site of the damage, removal of the bases in the damaged strand and resynthesis of the gap. The gene responsible for the DNA repair enzyme Fpg (formamydopyrimidine -DNA glycosylase) has been reported from *Synechococcus elongatus* and was suggested to be involved in photoprotection against oxidative damage [96]. It has been reported photoreactivation of UV-B induced inhibition of photosynthesis in *Anabaena* Sps [97]. The most common type of DNA damage induced by UVR is the formation of cyclobutyl pyrimidine dimmers (CPDs) and pyrimidine (6-4) pyrimidone photoproducts, leading to mutagenic, teratogenic or lethal effects in organisms, because these lesions prevent DNA replication and transcription. Organisms use different types of DNA repair mechanisms including photoenzymatic repair (PER), nucleotide excision repair (NER) and post replication repair [98].

Future Prospective

UVR reaching the earth surface has many detrimental effects on the physiological and biological processes of cyanobacteria. This result in the reduction in growth and survival, in addition to this photosynthesis, pigments, protein content, lipids, DNA, nitrogen fixation are adversely affected. However, these organisms have developed several defence mechanisms against the damage to attain success in sustaining growth and development. They sustain balance in damage and defence mechanisms to maintain their productivity and ecological importance. Several reports suggest damage by UVR, so further studies have to be done on various enzymatic and non- enzymatic defence mechanisms in cyanobacteria. DNA damage and repair mechanisms are to be studied in detail.Establishment of cyanobacteria under field condition may provide protective mechanism to other beneficial microorganisms from possible future threat of UV radiation to maintain the agricultural productivity sustainable.

ACKNOWLEDGMENT

SDS is thankful to UGC for providing one time grant (2013-14).

REFERENCES

[1] Kramlich, J. C., Linak, W. P., (1994). Nitrous oxide behaviour in the atmosphere, and in combustion and industrial systems. *Prog. Energy Combust. Sci.,* 20, 149.

[2] Singh, S. P., Kumari, S., Rastogi, R. P., Singh, K. L., Richa and Sinha, R. P. (2010b). Photoprotective and biochemical potentials of cyanobacterial sheath pigment, scytonemin. *Afr. J. Biotechnol.*, 9(5), 580-588.

[3] Crutzen, P. J., (1992). Ultraviolet on the increase. *Nature,* 356, 104-105.

[4] Smith, R. C., Pre´zelin, B. B., Baker, K. S., Bidigare, R. R., Boucher, N. P., Coley, T. Karentz, D., MacIntyre, S., Matlick, H. A., Menzies, D., Ondrusek, M., Wan, Z., Waters, K. J., (1992). Ozone depletion: ultraviolet radiation and phytoplankton biology in Antarctic waters. *Science,* 255, 952-959.

[5] Hader, D.-P., Kumar, H. D., Smith, R. C., Worrest, R. C., (2007). Effects of solar UV radiation on aquatic ecosystems and interactions with climate change. *Photochem. Photobiol. Sci.,* 6, 267-285.

[6] Fernanda Pessoa, M. (2012). Harmful effects of UV-radiation in algae and aquatic macrophytes- A review. *Emir. J. Food Agric.,* 24, 510-526.

[7] Norval, M., Cullen, A. P., de Gruijl, F. R, Longstreth, J., Takizawa, Y., Lucas, R. M., Noonan, F. P., van der Leun, J. C., (2007). The effects on human health from stratospheric ozone depletion and its interactions with climate change. *Photochem. Photobiol. Sci.,* 6, 232-251.

[8] Stanier, R. Y., Cohen-Bazire, G., (1977). Phototrophic prokaryotes: the cyanobacteria. *Annu. Rev. Microbiol.,* 31, 225-274.

[9] Rastogi, R. P., Sinha, R. P., (2009). Biotechnological and industrial significance of cyanobacterial secondary metabolites. *Biotechnol. Adv.,* 27, 521-539.

[10] Vaishampayan, A., Sinha, R. P., Hader, D.-P., Dey, T., Gupta, A. K., Bhan, U., Rao, A. L., (2001).Cyanobacterial biofertilizers in rice agriculture. *Bot. Rev.,* 67, 453-516.

[11] Fernanda Pessoa, M. (2012).Algae and aquatic macrophytes responses to cope to ultraviolet radiation- A review. *Emir. J. Food Agric.,* 24, 527-545.

[12] Gao, K., Yu, H., Brown, M. T., (2007). Solar PAR and UV radiation affects the physiology and morphology of the cyanobacterium *Anabaena* Sp. PCC 7120. *J. Photochem. Photobiol. B: Biol.,* 89, 117-124.

[13] Sinha, R. P., Hader, D.-P., (2002). Life under solar UV radiation in aquatic organisms. *Adv. Space Res.,* 30, 1547-1556.

[14] Kumar, A., Tyagi, M. B., Jha, P. N., Srinivas, G., Singh, A., (2003). Inactivation of cyanobacterial nitrogenase after exposure to ultraviolet-B radiation. *Curr. Microbiol.,* 46, 380-384.

[15] Dahms, H.-U., Lee, J.-S., (2010). UV radiation in marine ectotherms: molecular effects and responses. *Aquat. Toxicol.*, 97, 3-14.

[16] Dahms, H.-U., Dobretsov, S., Lee, J.-S., (2011). Effects of UV radiation on marine ectotherms in Polar Regions. Comp. *Biochem. Phys. C*, 153, 363-371.

[17] Donkor, V. A., Damian, H. A. K., Hader, D.-P., (1993). Effects of tropical solar radiation on the motility of filamentous cyanobacteria. *FEMS Microbiol. Ecol.,* 12, 143-148.

[18] Hader, D. P., Worrest, R. C., (1991). Effects of enhanced solar radiation on aquatic ecosystems. *Photochem. Photobiol.,* 53, 717-725.

[19] Tyagi, R., Srinivas, G., Vyas, D., Kumar, A., Kumar, H. D., (1992). Differential effect of ultraviolet-B radiation on certain metabolic processes in a chromatically adapting. *Nostoc. Photochem. Photobiol.,* 55, 401-407.

[20] Priyadarshini, I., Rath, B., (2012). Impact of UV-B radiation on two species of cyanobacteria of Odisha coast. *Asian J. Exp. Biol. Sci.,* 3(4), 733-739.

[21] Quesada, A., Vincent, W. F., (1997). Strategies of adaptation by Antarctic cyanobacteria to ultraviolet radiation. *Eur. J. Phycol.,* 32, 335-342.

[22] Vass, I., Sass, L., Spetea, C., Bakou, A., Ghanotakis, D., Petrouleas, V., (1996). UV-B induced inhibition of photosystem II electron transport studied by EPR and chlorophyll fluorescence. Impairment of donor and acceptor side components. *Biochemistry,* 35, 8964–8973.

[23] Vass, I., Turcsa'nyi, E., Touloupakis, E., Ghanotakis, D., Petroluleas, V., (2002). The mechanism of UV-A radiationinduced inhibition of photosystem II electron transport studied by EPR and chlorophyll fluorescence. *Biochemistry,* 41, 10200–10208.

[24] Blakefield, M. K., Harris, D. O., (1994). Delay of cell differentiation in *Anabaena aqualis* caused by UV-B radiation and the role of photoreactivation and excision repair. *Photochem. Photobiol.,* 59, 204-208.

[25] Sinha, R. P., Singh, N., Kumar, A., Kumar, H. D., Hader, M., Hader, D. P., (1996). Effect of UV radiation on certain physiological and biochemical processes in cyanobacteria. *J. Photochem. Photobiol. B: Biol.,* 32, 107-113.

[26] Kataria, S., Jajoo, A., and Guruprasad, K. N. (2014). Impact of increasing Ultraviolet-B (UV-B) radiation on photosynthetic processes. *J. Photochem. Photobiol.,* 137, 55-66.

[27] Sinha, R. P., Rastogi, R. P., Ambasht, N. K., Hader, D.-P., (2008). Life of wetland cyanobacteria under enhancing solar UV-B radiation. *Proc. Natl. Acad. Sci. India B,* 78, 53-65.

[28] Sinha, R. P., Hader, D.-P., (1998). Effects of ultraviolet-B radiation in three rice field cyanobacteria. *J. Plant Physiol.*, 153, 763-769.

[29] de Oliveira, C. A., de Castro Oliveira, W., and Rocha, S. M. (2014). Effect of light intensity on the production of pigments in *Nostoc* Sps. *Euro. J. Biol. Medi. Sci. Res.*, 2, 23-36.

[30] Jordan, B. R., James, P. E., Strid, A., Anthony, R. G., (1994). The effect of ultraviolet-B radiation on gene expression and pigment composition in etiolated and green pea leaf tissue: UV-B induced changes are gene-specific and dependent upon the developmental stage. *Plant Cell Environ.,* 17, 45–54.

[31] Pandey, R., Chauhan, S., Singhal, G. S., (1997). UVB-induced photodamage to phycobilisomes of Synechococcus Sp. PCC 7942. *J. Photochem. Photobiol.*, 40, 228–232.

[32] Bishop, N. I., (1961). The possible role of plastoquinone (Q-254) in the electron transport system of photosynthesis. *CIBA Symp.*, 1961; 385–404.

[33] Barbato, R., Frizzo, A., Friso, G., Rigoni, F., Giacometti, G. M., (1995). Degradation of the D1 protein of photosystem-II reaction centre by ultraviolet-B radiation requires the presence of functional manganese on the donor side. *Eur. J. Biochem.*, 227, 723-729.

[34] Gao, K., Li, P., Watanabe, T., Helbling, E. W., (2008). Combined effects of ultraviolet radiation and temperature on morphology, photosynthesis, and DNA of *Arthrospira (Spirulina) platensis* (Cyanophyta). *J. Phycol.*, 44, 777-786.

[35] Wu, H., Gao, K., Villafan˜ e, V. E., Watanabe, T., Helbling, E. W., (2005). Effects of solar UV radiation on morphology and photosynthesis of filamentous cyanobacterium *Arthrospira platensis*. *Appl. Environ. Microbiol.*, 71, 5004-5013.

[36] Rastogi, R. P., Singh, S. P., Incharoensakdi, A., Hader, D. P., Sinha, R. P., (2014). Ultraviolet radiation-induced generation of reactive oxygen species, DNA damage and induction of UV-absorbing compounds in the cyanobacterium *Rivularia* Sp. HKAR-4, *S. Afr. J. Bot.*, 90, 163-169.

[37] Rastogi, R. P, Richa, Kumar, A, Tyagi, M. B., Sinha, R. P., (2010). Molecular mechanisms of ultraviolet radiation-induced DNA damage and repair. *J. Nuc. Acids*, Article ID 592980, doi:10.4061/2010/592980 1-32.

[38] He, Y. Y., Hader, D. P., (2002). Reactive oxygen species and UV-B: effect on cyanobacteria. *Photochem. Photobiol. Sci.,* 1, 729-736.

[39] Sinha, R. P., Dautz, M., Hader, D.-P., (2001). A simple and efficient method for the quantitative analysis of thymine dimers in cyanobacteria, phytoplankton and macroalgae. *Acta Protozool.*, 40, 187-195.

[40] Bhargava, P., Atri, N., Srivastava, A. K., Rai, L. C., (2007). Cadmium mitigates ultraviolet-B stress in *Anabaena doliolum*: enzymatic and non-enzymatic antioxidants. *Biol. Plant,* 51, 546-550.

[41] Vladimirov, Y. A., Roshchupkin, D. I., Fesenko, E. E., (1970). Photochemical reactions in amino acids residues and inactivation of enzymes during U.V.-irradiation. A review. *Photochem. Photobiol.*, 11, 227–246.

[42] Creed, D., (1984). The photophysics and photochemistry of the near UV absorbing amino acids-II. Tyrosine and its simple derivates. *Photochem. Photobiol.*, 39, 563–575.

[43] Tevini, M., (1993). Molecular biological effects of ultraviolet radiation. In: *UV-B Radiation and Ozone Depletion. Effects on Humans, Animals, Plants, Microorganisms, and Materials* (Tevini M., Ed.), Boca Raton, FL: Lewis Publishers, pp. 1-15.

[44] Hifney, A. F., Issa, A. A., and Fawzy, M. A. (2013). Abiotic stress induced production of β-carotene, allophycocyanin and total lipids in Spirulina sp. *Journal of Biology and Earth Sciences*, 3, B54-B64.

[45] Kramer, G. F., Norman, H. A., Krizek, D. T., Mirecki, R. M., (1991). Influence of UV-B radiation polyamines, lipid peroxidation and membrane lipids in cucumber. *Phytochemistry,* 30, 2101–2108.

[46] Richa, Sinha, R. P., (2011). UV-mediated stress and is mitigation in cyanobacteria. *IJPAES,* 1(2), 155-166.

[47] Richa and Sinha,R. P.(2011).UV-mediated stress and is mitigation in cyanobacteria. *IJPAES,* 1(2):155-166.
[48] Bebout, B. M., Garcia-Pichel, F., (1995). UV B-induced vertical migrations of cyanobacteria in a microbial mat. *Appl. Environ. Microbiol.*, 61, 4215-4222.
[49] Reynolds, C. S., Oliver, R. L., Walsby, A. E., (1987). Cyanobacterial dominance: the role of buoyancy regulation in dynamic lake environments. *N. Z. J. Mar. Freshwater Res.,* 21, 379-390.
[50] Ramsing, N. B., Prufert-Bebout, L., (1994). Motility of *Microcoleus chthonoplastes* subject to different light intensities quantified by digital image analysis. In: (Stal L. J., Caumette, P., Eds), Microbial Mats, Structures, Development and Environmental Significance. Springer, Heidelberg, pp. 183-191.
[51] Vincent, W. F., and Quesada, A. (1994). Ultraviolet radiation effects on cyanobacteria: implications for Antarctic microbial ecosystems (pp. 111-124). American Geophysical Union.
[52] Quesada, A., Mouget, J. L., Vincent, W. F., (1995). Growth of Antarctic cyanobacteria under ultraviolet radiation: UVA counteracts UVB inhibition. *J. Phycol.,* 31, 242-248.
[53] Budel, B. (1999).Ecology and diversity of rock inhabiting cyanobacteria in tropical regions. *Eur. J. phycol.,* 34, 361-370.
[54] Miller, S. R., Wingard, C. E., Castenholz, R. W., (1998). Effects of visible light and ultraviolet radiation on photosynthesis in a population of the hot spring cyanobacterium, Synechococcus sp., subjected to high-temperature stress. *Appl. Environ. Microbiol.,* 64, 3893-3899.
[55] Quesada, A., Vincent, W. F., Lean, D. R. S., (1999). Community and pigment structure of Arctic cyanobacterial assemblages: the occurrence and distribution of UV-absorbing compounds. *FEMS Microbiol. Ecol.,* 28, 315–323.
[56] Sorensen, K. B., Canfield, D. E., Teske, A. P., Oren, A. (2005). Community composition of a hypersaline endoevaporitic microbial mat. *Appl. Environ. Microbiol.*, 71, 7352–7365.
[57] Adhikary, S. P., Sahu, J. K., (2000). Survival strategies of cyanobacteria occurring as crust in the rice fields under drought conditions. *Ind. J. Microbiol.*, 40, 53-56.
[58] Scherer, S., Ernst, A., Chen, T. W., Boger, P., (1984). Rewetting of drought-resistant blue-green algae: Time course of water uptake and reappearance of respiration, photosynthesis, and nitrogen fixation. *Oecologia,* (Berlin) 62, 418-423.
[59] Pattanaik, B., Adhikary, S. P., (2002). Blue green algal flora at some archeological sites and monuments of India. *Feddes Reportor.,* 113, 289-300.
[60] Mandal, S., Rath, J., Adhikary, S. P., (2011). Adaptation of the sheathed cyanobacterium. *Lyngbya majuscule* to ultraviolet-B. *J. Photoch. Photobio.,* B 102, 115-122.
[61] Banaszak, A. T., Trench, R. K., (2001). Ultraviolet sunscreens in dinoflagellates. *Protist.,* 152, 93-101.
[62] Singh, G., Piyoosh, K., Babele, Sinha, R. P., M. B., Tyagi, Kumar, A., (2013). Enzymatic and non-enzymatic defense mechanisms against ultraviolet-B radiation in two *Anabaena* species, *Process Biochemistry,* 48, 796-802.
[63] Mallick, N., Mohn, F. H., (2000). Reactive oxygen species: response of algal cells. *J. Plant Physiol.*, 157, 183-193.

[64] Wang, G., Hu, C., Li, D., Zhang, D., Li, X., Chen, K., Liu, Y., (2007). The response of antioxidant systems in *Nostoc sphaeroides* against UV-B radiation and the protective effects of exogenous antioxidants. *Adv. Space Res.*, 39, 1034-1042.

[65] Li, L., Zhao, J., Tang, X., (2010). Ultraviolet irradiation induced oxidative stress and response of antioxidant system in an intertidal macro algae *Corallina officinalis L. J. Environ. Sci.*, 22(5), 716-722.

[66] Malanga, G., Kozak, R. G., Puntarulo, S., (1999). N-Acetylcysteine-dependent protection against UV-B damage in two photosynthetic organisms. *Plant Sci.*, 141, 129-137.

[67] Santabarbara, S., Casazza, A. P., Ali, K., Economou, C. K., Wannathong, T., Zito, F., ... and Purton, S. (2013). The requirement for carotenoids in the assembly and function of the photosynthetic complexes in Chlamydomonas reinhardtii.*Plant Physiol.*, 161, 535–546.

[68] Jallet, D., Thurotte, A., Leverenz, R. L., Perreau, F., Kerfeld, C. A., and Kirilovsky, D. (2014). Specificity of the Cyanobacterial Orange Carotenoid Protein: Influences of Orange Carotenoid Protein and Phycobilisome Structures. *Plant Physiol.*, 164, 790-804.

[69] Castenholz, R. W., (1972). Low temperature acclimation and survival in thermophilic *Oscillatoria terebriforms*. In: *Taxonomy and* Biology of Blue-green algae (Desikachary, T. V. Ed), University of Madras, India, pp. 406-418.

[70] Tevini, M. Steinmuller, D., (1987). lnfluence of light, UV-B radiation, and herbicides on wax biosynthesis of cucumber seedlings. *J. Plant Physiol.,* 131, 111-121.

[71] Tevini, M., (2004). Plant responses to ultraviolet radiation stress. In: *Chlorophyll a Fluorescence a Signature of Photosynthesis* (Papageorgiou, G. C., Govindjee, Eds.), Springer, Dordrecht, pp. 605-621.

[72] Li, J., Ou-Lee, T. M., Raba, R., Amundson, R. G., Last, R. L., (1993). Arabidopsis flavonoid mutants are hypersensitive to UV-B irradiation. *Plant Cell,* 5, 171-179.

[73] Day, T. A., Vogelmann, T. C., (1995). Alterations in photosynthesis and pigment distributions in pea leaves following UV-B exposure. *Physiol. Plant*, 94, 433–440.

[74] Van de Staaij, J., de Bakker, N. V. J., Oosthoek, A., Broekman, R., van Beem, A., Stroetenga, M., Aerts, R., Rozema, J., (2002). Flavonoid concentrations in three grass species and a sedge grown in the field and under controlled environment conditions in response to enhanced UV-B radiation. *J. Photochem. Photobiol. B: Biol.,* 66, 21-29.

[75] Schmitz-Hoerner, R., Weissenböck, G., (2003). Contribution of phenolic compounds to the UV-B screening capacity of developing barley primary leaves in relation to DNA damage and repair under elevated UV-B levels. *Phytochemistry,* 64, 243–255.

[76] Agati, G., Brunetti, C., Di Ferdinando, M., Ferrini, F., Pollastri, S., and Tattini, M. (2013). Functional roles of flavonoids in photoprotection: new evidence, lessons from the past. *Plant Physiol.Biochem.: PPB/Societe francaise de physiologie vegetale*, 72, 35-45.

[77] Ziska, L. H., Teramura, A. H., Sullivan, J. H., (1992). Physiological sensitivity of plants along an elevational gradient to UV-B radiation. *Am. J. Bot.,* 79, 863-871.

[78] Rastogi R. P., Incharoensakdi, A., (2014). Characterization of UV-screening compounds, mycosporine-like amino acids, and scytonemin in the cyanobacterium Lyngbya Sp. CU2555. *FEMS Microbiol. Ecol.,* 87, 244–256.

[79] Rastogi, R. P., Singh, S. P., Incharoensakdi, A., Hader, D. P., Sinha, R. P., (2014). Ultraviolet radiation-induced generation of reactive oxygen species, DNA damage and

induction of UV-absorbing compounds in the cyanobacterium *Rivularia* Sp. HKAR-4. *S. Afr. J. Bot.,* 90, 163-169.

[80] Mushir, S. and Fatma, T., (2012). Monitoring stress responses in cyanobacterial scytonemin - screening and characterization. *Environ. Technol.,* 33, 153–157.

[81] Rastogi, R. P., Incharoensakdi, A., (2013). UV radiation-induced accumulation of photoprotective compounds in the green alga *Tetraspora* Sp. CU2551, *Plant physiol. Biochem.,* 70, 7-13.

[82] Conde, F. R., Churio, M. S., Previtali, C. M., (2000). The photoprotector mechanism of mycosporine-like amino acids. Excited state properties and photostability of porphyra-334 in aqueous solution. *J. Photochem. Photobiol. B,* 56, 139-144.

[83] Conde, F. R., Churio, M. S., Previtali, C. M., (2004). The deactivation pathways of the excited-states of the mycosporine-like amino acids shinorine and porphyra-334 in aqueous solution. *Photochem. Photobiol. Sci.,* 3, 960-967.

[84] Conde, F. R., Churio, M. S., Previtali, C. M., (2007). Experimental study of excited-state properties and photostability of the mycosporine-like amino acid palythine in water solution. *Photochem. Photobiol. Sci.,* 6, 669-674.

[85] Proteau, P. J., Gerwick, W. H., Garcia-Pichel, F., Castenholz, R., (1993). The structure of scytonemin, an ultraviolet sunscreen pigment from the sheaths of cyanobacteria. *Experientia,* 49, 825-829.

[86] Bultel-Poncé, V., Felix-Theodore, F., Sarthon, C., Ponge, J. F., Bodo, B., (2004). New pigments from the terrestrial cyanobacterium *Scytonema.* Sp. collected on the Mitaraka Inselberg, French Guyana. *J. Nat. Prod.,* 67, 678-681.

[87] Dillon, J. G., Tatsumi, C. M., Tandingan, P. G., Castenholz, R. W., (2002). Effect of environmental factors on the synthesis of scytonemin, a UV-screening pigment, in a cyanobacterium (*Chroococcidiopsis.* Sp.) *Arch. Microbiol.,* 177, 322-331.

[88] Ehling-Schulz, M., Bilger, W., Scherer, S., (1997). UV-B induced synthesis of photoprotective pigments and extracellular polysaccharides in the terrestrial cyanobacterium *Nostoc commune. J. Bacteriol.,* 179, 1940-1945.

[89] Fleming, E. D., Castenholz, R. W., (2007). Effects of periodic desiccation on the synthesis of the UV-screening compound, scytonemin, in cyanobacteria. *Environ. Microbiol.,* 9, 1448-1455.

[90] Fleming, E. D., Castenholz, R. W., (2008). Effects of nitrogen source on the synthesis of the UV-screening compound, scytonemin, in the cyanobacterium *Nostoc punctiforme PCC 73102. FEMS Microbiol. Ecol.,* 63, 301-308.

[91] Vincent, W. F., Downes, M. T., Castenholz, R. W., Howard-Williams, C., (1993). Community structure and pigment organisation of cyanobacteria-dominated microbial mats in Antarctica. *Eur. J. Phycol.,* 28, 213-221.

[92] Garcia-Pichel, F., Sherry, N. D., Castenholz, R. W., (1992). Evidence for an ultra-violet sunscreen role of the extracellular pigment scytonemin in the terrestrial cyanobacterium *Chlorogloeopsis.* Sp *.Photochem. Photobiol.,* 56, 17-23.

[93] Smith, K. C., (1977). Ultraviolet radiation effects on molecules and cells. In: *The Science of Photobiology* (Smith KC, Ed.), Plenum Press, New York, pp. 113–142.

[94] Britt, R. D., (1996). Oxygen evolution. In: *Oxygenic Photosynthesis: The Light Reactions.* (Ort, D. R., Yocum, C. F., Eds), Kluwer Academic Publisher, Dordrecht, The Netherlands, pp.137-164.

[95] Taylor, R. M., (1996). Ultraviolet-B induced DNA lesions and their removal in wheat (*Triticum aestivum* L.) leaves. *Plant Cell Environ.*, 19, 171–181.

[96] Mclennan, A. G., Eastwood, A. C., (1986). An endonuclease activity from suspension cultures of *Daucus carota* which acts upon pyrimidine dimers. *Plant Sci.*, 46, 151-157.

[97] Muhlenhoff, U., (2000). The FAPY-DNA glycosylase (Fpg) is required for survival of the cyanobacterium *Synechococcus elongatus* under high light irradiance. *FEMS Microbiol. Lett.*, 187, 127-132.

[98] Han, T., Sinha, R. P., Hader, D.-P., (2001). UV-A/blue light-induced reactivation of photosynthesis in UV-B irradiated cyanobacterium, *Anabaena* Sp. *J. Plant Physiol.*, 158, 1403-141.

[99] Vincent, W. F., Roy, S., (1993). Solar ultraviolet-B radiation and aquatic primary production: damage, protection and recovery. *Environ. Rev.,* 1, 1-12.

In: Microbes in Soil and Their Agricultural Prospects
Editors: K. Kishore Choudhary and D. Wattal Dhar
ISBN: 978-1-63482-824-6

Chapter 9

ISOLATION AND CHARACTERIZATION OF *BURKHOLDERIA* FROM THE RHIZOSPHEREOF DIFFERENT CROPS AND ITS PLANT GROWTH PROMOTING ACTIVITIES

S. Priyanka*[*]*, M. Sivaji and R. Sridar
Department of Plant Biotechnology, Centre for Plant Molecular Biology and Biotechnology, Tamil Nadu Agricultural University (TNAU), Coimbatore, Tamil Nadu, India

ABSTRACT

The bacterium *Burkholderia*was isolated from the rhizosphere soil of *Mimosa pudica*, lemon, maize, sugarcane, sunflower, rice, okraand sunhemp. The isolates were checked for phosphate solubilization, nitrogen fixation and antagonistic activity against plant pathogenic fungi like *Macrophoeminaphaesolina*and*Rhizoctoniasolani*. Biochemical and physiological characterization were done for the obtained isolates to screen for further studies. Plant growth promoting activities were accessed for the isolates through secondary metabolite production. Molecular screening was done using 16S rRNA primers, sequenced and analyzed using the NCBI BLAST tool. Pot culture studies were performed by treating groundnut seedswith*Burkholderia*isolates, to improve root and shoot growth.

Keywords: nitrogen fixation, phosphate solubilization; antagonism, hydrogen cyanide production, IAA production

[*]Corresponding author: S.Priyanka, Department of Plant Biotechnology, Centre for Plant Molecular Biology and Biotechnology, Tamil Nadu Agricultural University (TNAU), Coimbatore, Tamil Nadu-641003. Email: spriyankamsc@ymail.com.

INTRODUCTION

The genus *Burkholderia*currently comprises more than 60 species, most of them recently described. [1] They are widely distributed in nature and found in soil, water and rhizosphere in association with plants. Several *Burkholderia*species are involved in agronomical and biotechnological applications such as, production of antifungal substances for medical, veterinary and agricultural purposes, [2] biodegradation of polluting compounds such as trichloroethane; ability to fix nitrogen; [3] phosphate-solubilization; [4] production of phytohormones, [5] promoting plant growth and direct antagonism or induction of resistance against pathogens. [6] Thus, the study of this genus is relevant for bioprospecting of new species or even strains with new or more pronounced properties with potential application in biotechnology. [7, 8]

For a long time, N_2-fixing ability in bacteria of the genus *Burkholderia*was recognized only in the species *Burkholderiavietnamiensis*, which attracted interest because of its ability to promote rice plant growth and grain yield. Recently, the analysis of N_2-fixing bacteria associated with maize and coffee plants grown under field conditions revealed the presence of *B. vietnamiensis*, as well as the richness of novel diazotrophic bacterial species belonging to the genus *Burkholderia*. [9]Several *Burkholderia* species have developed intimate beneficial interactions with plants and colonize roots, stems and leaves.

The majority of *Burkholderia* species however, are known as soil bacteria, which exhibit different types of non-pathogenic interactions with plants. The ecological role of species like *Burkholderiaglathei*, *Burkholderiagraminis*, *Burkholderiaphenazinium*, *Burkholderiacaribensis*, *Burkholderiacaledonica*, *Burkholderiahospita*, *Burkholderiaterricola* and *Burkholderiasacchari* are largely unknown. [10] *Burkholderiaphenazinium* strains were isolated from soil and were characterized by the production of iodinin. [11] Strains belonging to *B. graminis* were first isolated from the rhizosphere of wheat, corn and pasture grasses during a field survey of French and Australian soils. [12] An efficient exopolysaccharide producing strain, *Burkholderiacaribensis* was isolated from a vertisol on the island of Martinique. [13] *B. caledonica* strains were isolated from the rhizosphere of different plants in Scotland. *B.glathei*was isolated from fossil lateritic soils in Germany. Some of these soil-borne *Burkholderia* species were detected in several soil types as transconjugants that acquired the plasmids pJP4 or pEMT1 (containing genes required for the degradation of 2, 4-dichlorophenoxyacetic acid) following inoculation of a donor strain. [14] A more detailed study of some of these transconjugants entailed the description of *B. hospita* and *B. terricola*. [15]. *B. sacchari* can produce polyhydroxyalkanoates from a wide range of carbon sources. [16]

The exceptional metabolic versatility of *B. cepacia* complex bacteria and strains belonging to other *Burkholderia* species can also be used for bioremediation purposes. Constituents of crude oils [including polycyclic aromatic hydrocarbon (PHA) compounds], herbicides (including 2, 4-dichlorophenoxyacetic acid and 2, 4, 5-trichlorophenoxyacetic acid) TCE and ether derivatives used as gasoline additives can be degraded by *Burkholderia* isolates. [17] Several *Burkholderia* strains with potential applications in bioremediation are extremely well-studied. The *B. cepacia* strains like G4, [18] and CRE-7, [19] are efficient biodegraders. *Burkholderiafungorum*strain LB400 and other closely related *Burkholderia* strains are good biodegraders of polychlorinated biphenyls (PCBs) with "unparalleled ability to destroy environmentally important PCB congeners". *B. kururiensis*, was isolated from an aquifer polluted with trichloroethylene. [20]

Some of the *Burkholderia* strains are capable of inducing nodule formation in plants like *Mimosa*. [21] Endosymbionts of leaf galls of *Psychotriakirkii* were classified as *CandidatusBurkholderiakirkii*. [22] Uncultured bacterial endosymbionts of the arbuscularmycorrhizal fungi belonging to the *Gigasporaceae* were recently formally classified as *Candidatus*Glomeribactergigasporarum. [23] These bacteria are close relatives of the genus *Burkholderia* and were also reported to contain putative *nif* genes. [24] Most remarkably, the microflora inhabiting a pouch-shaped organ at the junction of the midgut in the intestine of *Tetraponera* ants partially consists of *Burkholderia* species, which are most likely involved in the oxidative recycling of nitrogen-rich metabolic waste [25]. These organisms are phylogenetically closely related to *B. fungorum* and *B. caledonica*.

Phosphorus (P) is one of the major essential macronutrients for biological growth and development. Microorganisms play a central role in the natural phosphorus cycle. They solubilize the bound phosphate present in soil. The phenomena of fixation and precipitation of P in soil is generally highly dependent on pH and soil type. In addition, studies on the diversity of plant-associated bacteria may contribute to the discovery of new beneficial plant-microbe interactions. Plant root-associated phosphate solubilizing bacteria (PSB) have been considered as one of the possible alternatives for inorganic phosphate fertilizers for promoting plant growth and yield. [26] *Burkholderiaanthina*is an efficient phosphate solubilizer. They not only provide plants with phosphorus, but also facilitate the growth of plants by fixing atmospheric nitrogen; [27] accelerating the accessibility of other trace elements; producing plant hormones such as auxins, cytokinins, and gibberellins, [28] releasing siderophores, hydrogen cyanide, enzymes and/or fungicidal compounds such as chitinase, cellulose, protease, [29] which ensure antagonism against phytopathogenic microbes. Therefore, it is worth to believe that production of plant growth promoting substances by microbes may effectively contribute to their effect on the enhancement of the plant performance. [30]

The biological control of plant diseases, insects and nematodes by microorganisms have been proposed as an alternative or a supplement to chemical pesticides and the use of introduced biological control could have enormous ecological and economic benefits. [31, 32] However, the greatest progress towards biological control of soil-borne plant pathogens have been made with antagonists introduced with the planting material, i.e., biological control with plant-associated microorganisms. [33] *Burkholderiacepacia* complex strains can be used to control seedling and root diseases *in vitro* and hence can replace chemical alternatives. Field tests have shown that such strains can colonize the rhizosphere of several crops, including corn, maize, rice, pea, sunflower and radish, and thereby significantly increases the crop yield. [34] One of the most-studied biocontrol agent, *B. ambifaria* strain AMMD^T which was isolated from the rhizosphere of peas (*Pisumsativum*) in Wisconsin (USA) in 1985. It has activity against *Pythiumaphanidermatum* (responsible for pre- and post-emergence damping-off in peas) and *Aphanomyceseuteiches* (responsible for root rot in peas).

This chapter reveals the techniques of isolation and characterization of *Burkholderia* species from the rhizosphere soil and assessment of their ability to solubilize bound phosphate, nitrogen fixation and antagonistic effect on soil borne plant pathogens. This chapter also emphasizes the use of *Burkholderia* as a biofertilizer to crops.

Materialsand Methods

Burkholderia Isolation

Rhizosphere soil samples of plants like *Mimosa*, lemon, maize, sugarcane, sunflower, rice, sunhemp and okra were collected from the fields in TNAU farms, Coimbatore (Table 1). Malate medium was used for the isolation of *Burkholderia* sp. [35] Sterile suspensions of the soil sample was made and 1 ml of suspensions from dilutions 10^{-5} and 10^{-6} was inoculated in sterile nitrogen free Malate medium. They were incubated for 3-5 days and the colonies which transformed the color of medium from yellow to blue, which is an indicative of nitrogen fixation, were selected. They were further purified and each isolate was characterized by Gram reaction, motility, morphology, temperature, pH and salt tolerance tests, biochemical tests etc.

Characterization of the Obtained Isolates

The isolates were purified and checked for Gram reaction and characterized biochemically. Exocellular enzymatic tests like gelatin liquefaction, starch hydrolysis, lipid hydrolysis, caesin hydrolysis were performed. Also, catalase test, MR-VP, indole test, urease test, citrate utilization test, hydrogen sulfide production test, carbohydrate fermentation test with utilization of sugars like glucose, lactose were performed for all the isolates and the results were recorded. [36]

Phosphate Solubilization

Burkholderia isolates were inoculated in Sperber's Hydroxyl Apatite medium (HAP). [37] The HAP plates were incubated at 37°C for 3-5 days for determining the 'P' solubilizing ability of the isolates, which was indicated by zone formation around the colonies. Quantitative estimation of P solubilization was carried out as per the method described by Olsen et al. (1975) and the results were recorded. [38]

Nitrogen Fixation

The Nitrogen fixing ability of the isolates were identified by growing the cultures in a nitrogen free BMGM broth, [39] at an initial pH of 5.7 by using malate as a carbon source. The sterilized BMGM broth was inoculated with the test isolates and incubated at 37^0C. The alterations in pH were recorded at 3 days interval and results were tabulated. The isolates were further inoculated in nitrogen free BAZ medium and incubated for 3-5 days and color change of the medium was observed for confirmation of N_2 fixing activity.

From a total of 30 isolates, only 10 isolates which possessed the ability to solubilize phosphate as well as to fix nitrogen were selected for further study.

Table 1. List of selected *Burkholderia* isolates obtained from various fields of TNAU

S. NO.	CROPS	ISOLATE NAME
1.	Rice	R1
2.	Maize	MA2
3.	Maize	MA4
4.	Sunflower	SU6
5.	Sunflower	SU8
6.	okra	B1
7.	Sugarcane	SC5
8.	Lemon	L8
9.	Sunhemp	SH4
10.	*Mimosa*	BM5

IAA Production

Burkholderia isolates were screened for indole acetic acid production. [40] The bacterial cultures were inoculated in the respective medium (LB/nutrient broth) with tryptophan (2 mg/ml) and/or without tryptophan and incubated at 37^0C for 4 days. 10 ml culture was centrifuged at 3000 rpm for 20 minutes. To 2 ml of the supernatant, 2 drops of orthophosphoric acid and 4 ml of Salkowski's reagent (35% perchloric acid; 2% 0.5M $FeCl_3$) were added. Development of a pink colour indicates IAA production. O.D. was read at 530 nm using Spectrophotometer. The level of IAA produced was estimated by a standard IAA graph. [41]

Hydrogen Cyanide (HCN) Production

The *Burkholderia*isolates were inoculated in King's B medium amended with 4.4g/L of glycine. The Petriplate lids were covered with Whatmann No: 1 filter paper impregnated with 0.5% picric acid in 2% sodium carbonate. They were sealed with parafilm and incubated for 4 days. Development of brown color on the filter paper indicates positive results. [42]

Biocontrol Activity

The *Burkholderia* isolates were tested for their ability to inhibit plant pathogenic fungi like *Rhizoctoniasolani*(root rot disease*)* and*Macrophoeminaphaesolina,* which is responsible for charcoal rot disease in groundnut. The antagonistic activity of the isolates was determined by dual culture method. [43] The sterilized nutrient agar medium and potato dextrose agar medium were poured in Petriplate in the ratio 1:1. At one edge of the Petriplate test organism was streaked and at another end a 4mm disc of fungal pathogen was inoculated. The plates were incubated for 5 days at room temperature. The inhibition activity against plant pathogen, if any, was recorded by measuring the zone formed between the fungal and bacterial cultures.

Genomic DNA Isolation and Molecular Characterization of Isolates

Genomic DNA isolation was carried out by using phenol/chloroform/isoamyl alcohol method. [44] The total genomic DNA isolated from *Burkholderia*isolates was amplified by PCR, which was performed using the Eppendorf Master Cycler, Gradient (Eppendorf, Germany). PCR amplification was carried out using 16S rRNA primer.

Blast Sequencing Analysis

The amplified 16S rRNA gene PCR product of the isolatesB1 and R1, were sequenced in Sci-Genome Labs Pvt Ltd., Cochin, which was done through single pass analysis from forward and reverse direction. Sequencing was done by using Automated sequencer.

Pot Culture Studies

Surface sterilized seeds of *Arachis hypogea* (TNAU-Co6 variety) were treated with *Burkholderia* isolates (B1 and R1 isolates) at a rate of 10^6 CFU/ml per seed along with carboxy methyl cellulose as a binding agent. They were planted in sterile soil along with uninoculated control. For each treatment, three replications were maintained. They were placed in green house conditions and watered regularly. Observation was done after 65 days of sowing.

Results and Discussion

Characterization of the Isolates

The Gram negative *Burkholderia* isolates were characterized based on salinity, pH and temperature tolerance tests and biochemical tests. They were regular rods or slightly curved coccobacilli. They formed slightly yellow, smooth pinpoint colonies in nutrient agar. Also, Sudan Black B dye staining revealed the accumulation of polyhydroxy butyrate granules (PHB). The results of biochemical tests were presented in Table 2. [45]

Salt Tolerance

All the 10 *Burkholderia*isolates grow luxuriantly at a salt concentration of 0.1%-0.8% and optimum being 0.3%. The isolates like MA4, SU6, BM5, MA2, SH4, can tolerate high salt content *viz.*, 1%, 2% and 3%. [46]

Table 2. Biochemical characterization of *Burkholderia* isolates

S. No.	Test	R1	MA2	MA4	SU6	SU8	B1	SC5	L8	SH4	BM5
1.	Gram stainig	-	-	-	-	-	-	-	-	-	-
2.	Starch hydrolysis	-	-	-	-	-	-	-	+/-	+	+/-
3.	Caesin hydrolysis	-	-	+	-	-	-	+	-	-	+
4.	Lipid hydrolysis	+	-	+	-	-	+	-	-	+	+
5.	Gelatin liquefaction	-	-	-	-	+	-	-	+	+	+
6.	Urease	-	-	-	-	-	-	-	-	-	-
7.	Catalase	+	+	+	+	+	+	+	+	+	+
8.	Citrate	+	+/-	+	+/-	+/-	+	+	+	+	+
9.	H_2S production	-	-	-	-	-	-	-	-	-	-
10.	Glucose utilization	AG	AG	AG	AG	G	AG	AG	AG	AG	AG
11.	Lactose utilization	G	G	G	G	AG	G	AG	AG	AG	G
12.	MR	-	-	+	+	-	+/-	+	+/-	-	+/-
13.	VP	+	+	-	-	+	+	-	+	+	+
14.	Indole	-	-	-	-	-	-	-	-	-	-
15.	Oxidase	+	+	+	+	+	+	+	+	+	+
16.	Triple sugar iron test	Alkaline slant and alkaline butt	Acid slant acid butt	Acid slant acid butt	Alkaline slant acid butt	Acid slant acid butt	Acid slant acid butt	Alkaline slant acid butt	Acid slant acid butt	Acid slant acid butt ↑	Acid slant acid butt ↑

'+' = positive, '-'= negative, '+/-'= border, 'AG'=acid and gas production,' ,↑ 'G'=gas production.

Temperature & pH Tolerance

All the 10 *Burkholderia*isolates can tolerate a wide range of temperature like 20^0C-45^0C, while optimum being 35^0C. However temperature tolerance highly depends on the salt concentration. Also they grow luxuriantly in a wide pH range of 5.0-9.0. [47]

Phosphate Solubilization

Several species of *Burkholderia* are efficient P solubilizers. [48] The isolates were checked for their ability to solubilize potassium dihydrogen phosphate and rock phosphate in HAP medium. Zone of 'P' hydrolysis formed by the test organisms in HAP plates was measured following incubation of 2-4 days (Table 3). The isolate B1 formed better halozone which measured 0.40 cm followed by isolate MA4 (0.34cm) and R1 (0.32cm) (Figure 1). A graphical representation of the results of quantitative estimation of P solubilization was presented in Figure 2.

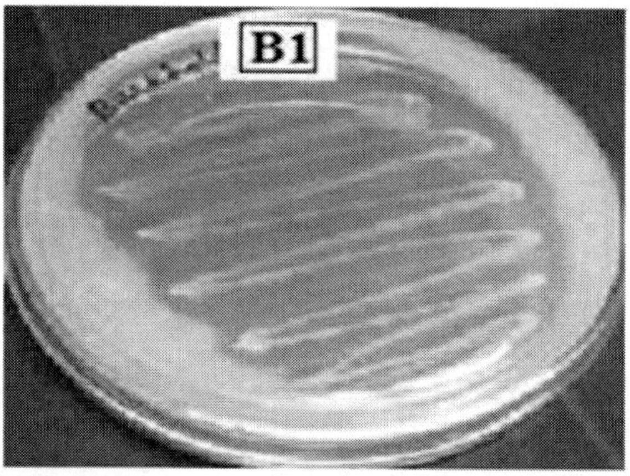

Figure 1. 'P' hydrolysis: Zone formation of B1 isolate in HAP medium.

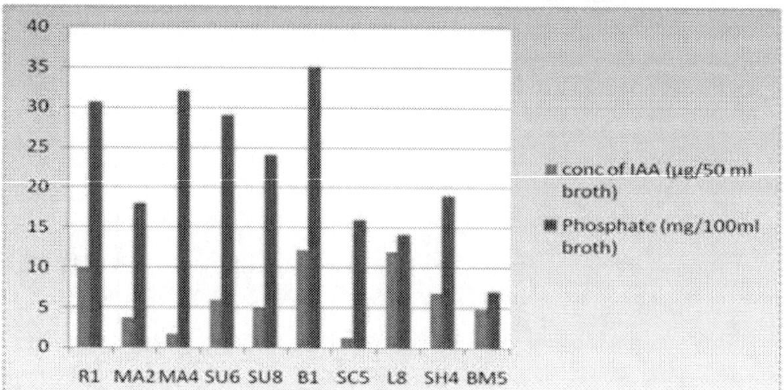

Figure 2. Quantitative estimation of phosphate solubilization and IAA production by *Burkholderia* isolates.

Table 3. Zone formation in HAP medium

S. No.	Name of isolate	Zone (in cms)
1.	**R1**	**0.32**
2.	MA2	0.10
3.	**MA4**	**0.34**
4.	**SU6**	**0.32**
5.	SU8	0.18
6.	**B1**	**0.40**
7.	SC5	0.30
8.	L8	0.14
9.	SH4	0.22
10.	BM5	0.16

Nitrogen Fixation

Nitrogen fixing ability of the test organisms was assessed using BMGM broth (pH 5.7). Among the10 isolates, 6 isolates *viz*, B1, MA4, R1, SU6, SU8 and SC5 showed more nitrogen fixing ability in BMGM broth. The alterations in pH of the mediumwere recorded and presented in Table. 4.

Table 4. Change of pH in BMGM broth

S. No.	Isolates	pH value							
		0 day	3rd day	6th day	9th day	12th day	15th day	18th day	21st day
1.	**R1**	**5.7**	**7.5**	**7.9**	**8.2**	**8.4**	**8.67**	**8.89**	**9.01**
2.	MA2	5.7	5.9	6.2	6.5	6.5	6.6	7.0	7.22
3.	**MA4**	**5.7**	**7.8**	**8.0**	**8.1**	**8.5**	**8.5**	**8.8**	**9.01**
4.	SU6	5.7	7.6	8.0	8.1	8.5	8.5	8.90	8.92
5.	SU8	5.7	6.6	6.8	7.0	7.0	7.7	8.1	8.10
6.	**B1**	**5.7**	**7.8**	**8.0**	**8.3**	**8.5**	**8.5**	**8.93**	**9.25**
7.	**SC5**	**5.7**	**7.5**	**7.9**	**8.0**	**8.5**	**8.55**	**8.85**	**9.0**
8.	L8	5.7	6.0	6.0	6.7	6.8	6.8	6.92	7.75
9.	SH4	5.7	5.8	6.0	6.1	7.0	7.0	7.17	7.39
10.	BM5	5.7	5.7	6.0	6.5	6.5	6.6	7.2	7.93
11.	Control	5.7	5.7	5.7	5.7	5.7	5.7	5.7	5.7

Thus B1, R1, MA4 and SC5 isolates increased pH of the broyhrom 5.7 to 9.0 after 21 days incubation, which was characterized by change in color of the medium from yellow to green and finally to deep blue (Figure 3). Isolates like SU6 and SU8 showed a pH value of 8.10 - 8.92. Also, when the isolates were inoculated in nitrogen free BAZ medium, similar results were obtained following incubation for 5-10 days which is a confirmation for N_2 fixation (Figure 4). [49]

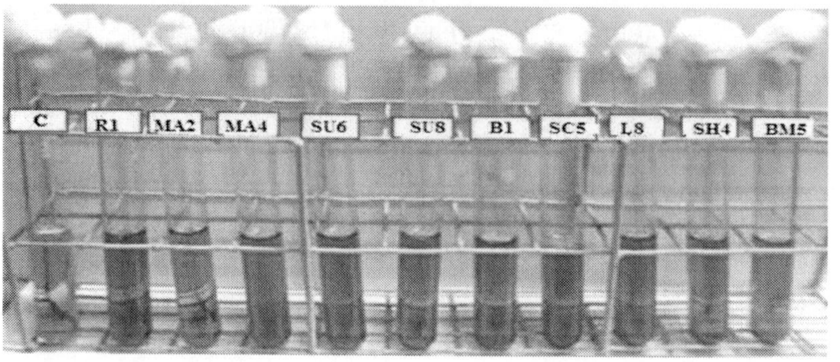

Figure 3. Nitrogen fixing activity of*Burkholderia* isolates in BMGM broth

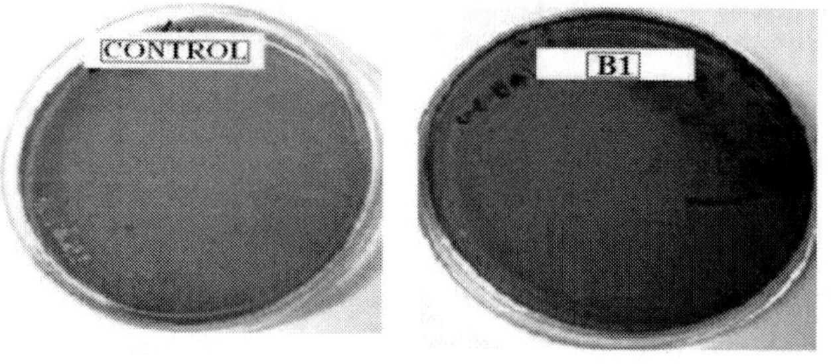

Figure 4. B1 isolate inoculated in BAZ medium.

HCN Production

The*Burkholderia* isolates were checked for Hydrogen cyanide production following incubation for 3-4 days in KB medium amended with glycine. Glycine acts as the precursor for the production of cyanide. The isolates R1, B1, MA4 and SU6 showed a characteristic change in color of the filter paper from yellow to orange and then to brown which indicated positive result. [50]

Biocontrol Activity

Antagonistic activity of *Burkholderia* was tested by dual culture method. Among the 10 isolates, B1 and R1 formed clear zone and inhibited the plant pathogen *Macrophoeminaphaesolina*. The zone formed by R1 isolate in dual culture plate was measured as 2cm and that of B1 is 1cm (Figure 5). No inhibition was found for *Rhizoctoniasolani.* [51]

Although many isolates like R1, MA4, SU6 and B1 produced significant results in all the above experiments, B1 and R1 were selected for molecular screening and pot culture experiments because of their higher antagonistic ability.

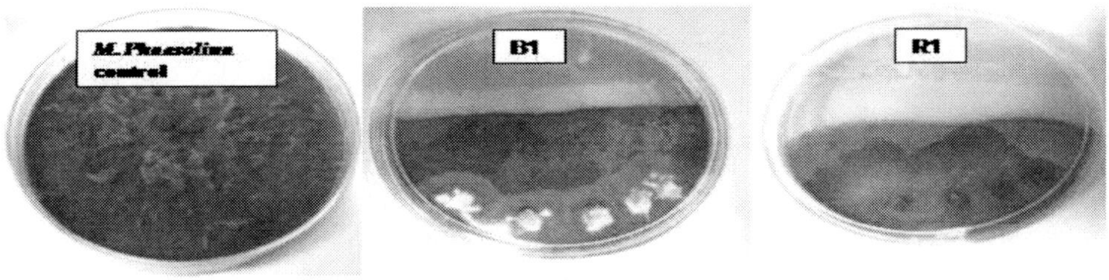

Figure 5. Antagonistic activity of B1 and R1 isolates against *M. phaesolina* indicated by zone formation.

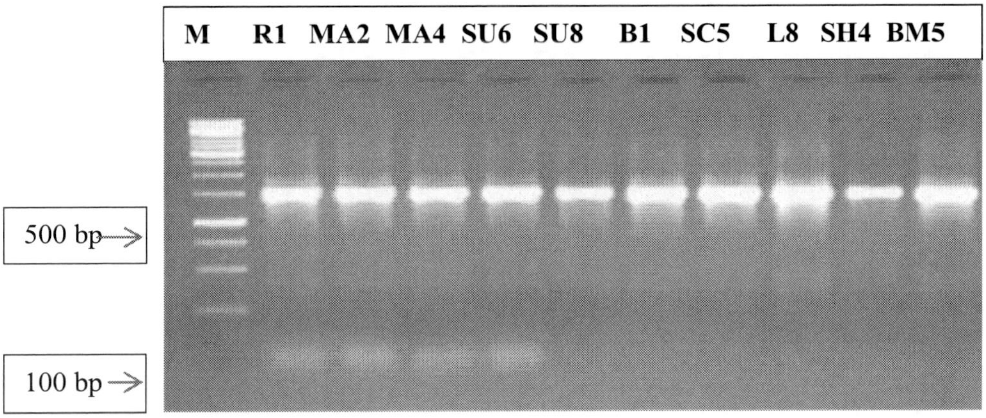

Figure 6.Amplification of 16S rRNA primers.

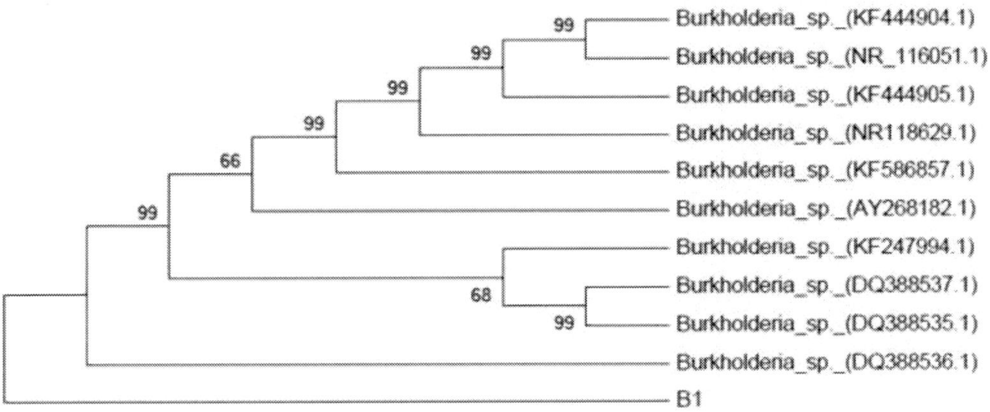

Figure 7.Dendrogram of B1 *Burkholderia* isolate.

Molecular Characterization of *Burkholderia* Isolates

When the genomic DNA of *Burkholderia* isolates were amplified by PCR using the16S rRNA primers, they yielded 500bp band which was notified by running the PCR product on 1.2% agarose gel.(Figure 6)

Nucleotide regionof 16S rRNAgenome of the isolates B1 and R1 were sequenced by using automated sequencer. The sequences were blasted against NCBI database. It was found that B1 isolateshowed 99% homology with ***Burkholderiathailandensis***(Gene Bank accession number: NC 007651.1) and the R1 isolate showed 100% homology with ***Burkholderiavietnamiensis*** (Gene Bank accession number: NC 009256.1).Dendrogram was also constructed for B1 and R1 isolates. (Figure 7 and 8)

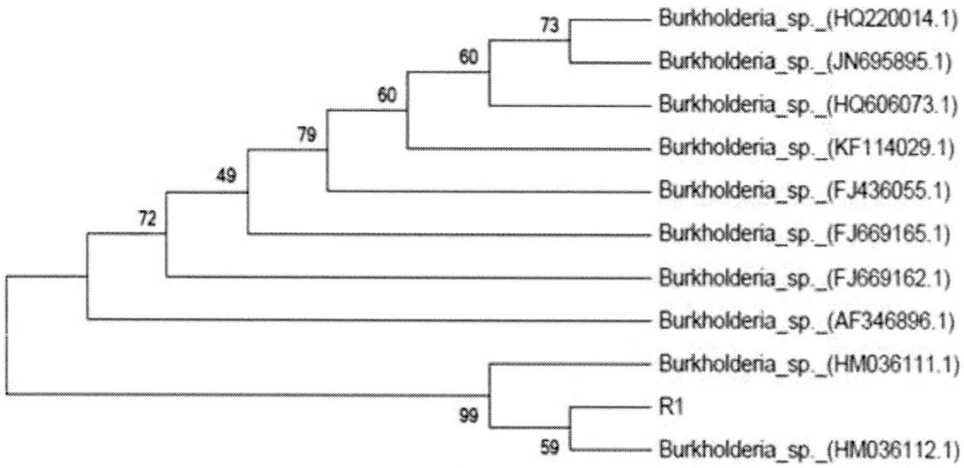

Figure 8. Dendrogram of R1 *Burkholderia* isolate.

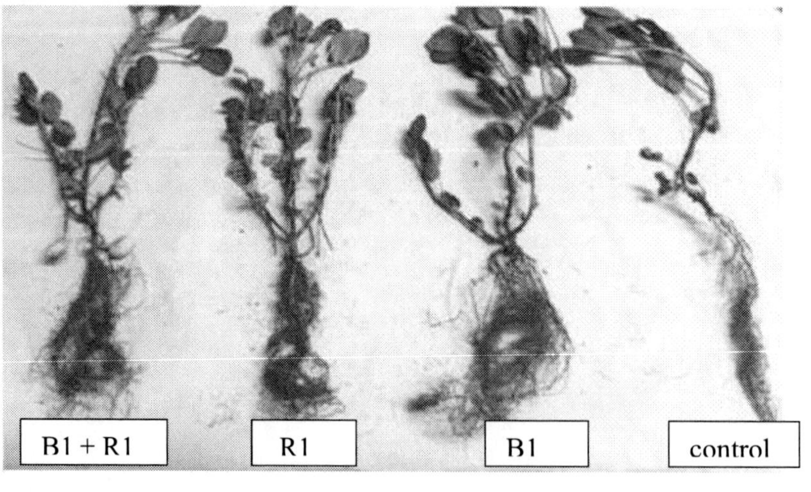

Figure 9. 65 days old Groundnut plants.

Pot Culture Studies

The results of the pot culture study of Groundnut revealed that *Burkholderia* treated plants showed more root and shoot growth when compared to uninoculated control. [52] Also, peg initiation was faster in treated plants and they resist pathogen attack. Similarly, the plants showed better growth, when B1 and R1 isolates were inoculated together. (Figure 9)

CONCLUSION

Based on the plant growth promoting activities of *Burkholderia,* the isolates B1 and R1 have a very high potential to be used as a bioinoculant for crop plants. They not only provide plants with essential nutrients like phosphorous, Calcium, iron and nitrogen, but also resist pathogen attack. Moreover, the ability of the isolates to withstand a wide range of temperature, pH and salinity, makes them more potent to be used as a biofertilizer in different geographical conditions irrespective of the nature of soil.

ACKNOWLEDGMENT

The authors are grateful to DBT, Govt. of India for providing financial support to carry out this research work.

REFERENCES

[1] Suarez-Moreno, Z. R., Caballero-Mellado, J., Coutinho, B. G., Mendonca-Previato, L., James, E. K. &Venturi, V. (2012). Common features of environmental and potentially beneficial plant-associated *Burkholderia*. *Microb. Ecol. 63*, 249-266.

[2] Sultan, Z., Park, K., Lee, S. Y., Park, J. K., Varughese, T. & Moon, S.S. (2008). Novel oxidized derivatives of antifungal pyrrolnitrin from the bacterium *Burkholderiacepacia*K87. *J. Antibiot. 61*, 420-425.

[3] Wong-Villarreal, A.& Caballero-Mellado, J. (2010). Rapid identification of nitrogen-fixing and legume-nodulating*Burkholderia*species based on PCR 16S rRNA species-specific oligonucleotides. *Syst. Appl. Microbiol. 33*, 35-43.

[4] Valverde, A., Delvasto, P., Peix, A., Velazquez, E., Santa Regna, I., Ballester, A., Rodriguez-Barrueco, C., Garcia-Balboa, C. &Igual, J.M. (2006). *Burkholderiaferrariae*sp. nov., isolated from an iron ore in Brazil. *Int. J. Syst. Evol. Microbiol. 56*, 2421-2425.

[5] Shaharoona, B., Jamro, G.M., Zahir, Z.A., Arshad, M.&Memon, K.S. (2007). Effectiveness of various *Pseudomonas* spp. and *Burkholderiacaryophylli*containing ACC-deaminase for improving growth and yield of wheat (*Triticumaestivum* L.). *J. Microbiol. Biotechnol. 17*, 1300-1307.

[6] Vandamme, P., Opelt, K., Knochel, N., Berg, C., Schomann, S., DeBrandt E., Eberl L., Falsen, E. & Berg, G. (2007). *Burkholderiabryophila*sp. nov. and *Burkholderia*

*megapolitana*sp. nov., moss-associated species with antifungal and plant-growth-promoting properties. *Int. J. Syst. Evol. Microbiol. 57*, 2228-2235.

[7] O'Sullivan, L.A.&Mahenthiralingam, E. (2005). Biotechnological potential within the genus *Burkholderia*. *Lett. Appl. Microbiol. 41*, 8-11.

[8] Payne, G.W., Vandamme, P., Morgan, S. H. & Lipuma, J.J. (2005). Development of a recA gene-based identification approach for the entire *Burkholderia*genus. *Appl. Environ. Microbiol.71*,3917-3927.

[9] Estrada-de los Santos P., Bustillos-Cristales R. & Caballero Mellado J. (2001).*Burkholderia*, a genus rich in plant-associated nitrogen fixers with wide environmental and geographic distribution. *Appl. Environ. Microbiol.67*, 2790–2798.

[10] Zolg, W. & Ottow, J.C.G. (1975). *Pseudomonas glathei* sp. nov., a new nitrogen scavenging rod isolated from acid lateritic relicts in Germany. *Z. Allg. Mikrobiol.15*, 287-299.

[11] Bell, S. C. & Turner, J.M. (1973). Iodinin biosynthesis by a pseudomonad. *BiochemSoc Trans. 1*, 751-753.

[12] Viallard, V., Poirier, I., Cournoyer, B., Haurat, J., Wiebkin, S., Ophel-Keller, K.&Balandreau, J. (1998) *Burkholderiagraminis* sp. nov., a rhizospheric *Burkholderia*species, and reassesment of [*Pseudomonas*]*phenazinium*, [*Pseudomonas*] *pyrrocinia*and [*Pseudomonas*]*glathei*as*Burkholderia*. *Int J. SystBacteriol.48*,549-563.

[13] Achouak, W., Christen, R., Barakat, M., Martel, M. H. &Heulin, T. (1999). *Burkholderiacaribensis* sp. nov., an exopolysaccharide-producing bacterium isolated from vertisolmicroaggregates in Martinique. *Int J. SystBacteriol. 49*, 787–794.

[14] Newby, D.T., Gentry, T.J.&Pepper, I.L. (2000). Comparison of 2,4-dichlorophenoxyacetic acid degradation and plasmid transfer in soil resulting from bioaugmentation with two different pJP4 donors. *Appl. Environ. Microbiol. 66*, 3399-3407.

[15] Goris, J., Dejonghe, W., Falsen, E., De Clerck, E., Geeraerts, B.&Willems, A. (2002) Diversity of transconjugants that acquired plasmid pJP4 or pEMT1 after inoculation of a donor strain in the A- and B-horizon of an agricultural soil and description of *Burkholderiahospita* sp. nov. and *Burkholderiaterricola* sp. nov. *Syst. Appl Microbiol. 25*, 340-352.

[16] Bramer, C.O., Vandamme, P., Da Silva, L.F., Gomez, J. G. C. & Steinbuchel, A.(2001). *Burkholderiasacchari* sp. nov., a Polyhydroxyalkanoate accumulating bacterium isolated from soil of a sugarcane plantation in Brazil. *Int J. SystEvol. Microbiol. 51*, 1709-1713.

[17] Bhat, M.A., Tsuda, M., Horiike, K., Nozaki, M., Vaidyanathan, C.S.&Nakazawa, T. (1994) Identification and characterisation of a new plasmid carrying genes for degradation of 2,4-dichlorophenoxyacetate from *Pseudomonas cepacia* CSV90. *ApplEnvMicrobiol. 60*, 307-312.

[18] Massol-Deya, A., Weller, R., Rios-Hernandez, L., Zhou, J.Z., Hickey, R. F. & Tiedje, J.M. (1997). Succession and convergence of biofilm communities in fixed-film reactors treatingaromatic hydrocarbons in groundwater. *Appl. Environ. Microbiol.63*, 270-276.

[19] Mueller, J.G., Devereux, R., Santavy, D.L., Lantz, S.E., Willis, S. G. & Pritchard, P.H.(1996).Phylogenetic and physiological comparsions of PAH-degrading bacteria from geographically diverse soils. *Ant Leeuwenhoek.71*, 329-343.

[20] Zhang H., Hanada S., Shigematsu T., Shibuya K., Kamagata Y., Kanagawa T. & Kurane R. (2000). *Burkholderiakururiensis* sp. *nov.*, a trichloroethylene (TCE)-degrading bacterium isolated from an aquifer polluted with TCE. *Int. J. Syst. Evol. Microbiol. 50*,743-749.

[21] Santini, A.C., Santos, H.R.M., Gross, E. & Correa, R.X. (2013).Genetic diversity of *Burkholderia*(Proteobacteria) species from the Caatinga and Atlantic rainforest biomes in Bahia, Brazil. *J. Genetics and Molecular Research. 12* (1), 655-664.

[22] Van Oevelen, S., De Wachter, R., Vandamme, P., Robbrecht, E. & Prinsen, E. (2002). Identification of the bacterial endosymbionts in leaf galls of *Psychotria* (*Rubiaceae*, angiosperms) and proposal of '*CandidatusBurkholderiakirkii*' sp. nov.*Int J. Syst.Evol.Microbiol. 52*, 2023-2027.

[23] Bianciotto, V., Lumini, E., Bonfante, P.&Vandamme, P. (2003). '*CandidatusGlomeribactergigasporarum*' gen nov., sp. nov., an endosymbiont of arbuscularmycorrhizal fungi. *Int J. SystEvol.Microbiol.53*,121-124.

[24] Minerdi, D., Fani, R., Gallo, R., Boarino, A. & Bonfante, P. (2001). Nitrogen fixation genes in an endosymbiotic*Burkholderia* strain. *Appl Environ Microbiol.67*, 725-732.

[25] Van Borm, S., Buschinger, A., Boomsma, J.J.&Billen, J. (2002). *Tetraponera* ants have gut symbionts related to nitrogen-fixing root-nodule bacteria. *Proc. R. Soc. Lond B. Biol. Sci.269*, 2023-2027.

[26] Thakuria D., Talukdar N. C., Goswami C., Hazarika S., Boro R. C. & Khan M. R. (2004). Characterization and screening of bacteria from rhizosphere of rice grown in acidic soils of Assam. *Curr. Sci. 86*, 978-985.

[27] Sahin, F., Cakmakci, R.& Kantar F. (2004). Sugar beet and barley yields in relation to inoculation with N_2-fixing and phosphate solubilizing bacteria. *Plant soil. 265*, 123-129.

[28] Gutierrez-Manero, F.J., Ramos-Solano, B., Probanza, A., Mehouachi, J., Tadeo F. R. & Talon M. (2001). The plant-growth-promoting rhizobacteria*Bacillus pumilus*and *Bacillus licheniformis*produce high amounts of physiologically active gibberellins. *Physiol.plant.111*, 206-211.

[29] Hamdali, H., Hafidi, M., Joe, M., Virolle, L. & Ouhdouch, Y. (2008). Growth promotion andprotection against damping-off of wheat by two rock phosphate solubilizing actinomycetes in a P deficient soil under greenhouse conditions. *Appl. Soil. Ecol. 40*, 510-517.

[30] Hameeda, B., Harini, G., Rupela, O. P., Wani, S. P. & Reddy, G. (2006). Growth Promotion of maize by Phosphate Solubilizing Bacteria Isolated from Composts and Macrofauna. *Microbiol. Res. 163*, 234-242.

[31] Parke, J. L. & Gurian-Sherman, D. (2001). Diversity of the *Burkholderiacepacia* complex and implications for risk assessment of biological control strains. *Annu Rev Phytopathol.39*, 225-258.

[32] Gerhardson, B. (2002).Biological substitutes for pesticides. *Trends Biotechnol. 20*, 338-343.

[33] Cook, R.J. (1990). Twenty-five years of progress towards biological control. In: *Biological Control of Soil-Borne Pathogens* (Hornby, D.,Ed.), Walingford, UK, CAB International, 1-14.

[34] Hebbar, P.K., Martel, M.H.&Heulin, T. (1998). Suppression of pre- andpostemergence damping-off in corn by *Burkholderiacepacia*. *Eur J. Plant Path.104*, 29-36.

[35] Coenye T., Laevens S., Willems A., Ohlen M., Hannant W., Govan J. R. W., Gillis M., Falsen E. & Vandamme P. (2001). *Burkholderiafungorum* sp. nov. and *Burkholderiacaledonica*sp.nov., two new species isolated from the environment, animals and human clinical samples. *Int J. Syst Evol. Microbiol.51*, 1099-1107.

[36] Kang, J.G., Shin, S.Y., Kim, M.J., Bajpai, V., Maheshwari, D.K.& Kang, S.C. (2004). Isolation and antifungal activities of 2-Hydroxy methyl-chroman-4-one produced by *Burkholderia* sp. MSSP.*J Antibiot. 57*(11), 726-731.

[37] Ayyakkannu, K.&Chandramohan, D. (1971). On the occurrence and distribution of Phosphobacteriain the marine environment at Porto Novo.*Curr. Sci.39*, 398-399.

[38] Olsen, S.R., Cole, C.V., Watanable, F.S.& Dean, L.A. (1954). Estimation of soil phosphorous in soil by extraction with sodium bicarbonate. USDA Circular no. 939.

[39] Govindarajan, M., Kwon, S.W.&Weon, H.Y. (2007). Isolation, molecular characterization and growth- promoting activities of endophytic sugarcane diazotroph*Klebsiella*sp. GR9. *World J. Microbiol. Biotechnol. 23*, 997-1006.

[40] Egamberdiyeva, D. (2005). Plant-growth promoting rhizobacteria isolated from a Calcisol ina semi-arid region of Uzbekistan: biochemical characterization and effectiveness. *J.PlantNutrit. Soil. Sci*, *168*, 94-99.

[41] Bric, J.M., Bostock, R.M.& Silverstone, S.E. (1991). Rapid *in situ* assay for indoleacetic acid production by bacteria immobilized on a nitrocellulose membrane. *Applied Environ. Microbiol. 57*, 535-538.

[42] Bakker, A. W. & Schippers, B. (1987). Microbial cyanide production in the rhizosphere in relation to potato yield reduction and *Pseudomonas* sp. mediated plant growth stimulation. *Soil Biol.Biochem.19*, 451-457.

[43] Rangeshwaran, R. & Prasad, R.D. (2000). Biological control of Sclerotium rot of sunflower. *Indian Phytopath. 53*, 444-449.

[44] Lazo, G.R., Roffey, R.& Gabriel, D.W. (1987). Conservation of plasmid DNA sequencesand pathovar identification of strains of *Xanthomonascampestris. J. Phytopathology*, *77*, 448-453.

[45] Pandey, P., Kang S. C. & Maheshwari, D.K. (2005). Isolation of endophytic plant growth promoting *Burkholderia*sp. MSSP from root nodules of *Mimosa pudica. Current Science. 89*, 1-10.

[46] Robertson,J., Levy, A., Sagripanti, J. L. & Inglis. T.J. (2010). The survival of *Burkholderiapseudomallei* in liquid media. *Am. J. Trop Med Hyg. 82* (1), 88-94.

[47] Theocharis, A., Bordiec, S., Fernandez, O., Paquis, S., Dhont-Cordelier, S., Baillieuel, F., Clement, C. &Barka, E.A. (2012). BurkholderiaphytofirmansPsJ Norimes Vitisvinifera L. and confers a better tolerance to low non freezing temperatures. *Mol. Plant Microbe Interactions. 25* (2), 241-249.

[48] Walpola, B.C. & Yoon, M. (2013). Isolation and characterization of phosphate solubilizing bacteria and their co-inoculation efficiency on tomato plant growth and phosphorous uptake. *African Journal of Microbiology Research. 7*, 266-275.

[49] Gillis, M., Van, T.V., Bardin, R., Goor, M., Hebbar, P., Willems, A., Segers, P., Kersters, K., Heulin, T. & Fernandez, M.P. (1995). Polyphasic taxonomy in the genus*Burkholderia* leading to an emended description of the genus and proposition of *Burkholderiavietnamiensis* sp. nov.for N2-fixing isolates from rice in Vietnam. *Int. J. Syst. Bacteriol.45*, 274-289.

[50] Voisard, C., Keel., C. D.&Defago, G. (1989). Cyanide production by *Pseudomonas fluorescens* helps suppress black root rot of tobacco under gnotobiotic conditions. *EMBO. J. 8*(2), 351-358.

[51] Heungens, K. & Parke, J.L. (2000). Zoospore homing and infection events: effects of the biocontrol bacterium *Burkholderiacepacia* AMMDR1 on two oomycete pathogens of pea (*Pisumsativum* L.). *Appl Env Microbiol. 66*, 5192-5200.

[52] Tran, V.V., Berge, O., Ngo Ke, S., Balandreau, J. & Heulin, T. (2000). Repeated beneficial effects of rice inoculation with a strain of *Burkholderiavietnamiensis* on early and late yield components in low fertility sulphate acid soils of Vietnam. *Plant Soil. 218*, 273-284.

SECTION B. AGRICULTURAL PROSPECTS

In: Microbes in Soil and Their Agricultural Prospects
Editors: K. Kishore Choudhary and D. Wattal Dhar
ISBN: 978-1-63482-824-6

Chapter 10

ADVANCES IN THE DEVELOPMENT OF MICROBIAL BIOFERTILIZERS AS A TOOL FOR SUSTAINABLE AGRICULTURE

***Vandana Bharti*[1], *Archana Mehta*[1] *and Ashwani Kumar*[1•]**
[1]Department of Botany, Dr. HarisinghGour University
(Central University), Sagar, MP, India

ABSTRACT

The wide-ranging application of chemical fertilizers adversely affected the soil fertility all over the globe. To resolve this problem it has become important to apply long term integrating management practices for maintaining sustainability in agriculture. In this regard, the promotion and development of biofertilizers may act as potential tool to solve this problem.

Presently the development of novel assay system, advances in genomics, proteomics, and transcriptomics and microscopic techniques, significant progress has been achieved in the field of identification of microbes as a potential biofertilizers. These microbial biofertilizers, commonly known as PGPR, promotes plant growth by solubilizing phosphate, siderophore production, biological nitrogen fixation, phytohormone production, induction of systemic resistance, and inhibition of ethylene synthesis. This chapter focuses on the concept of sustainable agriculture and various detection methods of PGPR, including integrative understanding of the role of PGPR in salinity tolerance and sustainable crop production.

Keywords: biofertilizers, PGPR, salinity tolerance, sustainable agriculture

INTRODUCTION

Salinity of soil is a major environmental constraint for production of crop. According to an estimate it covers 45 million hectares of irrigated ground, and is expected to increase due to global climate changes and of many irrigation practices [1, 2, 3]. In arid and semi arid regions, abiotic stresses such as salinity stress, water stress, high light intensity, flooding, temperature, toxic metals, wounding and biotic stress including pests and pathogens can reduce plant growth and yield [4]. Moreover, the deleterious effects of salt stress on agricultural crop yield are significant, in terms of slower growth rate, reduced tillering and reproductive development [5, 6].

Reduction in growth rate as well as changes in metabolic processes in plants due to low consumption of water is mainly affected by salinity of soils [7]. In contrast to crop plants, there are some wild plants that succeed in the saline environments (saline deserts, estuaries halophytes and seashore). Diverse anatomical and physiological adaptations respond the ion toxicity and water scarcity in saline conditions to overcome stress in different ways [8]. Therefore in this direction to maintain the yields on saline soils and damaged land, it is essential to use different salt tolerant crop.

There are many free-living bacteria in the soil, that can promote plant growth by different ways, and are known as Plant Growth Promoting Rhizobacteria (PGPR) [9]. These bacteria alleviate salt stress and promote plant growth by direct and indirect method. In indirect ways, these bacteria decrease some of toxic effects of fungus or other phytopathogenic organism by numerous mechanisms. However, the direct mechanisms of PGPR for the plant growth enhancement include stress hormone such as ethylene gaseous hormone. PGPR reduce the ethylene levels in stressed plants by breaking down the precursor of ethylene hormone,1-aminocyclopropane-1-carboxylate to α-ketobutyrate and ammonium through enzyme ACC deaminase [10, 11].

A novel plant growth promoting rhizobacteria, i.e., *Sphingobacterium pakistanensis* sp. has been isolated from rhizosphere of *Vigna mungo* [12]. The advantage of PGPR is not only to produce plant growth promoting substances but also providing resistant to different pesticides [13]. The maturity and yield of wheat plant in saline soil extensively influence by salt-tolerant plant growth-promoting rhizobacteria (ST-PGPR) [14]. This chapter focused on the concept of sustainable agriculture, various detection methods of PGPR, including integrative understanding of the role of PGPR in salinity tolerance and sustainable crop production.

The Concept of Sustainable and Alternative Agriculture

Sustainable agriculture has many definitions that include the similar elements of productivity, vigor, abundance, maintenance, protection and the environment, conflicting only within the extent of stress. The Agricultural Research Service [15] defines sustainable agriculture as 'Agriculture that for the foreseeable future will be productive, competitive and profitable, conserve natural resources, protect the environment, and enhance public health, food quality, and safety.

Sustainable agriculture is increasingly viewed as a long-term goal worldwide [16]. In addition, "sustainable" mean a time span and the capability of an agricultural system to develop and tolerate indefinitely [17, 18]. Alternative agricultural practices provide the best means of achieving this goal. It is also important to note that the single most important component of a sustainable farming system is skilled management.

Alternative agriculture is to optimize the use of internal production inputs (i.e., on-farm resources) and skilled management in ways that provide acceptable levels of sustainable crop yields and livestock production, and result in economically profitable net returns. This approach emphasizes such cultural and management practices as crop rotations, use of animal and green manures, and conservation tillage to control soil erosion and nutrient losses [19].

BIOFERTILIZERS

The term 'microbial inoculants' or Biofertilizers can be defined as live or dormant cells of efficient strain that have ability for nitrogen fixation, phosphate solubilization or cellulolytic activity, used for application of seed, soil or composting areas with the objective of increasing the numbers of such microorganisms and accelerate certain microbial process to enhance the amount of the accessibility of nutrients in a form which can absorb by plant. In large sense, the term may be used to include all organic resources (manure) for plant growth which are rendered in an available form for plant absorption through microorganisms or plant associations or interactions.

Organisms that are commonly used as biofertilizers component are nitrogen fixers (N-fixer), potassium solubilizer (K-solubilizer) and phosphorus solubilizer (P- solubilizer), or with the combination of fungi [20]. Biofertilizers activate the microorganisms that are found in the soil, thus restoring the soils' natural fertility and protecting it against soil diseases and droughts, which stimulates the growth of plants. There are many different types of biofertilizers and each type has a different effect on the soil, depending on the crop. The two types of bio-fertilizers that are commonly used are nitrogen and phosphorus biofertilizers.

India is an agricultural based country. In order to feed the ever growing populations, India has to increase the per unit area productivity. According to United Nations Food and Agriculture Organization (FAO) estimations, average demand used for the agricultural commodities in 2030 will be 60% higher than present time and more than 85% of this additional demand will be from developing countries. For greater than a half century, the world has relied on the concept of increasing crop yields to supply a constantly increasing requirement of food. Therefore, vertical expansion of food production is necessary. In order to increase the unit area productivity of agricultural land, the role of different crop nutrients in contributing increased crop yield is vital. Among the all crop nutrients, nitrogen as well as phosphorus plays an important role in increasing the crop productivity. Further, the nitrogenous chemical fertilizers are manufactured industrially using nonrenewable petroleum products under high temperature and high pressure. Increase in petroleum cost day by day effects the cost of the chemical fertilizers. In addition, more than 50% of the applied N-fertilizers are somehow lost through different agricultural processes which not only lead to economical loss to the farmers and polluted environment consequently. Feedstock/fossil fuels depletion and increasing fertilizer cost are making marginal farmer unaffordable. Growing concern about environmental hazards, increasing threat to sustainable agriculture are some of the other reliable reasons for the biofertilizers promotion. However, plant nutrients like N, P and K are highly essential for plant growth and metabolism. It is also evident that plants utilizes nutrients in greater amounts from soil in modern intensive cultivation and needs replenishment. Under such conditions microorganisms offer good alternative technology to replenish crop nutrients [21].

In agricultural eco-system, microorganisms have vital role in fixing/solubilizing/ mobilizing/nutrient recycling [22]. These microorganisms occur in soils naturally, but their populations are often scanty. In order to increase the crop yield, the desired microbes from rhizosphere are isolated and artificially cultured in adequate count and mixed with suitable carriers or as they are in suitable combinations (microbial consortium) by artificial culturing. The use of microbial biofertilizers to improve soil quality and health, to enhance crop yield and quality, and to reduce our dependence on agrichemicals is a promising new technology with great potential. However, we need to know more about their mode of action and interaction within the soil-plant ecosystem [23].

Plant Growth Promoting Rhizobacteria (PGPR)

Plant growth promoting rhizobacteria, or PGPR, are group of non-pathogenic, heterogeneous bacteria that are associated with plant roots by colonizing either the root itself, or the rhizosphere, along with mediate improvements in plant growth. While it is quite possible for a bacterium to benefit plant growth or health while colonizing the phyllosphere, and the phyllosphere to influence plant defense responses. Kloepper [24] was first to define PGPR as soil bacteria that enhance plant growth by colonizing the roots of plants..

A number of authors have reported significant increases in agricultural crops under green house and field conditions in response to PGPR [25, 26, 27, 28, 29, 30]. Besides plant growth promoting PGPR have different activity i.e., increase the effectiveness of fertilizers, alleviate abiotic stresses, control plant pathogens, and degradation of xenobiotic compounds [31, 32, 33, 34, 35, 36]. Role of PGPR on plant growth and development have been summarised in table 1.

The mechanisms employed by plant growth-promoting bacteria and its effect on plant growth (Table 2). But it's not always necessary that these biofertilizers work efficiently in the field and colonize the plants rhizosphere. The ineffectiveness of PGPR in the field has often been attributed to their inability to colonize plant roots [53, 54].

A variety of bacterial traits and specific genes contribute to this process, but only a few have been identified. These include motility, chemotaxis to seed and root exudates, production of pili or fimbriae, production of specific cell surface components, ability to use specific components of root exudates, protein secretion, and quorum sensing. The use of plant growth promoting rhizobacteria (PGPR) for sustainable agriculture has increased tremendously in various parts of the world. Significant increases in growth and yield of agronomically important crops in response to inoculation with PGPR have been repeatedly reported by Figueiredo et al. [28].

In addition to their proven effectiveness in agriculture, PGPR have potential in solving environmental problems such as soil erosion in arid zones is prevent by improving growth of desert plants it also reduces dust pollution.

Table 1. Role of PGPR on plant growth and development

S. No	PGPR	MECHANISMS	STRESS	PLANTS	REF.
1	*Kocuria turfanensis* 2M4	Phosphate solubilization	Salinity stress	*Arachis hypogaea*	37
2	*Bacillus subtilis* OSU142, *B. megaterium* M3, *Azospirillum brasilense*	Nitrogenenage activity, alternative as fertilizers	To improve seed growth under field condition	Wheat, Barley	38
3	*Pseudomonas fluorescens Pseudomonas putida*	ACC-deaminase	Water deficit stress	Maize Plants	39
4	*Azospirillum brasilense*	Partial biological N-fixation.		Wheat	40
5	*Bacillus* spp., *Azotobacter* spp., *Acetobacter* spp., *Pseudomonas* spp.	Solubilization of phosphorus and potassium	Deficient of nutrients (phosphorus andpotassium)	Chickpea, Wheat	41
6	*Brachybacterium* sp., *Zhihengliuella* sp., *B. casei, Vibrio* sp., *Haererehalobacter* sp., *Halomonas* sp., *Pseudomonas* spp., *Cronobacter sakazakii, R. radiobacter,* and *Mesorhizobium* sp.	Production of IAA, activity of ACC deaminase and phosphate solubilization	Salt stress	*Salicornia brachiata*	42
7	Bradyrhizobiaceae, Rhodospirillaceae, Pseudomonadaceae, and Aurantimonadaceae, and Chitinophagaceae and Flexibacteraceae	Aminocyclopropane carboxylate deaminase and β–propeller phytase.	Desert stress, to improve root biomass and total plant biomass	Cucumber plants	43
8	*Glomus mosseae* (AM)	Oxidative stress reduced As (V) toxicity and modulating antioxidative mechanisms in pea plants	Arsenic stress, damage chlorophyll pigments and oxidative stress induced	*Pisum sativum L.* (pea)	44
9	*Pseudomonas* sp., *Xanthomonas* sp., *Bacillus* sp., *Rhizobium* sp., *Sphingomonas* sp.	Nitrogenase activity.	To promote plant growth	*Brachiaria brizantha*	45
10	*Azospirillum brasilense* and *Pantoea dispersa*	higher stomatal conductance, increase photosynthesis, improved N nutrition	Salinity tolerance	Sweet pepper	46
11	*Pseudomonades* sp. and *Bacillus lentus*	ascorbate peroxidase (APX) and glutathione reductase (GR) enzymes activity	Salinity stress	Basil	47

Table 1. (Continued)

S. No	PGPR	MECHANISMS	STRESS	PLANTS	REF.
12	*Azotobacter chroococcum, Azospirillum* spp.,*Pseudomonas* spp., *Azospirillum* spp., *Pseudomonas fluorescens* and *Basillus subtilis*	Improve physiological and morphological responses	Flooding stress	Canola plants	48
13	*Gigaspora gigantean*	For enhancing the biomass production	To improve the growth and survival of plant seedlings in forestation	*Ruta graveolens*	49
14	*B.cereus*	Activitiy of peroxidase, superoxide dismutase, ascorbate peroxidase and catalase are Increased	Against salt stress	*Vigna radiata, Cicer arietinum* and *Oryza sativa*	50
15	*Pontibacter niistensis* NII-0905	Production of HCN, siderophore, IAA and Phosphate solubilization	Plant growth promoting potential	Cowpea	51
16	*Pseudomonas fluorescens, Bacillus subtilis*	Increase activity of stress-related enzymes	Water stress	Green gram	52

They contribute in different phytoremediation practice (bioaugmentation, Phytoextraction, phytodegradation, phytostabilization, rhizodegradation, phytotransformation and phycoremediation) to decontaminate soils and waters and maintain restoration of mangrove ecosystems that lead to improve fisheries [55].

The *Brevundimonas* sp. a multi-trait beneficial microbe for sustainable agriculture in arid regions have nitrogenase activity also enhanced the growth of Bt-cotton plants (plant height (68.41%), shoot dry weight (58.44%) and root dry weight (64.81%) are isolated from rhizosphere of *Saccharum L.* grown in arid region [56].

According to Baris et al. (2014), nitrogen requirements satisfy in seed inoculations with PGPR, and mixture inoculation but *Bacillus megaterium* M3 and MIX (*Bacillus subtilis* OSU142, *B. megaterium* M3, *Azospirillum brasilense* Sp. 245) inoculation provided higher PNE concentrations than mineral fertilizer application for wheat and barley under field conditions.

Interaction of PGPR with Plants

There are varieties of ways in which PGPR may improve the growth or health of plants. Mechanisms of plant growth promotion include increasing plant nutrient acquisition, modification of plant growth and development, modification of the soil environment to promote plant growth, and biocontrol of plant pathogens. Biocontrol can be via direct mechanisms, where the pathogen itself is attacked, or via indirect ones, where plant defense responses against the pathogen are induced (Figure 1).

Conceptually, PGPR can have an impact on plant growth and development in two different ways: indirectly or directly.

Table 2. Mechanisms employed by plant growth-promoting bacteria and its effect on plant growth

S.No.	Mechanisms	Effect on plant growth
1	Root-associated nitrogen fixation	Biomass and nitrogen content are Increase
2	Production of plant hormones (auxin, giberellin, cytokinin)	Stimulate root branching, increase shoot and root biomass, and induce the reproductive cycle
3	solubilization of Phosphate	Increase biomass and P content
4	Inhibition of ethylene synthesis	Increase root length
5	oxidation of Sulfur	Increase biomass and foliar nutrient content
6	Production of signal molecules and enhanced proton extrusion	Change in plant metabolism related to mineral absorption
7	By Increasing of root permeability	Increase biomass and nutrient uptake
8	improve mineral uptake	Increase biomass and nutrient uptake
9	By Increasing of nitrite production	Increase formation of lateral roots
10	Accumulation of nitrate	Increase biomass and nitrate content
11	Reduce toxicity of heavy-metal	Protection against nickel toxicity
12	By Increasing thesize of legume nodules	Increase biomass, N content, and reproductive yield
13	Increase alder root nodules or size	Increase biomass and N content
14	Increase frequency of infection by endomycorrhizal fungi	Increase biomass
15	Increase number of ectomycorrhizal root tips	Increase biomass
16	Increase temporary 'rain root' production in cacti	Improve survival of seedlings during drought
17	Increase resistance to adverse conditions (drought, salinity, compost toxicity)	Improve survival of seedlings and increase biomass
18	Additive hypothesis	Increase biomass as a result of several small-magnitude mechanisms working in concert or at the same time

The indirect promotion of plant growth occurs when bacteria decrease or prevent some of the deleterious effects of a phytopathogenic organism by one or more mechanisms. On the other hand, the direct promotion of plant growth by PGPR generally entails providing the plant with a compound that is synthesized by the bacterium or facilitating the uptake of nutrients from the environment [9]. At present, less than 20 different biocontrol PGPR strains are commercially available. However, this number should increase in near future due to the development of new screening procedures [58], and development of genetically engineered strains. Traits associated with the biocontrol of plant pathogens include:

- Synthesis of antibiotic
- Secretion siderophores forbinding of iron

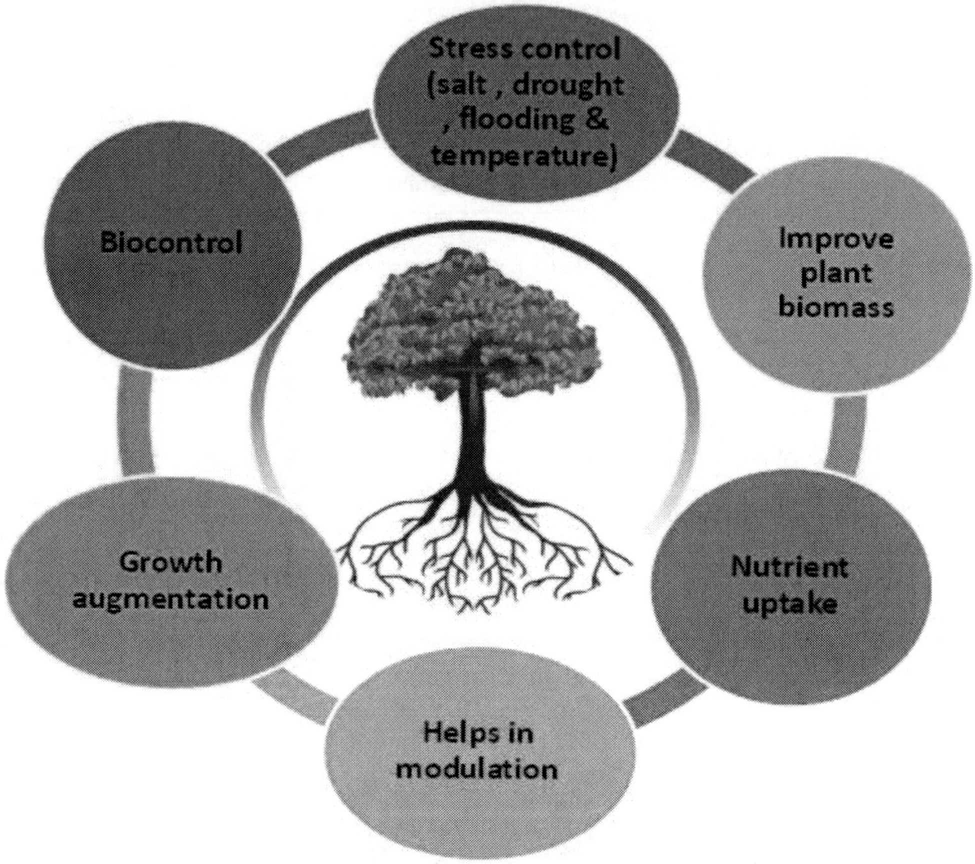

Figure 1. Different role of PGPR in stress tolerance, plant growth and development.

- Low molecular weight metabolites with antifungal activity are Produced
- Fungal cells lyse enzymes including chitinase, b-1, 3- glucanase, protease, or lipase are produced
- Out-competing phytopathogens for nutrients and niches on the root surface
- Lowering the production of ethylene in plants by controlling production of the enzyme ACC deaminase.

Different PGPR may directly facilitate the proliferation of their plant hosts in several ways. They may include atmospheric nitrogen fixing, synthesis of siderophores, production of phytohormones (auxins and cytokinins), and solubilization of minerals (phosphorus). A particular bacterium may promote plant growth and development using any one, or more, of these mechanisms. Moreover, a bacterium may utilize different traits at various times during the life cycle of the plant. For example, following seed germination a PGPR may lower the plant's ethylene concentration thereby decreasing the ethylene inhibition of seedling root length. Once the seedling has depleted the resources that are contained within the seed, the same PGPR may help to provide the plant with iron and phosphorus from the soil. The impact of the mechanisms by which the bacterium provides a compound or nutrient such as fixed nitrogen, phosphorus or iron to the plant, varies considerably depending upon the soil composition.

Thus, PGPR often have little or no measurable effect on plant growth when the plants are cultivated in nutrient-rich soil and grown under optimal conditions.

Detection Methods of PGPR

Progress in the identification of new, previously uncharacterized genes is being made using nonbiased screening strategies that rely on gene fusion technologies.

These strategies employ reporter transposons and *in vitro* expression technology CIVET to detect genes expressed during colonization. Using molecular markers such as green fluorescent protein or fluorescent antibodies it is possible to monitor the location of individual rhizobacteria on the root using confocal laser scanning microscopy. The approach has also been combined with an rRNA-targeting probe to monitor the metabolic activity of a rhizobacterial strain in the rhizosphere and showed that bacteria located at the root tip were most active [59].

Identification and Characterization of Beneficial Bacterial Strains for Agriculture

Morphological, physiological, and molecular characteristics of bacteria involves various methods such as G + C contents, mol (%), fatty acid analysis, 16SrRNA sequencing, and DNA–DNA hybridization.

Taxonomy of PGPR

Identify different bacterial strains divided into categories based upon:

(1) Morphological, physiological, and biochemical characters such as API kits, VITEK cards, and Biolog plate
(2) Chemotaxonomic characters (PAGE, FAME profiles)
(3) Genomic characters (16SrRNA gene sequencing, DNA–DNA relatedness, and other techniques). For overviews of modern taxonomy, paper can be referred, such as Logan et al. [60]. Zhang et al. [59] isolated bacteria from wheat roots by wheat germ agglutinin labeled with fluorescein isothiocyanate and characterize as PGPR.

Chemotaxonomic Characters

Some chemotaxonomic techniques for e.g., polar lipid analysis, quinone content, FAME profiling, PAGE analysis of whole-cell proteins, cell wall diamino acid content, pyrolysis mass spectrometry, Fourier transform infrared spectroscopy, and matrix-assisted laser desorption/ionization time-of-flight (MALDI-TOF) mass spectrometry, Raman spectroscopy etc. are applied for PGPR identification. FAME analysis is presently the only chemotaxonomic technique that is linked to a commercial database for identification purposes. Fatty acids are the major constituents of lipids and lipopolysaccharides showing variability in chain length, double-bond position, and substituent groups are perfectly suitable for taxon description and for comparative analyses of profiles.

For polyphasic taxonomy of aerobic endospore, forming bacteria Sodium dodecyl sulfate-PAGE of whole-cell proteins requires standardized conditions of growth with a rigorously standardized procedure for analysis, and normalization of the data for computer-assisted comparison of the results [60]. The taxonomic importance of polar lipids has now been demonstrated for some novel genera among the *Bacillaceae*, although many polar lipids detected have not yet been structurally characterized. Likewise, quinones (MK-7, MK-8, and MK-9) have so far been reported for representatives of Bacillaceae. Finally, pyrolysis mass spectrometry, fourier transform infrared spectroscopy and UV resonance raman spectroscopy are sophisticated analytical techniques that examine the total chemical composition of bacterial cells. These methods have been used for taxonomic studies of particular groups of bacteria, including the members of the family Bacillaceae [60].

Genetic Approaches

Genotypic methods are those that are directed toward DNA or RNA molecules. Undoubtedly, these methods have revolutionized the bacterial identification system and taxonomy. Different techniques are now available to subtype bacteria up to strain level, such as restriction fragment length polymorphism (RFLP), plasmid-profilling, ribotyping, amplified ribosomal DNA restriction analysis (ARDRA), pulsed field gel electrophoresis (PFGE), and randomly amplified polymorphic DNA (RAPD). Different PGPR have already been characterized by one or more of these methods [61, 62]. Determination of the moles percent guanosine plus cytosine is one of the classical genotypic methods. Generally, the range observed is not more than 3% within a well-defined species and not more than 10% within a well-defined genus [63]. DNA–DNA hybridization or DNA DNA reassociation technique is because at high temperatures, DNA can be denatured, but the molecule can be brought back to its native state by lowering down the temperature (reassociation). This technique considers the comparison between whole genome of two bacterial species. There are many different methods for DNA–DNA hybridization [64], but it is important to state that this technique gives the relative % of similarity but not the actual sequence identity.

DNA microarray is a method, which was lined up to overcome the short comings of DNA–DNA hybridization. However, it is still an expensive methodology. The 16SrRNA molecule comprises of variable and conserved regions, and universal primers for the amplification of full 16SrRNA gene are usually chosen from conserved region while the variable region is used for comparative taxonomy. The 16SrRNA gene sequence is deposited in databases such as Ribosomal Database Project II (http://rdp.cme.msu.edu/) and GenBank (http://www.ncbi.nlm.nih.gov/). Sequences of related species for comparative phylogenetic analysis can also be retrieved from these databases. Thereafter, sequence comparing software packages such as BLAST and CLUSTAL X are used for alignment of 16SrRNA gene sequence. The extent of relatedness between bacterial species can be scrutinized by the construction of phylogenetic trees or dendrograms.

Finally, sequences of other highly conserved housekeeping or other protein encoding genes, such as rpoB, gyrB, recA, have also great potential for taxonomic analysis at the species level. For example, Mota et al., [65] obtained clustering patterns for *Paenibacillus* based upon rpoB sequence comparisons that were similar to those obtained with 16SrRNA gene sequences. Moreover, Cerritos et al,. [66] included recA sequence comparisons in the work that led to the proposal of a new *Bacillus* species.

Salt Tolerance Traits in Plants

In plant taxa, the widespread genetic diversity is distributed that exists more than several genera for salt tolerance [67]. Halophytes have the capacity to accommodate extreme saline environments because of very special anatomical, morphological and adaptations mechanisms while glycophytes are salt sensitive plants [68].

However, they have unique characteristics for which the genes are not likely to be introgressed easily into crop plants. Research of recent decades has established that most halophytes and glycophytes tolerate salinity by rather similar strategies often using analogous tactical processes [69]. The cytotoxic ions in saline environments, typically Na^+ and Cl^- are compartmentalized in vacuole as osmotic solutes [70]. Many of the molecular entities that mediate ion homeostasis and salt stress signaling are similar in all plants.The fact that cellular ion homeostasis is controlled and effected by common molecular entities and made it feasible to use of model genetic organismal systems for the dissection of the plant salt stress response [71, 72, 73, 74]. Research on the plant genetic model *Arabidopsis* has increased greatly our understanding of how cellular salt tolerance mechanisms are integrated and coordinated in an organism context, and are linked to essential phenological adaptations. Since *Arabidopsis* is a glycophyte, a salt tolerant genetic model will be required to delineate if salt tolerance is affected most by form or function of genes or more by differences in the expression of common genes due to transcriptional or post-transcriptional control [73].

Antioxidant Enzyme Activation

Salt stress causes to accumulate reactive oxygen species (ROS) on the cell, that can damage the membrane and membrane processes of lipids, proteins and nucleic acids [75]. Plant defense against ROS is related to antioxidant defense systems including catalase, peroxidase, superoxide dismutase, glutathione reductase, ascorbate peroxidase and non-enzymatic compounds that includes ascorbate, α-tocopherol and carotenoides, ascorbate, compounds have a vital role to protect plants against ROS effects [76]. Due to salt stress, antioxidant enzymes are activated as part of protection mechanism [77]. Salinity toxicity is a global agricultural and eco-environmental problem. Salt stress is the major abiotic stresses faced by plants, which adversely affect their productivity. Salt stress causes reduction of crop yield and alterations in plant metabolism, including a reduced water potential, ion imbalances and toxicity and sometimes severe salt stress may even threaten survival. Salinity also leads to oxidative stress in plants due to the production of ROS such as the super oxide radical, hydrogen peroxide, and hydroxyl radical [78]. Attempts to reduce oxidative damages under the salt stress conditions have included the manipulation of ROS scavenging enzymes by gene transfer technology. It is important to maintain and/or increase the productivity (photosynthetic capacity) under stressful environment by developing plants that arewell adapted to environmental stress through manipulating antioxidant system [79].

Phosphatase and Solute Accumulation

At low water potentials, stress tolerant plants maintain their turgor by increasing the number of solute molecules in the cell. Several plants accumulate organic osmolytes in response to the imposition of environmental stresses that cause cellular dehydration. The adaptive role of compounds in stress physiology is to mediate osmotic adjustment and protecting subcellular structure in plants. Transgenic plants are engineered in such a way that they could subsidiary improve drought or salt tolerance by accumulatation of proline, mannitol, fructans, trehalose, glycinebetaine, and ononitol. The proline and glycine betaine synthesis may buffer cellular redox potential and functional significance of compatible solute accumulation.In transgenic-engineered plants the disturbances of hexose sensing to produce trehalose, fructans, or mannitol was observed that mightbe major contributory factor to the stress-tolerant phenotypes [80].

Induction of Plant Hormones

The plant hormones play vital roles in plants to adapt in changing environments, by mediating growth, development, nutrient allocation, and source/sink transitions. Some plant hormones are involved in multiple processes while ABA is the stress-responsive hormone; the role of cytokinins, brassinosteroids, and auxins during environmental stress is emerging.

Cross talk between the different plant hormones results in synergistic or antagonic interactions that play crucial roles in response of plants to abiotic stress [81]. The characterization of the molecular mechanisms regulating hormone synthesis, signaling, and action are facilitating the modification of hormone biosynthetic pathways for the generation of transgenic crop plants with enhanced abiotic stress tolerance. Plant hormones regulate every aspect of plant growth and development and the responses of plants to biotic and abiotic stresses. Plant growth regulators include the five classical phytohormones: abscisic acid (ABA), ethylene, cytokinin (CK), auxin (IAA), gibberellin (GA), jasmonate (JA), as well as brassinosteroids (BR), salicylic acid (SA), nitric oxide (NO), and strigolactone (SL), and it is likely that additional growth regulators are yet to be discovered [82].

Lipids and Fatty Acids

Plants respond to abiotic and biotic stress by change in membrane fluidity or releasing by α-linolenic (18:3) from lipids membrane [83]. The change of membrane fluidity is mediated by changes in unsaturated fatty acid levels, a function provided in part by the regulated activity of fatty acid desaturases. Adjustment of membrane fluidity maintains an environment suitable for the function of critical integral proteins during stress. α-linolenic acid, released from membrane lipid by regulated lipase activity, is the precursor molecule for phyto-oxylipin biosynthesis. The modulation of chloroplast oleic acid (18:1) levels is central to the normal expression of defense responses to pathogens in *Arabidopsis*. Oleic (18:1) and linolenic (18:2) acid levels, in part, regulate development, seed colonization, and mycotoxin production by *Aspergillus* spp.

Osmotic Stress Induced Proteins

Plant tolerance against osmotic stress is a complex multigenic trait, a demand exists for genome wide analysis, including 'omics' approaches suitable for uncovering important gene sets involved in this important process [84]. Osmotic stress greatly affects cells both at the micro (i.e., membrane structure), and at the macro level (i.e., the physiology of the whole plant), with results that reflect the variety of responses involved in the acquirement of tolerance. At the microcellular level activation of genes in the categories 'cellular structure, organization and biogenesis' and 'transport protein ion channel carrier' was observed, showing the importance of the maintenance of cellular structures and of the control of ion exchange with the environment [85, 86]. For different cellular processes including stress responses, plant development, pathogen defense and transcriptional regulation proteins such as zinc finger are required [87]. A current workon rice plant showed that the zinc finger family (C2H2-type) was represented by 189 members out of them 26 members are respond to different environmental stresses [88]. Plants environmental stresses tolerance is a very complex phenomenon. Soares-Cavalcanti et al., [89] evaluates a set of genes involved in osmotic response, comparing soybean and medicago with the well-described *Arabidopsis thaliana* model plant.

Some Other Abiotic Stress

Temperature Stress

Changes in common physiological system like photosynthesis, dark respiration, membrane stability, and mitochondrial respiration are induced by heat stress. There are considerable losses in pre-harvest and post-harvest crop are arise due to high temperature. In stress condition plants have developed a sequence of method for detoxification that break down the generation and reactions of highly toxic reactive oxygen species (ROS) that cause injury in plants [90, 91]. Antioxidant enzymes such as ascorbate peroxidase, superoxide dismutase, glutathione reductase, catalase, and some metabolites like ascorbic acid, carotenoids, α-tocopherol and glutathione protect cell and subcellar systems in plants by lowering the cytotoxic effects of ROS [92]. The action of antioxidant enzymes superoxide dismutase (SOD), ascorbate peroxidase (APX) and catalase (CAT) are increases at high temperature [93].

Logan et al. [94] discussed the defensive mechanisms of antioxidant enzymes, which provide protection in plants against assimilation of excessive sun light. They give an idea of transgenic plants with high antioxidant enzyme activities will help to improve stress; the effect of such transgenic manipulation strongly depends on the manner in which the stress is imposed.

Water Drought Stress

Drought stress reduces the growth parameters as well asamplified the ajmalicine content. In *Catharanthus roseus* plant the growth parameters are enhanced under drought stress environment after treatment with *Pseudomonas fluorescens* and partially ameliorated the drought induced growth inhibition by increasing the fresh and dry weights significantly. PGPR provides an eco-friendly approach and can be used as an agent in stress amelioration [95]. Arzanesh et al., hypothesize that the isolated strains of *Azospirillum* sp. may alleviate the adverse effects of drought stress on wheat (*Triticum aestivum* L.) growth [96].

CONCLUSION

It is a well-known fact that, PGPR are a group of bacteria that play an important roles in plant growth promotion. The wide variability in the performance of PGPR all over the globe may be due to various environmental factors (climate, weather conditions, soil characteristics or the composition or activity of theindigenous microbial flora).These factors may affect their growth, performance and effects on plants.The exact mechanisms by which PGPR promote plant growth are not fully understood and need more systematic research. The application of 'omics' technologies are enhancing our capability to understand and manage the rhizosphere that will lead to the discovery of new products with improved effectiveness. The application of PGPR offers an environmentally sustainable approach to increase crop production and health.

REFFERENCES

[1] Munns, R., Tester, M., (2008). Mechanisms of salinity tolerance. *Annu. Rev. Plant Biol.,* 59, 651-81.

[2] Rengasamy, P., (2010) Soil processes affecting crop production in salt-affected soils. *Funct. Plant Biol.*, 37(7), 613-620.

[3] Kumar, A., Gupta, A., Azooz, M. M., Sharma, S., Dames, J., Ahmad, P., (2013). Genetic approaches to improve salinity tolerance in plants. In: Salt stress in plants:

Omics, signaling and adaptations, (Ahmad P., Azooz, M. M., Prasad, M. N. V., Eds.), New York: Springer.

[4] Glick, B. R., Todorovic, B., Czarny, J., Cheng, Z., Duan, J., McConkey, B., (2007). Promotion of plant growth by bacterial ACC deaminase. *Critic. Rev. Plant Sci.,* 26, 227-242.

[5] Ahmad, P., Hakim, R. K., Kumar, A., Ashraf, M., Akram, N. A.,(2012). Salt induced changes in photosynthetic activity and oxidative defense system of three cultivars of mustard (*Brassica juncea* L.). *Afr. J. Biotechnol.*, 11, 2694-2703.

[6] James, R. A., Blake, C., Zwart, A. B., Hare, R. A., Rathjen, A. J., Munns, R., (2012). Impact of ancestral wheat sodium exclusion genes *Nax1* and *Nax2* on grain yield of durum wheat on saline soils. *Funct. Plant Biol.,* 39, 609-618.

[7] Munns, R., (2002). Comparative physiology of salt and water stress. *Plant Cell Environ.,* 239-250.

[8] Roy, S. J., Negrao, S., Tester, M., (2014). Salt resistant crop plants, *Cur. Opin. Biotechnol.,* 26, 115-124.

[9] Glick, B. R., (1995). The enhancement of plant growth by free-living bacteria. *Can. J. Microbiol.,* 1995, 41, 109-117.

[10] Glick, B. R., Cheng, Z., Czarny, J., Duan, J., (2007). Promotion of plant growth by ACC deaminase-producing soil bacteria. *Eur. J. Plant Pathol.*, 119, 329-339.

[11] Hontzeas, N., Hontzeas, C. E., Glick, B. R., (2006). Reaction mechanisms of the bacterial enzyme 1-aminocyclopropane-1-carboxylate deaminase. *Biotechnol. Adv.,* 24, 420-426.

[12] Ahmed, I., Muhammad E., Yeseul, S., Jayoung, P., Nauman K., Rifat, H., Young, H. C., (2014). *Sphingobacterium pakistanensis* sp. nov., a novel plant growth promoting rhizobacteria isolated from rhizosphere of *Vigna mungo. Antonie van Leeuwen,* 105, 325-333.

[13] Chennappa, G., Adkar-Purushothama, C. R., Naik, M. K., Suraj, U., Sreenivasa, M. Y., (2014). Impact of pesticides on PGPR activity of *Azotobacter sp.*isolated from pesticide flooded paddy soils. *Greener J. Agric. Sci.,* 4, 117-129.

[14] Upadhyay, S. K., Singh, D. P., (2014). Effect of salt-tolerant plant growth-promoting rhizobacteria on wheat plants and soil health in a saline environment. *Plant Biol.,* 10.1111/plb.12173.

[15] U S Department of Agriculture. Report and recommendations on organic farming.U.S, Government Printing Office, Washington, D.C., USA, 1980, p. 95.

[16] Reganold, J. P., Papendick, R. I., Parr, J. F., (1990). Sustainable agriculture. *Scientific American,* 262, 112-120.

[17] Parr, J. F., Papendick, R. I., Hornick, S. B., Meyer, R. E., (1992). Soil quality: attributes and relationship to alternative and sustainable agriculture. *Amer. J. Alternat. Agric.*, 7, 5-11.

[18] National Research Council. Alternative Agriculture. Committee on the role of alternative fanning methods in modern production Agriculture. Board on Agriculture, National Academy Press, Washington, D.C., USA. 448, 1989.

[19] Parr, J. F., Hornick, S. B., Kaufman, D. D., (1994). Use of microbial inoculants and organic fertilizers in agricultural production. Extension Bulletin 394. Food and Fertilizer Technology Center, Taipei, Taiwan. 16.

[20] Mohammadi, K., Sohrabi, Y., (2012). Bacterial biofertilizers for sustainable crop production: A Review. *ARPN J. Agric. Biol. Sci.,* 7, 307-316.

[21] Pindi P. K., (2012). Liquid Microbial Consortium- A potential tool for sustainable soil health. *J. Biofertil. Biopestic.,* 4, 1-9.

[22] Mia, M. A. B., (2010). Shamsuddin, ZH. *Rhizobium* as a crop enhancer and biofertilizer for increased cereal production. *Afr. J. Biotechnol.,* 9, 6001-6009.

[23] Pindi, P. K., Satyanarayana, S. D. V., (2012). Liquid microbial consortium- A potential tool for sustainable soil health. *J. Biofertil. Biopestici.,* 3, 1-9.

[24] Kloepper, J. W., (1993). Plant growth-promoting rhizobacteria as biological control agents. Soil Microbial Ecology: Applications in Agricultural and Environmental Management. (Metting, Jr., F. B., Ed.), Marcel Dekker Inc., New York, USA. Pp. 255-274.

[25] Arshad, M., Shaharoona, B., Mahmood, T., (2008). Inoculation with plant growth promoting rhizobacteria containing ACC-deaminase partially eliminates the effects of water stress on growth, yield and ripening of *Pisum sativum* L. *Pedosphere,* 18, 611-620.

[26] Banchio, E., Bogino, P. C., Zygadlo, J., Giordano, W., (2008). Plant growth promoting rhizobacteria improve growth and essential oil yield in *Origanum majorana* L. *Biochem. Syst. Ecol.*, 36, 766-771.

[27] Contesto, C., Desbrosses, G., Lefoulon, C., Bena, G., Borel, F., Galland, M., Gamet, L., Varoquaux, F., Touraine, B. (2008). Effects of rhizobacterial ACC deaminase activity on *Arabidopsis* indicate that ethylene mediates local root responses to plant growth-promoting rhizobacteria. *Plant Sci.*, 175, 178-189.

[28] Figueiredo, M. V. B., Martinez, C. R., Burity, H. A., Chanway, C. P., (2008). Plant growth-promoting rhizobacteria for improving nodulation and nitrogen fixation in the common bean (*Phaseolus vulgaris* L.). *World J. Microbiol. Biotechnol.,* 24, 1187-1193.

[29] Mubeen, F., Aslam, A., Radl, V., Schloter, M., Malik, K. A., Hafeez, F. Y., (2008). Role of nature's fertility partners with crop protectants for sustainable agriculture. In: Biological nitrogen fixation: towards poverty alleviation through sustainable agriculture (Dakora, F.D, et al. Eds.). Springer, the Netherlands, pp.153-154.

[30] Naiman, A. D., Latronico, A., Garca de Salamone, I. E., (2009). Inoculation of wheat with *Azospirillum brasilense* and *Pseudomonas fluorescens*: impact on the production and culturable rhizospheremicroflora. *Eur. J. Soil Biol.,* 45, 44-51.

[31] Ahmad, R., Arshad, M., Khalid, A., Zahir, Z. A., (2008). Effectiveness of organic-/bio-fertilizer supplemented with chemical fertilizers for improving soil water retention, aggregate stability, growth and nutrients uptake of maize (*Zea mays* L.). *J. Sust. Agri.,* 31, 57-77.

[32] Amor, F. M. D., Martı´nez, A. S., Fortea, M. I., Legua, P., Delicado, E. N., (2008). The effect of plant-associative bacteria (*Azospirillum* and *Pantoea*) on the fruit quality of sweet pepper under limited nitrogen supply. *Sci. Hortic.*, 117,191-196.

[33] Dell'Amico, E., Cavalca, L., Andreoni, V., (2008). Improvement of *Brassica napus* growth under cadmium stress by cadmium-resistant rhizobacteria. *Soil Biol. Biochem.,* 40, 74-84.

[34] Lebeau, T., Braud, A., Je´ze´quel, K., (2008). Performance of bioaugmentation-assisted phytoextraction applied to metal contaminated soils: a review. *Environ. Pollut.,* 153,497-522.

[35] Masoud, A., Abbas, S. T., (2009). Evaluation of fluorescent pseudomonads for plant growth promotion, antifungal activity against *Rhizoctonia solani* on common bean, and biocontrol potential. *Biol. Control,* 48, 101-107.

[36] Kohler, J., Herna´ndez, J. A., Caravaca, F., Rolda´n, A. (2009). Induction of antioxidant enzymes is involved in the greater effectiveness of a PGPR versus AM fungi with respect to increasing the tolerance of lettuce to severe salt stress. *Environ. Exp. Bot.,* 65, 245-252.

[37] Goswamia, D., Pithwaa, S., Dhandhukiab, P., Thakker, J. N., (2014). Delineating *Kocuria turfanensis* 2M4 as a credible PGPR: Novel IAA-producing bacteria isolated from saline desert. *J. Plant Interact.,* 9, 566-576.

[38] Baris, O., Sahin, F., Turan, M., Orhani, F., Gulluce, M. (2014).Use of plant growth promoting rhizobacteia (pgpr) seed inoculation as an alternative fertilizer inputs in wheat and barley production. *Commun. Soil Sci. Plant Analysis*, (In press).

[39] Ghanbari Z. S., Moosavi S. G., Zabihi H. R., Seghateslami M. J., (2013). The effect of plant growth promoting rhizobacteria (PGPR) and zinc fertilizer on forage yield of maize under water deficit stress conditions, *Tech. J. Engin. App. Sci.,* 3, 3281-3290.

[40] Piccinin, G. G., Braccini, A. L., Dan, L. G. M., Scapim, C. A., Ricci, T. T., (2013). Efficiency of seed inoculation with *Azospirillum brasilense* on agronomic characteristics and yield of wheat. *Ind. Crops Products*, 43, 393–397.

[41] Patil, N. N., Jadhav, S., Ghorpade, S. S., Sharma, A. B., (2013). Isolation and enrichment of sugar press mud (spm) adapted microorganism for production of biofertilizer by using sugar press mud. *Internat. J. Adv. Biotechnol. Res.,* 4, 96-104.

[42] Jha, B., Gontia, I., Hartmann, A., (2012). The roots of the halophyte *Salicornia brachiata* are a source of new halotolerant diazotrophic bacteria with plant growth-promoting potential. *Plant Soil,* 356, 265–277.

[43] Jorquera, M. A., Shaharoona, B., Nadeem, S. M., Mora, M. L., Crowley, D. E., (2012). Plant growth-promoting rhizobacteria associated with ancient clones of creosote bush (*Larrea tridentata*). *Microb. Ecol.*, 1-5.

[44] Garg, N., Singla, P., (2012). The role of *Glomus mosseae* on key physiological and biochemical parameters of pea plants grown in arsenic contaminated soil. *Sci.Hortic.,* 143, 92-101.

[45] Silva, M. C. P., Figueiredo, A. F., Andreote, F. D., Cardoso, E. J. B. N., (2013). Plant growth promoting bacteria in *Brachiaria brizantha. World J. Microbiol. Biotechnol.,* 29, 163-71.

[46] Francisco, M. A., Paula, C. C., (2011). Plant growth-promoting bacteria as a tool to improve salinity tolerance in sweet pepper. *Functional Plant Biol.,* 39, 82-90.

[47] Golpayegani, A., Tilebeni, H. G., (2011). Effect of biological fertilizers on biochemical and physiological parameters of basil (*Ociumum basilicm* L.) Medicine Plant. *American-Eurasian J. Agric. Environ. Sci.*, 11, 445-450.

[48] Habibzadeh, F., Sorooshzadeh, A., Pirdashti, H., Mohammad, S. A., Sanavy, M. A., (2012). comparison between foliar application and seed inoculation of biofertilizers on canola (Brassica napus L.) grown under waterlogged conditions. *AJCS*, 6, 1435-1440.

[49] Parkash, V., Aggarwal, A., Sharma, V., (2011). Rhizospheric effect of *vesicular arbuscular mycorrhizal* inoculation on biomass production of *Ruta graveolens* L.: A potential medicinal and aromatic herb. *J. Plant Nutr.,* 34, 1386-1396.

[50] Chakraborty, U., Roy, S., Chakraborty, A. P., Dey, P., (2011). Chakraborty, B. Plant growth promotion and amelioration of salinity stress in crop plants by a salt-tolerant bacterium. *Recent Res. Sci. Technol.,* 3, 61-70.

[51] Dastager, S. G., Deepa, C. K., Pandey, A., (2011). Plant growth promoting potential of *Pontibacter niistensis* in cowpea (*Vigna unguiculata* (L.) Walp.). *Appl. Soil Ecol.,* 49,250-255.

[52] Saravanakumar, D., Kavino, M., Raguchander, T., Subbian, P., Samiyappan, R., (2011). Plant growth promoting bacteria enhance water stress resistance in green gram plants. *Acta Physiol. Plant*., 33, 203-209.

[53] Benizri, E., Baudoin, E., Guckert, A., (2001). Root colonization by inoculated plant growth promoting rhizobacteria. *Biocontrol Sci. Techno*., 11, 557-574.

[54] Bloemberg, G. V., Lugtenberg, B. J. J., (2001). Molecular basis of plant growth promotion and biocontrol by rhizobacteria. *Curr. Opin. Plant Biol*., 4, 343-350.

[55] De-Bashana, L. E., Hernandeza, J. P., Bashan, Y., (2012). The potential contribution of plant growth-promoting bacteria to reduce environmental degradation- A comprehensive evaluation. *Appl. Soil Ecol.,* 61, 171-189.

[56] Kumar, V., Gera, R., (2014). Isolation of a multi-trait plant growth promoting *Brevundimonas* sp. and its effect on the growth of *Bt*-cotton. *Biotech*., 4, 97-101.

[57] Baris, O., Sahin, F., Turan, M., Orhan, F., Gulluce, M., (2011). Use of Plant growth promoting rhizobacteria (PGPR) seed inoculation as an alternative fertilizer inputs in wheat (*Triticum aestivum* spp. *vulgare*) and barley (*Hordeum vulgare*) production. *J. Plant Nutr.*

[58] Putcha, V. S., Allen, S. J., (1997). A technique for screening bacterial inoculants in the field. In: *Plant growth promoting rhizobacteria: Present Status and Future Prospects*, (Ogoshi, A., Kobayashi, K. K., Homma, Y., Kodama, F., Kondo, N., Akino, S., Eds.), OECD, Paris, pp. 221-222.

[59] Zhang, J., Liu, J., Meng, L., Ma, Z., Tang, X., Cao, Y., Sun, L., (2012). Isolation and characterization of plant growth-promoting rhizobacteria from wheat roots by wheat germ agglutinin labeled with fluorescein isothiocyanate. *The J. Microbiol.,* 50, 191-198.

[60] Logan, N. A., Berge, O., Bishop, A. H., Busse, H. J., De Vos, P., Fritze, D., (2009). Heyndrickx M., Kampfer P., Rabinovitch L., Salkinoja-Salonen M. S., Seldin, L., Ventosa, A. Proposed minimal standards for describing new taxa of aerobic, endospore-forming bacteria. *Int. J. Syst. Evol. Microbiol.,* 59, 2114-2121.

[61] Oliveira, I. A., Vasconcellos, M. J., Seldin, L., Paiva, E., Vargas, M. A. T., Sa, N. M. H., (2000). Random amplified polymorphic DNA analysis of effective *Rhizobium sp.* associated with beans cultivated in Brazilian cerrado soils. *Braz. J. Microbiol*., 31, 39-44.

[62] Depret, G., Laguerre, G., (2008). Plant phenology and genetic variability in root and nodule development strongly influence genetic structuring of *Rhizobium leguminosarum biovar viciae* populations nodulating pea. *New Phytol*., 179, 224-235.

[63] Stackebrandt, E., Goebel, B. M. .Taxonomic Note: A place for dna-dna reassociation and 16s rRNA sequence analysis in the present species definition in bacteriology. *Int. J. System. Bacteriol*., 44, 846-849.

[64] Mora, R. R., (2006). DNA-DNA reassociation methods applied to microbial taxonomy and their microbial taxonomy and their critical evolution. In: Molecular identification,

systematic, population structure of prokaryotes, (Stackebrandt, E., Eds) Springer, Berlin, Heidelberg, pp. 23-49.

[65] Mota, F. F., Gomes, E. A., Paiva, E., Seldin, L., (2005). Assessment of the diversity of *Paenibacillus* species in environmental samples by a novel rpoB-based PCR-DGGE method. *FEMS Microbiol. Ecol.*, 53, 317-328.

[66] Cerritos, R., Vinuesa, P., Eguiarte, L. E., Herrera-Estrella, L., Alcaraz-Peraza, L. D., Arvizu-Go′mez, J. L., Olmedo, G., Ramirez, E., Siefert, J. L., Souza, V., (2008). *Bacillus coahuilensis* sp. nov., a moderately halophilic species from a desiccation lagoon in the Cuatro Cie′negas Valley in Coahuila, Mexico. *Int. J. Syst. Evol. Microbiol.*, 58, 919-923.

[67] Greenway, H., Munns, R., (1980). Mechanisms of salt tolerance in nonhalophytes. *Annu. Rev. Plant Physiol.,* 31, 149-190.

[68] Flowers, T. J., Hajibagheri, M. A., Clipson, N. J. W., (1986). Halophytes. *The Quart. Rev. Biol.*, 61, 313-337.

[69] Hasegawa, P. M., Bressan, R. A., Zhu, J. K., Bohnert, H. J., (2000a). Plant cellular and molecular responses to high salinity. *Annu. Rev. Plant Physiol.*, 51, 463-499.

[70] Blumwald, E., Aharon, G. S., Apse, M. P., (2000). Sodium transport in plant cells. *Biochemica et Biophysica Acta,* 1465, 140-151.

[71] Sanders, D., (2000). Plant biology: The salty tale of *Arabidopsis*. *Curr. Biol.*, 10, 486-488.

[72] Zhu, J. K., (2000). Genetic analysis of plant salt tolerance using *Arabidopsis*. *Plant Physiol.,* 124, 941-948.

[73] Zhu, J. K., (2001). Plant salt tolerance. *Trends in Plant Sci.*, 6, 66-71.

[74] Hasegawa, P. M., Bressan, R. A., Pardo, J. M., (2000b). The dawn of plant salt to tolerance genetics. *Trends Plant Sci.,* 5, 317-319.

[75] Sharma, P., Jha, A. B., Dubey, R. S., Pessarakli, M., (2012). Reactive oxygen species, oxidative damage, and antioxidative defense mechanism in plants under stressful conditions. *J. Bot.*, 1-26.

[76] Gill, S. S., Tuteja, N., (2010). Reactive oxygen species and antioxidant machinery in abiotic stress tolerance in crop plants. *Plant Physiol. Biochem.*, 48, 909-930.

[77] Foyer, C. H., Noctor, G., (2005). Oxidant and antioxidant signalling in plants: A re-evaluation of the concept of oxidative stress in a physiological context. *Plant Cell Environ.*, 28, 1056-1071.

[78] Shahbazi, M., Arzani, A., Saeidi, G., (2011). Effect of NaCl treatments of seed germination and antioxidant activity of Canola (*Brassica Napus*) cultivars, *J. Bot.*, 41, 67-73.

[79] Amirjani, M. R., (2010). Effect of salinity stress on growth, mineral composition, poline content, antioxidant enzymes of soybean. *Am. J. Plant Physiol.,* 5, 350-360.

[80] Bray, E. A., Bailey-Serres, J., Weretilnyk, E., (2000). Responses to abiotic stresses.In: Biochemistry and Molecular Biology of Plants, (Buchanan, B. B., Gruissem, W., Jones, R. L., Eds.), *Am. Soc. Plant Physiologists*, Rockville, MD, pp. 1159-1175.

[81] Zvi, P., Eduardo, B., (2011). Hormone balance and abiotic stress tolerance in crop plants. *Current Opin. Plant Biol.,* 14, 290-295.

[82] Peleg, Z., Blumwald, E., (2011). Hormone balance and abiotic stress tolerance in crop plants. *Curr. Opin. Plant Biol.,* 14, 290-295.

[83] Robert, G. U., (2008). Fatty acid unsaturation, mobilization, and regulation in the response of plants to stress. *Biotechnol. Lett.,* 30, 967-977.

[84] Hirayama, T., Shinozaki, K., (2010). Research on plant abiotic stress responses in the post-genome era: Past, present and future. *J. Plant,* 61, 1041-1052.

[85] Maurel, C., Javot, H., Lauvergeat, V., Gerbeau, P., Taournaire, C., Santoni, V., Heyes, J., (2002). Molecular physiology of aquaporins in plants. *Int. Rev. Cytol.*, 215, 105-148.

[86] Estrella, R. V., Barkla, B. J., Bohnert, H. J., Pantoja, O., (2004). Novel regulation of aquaporins during osmotic stress. *Plant Physiol.,* 2318-2329.

[87] Ciftci-Yilmaz, S., Mittler, R., (2008). The zinc finger network of plants. *Cell Mol. Life Sci.,* 65, 1150-1160.

[88] Agarwal, P., Arora, R., Ray, S., Singh, A. K., Singh, V. P., Takatsuji, H., Kapoor, S., Tyagi, A. K., (2007). Genome-wide identification of C (2) H (2) zinc-finger gene family in rice and their phylogeny and expression analysis. *Plant Mol. Biol.*, 65, 467-485.

[89] Soares-Cavalcanti, N. M., Belarmino, L. C., Kido, E. A., Wanderley-Nogueira, A. C., Bezerra-Neto, J. P., Cavalcanti-Lira, R., Pandolfi , V., Nepomuceno, A. L., Abdelnoor, R. V., Nascimento, L. C., Benko-Iseppon, A. M., (2012). *In-silico* identification of known osmotic stress responsive genes from *Arabidopsis* in soybean and Medicago. *Genet. Mol. Biol.,* 35, 315-321.

[90] Liu, X., Huang, B., (2000). Heat stress injury in relation to membrane lipid peroxidation in creeping bentgrass, *Crop Sci.,* 40, 503-510.

[91] Larkindale, J., Huang, B., (2004). Thermo-tolerance and antioxidant systems in *Agrostis stoloifera*: involvement of salicylic acid, abscisic acid, calcium, hydrogen peroxide, and ethylene. *J. Plant Physiol.,* 161, 405-413.

[92] Sairam, R. K., Shrivastava, G. C., Saxena, D. C., (2000). Increased antioxidant activity under elevated temperature: a mechanism of heat stress tolerance in wheat genotypes. *Biol. Plant*, 43, 245-251.

[93] Almeselmani, M., Deshmukh, P. S., Sairam, R. K., Kushwaha, S. R., Singh, T. P., (2006). Protective role of antioxidant enzymes under high temperature stress. *Plant Sci.,* 171, 382-388.

[94] Logan, B. A., Kornyeyev, D., Hardison, J., Holaday, A. S., (2006). The role of antioxidant enzymes in photoprotection. *Photosynthesis Res.,* 88, 119-132.

[95] Jaleel, A. C., Manivannan, P., Sankar, B., Kishorekumar, A., Gopi, R., Somasundaram, R. Paneerselvam, R., (2007). *Pseudomonas fluorescens* enhances biomass yield and ajmalicine production in *Catharanthus roseus* under water deficit stress. *Colloids Surf. B: Biointerfaces*, 60, 7-11.

[96] Arzanesh, M. H., Alikhani, H. A., Khavazi, K., Rahimian, H. A., Miransari, M., (2011). Wheat (*Triticum aestivum* L.) growth enhancement by *Azospirillum* sp. under drought stress. *World J. Microbiol. Biotechnol.*, 27, 197-205.

In: Microbes in Soil and Their Agricultural Prospects
Editors: K. Kishore Choudhary and D. Wattal Dhar
ISBN: 978-1-63482-824-6

Chapter 11

PROSPECTS OF SILICATE SOLUBILIZING BACTERIAL INOCULANTS IN SUSTAINABLE AGRICULTURE

***M. Sahul Hameed*[1]*, R. Anandham*[1]*, N. Premalatha*[1]*, M. Karthik*[1]*, M. Ranjith*[1]*, P. Arul Jose*[1]*, J. Prabaharan*[2]*, S. Thiyageshwari*[2] *and N. O. Gopal*[1]**

[1]Department of Agricultural Microbiology, Agricultural College and Research Institute, Tamil Nadu Agricultural University, Madurai, India
[2]Department of Soils and Environment, Agricultural College and Research Institute, Tamil Nadu Agricultural University, Madurai, India

ABSTRACT

Silicon (Si) is the second most abundant element in the Earth's crust. Although biologically common, it is a minor element in the majority of living organisms. In nature, silica scarcely occurs as a monomer but is mostly present in a more or less polymerized state and cannot be absorbed unless depolymerized. Silicon is considered essential for rice, maize, sugarcane and other graminaceous plants. It is important for maintaining leaves erect, increasing photosynthesis and conferring resistance against pests and diseases and also plays a role in the husk formation and in oxygen transport to roots. Silicon was found to suppress the activity of certain enzymes; in particular Si suppression of invertase resulted in the reduction in phosphatase, which provided a greater supply of essential high-energy precursors needed for optimum growth. The current trend of cultivating high-yielding, nitrogen-responsive varieties warrants increased application of nitrogen fertilizers, which in turn results in a higher demand for silica. Agricultural soils are largely composed of silicate minerals, but many soils contain an inadequate supply or are naturally low in available silica. It is likely that the silica content in some regions may be a limiting factor for the sustainable crop production. Naturally occurring silica can be depolymerized and solubilized by microorganisms. The mechanisms of bio-dissolution of silica by silicate solubilizing bacteria and their effects on various crop plants are discussed.

Keywords: bioinoculants, *Bacillus*, potassium solubilizers, *Pseudomonas*, silicate solubilizing bacteria

INTRODUCTION

Soil consists of sand, silt, and clay, all of which contain either a primary or secondary form of silicate minerals. Some silicate minerals contain potassium as well. Weathering and organic agents resulting from decomposition of plant and animal residues, root exudates and/or metabolites of rhizosphere and non-rhizosphere microorganisms can create an environment favorable for the breakdown of silicate minerals and the release of the ionic component [1]. Some wastes from mining and mineral processing may be useful for low-input agriculture, e.g., wastes from a granite crusher yard and basic slag from the steel industry can be reutilized as potential silicate sources providing silicon nutrition for agricultural purposes. Farmers, agricultural scientists and policy planners have searched for eco-friendly and environmentally safe on-farm inputs to minimize environmentally destructive and economically disruptive off-farm inorganic inputs so that the resources available today are passed on to the posterity without exhaustion or a loss in productivity. Recycling of industrial wastes to replenish what has been removed from soil and addition of biomass and biofertilizers are gaining momentum among farmers to preserve soil health.

Biofertilizers are an eco-friendly and low cost technology, and their application plays a major role in maintaining soil fertility, nutrient transformation, crop sanitation and sustainability. Thus, the diazotrophic rhizobiocoenosis is an important biological process that plays a major role in satisfying the nutritional requirements of crops. Microbiologists have turned to investigating bio-dissolution and generation of plant nutrients such as phosphorus and potassium *in situ.* Such organisms can be used to hasten the dissolution of inorganic silicate minerals, such as a granite crusher yard powder, sand and basic slag, to release silicon (Si) and potassium (K). This chapter mainly focuses on the forms and importance of Si and K. Further, microbial solubilization of Si and K from ores and their effects on crop plants in sustainable agriculture are discussed.

SILICON IN SOIL

Silicon is abundant in the Earth's crust (constituting up to 28%) and occurs in various complex mineral forms. The main forms of silicon in soil are soluble silica, mainly monosilicic acid (H_4SiO_4) and polysilicic acid ($SiO_2 \cdot nH_2O$) [2], crystalline and amorphous silica, $Si(OH)_4$ and Mn-silicates [3]. Silicon is believed to be beneficial to rice plants as it increases available phosphorus in soil, enhances phosphorus uptake, improves water use efficiency, reduces toxicities associated with Mn, Fe and Al, increases mechanical strength of stems, improves growth habits, reduces shattering of grains and controls insect pests [4-7]. Yeo et al. [8] found that Si reduced sodium uptake by rice in saline conditions by reducing the transpiration by-pass flow. Compartmentation of Si increases Mn binding to the cell wall [9] or results in a more homogeneous distribution of Mn in the leaves [10, 11].

In rice leaves, 90% or more of the silicic acid exists as silica gel (polysilicic acid). The deposition of silica renders the plants stiff and erect. The major nutrient, nitrogen, favors succulence, tenderness and lodging, predisposing the plants to pests and diseases. On the other hand, silica reverses the unfavorable effects of higher doses of nitrogen. The current trend of cultivating high-yielding, nitrogen-responsive varieties warrants increased application of nitrogen fertilizers, which in turn results in a higher demand for silica, whose supply in soil is not adequate. This necessitates using silicon additives for crop nutrition, including rice, sugarcane, banana and potato.

Silicon combines with four oxygen atoms to form the silica tetrahedral structure. Besides, several other elements, namely, Al, Fe, Na, K, Mg, Mn, Br and Zn, also bind to silicon. The size of the unit cell is determined by the sizes of these constituent ions. Aluminum is substituted for silicon in the central position of a tetrahedron in a large number of silicates. The levels of silica in soil vary widely depending on the nature of the parent material and the transformation it has undergone. More than a hundred relatively common silicate minerals have been listed by Dana [12]. Various silicate minerals found in soil are given in the following table (Table 1).

Table 1. Silicate minerals found in soil

S. No.	Name of the mineral	Molecular formula
Potassium containing silicate minerals		
1.	Adularia	$KAlSi_3O_8$
2.	Anorthoclase	$(KAlSi_3O_8 - NaAlSi_3O_8)$
3.	Apophyllite	$K Ca_4 (Si_4 O_{10})_2 F. 8H_2O$
4.	Astrophyllite	$(K,Na)_3(Fe^{++}, Mn)_7Ti_2Si_8O_{24}(O,OH)_7$
5.	Biotite	$K(Mg, Fe)_3 (AlSi_3O_{10}) (OH)_2$
6.	Glauconite	$(K, Na, Ca)_{0.5-1} (Fe^{3+},Al, Fe^{2+}, Mg)_2 (Si, Al)_4 O_{10} (OH)_2. 4H_2O)$
7.	Hyalophane	$(K, Ba) (Al, Si)_2 Si_2O_8$
8.	Kalsilite	$KAlSiO_4$
9.	Lepidolite	$K(Li, Al)_{2-3} (AlSi_3O_{10}) (O, OH, F)_2$
10.	Leucite	$KAlSi_2O_6$
11.	Microcline	$KAlSi_3O_8$
12.	Muscovite	$KAl_2 (Al Si_3 O_{10}) (OH)_2$
13.	Orthoclase	$KAlSi_3O_8$
14.	Phillipsite	$(Ca,Na_2,K_2)_3Al_6Si_{10}O_{32}\cdot 12H_2O$
15.	Phlogopite	$K Mg_3 (Al_3 Si_3 O_{10}) (OH)_2$
16.	Roscoellite	$KV_2 (Al Si_3 O_{10}) (OH)_2$
17.	Sanidine	$(K,Na)(Si,Al)_4O_8$
Other silicate minerals		
18.	Actinolite	$Ca_2(Mg,Fe)_5Si_8O_{22}(OH)_2$
19.	Albite	$Na Al Si_3 O_8$
20.	Andalusite	$Al_2 SiO_5$
21.	Anorthite	$Ca Al_2 Si_2 O_8$
22.	Augite	$Ca (Mg, Fe, Al) Si, Al)_2 O_6$
23.	Anthophyllite	$(Mg,Fe)_7Si_8O_{22}(OH)_2$
24.	Benitoite	$Ba Ti Si_3 O_9$

Table 1. (Continued)

S. No.	Name of the mineral	Molecular formula
25.	Danburite	$CaB_2(SiO_4)_2$
26.	Diopside	$Ca\ Mg\ Si_2\ O_6$
27.	Enstatite	$Mg\ SiO_3$
28.	Epidote	$Ca_2Al_2(Fe^{3+};Al)(SiO_4)(Si_2O_7)O(OH)$
29.	Fayalite	$Fe_2\ SiO_4$
30.	Forsterite	$Mg_2\ SiO_4$
31.	Halloysite	$Si_4\ Al_4\ O_{10}\ (OH).\ 4H_2O$
32.	Hemimorphite	$Zn_4\ Si_2\ O_7\ (OH)_2.\ H_2O$
33.	Hornblende	$Ca_2(Mg, Fe, Al)_5\ (Al, Si)_8O_{22}(OH)_2$
34.	Hypersthene	$(Mg, Fe)\ SiO_3$
35.	Illite	$Ky\ (Al_4\ Fe_4\ Mg_4\ Mg_6)\ Si_{8\text{-}Y}\ Al_Y\ O_{20}\ (OH)_4$

ROLE OF SILICON IN PLANTS

Silicon in a chemically combined form is ubiquitous in nature. The Si content of soils can vary dramatically, from <1% to 45% of dry weight [13], and its presence in the form of silicic acid, $Si(OH)_4$, or in its ionized form, $Si(OH)_3O^-$, which predominates at pH >9, allows its uptake by plants. Silicic acid is generally found in soils at concentrations ranging from 0.1 to 0.6 mM [14]. The rice plant absorbs silicon in the form of orthosilicic acid (H_4SiO_4 or $Si(OH)_4$), along with water from the growth medium. As water is lost through transpiration, the silica concentration increases, and at higher levels orthosilicic acid polymerizes into silica gel ($SiO_2{\cdot}nH_2O$) through non-enzymatic reactions [15].

It was suggested that silica in plants reaches leaf surface and filters harmful ultraviolet radiation, with leaf cells acting as 'windows' transmitting the light energy to photosynthetic mesophyll and cortical tissues beneath the epidermis. Silicon is not a constituent of cellular components because Si - Si bonds are considerably weaker than C - C bonds. Silicon is primarily deposited on the walls of the epidermis and vascular tissues, conferring strength, rigidity and resistance to pests and diseases. Although silicon has no biochemical role in plant development, silicon-enzyme complexes are found to protect or regulate photosynthesis [16].

Van Hoest [17] reported that different parts of the rice plant can have large differences in Si accumulation, varying from 0·5 g/kg^{-1} in polished rice, 50 g/kg^{-1} in rice bran, 130 g/kg^{-1} in rice straw, 230 g/kg^{-1} in rice hulls to 350 g/kg^{-1} in rice joints (found at the base of the grain). Si is found in plants at concentrations ranging from 0·1% to 10% on the dry weight basis, an amount equivalent to, or even exceeding, several macronutrients [14]. Plants deprived of Si are often weaker structurally and more prone to abnormalities of growth, development and reproduction, and it is the only nutrient that is not detrimental when absorbed in excess [18].

Hodson et al. [19] showed that Si accumulation mainly, but not exclusively, occurs in monocotyledonous plants. Plants of the families *Poaceae*, *Equisetaceae* and *Cyperaceae* show high Si accumulation (>4%), the *Cucurbitales*, *Urticales* and *Commelinaceae* show intermediate Si accumulation (2–4%), while most other species demonstrate little accumulation. Si concentrations in shoots tend to decrease in following order: liverworts > horsetails > clubmosses > mosses > angiosperms > gymnosperms > ferns. Gallo et al. [20]

observed that in rice, oat, rye and wheat the seed coat accumulated most of the silica compared to the grain. They also observed that leaves and stems of maize and sorghum and leaves of sugarcane and bamboo had the highest Si levels compared to other plant parts.

The application of silicon generally increased the uptake of phosphorus, calcium and magnesium and the formation of carbohydrates and decreased the uptake of nitrogen, potassium, iron and manganese in transplanted rice [21]. The application of silicates not only augmented their absorption by rice plants but also had a significant interrelationship with other nutrients. Silica nutrition has direct and indirect beneficial effects on the rice growth, largely due to its unique physiological role [15, 22]. Silica application to rice was found to increase the grain yield under both upland and waterlogged conditions [23]. Addition of 100 mg of SiO_2 kg^{-1} to soil increased plant height, tiller numbers, the number of leaves, leaf area, dry matter production and the leaf area ratio in rice. The transpiration rates in plants supplied with silica were reduced. The supply of silica also altered the levels of carbohydrates, protein and phenol, and these changes varied slightly with the stage of the rice crop and the plant part.

Silicon application significantly increased the grain yield by increasing the number of panicles per plant, spicklets per panicle, mature grains per panicle, grain weight and harvest index in rice [24], and silica promoted the photosynthetic activity and P utilization in rice [25] and water use efficiency of the plants [26].

The application of silicate materials increased leaf chlorophyll and decreased leaf freckling in sugarcane [27]. Boylsone et al. [28] reported that in cotton silicon was necessary for the development of fibers. Silicon is not only important structurally but also plays a metabolic role in the synthesis of protein, chlorophyll, DNA, RNA, xanthophylls, lipids and thymidylate kinase [29].

Savant et al. [30] reported that sugarcane absorbs more Si than any other mineral nutrient, accumulating approximately 380 kg ha^{-1} of Si. Sugarcane responses to silicon fertilization have been documented, and silicon applications on commercial fields are routine in certain areas of the world. The reasons for this plant response or yield increase are not fully understood, but several mechanisms have been proposed. Some studies indicated that sugarcane yield responses to silicon might be associated with induced resistance to biotic and abiotic stresses, such as disease and pest resistance, Al, Mn, and Fe toxicity alleviation, increased P availability, reduced lodging, improved leaf and stalk erectness, freeze resistance, and improved plant water economy. Elawad et al. [27] revealed that application of 15 metric tons ha^{-1} of silicates (Florida slag) increased leaf chlorophyll by 78% and 65% and decreased leaf freckling by 46% and 41% in the first and ratoon crops of sugarcane, respectively. It also increased the levels of Si, P, Ca, and Cu and reduced the levels of N, K, Mg, Fe, Mn, and Zn in the leaves and also increased soil pH and Si, P, Ca, and Mg content. The application of black or grey ash of rice hulls made rice seedlings healthy and strong and increased their biomass [31]. Talashilkar and Chavan [32] reported that application of rice hull ash as a source of silicon resulted in a significant increase in rice grain and straw yields.

Lee [33] found that the rice husk had the most influence on the availability of silica in soil, which was attributed to fast release of soluble silica by the rice husk. Addition of silicate materials, *viz.,* rice husks, furnace slag and sodium metasilicate at the 500 ppm SiO_2 level increased the available silicon status in soil [34]. Calcium silicate fertilizers, which are made from slags, have been applied to paddy fields as a source of silicon to increase rice production

[35]. Rice yield increases due to continuous application of siliceous slag at 1.5 to 3.0 t ha^{-1} were observed in pot and field experiments [36, 37].

Silva [38] reported that application of Si as calcium silicate, sodium silicate or basic slag increased the solubility of adsorbed soil phosphorus, decreased phosphorus fixation, increased cation exchange capacity and decreased aluminum activity. Vyas and Motiramani [39] reported that addition of organic matter and silicate increased the available phosphorus in the soil. Chinnaswami and Chandrasekaran [40] reported that addition of silica as silica gel to soils in Tamil Nadu, India released more phosphorus, with its levels higher in lateritic sandy loam than in clay and non-lateritic sandy loam soil. The application of calcium silicate to highly weathered savanna soils enhanced the response of wetland rice to applied phosphate [41]. Ma et al. [42] reported that rice straw application to the soil is one of the means to supply a sufficient amount of silicon to rice plants for healthy growth. Silicon is released from two sources in this case, from the straw itself and from the soil silicon liberated by straw decomposition. The same field experiments showed that straw application increased the silica content of rice shoots. Kamimura et al. [43] reported that a paddy soil derived from glassy volcanic ash was improved by application of rice straw, which resulted in a higher percentage of ripened grains and an increased yield.

INTERACTION OF SILICON WITH OTHER NUTRIENTS

Nitrogen

Silicon forms important interrelationships with other nutrients. Applied Si seems to interact favorably with N, P, and K and has the potential to improve their efficiency in terms of the plant yield response. Silicon helps rice plants resist biotic stresses (insect pests and fungal diseases) and tolerate abiotic stresses (Al, Fe, and Mn toxicities). It also helps to reduce cuticular transpiration and, to some extent, crop lodging caused by an excessive N supply [44]. The application of silicon has the potential to raise the optimum nitrogen rate and to increase productivity of existing low-land rice fields [45]. The maintenance of erect leaves by proper silicon fertilization for a higher photosynthetic efficiency becomes more important when rice is grown with liberal application of nitrogen fertilizers in low-land rice fields, which have highly weathered tropical soils with a low silicon supplying capacity [46]. Islam and Saha [21] reported that the addition of silicon to culture media decreased the nitrogen content of rice plants with increasing rates of silicon application. The decrease in nitrogen uptake may be a reason for reduced plant lodging due to silicon application.

Phosphorus

Yong et al. [47] showed that appropriate Si application to a low P solution could enhance phosphorus absorbability and utilization ability in roots of maize seedlings, increase the content and accumulation of phosphorus and silicon, as well as dry matter accumulation in different organs, improve the chlorophyll content (chlorophyll a, 12.0%; chlorophyll b, 34%; total chlorophyll, 18.1%) and the net photosynthetic rate (20.0%) of leaves, along with the root/shoot ratio. Finally, the authors concluded that 1.5 mmol/L Si was considered optimal for maize.

Birch [48] reported that there was a relationship between the amount of water-soluble silica in soil and the soil phosphorus status. It has been suggested that adding water-soluble silicate to soil may increase the amount of phosphate the crop can take up from the soil [49]. This increased phosphorus content had been attributed to better availability of soil P and/or an enhanced mobility of P from roots to stems. Silicon addition increased the phosphorus concentration in rice plants and promoted the phosphorus uptake [21]. Roy et al. [50] found that silicate application increased the phosphorus concentration in green tops of sugarcane (metabolically more active tissue) and decreased phosphorus in the stalk (metabolically less active tissue) when P nutrition was low. The addition of soluble silica to a soil low in available phosphorus was found to increase the available P by displacement of adsorbed phosphorus ions by silicate ions [51]. Jones and Handreck [52] reported that the release of phosphorus anions by silicate anions was not due to anion exchange, but possibly due to an increase in pH, which is known to free phosphate from its complexes with iron and aluminum.

Potassium

Potassium is required for many plant system functions. It regulates and activates photosynthesis, metabolism of carbohydrates, organic acids, fats and nitrogenous compounds, protein synthesis, water use efficiency, elongation, resistance to drought, frost, lodging, pests and diseases, and physiological disorders [53]. Mitsui and Takatoh [54] and other authors [55, 56] reported that supplying 100 ppm of silicon to a culture solution not only increased the formation of new roots and tillers, advanced heading, suppressed silicon deficiency, and increased the number of seeds but also increased the level of potassium in plants. However, few investigators [21, 57] also reported a general decrease in the plant potassium content due to silicon in a solution culture. The K removal by a crop is either equal to or higher than that of N in many cases. After the introduction of high-yielding varieties and crop hybrids, the K removal from the soil increased several fold, and thus the application of K fertilizer is required, along with N and P applications, to replenish the soil supply so as to achieve nutrient balance. K is usually the most abundant nutrient element in soil [58]. Igneous rocks of the Earth's crust have higher K content than sedimentary rocks. Among the former, granites and syenties contain 46 to 54 g/kg^{-1}, basalt contains 7 g/kg^{-1} and peridotites contain 2.0 g/kg^{-1} of K [59].

The total K content in soil ranges between 3,000 and 100,000 kg ha^{-1} in the upper 0.2 m of the soil profile, of which 98% is bound in the mineral form whereas 2% is in the soil solution [60, 61]. Potassium is present in soil in four forms which are in dynamic equilibrium. They represent exchangeable K which is held by the negative charges of organic matter and clay minerals. It is easily exchanged with other cations and is readily available to plants. Non-exchangeable K ions are held in interlayers and gaps [62]. Non-exchangeable K is released to the exchangeable form when levels of exchangeable and soil solution K are decreased by crop removal or by leaching and, perhaps, by a significant increase in microbial activity [63, 64].

Common soil K-bearing minerals, in the order of availability of their K to plants, are biotite, muscovite, orthoclase and microcline [65], which are also silicate minerals. Potassium is depleted from the minerals in a centripetal fashion [66]. Arnold [67] stated that the weathering ratio of biotite and muscovite were similar. Potassium weathering occurs more rapidly in biotite than in muscovite. Feldspars and micas are important reserves of available K in soils [68]. Sparks and Huang [69] reported that the release of K from micas can proceed either by ion exchange and the consequent transformation of micas to expansible 2:1 layer silicates or by dissolution.

Interactions of applied K and Si in soil seem to have beneficial effects on rice yields. Si application (140 and 280 kg ha^{-1}) increased the upland rice yield response to applied K on an ultisol in West Sumatra [70]. Baba et al. [71] reported that low soil moisture and high humidity, which are common in upland rice regions, reduced Si and K absorption by rice plants and increased their susceptibility to biotic and abiotic stress. Park and Kim [72] reported that application of silica slag or wollastonite to soils low in Si and K improved the efficiency of applied K. Nogushi and Sugawara [73] reported that K deficiency reduced the accumulation of Si in epidermal cells of leaf blades, thus increasing the susceptibility of the plant to rice blast.

Other Elements

Nwugo et al. [74] investigated effects of silicon nutrition on the symptoms of low-level cadmium (Cd) toxicity in hydroponically grown rice seedlings. The results showed that low-level Cd treatment generally inhibited plant growth and photosynthesis. However, the addition of 0.2 or 0.6 mM Si significantly reduced the root and leaf Cd content. Kaufman et al. [75] found that foliar-applied gibberellic acid (5.0 mg of GA_3 into the leaf whorl) increased 2- to 3-fold the internodal length and fresh weight and decreased the silicon content (as much as 50%) of sugarcane. Gibberellic acid had little or no effect on growth or silicon content of leaf laminae or sheaths.

Liang [76] conducted a study with two barley cultivars, Kepin No.7 (salt-sensitive) and Jian 4 (salt-tolerant), grown in a hydroponic system containing 120 mol m^{-3} NaCl only or 120 mol m^{-3} NaCl and 1.0 mol m^{-3} Si (as potassium silicate). Compared with the plants treated with salt alone, the superoxide dismutase (SOD) activity in plant leaves and HC-ATPase activity in plant roots increased and the malondialdehyde (MDA) concentration in plant leaves decreased significantly for both cultivars when treated with salt and Si. The addition of Si was also found to reduce sodium but increase potassium concentrations in shoots and roots of salt-stressed barley. Sodium uptake and transport into shoots from roots were strongly inhibited by added Si under salt-stress conditions. However, Si addition had little effect on calcium concentrations in shoots of salt-stressed barley. Thus, Si-enhanced salt tolerance is attributed to selective absorption and transport of potassium and sodium by plants.

Islam and Saha [21] reported that calcium and magnesium concentrations in rice plants increased gradually with increasing rates of silicon application. The inhibition of calcium uptake caused by aluminum was eliminated by silicon application [77]. On the other hand, Inanaga and Okasaka [78] suggested that silicon might suppress the absorption of Ca^{2+} ions by the rice plant by possibly competing with calcium for binding sites on the cell wall. Silicic acid in the soil solution may have an important role in reactions that control the availability of trace elements.

Ponnamperuma [79] reported that a sufficient supply of silica makes oxygen transport from the plant top to roots more efficient through the enlargement of gas channels, which increases oxidation and subsequent deposition of iron and manganese on the root surface. Silicon fertilization alleviates aluminum, iron and manganese toxicities in certain acid soils [44]. Okuda and Takahashi [80] reported that silica promoted the oxidation power of roots and thus decreased the solubility and uptake of Fe and Mn. Silica ameliorates aluminum toxicity to plant growth by decreasing the activity of free Al^{3+} in solution and also by reducing the internal toxicity of aluminum.

The aluminum toxicity effect on growth of sorghum and maize was decreased when silica was added to the nutrient solution [77, 81]. Hodson and Wilkins [82] found that exposure to aluminum markedly increased silica levels in the cortical cell wall of roots of spruce seedlings in an acid soil, and the formation of aluminosilicate compounds in the cell wall inhibited the uptake of aluminum into the protoplast, thus reducing the aluminum toxicity.

Silicate Solubilizing Bacteria (SSB)

Norkina and Pumpyanskaya [83] reported that the silicate solubilizing bacterium *Paenibacillus musiliaginosus* (formerly *Bacillus musiliaginosus*) var. *siliceous* liberated potassium from feldspar and aluminosilicates *in vitro*. They also showed that silicate solubilizing bacteria can fix small amounts of nitrogen. The ubiquitous occurrence of silicate solubilizing bacteria in the rhizosphere of rice has been previously described [84]. Potassium-solubilizing bacteria are normally capable of degrading silicate minerals and releasing potassium and other elements in plant available forms [85]. The maximum counts of silicate solubilizing bacteria were found in rhizosphere soils of system of rice intensification (SRI) fields (8.51×10^3 cfu g^{-1} of dry soil) and aerobic rice fields (8.26×10^3 cfu g^{-1} of dry soil) 15 and 45 days after transplanting, respectively [86].

Isolation of Silicate Solubilizing Bacteria

Silicate solubilizing bacteria were isolated from rhizosphere soils. Rhizosphere soil samples were serially diluted up to 10^{-3} dilution, and a 0.1 mL aliquot of an appropriate dilution was plated on agar medium containing 0.5% of insoluble magnesium trisilicate [87]. Halo zones around bacterial colonies indicated silicate solubilizing activity (Figure 1).

Screening of Silicate Solubilizing Bacteria

In vitro Determination of Silicate Solubilizing Ability of Bacterial Isolates

Plate Assay

To test for silicate solubilizing activity, 10 μL of a bacterial isolate culture (1×10^7 cfu mL^{-1}) was spot inoculated in the center of a Bunt and Rovira agar medium plate [87], supplemented with 0.5% insoluble $MgSiO_3$, and incubated for 48 h. The diameter of the clearing zone around the colony was measured. The solubilizing efficiency of silicate solubilizing bacteria was calculated as follows [88]: (Diameter of solubilization zone – Colony diameter) / Colony diameter × 100.

Quantification of *in vitro* Solubilization of Insoluble Silicate and Potassium

The solubilization of insoluble inorganic silicate sources, *viz.,* vermiculite, potassium-aluminum silicate and tricalcium phosphate, by silicate solubilizing bacteria was assessed according to the methods described earlier [89, 90]. The silicate, zinc and phosphate solubilization efficiency was estimated in all seven silicate solubilizing bacterial isolates. Among the isolates, SSB4 exhibited the highest silicate (158.06%) and maximum zinc (106.25%) solubilization efficiencies, while isolate SSB2 exhibited the highest phosphate solubilization efficiency (106.38%). Molecular characterization was performed by 16S rRNA gene sequencing, and strains SSB2, SSB4 and SSB7 were identified as *Achromobacter xylosoxidans*, *Bacillus altitudinis* and *Pseudomonas gessardii*, respectively.

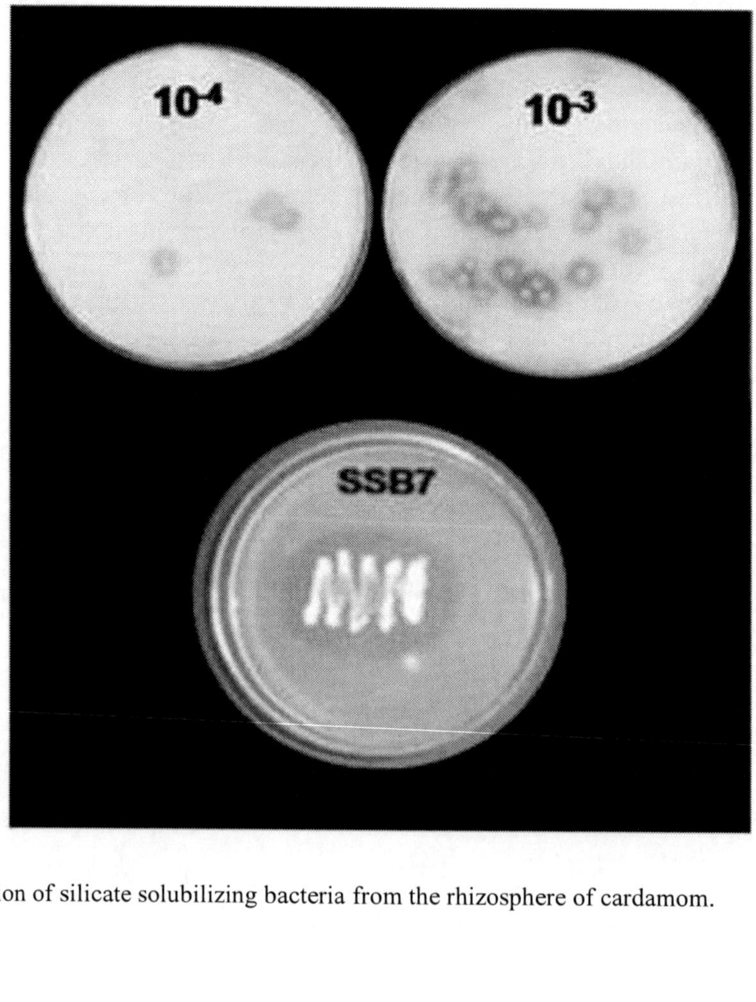

Figure 1. Isolation of silicate solubilizing bacteria from the rhizosphere of cardamom.

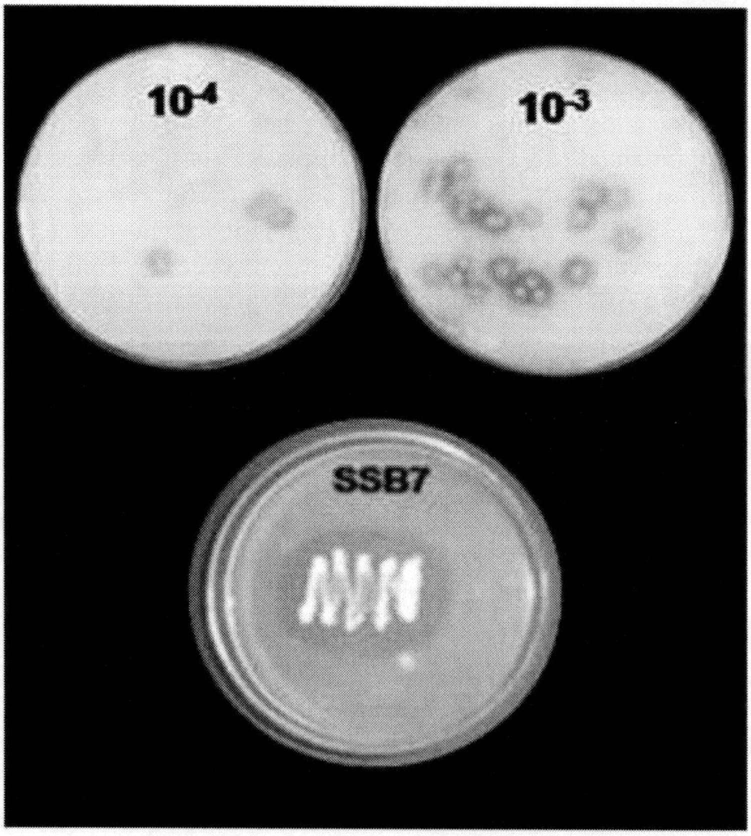

POSSIBLE MECHANISMS OF SILICATE SOLUBILIZATION

The growth of certain bacteria in media containing silicates has been well established [91]. Ehrlich [92] reported that the growth of *Penicillium simplicissimus* in glucose mineral salts medium in the presence of basalt, granite, granodiorite and quartzite released silica. Sterilization by autoclaving releases some amount of silica from these minerals, suggesting a role for high temperature, similar to weathering, as reported by Hall [93].

The solubilization of silica from silicate minerals may be due to several bacterial metabolites or products, *viz.*, carbon dioxide, organic acids, alkali, H_2S and exopolysaccharides.

It is well known that CO_2 is a product of microbial decomposition of organic matter and root respiration. Waksman and Starkey [94] showed the reaction of orthoclase degradation by carbon dioxide:

$$\underset{\text{Orthoclase}}{2\,KAlSi_3O_8} + 2\,H_2O + CO_2 \rightarrow \underset{\text{Kalonite}}{H_2Al_2Si_2O_9} + K_2CO_3 + 4\,SiO_2$$

Duff and Webley [95] reported the dissolution of silicates due to complexing of 2-keto-gluconate with cationic components, as these complexes are more stable than silicates:

$$CaSiO_3 \rightleftharpoons Ca^{2+} + SiO_3^{-}$$

$$Ca^{2+} + \text{2-keto-gluconate} \rightleftharpoons \text{Ca 2-keto-gluconate}$$

Organic acids are inherently present in biological materials, and microorganisms synthesize them in the process of cellular metabolism and also secrete organic acids during fermentation of organic substrates. Organic acids play a role in weathering, and this mechanism may operate both in the *in vitro* breakdown and in soil. Field observations and laboratory experiments have demonstrated that microbes accelerate weathering reactions of aluminosilicate minerals, especially when in direct contact with a mineral surface, by providing metal-complexing ligands, changing redox conditions or mediating formation of secondary mineral phases. Barker et al. [96] have reported that several strains of bacteria produced organic and inorganic acids and extracellular polymers which increased the release of cations from biotite (Si, Al). These bacteria colonize all mineral surfaces, preferentially along the cleavage plane and edge of minerals. Organic acids can supply H^+ ions to the medium and promote silicate hydrolysis, on one hand, and make complexes with cationic components, on the other hand, as they are potent complex-forming agents [97]. Citric, oxalic, keto, hydroxy and carboxylic acids usually form complexes with cations and promote their removal and retention in the medium in a dissolved state. The breakdown of silicate concurrently leads to the release of potassium from silicate minerals that have potassium in their crystal structure.

Effect of Inoculation of Crop Plants with SSB

Growth chamber studies were conducted to investigate the effects of silicon on the distribution of cadmium in rice seedlings grown hydroponically in the presence of toxic levels of Cd. Added Si significantly alleviated the toxicity of Cd in aerobically grown rice seedlings and partly overcame the reduction in growth due to Cd. This amelioration was correlated with a reduction in Cd uptake. Si increased Cd accumulation in the roots and restricted the transport of Cd from roots to shoots, so the accumulation of Cd in the shoots decreased by 33% [98].

Chandramani et al. [99] carried out field experiments to analyze induced resistance to certain major pests of rice, *viz.*, leaf folder, stem borer and gall midge. Populations of these insects were assessed at regular intervals on plants treated with neem cake, farm yard manure (FYM), *Azospirillum*, a phosphobacterium, silicate solubilizing bacteria and lignite fly ash. The results revealed that the combination of FYM, the three biofertilizers, lignite fly ash and neem cake, applied in splits, significantly reduced the incidence of leaf folder (by 76.69%), stem borer (by 58.66%) and gall midge (by 66.81%) as compared to inorganic NPK. The insect populations were correlated with silica content, which was analyzed in the leaf sheath and lamina of plants treated with the organic nutrients. As the age of the plants increased, silica content proportionally increased in the plants. The treatment with biofertilizers, lignite fly ash, FYM and neem cake in splits resulted in a significantly higher grain yield (5.49 t/ha^{-1}), which was 2% higher than that in NPK-applied plots (5.38 t/ha^{-1}). It was concluded that application of organic sources of nutrition reduced the incidence of leaf folder, stem borer and gall midge and increased the rice yield.

The release of K and P from feldspar and rock phosphate was substantially enhanced in pure cultures inoculated with silicate dissolving bacteria. *Bacillus cereus* had a much higher potential to release K at pH 6.5-8.0 than at other pH values. More K and P was released under more aerobic conditions [100]. The mean effect of silicate bacteria on feldspar dissolution was appreciably greater in the soils tested (62% of total K) than in the bacterial cultures (53% of total K). The bacteria were found to solubilize insoluble magnesium trisilicate *in vitro*. The dissolved silicon content increased with bacterial growth, and after 4 weeks varying concentrations of soluble silicon were present in the media in which *Azotobacter*, *Azospirillum*, *Bacillus megaterium* var. *phosphaticum* and *Rhizobium* were grown [7]. *Bacillus edaphicus* strain NBT was able to efficiently mobilize potassium in cotton and rape when illite was added to the soil. The potassium content increased in both plants (by 30% and 26%, respectively) grown in soils treated with insoluble potassium and inoculated with *B. edaphicus* strain NBT. Inoculation with this bacterial strain also resulted in higher N and P contents in above-ground plant components. The bacterial isolate was also able to colonize and develop in the rhizosphere soil of cotton and rape after root inoculation [101].

Inoculation of SSB7, along with organic siliceous amendments (lignite), to cardamom under field conditions resulted in enhanced crop growth and yield. The number of tillers per clump, tiller height, number of panicles per clump and number of capsules per panicle as well as green pod yield also significantly increased, while clump rot disease (*Rhizoctonia solani*) and stem borer (*Conogethes punctiferalis*) incidences were significantly reduced in the cardamom plantations (Figure 2) [102].

Figure 2. Effect of inoculation of the silicate solubilizing bacterium Bacillus sp. SSB7 on the number of tillers per clamp of cardamom (*Elettaria cardamomum*).

Combined inoculation of ashwagandha with *Azospirillum*, phosphobacteria and silicate solubilizing bacteria, along with rock phosphate, under field conditions produced similar results with application of Azophos + SSB + 100 kg of N + 25 kg of P as superphosphate + 25 kg of P as rock phosphate +100 kg of K and recorded maximum chlorophyll content (4.12 mg/g^{-1}), shoot length (76.50 cm), leaf area (1,350.42 cm^2/plant^{-1}), root length (26.19 cm/plant^{-1}), root girth (2.38 cm/plant^{-1}), root fresh weight (21.33 g/plant^{-1}), root dry weight (6.18 g plant^{-1}) and the total alkaloid content of 1.46% in ashwagandha roots [86]. Application of silicate solubilizing bacterial strains *Pseudomonas gessardii* SSB7 and *Achromobacter xylosoxidans* SSB2 recorded the maximum seed germination (88% and 91%), plant height (85.6 and 55.1 cm) and biomass (3.10 and 3.60 g/plant^{-1}) in rice plants (cv. ADT 39 and Anna (R) 4, respectively) in wet-land and aerobic conditions [86].

Mineral Solubilization by Silicate Solubilizing Bacteria

Phosphate Solubilization

Kapoor et al. [103] showed that solubilization of inorganic phosphate depends on the type of organic acids produced by phosphate solubilizing microorganisms, which chelate phosphate ions. Tri-and dicarboxylic acids were more effective chelators compared to monocarboxylic and aromatic acids. Raj et al. [104] observed an augmentation in the grain yield of rice due to inoculation of silicate solubilizing bacteria or phosphobacteria, along with 75% of the recommended 50 kg/ha^{-1} P_2O_5.

A phosphate solubilizing bacterium, *Enterobacter intermedium,* isolated from the rhizosphere, exhibited a strong ability to solubilize insoluble phosphate, which was mediated by the production of organic acids, especially 2-ketogluconic acid (2-KGA), in culture media

[105]. Various bacterial isolates from the rhizosphere of field-grown *Trigonella* sp., which solubilized phosphates, were identified as *Bacillus*, *Pseudomonas*, *Azotobacter* and *Azospirillum* [106]. Chavada et al. [107] screened diazotrophic bacteria from a saline desert soil for their phosphorus solubilization potential, and the results revealed that all isolates tested produced large clear zones on Pikovskaya's medium, indicative of the potential of the isolates to be used as a phosphate biofertilizer even under saline conditions.

Potassium Solubilization

Solubilization of potassium from potassium-aluminum silicate minerals by selected bacterial strains was shown to be due to the action of different organic acids, such as citric, oxalic, malic, succinic and tartaric acids. Also, potassium solubilizing fungi, *Aspergillus niger* and *Aspergillus terreus*, were isolated [108,109]. Among 137 cultures tested, 20 bacterial strains showed significant potassium solubilizing activity on mica powder-supplemented plates, and the amount of K released by different strains varied from 15 to 48 mg L^{-1}. In glucose-amended medium broth, bacterial strains WPS73 and NNY43 caused significant K solubilization (41.0 and 48.0 mg L^{-1}, respectively). Bacterial strain WPS73 caused the maximum solubilization (49.0 mg L^{-1}) at 25 °C, whereas bacterial strain NNY43 caused the maximum solubilization at 30 °C [110]. The potassium content increased in okra grown in soils treated with insoluble potassium and inoculated with a strain of *Enterobacter hormaechei* and the fungus *A. terreus* [111].

Zinc Solubilization

The term zinc solubilizing bacteria (ZSB) was created for those bacteria that are capable of solubilizing insoluble zinc compounds and minerals in agar plates as well as in soil [112,113]. Zinc is an essential micronutrient that plays a vital role in various metabolic processes in plants, and its deficiency adversely affects the growth and development of crop plants [114]. The available zinc content in Indian soils is low; however, the total zinc content is substantially high as zinc exists in fixed forms such as smithsonite ($ZnCO_3$), sphalerite (ZnS), zincite (ZnO), franklinite ($ZnFe_2O_4$), wellemite (Zn_2SiO_4), and hopeite ($Zn_3(PO_4)_2 \cdot 4H_2O$), which are sparingly soluble. It is plausible to assume that exploitation of zinc solubilizing bacteria may aid in overcoming zinc deficiency.

Senthilkumar [115] isolated zinc solubilizing organisms from rice rhizosphere and ore sources, and the studies on inoculation of zinc solubilizing bacteria to rice in a plot culture resulted in the correction of zinc deficiency, in addition to an increase in growth parameters. Zinc solubilizing microorganisms can solubilize zinc from inorganic and organic pools of total soil zinc and can be utilized to increase zinc availability to plants. However, only some bacterial species of the genera *Acinetobacter*, *Bacillus*, *Gluconacetobacter* and *Pseudomonas* have been reported to solubilize zinc [116-118]. Inclusion of a zinc solubilizing bacterium as a bioinoculant in the crop production technology is really beneficial for a country like India having a high incidence of zinc deficiency (more than 70%) [113]. Iqbal et al. [119] examined the application of zinc solubilizing bacterial isolates, along with zinc phosphate, and demonstrated an increase in plant growth parameters, including seedling length, in *Vigna radiata*.

Development of Pest and Disease Resistance due to Silicate Solubilizing Bacteria

Jeffery et al. [120] reported that incorporation of silicate in a potting mixture reduced the incidence of damping off caused by *Pythium ultimum* on cucumber. There appeared to be a minimum time for accumulation of silicon by seedlings to obtain protection from challenges imposed by the pathogen. Belanger et al. [121] strongly suggested that Si mediated active localized cell defenses in wheat against a *Blumeria graminis* f. sp. *tritici* attack. Histological and ultrastructural analyses revealed that epidermal cells of Si-supplied plants reacted to the *B. graminis* f. sp. *tritici* attack with specific defense reactions, including papilla formation, production of callose, and release of an electron-dense osmiophilic material identified by cytochemical labeling as containing glycosylated phenolics. The phenolic material was not only accumulated along the cell wall but also associated with altered integrity of haustoria in a manner similar to that of localized phytoalexins as reported for other pathosystems. Fawe et al. [122] reported that silicon was involved in the increased resistance of cucumber to powdery mildew (*P. ultimum*) by enhancing antifungal activity of infected leaves. This antifungal activity was attributed to the presence of low molecular weight metabolites and a phytoalexin.

Much of the work on possible mechanisms by which Si affects susceptibility to diseases has been done on cucumber, but the insights gained from these investigations may also apply to other plants. Soluble Si taken up by plants accumulated in the apoplast, particularly in epidermal cell walls [14, 123, 124]. This observation has led many investigators to hypothesize that Si inhibits fungal diseases by physically inhibiting fungal germ tube penetration through the epidermis [125, 126]. However, subsequent investigations have found that only trichome bases of the cucumber epidermis were silicified [123, 125].

Conclusion

To date, the application of Si nutrition in crop production remains unexplored. Hence, comprehensive applied research is needed to optimize the doses of silica for various crop plants and also explore its availability in various types of soil. In the future, crop-specific, efficient silicate solubilizing bacteria should be identified and molecular mechanisms and enzymes responsible for silica solubilization should be investigated.

Acknowledgment

This work was supported by Tamil Nadu Agricultural University, Coimbatore 641 003, Tamil Nadu, India.

References

[1] Boyle, J. R. & Voigt, G. K. (1973). Biological weathering of silicate minerals. Implications for tree nutrition and soil genesis. *Plant Soil.*, *38*, 191-201.

[2] Mckeague, J. A. & Cline, M. G. (1963). Silica in soil. *Advan. Agron.*, *15*, 339-396.

[3] Ponnamperuma, F. S. (1972). The chemistry of submerged soils. *Adv. Agron.*, *24*, 29-96.

[4] Reifenberg, A. & Buckwold, S. J. (1954). The release of silica from soils by the orthophosphate anions. *J. Soil Sci.*, *5*, 106-115.

[5] Elawad, S. H. & Green, V. E. (1979). Silicon and the rice plant environment: a review of recent research II. *Riso.*, *28*, 235-253.

[6] Ma, J. F., Miyake, Y. & Takahashi, E. (2001). Silicon as a beneficial element for crop plants. In: *Silicon in Agriculture* (Datonoff, L., Korndorfer, G., Synder, G., Eds.), Elsevier Science Publishing, New York, 17-39.

[7] Muralikannan, N. (1996). Biodissolution of silicate, phosphate and potassium by silicate solubilizing bacteria in rice eco system. *M.SC. (Ag.) Thesis*, TNAU, Coimbatore.

[8] Yeo, A. R., Flowers, S. A., Rao, G., Welfare, K., Senanayake, N. & Flowers, T. J. (1999). Silicon reduces sodium uptake in rice (*Oryza sativa* L.) in saline conditions and this is accounted for by a reduction in the transpirational bypass flow. *Plant Cell Environ.*, *22*, 559-565.

[9] Rogalla, H. & Romheld, V. (2002). Role of leaf apoplast in silicon mediated manganese tolerance of *Cucumis sativus* L. *Plant Cell Environ.*, *25*, 549-555.

[10] Horst, W. J. & Marschner, H. (1978). Effect of silicon on manganese tolerance of bean plants (*Phaseolus vulgaris* L.). *Plant Soil.*, *50*, 287- 303.

[11] Iwasaki, K. & Matsumura, A. (1999). Effect of silicon on alleviation of manganese toxicity in pumpkin (*Cucurbita moschata* Duch cv. *Shintosa*). *Soil Sci. Plant Nutr.*, *45*, 909-920.

[12] Dana, J. D. (1959). Dana's manual of mineralogy rev. by C. S. Hurlbut, Jr., Wiley, New York.

[13] Sommer, M., Kaczorek, D., Kuzyakov, Y. & Breuer, J. (2006). Silicon Pools and Fluxes in Soils and Landscapes - A Review. *J. Plant Nutr. Soil Sci.*, *169*, 310-329.

[14] Epstein, E. (1994). The anomaly of silicon in plant biology. *Proceedings of the National Academy of Sciences., USA. 91*, 11-17.

[15] Yoshida, S. (1975). The physiology of silicon in rice. Food fertilizer tech. *Centre Technical Bull. 25*.

[16] Tisdale, S. L., Nelson, W. L. & Beatson, J. D. (1985). *Soil fertility and fertilizers.* MacMillan Publ., New York, p. *754*.

[17] Van Hoest, P. J. (2006). Rice straw, the role of silica and treatments to improve quality. *Animal Feed Sci. Tech.*, *130*, 137–171.

[18] Epstein, E. (1999). Silicon. *Ann. Rev. Plant physiol. Plant Mol. Biol.* 50, 641-644.

[19] Hodson, M. J., White, P. J., Mead, A. & Broadley, M. R. (2005). Phylogenetic variation in the silicon composition of plants. *Ann. Bot.*, *96*, 1027–1046.

[20] Gallo, J. R., Furlani, P. R., Bataglia, O. C. & Hiroce, R. (1974). Silicon content in grass and forage crops. *Cienciae culture.*, *26*, 282-293.

[21] Islam, A. & Saha, R. C. (1969). Effects of silicon on the chemical composition of rice plants. *Plant soil.*, *30*, 446-458.

[22] Okuda, A. & Takahashi, E. (1965). The role of silicon. In: *"The mineral nutrition of the rice plant"*. John Hopkins press. Baltimore. Maryland, 123-146.

[23] Datta, N. P. & Shinde, J. E. (1965). Yield and nutrition of rice under upland and waterlogged conditions - Effect of nitrogen, phosphorus and silica. *J. Indian Soc. Soil Sci.*, *13*, 53-60.

[24] Rani, Y. A., Narayanan, A., Devi, V. S. & Subbaramamma, P. (1997). The effect of silicon application in growth and yield of rice plants. *Ann. Plant. Physiol.*, *11*, 123-128.

[25] Ma, J. F. & Takahashi, E. (1990). Effect of silicon on the growth and phosphorus uptake of rice plants under different phosphorus supplies. *Plant Soil.*, *126*, 115-119.

[26] Vandervorm, P. D. J. (1980). Uptake of Si by five plant species as influenced by variations in Si-supply. *Plant Soil,*. *56*, 153-156.

[27] Elawad, S. H., Gascho, G. J. & Street, J. J. (1982). Response of Sugarcane to silicate source and rate. I. Growth and Yield. *Agron.*, *J. 74*, 481-484.

[28] Boylsone, E. K., Hebert, J. J., Hensarling, T. P., Browdow, J. M. & Thibodeaux, D. P. (1990). Role of silicon in developing cotton fibres. *J. Plant Nutr.*, *13*, 131-148.

[29] Lewin, J. C. (1955). Silicon metabolism in diatoms III. Respiration and silicon uptake in *Novicula pelliculosa*. *J. Gen. Physiol.*, *39*, 1-10.

[30] Savant, N. K., Korndörfer, G. H., Datnoff, L. E. & Snyder, G. H. (1999). Silicon nutrition and sugarcane production: A Review. *Plant Nutr.*, *22*, 1853-1903.

[31] Savant, N. K. & Sawant, A. S. (1995). Nutrition composition of rice seedlings as influenced by rice hull ash application to seedbed. *Oryza.*, *43*, 62-71.

[32] Talashilkar, S. C. & Chavan, A. S. (1996). Effects of rice hull ash on yield and uptake of silicon and phosphorus by rice cultivars at different growth stages. *J. Ind. Soc. Soil Sci.*, *44*, 340-342.

[33] Lee, T. S. (1983). Improvement of latosolic paddy fertility. *Soils and Fertilizers in Taiwan*, 1-20.

[34] Subramannian, S. & Gopalswamy, A. (1990). Influence of silicate and phosphate materials on availability and uptake of silicon and phosphorus in acid soils. *Oryza.*, *27*, 267-273.

[35] Kato, N. & Owa, N. (1997). Dissolution of slag fertilizers in a paddy soil and Si uptake by rice plant. *Soil Sci. Plant. Nutr.*, *43*, 329-341.

[36] Lian, S. & Wang, C. H. (1983). Effect of long term applications of siliceous slag on paddy yield and chemical properties of soil. *J. Agrl. Res. China.*, *32*, 185-199.

[37] Masaoka, Y. & Takawan, N. (1987). Effects of silicon in warm season forage crops on cell wall digestion by cellulose. *J. Japanese Soc. Grassland Sci.*, *32*, 417-418.

[38] Silva, J. A. (1971). Possible mechanism of crop response to silicate application. Proc. Int. Symp. *Soil Fert. Evaluation.*, *1*, 805-814.

[39] Vyas, M. K. & Motiramani, D. P. (1971). Effect of organic matter, silicates and moisture levels on availability of phosphate. *J. Ind. Soc of Soil Sci. 19*, 39-43

[40] Chinnaswami, K. N. & Chandrasekaran, S. (1978). Silica status in certain soils of Tamil Nadu. *Madras Agric J. 65*, 743-746.

[41] Friesen, D. K., Thomas, R. J., Rivera, M., Asakawa, N. & Bowen, W. (1998). Nitrogen dynamics under monocultures and crop rotations on a Colombian savanna oxisol. In:

Proceedings of the *16th World Congress of Soil Science*, Symposium *14*, AFES, Montpellier, France.
[42] Ma, J., Nishimura, K. & Takahashi, E. (1989). Effect of silicon on the growth of rice plant at different growth stages. *Soil Sci. Plant Nutr.*, *35*, 347-356.
[43] Kamimura, Y, Furue, K. & Nishizono, N. (1994). Improvement of paddy soil derived from glassy volcanic ash (shirasu) by successive applications of organic and inorganic amendments. *Soil Sci. Pl. Nu.*, *4*0, 39-48.
[44] Savant, N. K., Synder, G. H. & Datnoff, L. E. (1997). Silicon management and sustainable rice production. *Advan. Agron.*, *58*, 157-189.
[45] Ho, D. Y., Zang, H. L. & Zhang, X. P. (1980). The silicon supplying ability of some important paddy soil in South China. In: *Proceedings of the symposium on Paddy Soils*, held during 19-24th October 1980 at Nanjing, China, p. *95*.
[46] Yoshida, S., Navasero, S. A. & Ramirez, E. A. (1969). Effects of silica and nitrogen supply on some leaf characters of the rice plant. *Plant Sci.* ,*31*, 48-56.
[47] Yong, Y., Li, J., Shi, H., Yongpei, J. & Zhao, Y. (2008). Alleviation of silicon on Low-P stressed maize (*Zea mays* L.) seedling under hydroponic culture conditions. *World J Agri. Sci.*, *4*, 168-172.
[48] Birch, H. F. (1953). Phosphate response, base saturation and silica relationship in acid soils. *J. Agric. Sci.*, *43*, 229-235.
[49] Fisher, R. A. (1929). A preliminary note on the effect of sodium silicate in increasing the yield of barley. *J. Agric. Sci. 19*, 132-139.
[50] Roy, A. C., Ali, M. Y., Fox, R. L. & Silva, J. A. (1971). Influence of calcium silicate on Phosphate solubility and availability in Hawaiian latosols. *Proc. Int. Symp. Soil Fert. Evaluation.*, *1*, 757-765.
[51] Russell, E.W. (1973). *Soil conditions and plant growth.* 10th ed. Longman London. p. 756.
[52] Jones, L. H. P. & Handreck, K. A. (1963). Effect of iron and aluminium oxides on silica in solution in soils. *Nature.*, 198, 852-353.
[53] Perrenoud, S. (1990). Potassium and plant health. Research Topic No.3. IPI Berene. p*365*.
[54] Mitsui, S. & Takatoh, H. (1963). Nutritional study of silicon in graminaceous crops. *Soil Sci. Plant nutr.*, *9*, 49-58.
[55] Okamoto, Y. (1969). Effect of silicic acid upon rice plants. 9. on growth under high and low temperature of the culture solution. 10. On growth under high and low air temperatures *Proc. Crop Sci. Soc.*, *38*, 743-752.
[56] Kwon, H. G., Kim, V. G., Park, J. K. & Kim, Y. S. (1971). Effect of spacing, nitrogen levels and silica on photosynthesis of the rice plant. *Res. Rept. Rural Dev. Plant Environ.*, *14*, 73-81.
[57] Horiguchi, T. (1988). Effect of silicon on alleviation of Mn toxicity of rice plants. *Soil Sci. Plant Nutr.*, *34*, 65-73.
[58] Reitemeier, R. F. (1951). Soil Potassium. In. *Advances in Agronomy*, Vol. III. (Norman, A.G., Ed), American Society of Agronomy Academic Press, New York, 113-114.
[59] Malavolta, E. (1985). Potassium status of tropical and subtropical region soils. In *Potassium in agriculture* (Munson, R.D., Ed.), American Society of Agronomy. Madison, WI. 163-200.

[60] Schroeder, D. (1979). Structure and weathering of potassium containing minerals. *Proceedings of Congress of International Potash Institute.*, *11*, 43-63.

[61] Bertsch, P. M. & Thomas, G. W. (1985). Potassium status of temperate region soils. In: *Potassium in agriculture* (Munson, R.D., Ed.), American Society of Agronomy. Madison, WI. 131-162.

[62] Kittrick, J. A. (1966). Forces involved in ion fixation by vermiculite. *Soil Sci. Soc. Am. Proc.*, *30*, 801-803.

[63] Doll, E. C. & Lucas, R. E. (1973). Testing soils for potassium, calcium, and magnesium. In: *Soil testing and plant analysis* (Walsh, L.M., Beaton J.D., Eds.), Revised eds. Madison, WI: Soil Science Society of America, 133-152.

[64] Sparks, D. L., Zelazny, L. W. & Martens, D. C. (1980). Kinetics of potassium exchange in a paleudult from the Coastal Plain of Virginia. *Soil Sci. Soc. Am. J.*, *44*, 37-40.

[65] Huang, P. M., Crossan, L. S. & Rennie, D. A. (1968). Chemical dynamics of K release from potassium minerals common in soils. *Transactions of 9th International Congress of Soil Science*, *2*, 705-712.

[66] Mortland, M. M. (1958). Kinetics of potassium release from biotite. *Soil Sci. Soc. Am. Proc.*, *22*, 503-508.

[67] Arnold, P. W. (1960). Nature and mode of weathering of soil-potassium reserves. *J. Sci. Food Agy.*, 11, 285-92.

[68] Sparks, D. L. (1987). Potassium dynamics in soils. *Adv.Soil Sci.*, *6*, 1- 63.

[69] Sparks, D. L. & Huang, P. M. (1985). Physical chemistry of soil potassium. In: *Potassium in agriculture* (Munson, R.D., ed.). *Americ. Soc. Agron.* Madison, WI, 201-276.

[70] Burbey, A., Rizaldi, B. & Yulizar, Z. (1988). Response of upland rice to potassium and silicate application on Ultisol. *Pemberitaan Penelitian Sukurami*, *15*, 26-31.

[71] Baba, I., Iwata, I. & Takahashi, Y. (1956). Studies on the nutrition of rice plants with reference to *Helminthosporium* leaf spot (preliminary report). XI. Absorption and translocation of nutrients as influenced by soil moisture and air humidity. *Proc. Sci. Soc. Jpn.*, *24*, 169-172.

[72] Park, Y. D. & Kim, Y. S. (1970). The effect of wollastonite and manganese dioxide on rice grown on a flooded acid sulphate soil. *J. Korean society Soil Sci. Fert.*, *3*, 23-28.

[73] Nogushi, J. & Sugawara, T. (1966). Potassium and Japonica Rice: Summary of twenty five years' research, International Potash Institute Berne, Switzerland, p. *102*.

[74] Nwugo, C. C. & Huerta, A. J. (2008). Effects of silicon nutrition on cadmium uptake, growth and photosynthesis of rice plants exposed to low-level cadmium. *Plant soil.*, *311*, 73-86.

[75] Kaufman, P. B., Najati, S., Myong, Lee, Carlson, T. J., Jones, J. D., Rigot, W., Bigelow, W. C. & Kraus, S. (1981). Effect of gibberellic acid on silica content and distribution in sugarcane. *Plant Physiol.*, *68*, 314-317.

[76] Liang, Y. (1999). Effects of silicon on enzyme activity and sodium, potassium and calcium concentration in barley under salt stress. *Plant Soil. 209*, 217-224

[77] Barcelo, J., Guevara, P. & Poschenrieder, C. (1993). Silicon amelioration of aluminium toxicity in teosinte. (*Zea mays* L. spp. *mexicana*). *Plant soil.*, *154*, 249-255

[78] Inanaga, S. & Okasaka, A. (1995). Calcium and silicon binding compounds in cell walls of rice shoots. *Soil Sci. Pl. Nutr.* 41, 103-110.

[79] Ponnamperuma, F. S. (1965). Review of the symposium o,n the Mineral nutrition of the rice plant. In "*The Mineral Nutrition of the rice plant*" John Hopkins press, Baltimore, MD, 295-328.

[80] Okuda, A. & Takahashi, E. (1962). The physiological role of Silicon in crop plants. *J. Soil Sci. Plant Nutr.*, *9*, 37-58.

[81] Galvez, L., Clark, R. B., Gourley, L. M. & Maranvilla, J. W. (1987). Silicon interactions with manganese and aluminium toxicity in sorghum. *J. Plant Nutr.*, *10*, 1139-1147.

[82] Hodson, M. J. & Wilkins, D. A. (1991). Localization of aluminium in the roots of norway spruce (*Picea abies* L. Karst) inoculated with *Paxillus involutus* Fr. *New Phytol.*, *118*, 273-238.

[83] Norkina, S. P. & Pumpyanskaya, L. V. (1956). Certain properties of silicate bacteria. *Dokl. Akal. Skh. Nauk.*, *3*, 27-31.

[84] Dwivedi, R. S., Mathew, C., Michael, K. L., Ray, P. K. & Amma, B. S. K. (1978). Carbonic anhydrase, carbon assimilation and canopy structure in relation to gut yield of coconut. In: *On patosynthesis and productivity*, *Indian Nat. Sci. Acad. Symp*. Lucknow, 44-46.

[85] Lü, C. & Huang, B. (2010). Isolation and characterization of *Azotobacteria* from pine rhizosphere. *Afr. J. Microbiol Res.*, *4*, 1299-1306.

[86] Karthik, M. (2012). Dynamics of diazotrophs and silicate solubilizing bacteria in three different rice (*Oryza sativa* L.) Ecosystems. *M. Sc. (Ag.) Thesis*, TNAU, Coimbatore.

[87] Bunt, J. S. & Rovira, A. D. (1955). Microbiological studies of some Antarctic soils. *J. Soil Sci.*, *6*, 119-128.

[88] Srivastav, S., Yadav, K. S. & Kundu, B. S. (2004). Prospects of using phosphate solubilizing *Pseudomonas* as biofungicide. *Indian J. Microbiol.*, *44*, 91-94

[89] Saxena, M. M. (1989). *Environmental analysis of water, soil and air.* Agro botanical publ., India. p.786.

[90] Stanford, S. & English, L. (1949). Use of flame photometer in rapid soil tests for K and C. *Agron. J. 41*, 446-447.

[91] Webley, D. M., Henderson, M. E. K. & Taylor, I. F. (1963). The microbiology of rocks and weathered stones. *J. Soil. Sci. 14*, 102-112.

[92] Ehrlich, H. I. (1981). *Geomicrobiology.* Marcel Dekker Inc. New York. P. *393*.

[93] Hall, F. R. (1972). Silica cycle. In: *The Encyclopedia of Geochemistry and Environmental sciences. Encyclopedia of Earth sciences series*, Vol. IVA (Fairbridge, R.W., Ed.), van Nostrand-Reinhold Publishing Co., New York, 1082-1085.

[94] Walksman, S. A. & Starkey, R. L. (1924). Microbiological analysis of soil as an index of soil fertility. *Soil Sci.*, *17*, 141-161.

[95] Duff, R. B. & Webley, D. M. (1959). 2-keto-gluconic acid as a neutral chelator produced by soil bacteria. *Chem. Ind.*, *10*, 1376-1377.

[96] Barker, W. W., Welch, S. A. & Banfield, J. F. (1997). Geomicrobiology of silicate weathering. In: *Minerological society of America Reviews in Mineralogy.*, *35*, 397-428.

[97] Avakyan, Z. A., Pivovara, T. A. & Karavaiko, G. I. (1986). Properties of a new species. *Bacillus mucilaginosus*. *Mikrobiologiya*, *55*, 477-482.

[98] Shi, X., Zhang, C., Wang, H. & Zhang, F. (2005). Effect of Si on the distribution of Cd in rice seedlings. *Plant Soil.*, *272*, 53–60.

[99] Chandramani, P., Rajendran, R., Muthiah, C. & Chinniah, C. (2010). Organic source induced silica on leaf folder, stem borer and gall midge population and rice yield. *J. Biopest. 3*, 423 – 427.

[100] Badr, M. A., Shafei, A. M. & Sharaf El-Deen, S. H. (2006). The Dissolution of K and P-bearing Minerals by Silicate Dissolving Bacteria and their effect on Sorghum Growth. *Res. J. Agric. Biol. Sci.*, *2*, 5-11.

[101] Sheng, X. F. (2005). Growth promotion and increased potassium uptake of cotton and rape by potassium releasing strain of *Bacillus edaphicus*. *Soil Biol. Biochem.*, *37*, 1918-1922.

[102] Anandkumar, P. (2009). Influence of silicate solubilizers on silicon nutrition of cardamom (*Elettaria cardamomum* (L.) Maton). *M.SC. (Ag.) Thesis*, TNAU, Coimbatore.

[103] Kapoor, K. K., Mishra, M. M. & Kukreja, K. (1989). Phosphate solubilization by soil microorganisms. *Ind. J. Microbiol.*, 29, 119-128.

[104] Raj, S. A., Gopalaswamy, G. & Kareem, A. A. (1994). Effect of graded levels of phosphorus and phosphobacteria inoculation on rice yield. *Madras Agric.*, *J. 81*, 457-458.

[105] Kim, K. Y., Hwangbo, H., Park, R. D., Ieong, K. Y., Kim, V. W. & Park Krishnan, H. B. (2002). 2-ketogluconic acid production and phosphate solubilization by *Enterobacter intermedium*. *17th WCSS*, 14-21, August 2002, Thailand, Symposium no. 06, Paper no. 919.

[106] Tank, N. & Saraf, M. (2003). Phosphate solubilization, exopolysaccharide production and indole acetic acid secretion by rhizobacteria isolated from *Trigonella foenumgraceum*. *Ind. J. Microbiol.*, *43*, 37-40.

[107] Chavada, N. B., Patel, R., Vanpuria, S., Raval, B. P. & Thakkar, P. V. (2010). A Study on isolated diazotrophic (Non Symbiotics) bacteria from saline desert soil as a biofertilizer. *Int. J. Pharm.*, *1*, 52-54.

[108] Prajapati, K. B. & Modi, H. A. (2012). Isolation and characterization of potassium solubilizing bacteria from ceramic industry soil. *CIB Tech. J. Microbiol.*, *1*, 8-14.

[109] Prajapati, K., Sharma, M. C. & Modi, H. A. (2012). Optimization of medium components for potassium solubilizing fungus - *Aspergillus terreus* (ksf 1) by Response surface methodology. *Ind. J. Fund. Appl. Life Sci.*, *2*, 54-60.

[110] Parmar, P. & Sindhu, S. S. (2013). Potassium Solubilization by Rhizosphere Bacteria: Influence of Nutritional and Environmental Conditions. *J. Microbiol. Res.*, *3*, 25-31

[111] Prajapati, K., Sharma, M. C. & Modi, H. A. (2013). Growth promoting effect of potassium solubilizing microorganisms on *Abelmoscus esculentus*. *Inter. J. Agri. Sci.*, *3*, 181-188.

[112] Raj, S. A. (2002). Biofertilizers for micronutrients. *Biofert. News Lett.* ,*10*, 8-10.

[113] Saravanan, V. S., Madhaiyan, M. & Thangaraju, M. (2007). Solubilization of zinc compounds by the diazotrophic, plant growth promoting bacterium *Gluconacetobacter diazotrophicus*. *Chemos. 66*, 1794-1798.

[114] Cakmak, I. (2008). Enrichment of cereal grains with zinc: Agronomic or genetic biofortification. *Plant Soil. 302*. 1-17.

[115] Senthilkumar, R. (1997). Biodissolution of zinc in rice ecosystem. *M.SC. (Ag.) Thesis*, TNAU, Coimbatore.

[116] Di Simine, C. D., Sayed, J. A. & Gadd, G. M. (1998). Solubilization of zinc phosphate by a strain of *Pseudomonas fluorescens* isolated from a forest soil. *Biol. Fertil. Soils.*, *28*, 87-94.

[117] Saravanan, V. S., Subramoniam, S. R. & Anthoni Raj, S. (2003). Assessing in vitro solubilization potential of different zinc solubilizing bacterial (ZSB) isolates. *Braz. J. Microb. 34*, 121-125.

[118] Sachdev, D., Nema, P., Dhakephalkar, P., Zinjarde, S. & Chopade, S. (2010). Assessment of 16S rRNA gene-based phylogenetic diversity and promising plant growth-promoting traits of *Acinetobacter* community from the rhizosphere of wheat. *Microbiol. Res.*, *165*, 627-638.

[119] Iqbal, U., Jamil, N., Ali, I. & Hasnain, S. (2010). Effect of zinc-phosphate solubilizing bacterial isolates on growth of *Vigna radiata*. *Ann. of Microbiol,. 60*, 243-248.

[120] Jeffery, S., Williams, E., Whicker, C., Khoo, C., Summerell, B., Offord, C., Fahy, P. & Noble, D. (2006). Some effects of silicate on *Pythium* Damping-off of cucumbers in potting mixes. *Acta. Horticul.*, *450*, 467–473.

[121] Belanger, R. M., Smith, C. M., Corkum, L. D. & Zielinski, B. (2003). Morphology and histochemistry of the peripheral olfactory organ in the round goby, *Neogobius melanostomus* (Teleostei: Gobiidae). *J. Morphol.*, *257*, 62–71.

[122] Fawe, A., Abou-Zais, M., Menzies, J. G. & Belanger, R. R. (1998). Silicon-mediated accumulation of flavonoid phytoalexins in cucumber. *Phytopathol.*, *88*, 396–401.

[123] Samuels, A. L., Glass, A. D. M., Ehret, D. L. & Menzies, J. G. (1993). The effects of silicon supplementation on cucumber fruit: changes in surface characteristics. *Annal. Bot.*, *72*, 433-440.

[124] Marschner, H. (1995). Nutritional physiology. In: *Mineral Nutrition of Higher Plants.*, (Marschner, H., ed.), Academic Press Limited, London, pp. 417–426

[125] Belanger, R. R., Bowen, P. A., Ehret, D. L. & Menzies, J. G. (1995). Soluble silicon: its role in crop and disease management of greenhouse crops. *Plant Disease.*, *79*, 329-336.

[126] Datnoff, L. E., Deren, C. W. & Snyder, G. H. (1997). Silicon fertilization for disease management of rice in Florida. *Crop Protec.*, *16*, 525-531.

In: Microbes in Soil and Their Agricultural Prospects
Editors: K. Kishore Choudhary and D. Wattal Dhar
ISBN: 978-1-63482-824-6

Chapter 12

BACTERIOPHAGES – NEXT GENERATION BIOCONTROL AGENT FOR PLANT DISEASE MANAGEMENT

R. Krishnamoorthy[1], V. Venkateswaran[2] and M. Senthilkumar[3*]

[1]Department of Environmental and Biological Chemistry, College of Agriculture and Life Sciences, Chungbuk National University, Cheongju, Chungbuk, Republic of Korea
[2]Ministry of Agriculture, New Delhi, India
[3*]Department of Agricultural Microbiology, TamilNadu Agricultural University, Coimbatore, TamilNadu India

ABSTRACT

Plant diseases caused by bacterial pathogens are one of the major economic problems in sustainable agriculture. Due to the development of antibiotic and pesticide resistances, control of these phytopathogens are more challenging in food production. Bacteriophages are viruses, which could offer an eco-friendly method of controlling phytopathogens. The concept of using bacteriophages in controlling plant diseases were started in the second decade of the 20th century, but it was not popularized due to the discovery of antibiotics at the same decade. Bacteriophages are easy to use and can enhance food safety without affecting the quality of the foodstuffs and without any side effects. The worldwide development of multi drug resistance plant pathogens resulted in switchover from chemotherapy to pre antibiotic era in agriculture. In the past two decades, there were numerous studies which focused on the isolation, formulation and application of bacteriophages against economically important bacterial phytopathogens. In this chapter, we consolidated the history and the phage formulation developed against

* Corresponding Author address: Department of Agricultural Microbiology, TamilNadu Agricultural University, Coimbatore -641003, TamilNadu India. E-mail:msenthilkumar@tnau.ac.in

important phytopathogens. Considering its application in agriculture, many phage strains were characterized and successfully applied in plant disease control.

However, the low success rate under field condition hammered the commercialization of phage products. In field condition, phage survival and multiplication were affected by many environmental factors which is a major constraint in the application of biological control agent in plant disease management. Future research should be focused on the development of formulations that can enhance the phage applicability and survival under field condition. Bacteriophage therapy may substantially improve the bacterial pathogen control and may play an important role in sustainable food production in near future.

Keywords: antibiotic resistance, bacteriophage therapy, formulation, phytopathogens, plant disease control

INTRODUCTION

Plant diseases are considered as an important biotic factor, which leads to significant crop losses worldwide. Crop losses in developing countries are often higher than the developed countries due to lack of farmers' knowledge on adopting appropriate control measure and resources. It is projected that 10-15 percent of the yield loss in developing countries are due to the disease attack, and the losses would be higher if the post harvest diseases are taken into the account. Presently in developing countries, more than 800 million peoples do not have sufficient food and around 1.3 billon people live on less than a dollar per day [1]. A recent study on the production constrains for six major crops (wheat, rice, sorghum, chickpea, cassava, and cowpea) in 13 Asian and African farming system revealed that the losses due to diseases ranges from 3 to 14 percent, while the yield losses from all biotic factors ranged from 16 to 37 percent [2].

Use of pesticides was the major form of disease control in many of the developing countries. However, there are several problems emerged with such approach like emergence of pesticide resistance in pathogen and harmful effect of chemicals on human health and environment. Integrated Disease Management (IDM) concept was popular among the farmers and researchers, because the diseases are managed by incorporating a range of control measures. IDM calls for minimal use of pesticide and give preferences to other control measures like cultural practices, host plant resistance and biological control. Even though the IDM is well established and effective, majority of the disease control strategies depends on single technology, may it be host-plant resistance, pesticide applications or cultural practices and rarely the combination of all [3].

Use of non chemical products for controlling plant diseases are of considerable interest because of the human health concerns and environmental problems [4]. Currently there are strict regulations existing on the chemical pesticides and the government pressure to remove the hazardous chemicals from the market is increasing. Therefore, the use of chemical pesticides will possibly be reduced in near future and this reduction will go along with the reliance on the use of microorganisms to control plant pathogens [4]. Nonetheless, the use of microorganisms to control wide spectrum of pathogens remains an unfulfilled goal in modern agriculture.

The major shortcomings of this microbial control of pathogens are not highly reliable, and several other reasons accounts for this limitation. Firstly, the population dynamics of the biocontrol agent, the natural population of the pathogens and the climatic conditions are much variable than those used in the laboratory to study the mode of action of biocontrol agent. Secondly, the climatic conditions that favor the multiplication of the pathogen might not be the same as the conditions required for the expression of the biocontrol agent's antagonistic activities [5]. Therefore, more research should be focused on identifying the environmental conditions that are required for the maximum expression of biocontrol agent beneficial traits under field condition for making microbial control of pathogens more reliable.

CURRENT STATUS OF PLANT DISEASE MANAGEMENT

Plant Disease Management Strategies

After the discovery of plant disease causal agent in the early nineteenth century, different strategies were developed to control phytopathogens. Wide array of control measures for specific plant diseases were developed with the knowledge on the plant - pathogen interaction. During late 1920sWhetzel [6] developed the basic principles for plant disease control as follows:

i) Avoidance - Selecting the site or time of the year where there are no inoculum or the environment is not suitable for the pathogen infection.
ii) Eradication - Complete removal of pathogen inoculum from the plant or from the area.
iii) Exclusion - Protecting the area from the pathogen inoculum by using proper quarantine measures.
iv) Protection - The process of reducing the effectiveness of inoculum level by using chemical or physical or biological barriers.
v) Resistance - Developing the plant cultivars resistance to pathogens.
vi) Therapy - Minimizing the yield loss due to the pathogen by manipulating the environment or using chemotherapy or biological therapy

Some short coming where found in successful disease management by using these basic principles. First, these principles were aimed on the complete removal of diseases from the plant. However, it is not possible in practical point of view; instead it is necessary to reduce the development progress of a pathogen and keeping them below the thresh hold level. Second drawbacks in these basic principles are - not considering the dynamics of plant diseases: that is diseases incidence and severity in time and space. Third, it emphasized the tactics without fitting them into an adequate strategy.

Considering this drawback, Arneson [7] rearranged the basic principle based on the epidemiological principles where the author developed three tactics such as reduction of initial inoculum, infection rate and the duration of the epidemic.

Reduction of Initial Inoculum

i) Avoidance - Reduce the level of disease by selecting a season or a site where the amount of inoculum is low or where the environment is unfavorable for infection.
ii) Eradication - Reduce the production of initial inoculum by destroying or inactivating the sources of initial inoculum (sanitation, removal of reservoirs of inoculum, removal of alternate hosts, etc.).
iii) Exclusion - Reduce the amount of initial inoculum introduced from outside sources.
iv) Protection - Reduce the level of initial infection by means of a toxicant or other barrier to infection
v) Resistance - Use cultivars that are resistant to infection, particularly the initial infection.
vi) Therapy - Use thermotherapy, chemotherapy and/or meristem culture to produce certified seed or vegetative planting stock.

Reducing the Infection Rate

i) Avoidance - Reduce the rate of inoculum production, infection rate, or rate of development of the pathogen by selecting a season or a site where the environment is not favorable.
ii) Eradication - Reduce the rate of inoculum production during the course of the epidemic by destroying or inactivating the source of inoculum (roguing).
iii) Exclusion - Reduce the introduction of inoculum from external sources during the course of epidemic.
iv) Protection - Reduce the rate of infection by means of a toxicant or some other barrier to infection.
v) Resistance - Plant cultivars that can reduce the rate of inoculum production, rate of infection, or rate of pathogen development.
vi) Therapy - Cure the plants that are already infected or reduce their inoculum production.

Reducing the Epidemic Duration

i) Avoidance - Plant early maturing cultivars or plant at a time that favors rapid maturation of the crop.
ii) Exclusion - Delay the introduction of inoculum from external sources by means of plant quarantine.

Two important things that must be considered in disease management strategies are disease triangle and disease cycle. Knowledge on the interaction between plant, pathogen and environment (Figure 1.) are important for an effective disease control. The basic principle is that the disease is the result of an interaction between a host, a potential pathogen and the environment [8]. If any one of these factors are lacking then disease will not occur.

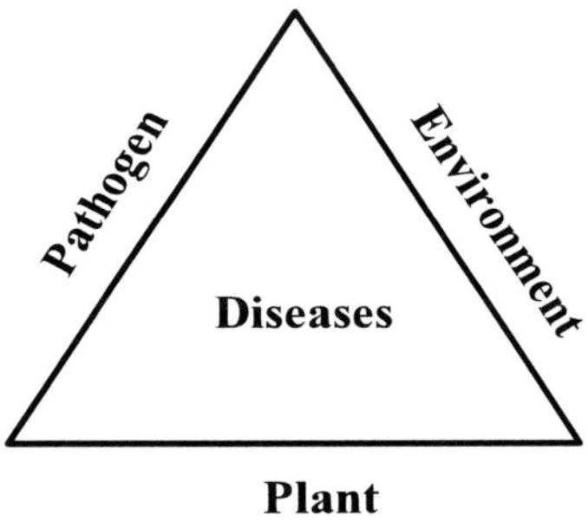

Figure 1. Disease triangle.

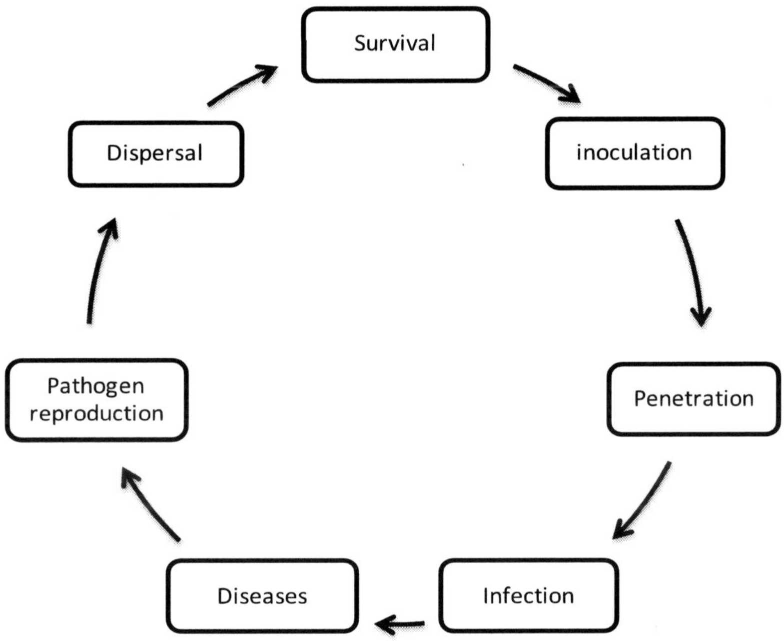

Figure 2. Disease cycle of a pathogen.

Knowledge on disease cycle is helpful to know how a particular pathogen goes through different stages during the disease development. Most of the plant pathogens follow seven different stages for successful infection and survival, as given in Figure 2.

Control of Plant Diseases – Prospects and Future

Plant diseases are defined as an irregularity in the function or structure on the plant cells or tissue caused by continuous irritation by a phytopathogen or environmental factors [9]. Plant disease control is a broad area of plant production, because pathogen substantially reduces the yield or the quality of the product. First and important step in controlling plant disease is diagnosis: accurate detection of causal agent. Diseases diagnosis is an art, this helps to differentiate the diseases caused by infectious agent (Phytopathogen) or by an abiotic factor. Diseases caused by environmental (abiotic) factors can be divided into 9 category (1) low and high temperature, (2) less or high moisture level, (3) unbalanced oxygen and carbondioxide concentration, (4) toxic compounds in the air, (5) nutrient deficiency, (6) mineral toxins, (7) unfavorable soil pH, (8) pesticide toxicity, and (9) inappropriate cultural practices. Whereas biotic factors (only microorganisms) can be divided into five groups: (1) Bacteria, (2) Fungi, (3) Nematodes, (4) Viruses and Virods, and (5) Phytoplasma. Second, step in plant disease control is the identification of causal agent under laboratory conditions by microscopic, biochemical methods and at molecular level. This step is the crucial one because accurate identification of the pathogen will be helpful to know the life cycle of the pathogen and its controlling strategy. Third step is the application of pathogen controlling agent that may be a chemical or biological agent. Even though application of some chemicals showed faster control of pathogens, it causes environmental pollution and pathogen resurgence. Application of biological agents may be helpful in sustainable agriculture production.

Bacterial disease control is achieved by chemical, biotechnology, breeding, biological agents and cultural practices. During 1880's copper based bactericides were introduced and still it is used in plant disease control. A major risk of the exclusive and regular use of copper spray for control of bacterial diseases may lead to the development of strains resistant to copper, which hampers disease management. Once resistance genes are acquired by the targeted plant pathogen, the frequency of the resistant strains in the pathogen population increases rapidly with the copious sprays and further applications are gradually less effective for disease control [10]. Copper resistance in bacteria is regulated by several genes [11]. Horizontal gene transfer through conjugation is the primary mechanism for acquisition of copper resistance by bacteria [12]. However, plasmid and chromosomal based copper resistance in bacteria were reported against many copper based bactericides. After the invention of antibiotics in 1928, antibiotic usage in both medical and agricultural field increased. Nonetheless, continuous and extensive use of antibiotic against bacterial pathogen increased the bacterial genes resistance to streptomycin antibiotic and reduced the efficiency in controlling bacterial spot of tomato, red pepper, fire blight of apple and pear [13], as well as other bacterial phytopathogens [11]. The plasmid encoded streptomycin resistance resulted in alternate approaches for disease control, including the use of other antibiotics and systemic acquired resistance (SAR). SAR is the plant based mechanism and it is exhibit against several pathogens. Activation of SAR requires the accumulation of endogenous salicylic acid (SA) in the tissues [14]. Although it reduces the pathogen resistance, they also showed yield loss in many crops [15, 16]. Development of plants with pathogen resistance gene is one of the major outcome of plant biotechnology, but the technique is expensive and successful only in few crops [17, 18]. Insertion of pathogen resistance gene into a commercially acceptable cultivar is difficult. Because of the drawbacks in chemicals, antibiotics, SAR and pathogen resistance gene, biological control gained more attention. In practical point of view, we cannot fully control the yield loss due to bacterial diseases but we can reduce the level of yield loss with the help of methods mentioned above. One of the best and ecologically safe ways of controlling bacterial pathogens is using biological weapon against them. Biological control of plant diseases were attempted using non pathogenic strains [19], saprophytic [20], and plant growth promoting bacteria [21]. All these disease controlling mechanisms showed variable outcome, hence, very recently people restarted to use bacteriophage for bacterial disease management [22, 23, 24]. Bacteriophages are the valuable resources, which can acts as a "biological magic bullet", where it will target only the particular bacterial pathogen in the plant ecosystem.

Biology and Ecology of Bacteriophages

Bacteriophages and Its Biology

Bacteriophages are the most abundant life on the planet, which have 13 families with 31 genera. It was estimated that the total number of species in this group is ~$1x10^8$ comprised of an estimated total of $1x10^{30}$ to $1x10^{32}$ phage particles; the volume approximately equal to $1x10^9$ metric tons [25]. Bacteriophages are present everywhere, we can say there are no place in this earth without phage and it is estimated that $1x10^6$ particles per drop of seawater and as many as $1x10^8$ particles per gram of soil. It acts as predators to control bacterial population with one half of the bacterial population being destroying every 48h. These bacteriophages are playing a vital role in maintaining bacterial population in an ecosystem [26]. More than a billion-year evolution emanating from this predator - prey relationship has made the bacteriophages as effective candidate against the bacterial pathogens [27]. In aquatic environments, phages are known to regulate bacterial population directly through host cell lysis and indirectly through the lytic release of nutrients to other bacterial communities, whereas the influence of phages on soil biogeochemical; processws remains unanswered.

Bacteriophages composed of nucleic acid covered by a protein coat called capsid and a nucleic acid called as genome. Capsid structure varied from small hexagonal structure to filaments, to highly complex structure consist of a head and tail. Genome can be a double stranded DNA or single stranded DNA or single stranded RNA. Some phages have only few thousand base pairs in their genome however; the phage G has the largest genome of 480,000bp. It is like an average bacterium, but still it lacks the genes essential for the replication. Bacteriophages are metabolically inactive until it gets the host, however the metabolism will be activated when it finds a prey. The process begins with the adsorption during the time capsid binds to the cell surface receptors of bacteria and subsequently the phage genome is inserted and it takes the control of bacterial cell. Generally phages use two kind of controlling mechanism (i) after infection it will multiply and the particles will be assembled within the cytoplasm of an infected bacterium, finally it will lyses the bacteria - process called as active infection (lytic cycle) (ii) another one is lysogeny process where phage genome integrates (as a prophage) into the bacterial chromosome as a giant gene complex and the bacteria will not lyses immediately. During lytic cycle the bacteria releases 10-100 virion per bacterial cell. If a phage kills bacteria, it usually accomplished with activation of a 2-component lysis system, consisting of a pore-forming hydrophobic protein called holins and a murein degrading enzyme (endolysins or lysins) [28]. Concerning the bacteriophage therapy the active infection (lytic cycle of phage infection) is the most preferable one, because the aim of this process is to get the bacteria killed. Therefore, it is likely preferred that phages with lytic property will act as a good candidate for any phage-based biocontrol method.

Classification of Bacteriophage

Bacteriophages that infect the eubacteria and archaea bacteria are called as "prokaryotes eating virus". Felix d'Herelle, one of the two discoverers of bacteriophages believed that there was only one bacteriophage with many races, the "Bacteriophagum intestinale" [29]. Subsequent research showed phages that are active against many bacterial species are found in numerous habitats, this disproved the single bacteriophage concept developed by d'Herelle. Until now scientist discovered that bacteriophages can infect more than 140 bacterial genera, including those consisting of (i) aerobes and anaerobes, (ii) exospore and endospore formers, (iii) cyanobacteria, spirochetes, mycoplasmas, and chlamydias, (iv) budding, gliding, ramified, stalked, and sheathed bacteria, (v) extreme halophiles and methanogens, and (vi) hyperthermophilic archaea. In addition, phages are also active against the endosymbiotic bacteria of paramecia and insects. The structure of phages is highly heterogeneous in physicochemical and biological properties suggesting that they are polyphyletic in origin.

Table 1. Classification of bacteriophages [30]

Order	Family	Nucleic acid
Caudovirales	*Myoviridae* *Siphoviridae* *Podoviridae*	Double-stranded DNA
Ligamenvirales	*Lipothrixviridae*[b] *Rudiviridae*	Double-stranded DNA
Unassigned	*Tectiviridae*[a] *Corticoviridae*[a] *Plasmaviridae*[b] *Fuselloviridae*	Double-stranded DNA
Unassigned	*Inoviridae* *Microviridae*	Single-stranded DNA
Unassigned	*Leviviridae*	Single-stranded RNA
Unassigned	*Cystoviridae*[b]	Segmented, double-stranded RNA

[a] Lipid containing; [b]Enveloped.

Structure varies from tailed, polyhedral, filamentous and pleomorphic. Majority of the phages has dsDNA as their genome; however, ssDNA, ssRNA, or dsRNA are also found among different phage groups. To till date phage structure and morphology for more than 5000 bacteriophages classified into 13 family and 31 genera have been examined by electron microscopy (Table 1).

Order *Caudovirales* (Tailed Phage)

Caudovirales orders are tailed phages with three families namely *Myoviridae*, *Siphoviridae* and *Podoviridae* [31]. This is the largest group of bacteriophages containing approximately 4950 phages. They may be the oldest virus group, dating back to 3.5 billion years ago which is more close to the bacterial evolution. Tailed phages infect both eubacteria and archaea, which may suggest that bacteriophages appeared before the bacterial kingdom separation. This group has linear dsDNA, a protein shell, non-envelop and binary symmetry

[32]. Capsid proteins (protein shell) are smooth and organized into capsomers (5 or 6 protein subunit). These groups also contain tails, which are helices or consist of staked disks and possess terminal adsorption structures like base plate, spikes and fibers. DNA composition generally resembles that of host bacteria, some unusual bases like 5-hydroxymethylcytosine was found in T_4 coliphage. Under this order largest bacteriophage, Phage G was grouped, which comprises of 498kbp and 684 genes.

Myoviridae

Phages with contractile tails consisting of a sheath and a central tube (at least 1250 observations, ca. 25% of tailed phages) [33].

Siphoviridae

Long, non-contractile tails (3000 observations, ca. 61% of tailed phages).

Podoviridae

Short, non-contractile tails (700 observations, ca. 14% of tailed phages).

These three families shows overlapping in the physicochemical properties and hence differentiation is very difficult. However,*Myoviridae* is differentiated from other two families by the size of the capsids, which is larger than other two groups. *Siphoviridae* and *Podoviridae* can be differentiated based on the tail length.

Polyhedral, Filamentous and Pleomorphic Phages (PFP Pages)

These are tailless phages comprising of 190 known viruses and accounts for only 4% of the known bacterial virus population. They are grouped into 10 families, some family have only one genus and a species. These 10 families are differing in their fundamental properties like enveloped and non enveloped. Based on the nature it is grouped into three types –first is polyhedral phages, which are icosahedra with cubic symmetry, second is filamentous phages with helical symmetry and third is few variable shape phages without symmetry axes. Some phages in this group contain lipids, which results in chloroform sensitive and low floating capacity. Mostly all these 10 families have narrow host range and many phages from *Fuselloviridae*, *Inoviridae*, *Lipothrixviridae*, and *Plasmaviridae* families are lysogenic which is not suitable for phage-mediated biocontrol of bacterialphytopathogens.

Polyhedral DNA Phages

Microviridae (ssDNA)

Phages are unenveloped icosahedra (ca. 27 nm in diameter) with 12 capsomers and contain circular ssDNA. Rolling circle model of DNA replication was found in these phages.

Corticoviridae (dsDNA)

Phage PM2 is the only member in this group, it was isolated from seawater and it contains lipids.

Tectiviridae (dsDNA)

Phages with rigid protein capsid containing thick, flexible, lipoprotein vesicles and apical spike (only in *Bacillus* tectiviruses)

Polyhedral RNA Phages

Leviviridae(ssRNA)

This is unenveloped phage and morphologically similar to polioviruses. Most of the phage belongs to this family are plasmid specific, it is divided into two genera based on serology and genome structure.

Cystoviridae (dsRNA)

Like *Corticoviridae* this family also has only one member. Mindich et al., [34] has found eight related viruses, this 8 phages are against phytopathogenic bacterium *Pseudomonas syringae*. This RNA phage enters the space between the cell wall and cytoplasmic membrane of the host bacterial and kills.

Filamentous Phages

Inoviridae (ssDNA)

It consists of two genera with rolling circle mechanism of DNA replication like *Microviridae* family. Genus Inovirus classified into 29 species with 42 phages [35] and exhibit the rigid or flexible nature of DNA.

Lipothrixviridae (dsDNA)

This family phages are against thermophilic archeabacteria, consists of six viruses. Rod shaped and contain lipoprotein envelop.

Rudiviridae (dsDNA)

Straight, rod shaped without envelop. These phages are against *Thermophilic sulfolobus*. It has two viruses Rudiviruses and Lipothrixviruses having same genome homologies, which suggest that they form a superfamily [36].

Fuselloviridae (dsDNA)

Fuselloviridae infect the *Archaeon sulfolobus*, which inhabitats at high-temperature (>70°C) and acidic (pH of <4.0) environments. Phages are spindle-shaped, flexible, and have protrusions that extend through the envelope.

Plasmaviridae (dsDNA)

Circular shape with enveloped and no division of head-tail structure. The capsid is about 80 nm in diameter and genome size is 12 kb.

ECOLOGY OF BACTERIOPHAGE

Phage ecology is the study of interaction between phage and their biotic and abiotic environments. Concerning the study about the ecology of phage, it has four subdivisions (i) adaptation ecology - concentrate on how it interact and adapt to the harsh environment like desiccation and ultraviolet (UV) rays from sunlight. (ii) Population ecology - study about the life history, growth, competition and growth in different environments (iii) community ecology- deals with stability of phage containing environment, this is complicated by the continuous co- evolution of bacteria with their phage predator (iv) ecosystem ecology - concentrate on the energy flow and nutrient cycling within the ecosystem - for example phage can disturb the bacteria responsible to phosphorus or nitrogen cycle.

Phages can be isolated from various sources like sewage water, human body, soil etc. According to Gill and Abedon [37], targeting the rhizosphere soil of the plant might be easier to isolate the phage compared to the phyllosphere of the plant. Considering the nature of soil, which is highly heterogeneous, partially hydrated matrix and diffusion of bacteriophages through the heterogeneous rhizosphere is a limitation in controlling soil borne phytopathogen. The phages should have the ability to diffuse through the hydrated matrix of the soil to infect and destroy the phytopathogen, which are often present in non-homogenous masses that are sometime protected by the polysaccharides. Therefore, the successes relay on the ability of the phage to target the bacteria in soil and breakthrough the barriers and infect before the phages are destroyed by environmental stresses. The problem of diffusing can be overcome by inundating the plant environment with more effective phages.

Interaction of phage on the phyllosphere is still not fully understood than the rhizosphere phage interaction. This is may be due to the nature of phyllosphere environment, which is more often exposed to the UV rays of sunlight and desiccation of leaf surface. Isolation of phage is very low in aerial portion of plants even with active infection of pathogens like *Erwinia amylovora*. Mostly the phages against phyllosphere bacterial pathogens were isolated from the soil around the infected plants. This may suggest that the phages have more favorable condition in soil for their survival and can call it as the reservoir of phage. If soil acts as reservoir for phage one question must be answered, how phage will reach the phyllosphere to control the bacterial pathogens, which infect the plant leaf? One possible explanation is that the phages attaches to the phyllosphere upon plant germination or it may transfer from phyllosphere of one plant to another by physical contact.

BACTERIOPHAGE THERAPY

Historical Development of Bacteriophage Therapy

The observation of filterable agent against cholera bacteria (*Vibrio cholera*) by Hankin in 1896 was the first findings of bacteriophages. Hankin isolated some filterable agent from Indian River Ganges and he observed that the filterable agent inhibited the infective agent of cholera. After 20 years of his finding the first official proof of bacteriophage was done by an Englishman Edward Twort [38] and a French Canadian Felis d'Herelle [39]. Later these two scientists were considered as bacteriophage discoverers, both of them spend many years to control many plant, animal and human diseases.

Even though scientific community did not accept the use of bacteriophage over 30 years from the time it was proved by d'Herelle, there was a huge expectation in medical field to use phage as a predator against the human pathogens. Four plague patients were successfully treated with phages by d'Herelle and as a consequence the British government invited him to India to treat the plague patients using phage. During 1921, Brunoghe and Maisin reported the control of *Bacillus anthracis* and *Staphylococcus* pathogens using the bacteriophage, followed by Beckerich and Hauduroyin (1922) against typhoid [40] and dysentery [41] causing bacteria. In 1924 Mallman and Hemstreet [42] collected the filtrate from cabbage tissue affected by *Xanthomonas campestris pv. campestris*, the filtrates showed inhibition of disease causing bacteria *invitro*. This was considered as the first successful attempt in plant disease control using bacteriophage. The next year another proof was given by Coons and Kotila [43], where they incubated the carrot disk with *Erwinia carotovora subsp. carotovora* alone and with phage. The authors observed rotting in the carrot disk inoculated with pathogen alone, and no symptom was observed with phage treatment. The same year Kotila and Coons isolated bacteriophage against the blackleg disease of potato [44].

In 1934, Massey [45] found that the flooding of soil with Nile river water reduced the incidence of bacterial blight disease. He isolated phage against blight causing bacteria *Xanthomonas malvacearum* from the same field, but not from the field, which was not flooded with Nile water. Thomas [46] used phage in seed treatment to control the Stewart's wilt of corn, caused by *Pantoea stewartii*. Stewart's wilt was the most important disease during that time where majority of the corn fields were affected. The author observed that the disease incidence was reduced from 18% to 1.4%. More than 800 research papers were published over the next 40 years concerning disease control in plants, animal and humans using phages. Even though phage therapy was successful against many bacterial diseases, controversial results was also obtained from all over the world. The major reasons for this report was the lack of knowledge regarding the biology of phage, poor scientific design of the commercial preparation, poor or uncharacterized phages in unknown concentration, no match was found between phage and target bacterium and commercial preparations containing bacterial toxins. Before the start of World War II (1939-1945) an extensive study was conducted about the possible therapeutic use of phage. Because of the war and the introduction of antibiotics made this research as unfocused one. Antibiotics are easy to produce and had broad spectrum activity against most of the pathogens. Hence, chemotherapy was considered as superior over phage therapy. Nevertheless, the bacteriophage research work was continued in former Soviet Union, to treat the war wounds of soldiers. During 1963, Okabe state that "the phage therapy was ineffective against the bacterial diseases" [47] and Goto [48] concluded that "controlling bacterial plant diseases in field level was not successful by using bacteriophage". Due to these conclusions and the invention of antibiotics hammered the use of phage against bacterial diseases.

Need for Phage Therapy

Bacterial resistance to antibiotic in agriculture industry makes problem to human health, because these resistance bacteria can enter into the food chain and leads to the healthcare problems. Over the last two decades, the public has alarmed by scientific data that the misuse of antibiotics in agriculture leads to the emergence of antibiotic resistance bacteria. Other than misuse, inappropriate use of antibiotics in agriculture is the major contribution for the emergences of antibiotic resistance bacteria. The first repot on antibiotic resistance in agricultural feedstock was documented in 1963 with increased level of antibiotic resistance was observed in *Salmonella typhimurium*, subsequently several antibiotic resistant bacterial isolates were found on the British feed lots.

There are two conditions needed for the antibiotic resistance development in bacteria - (i) the organism should constantly be exposed to the antibiotics (ii) the organism must develop the mechanism to transfer the resistance to next generation or to the other members of the same family. Each antibiotic have specific target site on bacterial cell. Some antibiotics target only the cell wall components, some others targets the cell membrane, RNA, DNA and enzymes in the cell. When resistance arises in bacteria to a particular antibiotic, it evolves diverse mechanisms and transmit to other members of the same genus. Chromosome and extrachromosomal elements (Plasmids) mediated antibiotic resistance were developed by mutations in many bacteria.

Streptomycin is the first antibiotic to be used in agriculture in late 1950's, against gram negative phytopathogens particularly against the fire blight (*Erwinia amylovora*), soft rot of cut flowers and potato seeds (*Pectobacterium* spp.), fruit-spotting or blossom-blast symptoms on apple, pear and related landscape trees (*Pseudomonas syringae*) and bacterial spot of pepper and tomato (*Xanthomonas campestris pv. vesicatoria*) [49].Streptomycin resistance was reported in phytopathogens such as *Erwinia amylovora* [50, 51], *Erwinia carotovora* [52], *Pseudomonas lachrymans* [53], *P. syringae* [54], and *X. campestris pv. Vesicatoria* [55]. The occurrence of copper resistance in phytopathogens such as *P. syringae* [56, 57] and *Xanthomonas campestris pv. Vesicatoria* [58, 59] was also observed. Presence of copper resistance gene was present in the plasmids of phytopathogens, for example in *P. syringae* pv. *tomato*, the copper resistance genes reside on a 35-kilobasepair (35-kb) plasmid [60]. In *X. campestris* pv. *vesicatoria* copper resistance resides on large plasmids of strains isolated in Florida and Oklahoma [61, 62]. Genetic studies of bactericide resistance (both antibiotics and copper containing bactericides) in phytopathogenic bacteria have increased in recent years. The intensive study on streptomycin resistance in bacteria, showed four types of streptomycin inactivating enzymes belongs to aminoglycoside phosphor transferase [APH(6) and APH(3")] and aminoglycoside nucleotidyltransferase [ANT(6) and ANT(3")] [63]. These four enzymes found both in gram positive and gram negative phytopathogens suggests that the gene transfer has played an important role in the dissemination of Sm^r (codes for resistance). The Sm^r are encoded by two genes - *strA* and *strB*[64] was isolated from clinical isolates of *Escherichia coli*, *Proteus* spp., *Providencia* sp., *Pseudomonas aeruginosa*, *Salmonella* spp. and *Shigella flexneni*[65].

The *strA* genes is responsible for the production of aminoglycoside phosphor transferase [APH(3")] in phytopathogen [52 53].The streptomycin-resistant strains with *strA* and *strB* genes in *Pseudomonas marginalis* and *P. syringae* pv. *actinidiae* was also reported from kiwifruit orchards of Korea and Japan [66]. Twenty five isolates of *P. syringae* pv. *tabaci* [67], four strains of *Xanthomonas vesicatoria* and one strain of *X. oryzae* pv. *oryzicola* showed high level of streptomycin resistance [68]. In several studies, copper and streptomycin resistance was observed in different strains [55, 69, 70], but Ritchie and Dittapongpitch [71] was first to report a single plasmid with resistances gene for both copper and streptomycin. Emergence of antibiotic resistance plasmids may be the consequence of combined and improper use of copper and streptomycin containing bactericides against the phytopathogens particularly against the *P. syringe* pv. *syringae*.

The bacteriology committee of the American Phytopathological society dispirited the use of medically important antibiotics in agriculture during 1978, because of the threat created by the antibiotic resistance phytopathogen in human health. The use of alternative bactericidal like tetracycline may create additional gene transfer events resulting in further development of antibiotic resistance determinants in phytopathogenic bacteria.

Advantages of Phage Therapy

With the increasing evidence of antibiotic resistant bacteria and lack of development of new generation of antibiotics a greater potential is available for phage therapy to treat numerous diseases. Phage therapy is certainly effective over antibiotics and has unique advantages. Here the advantages of phage therapy are framed over the use of antibiotics.

Act As a Bactericidal Rather Than Bacteriostatic

Some antibiotics arrest bacterial multiplication (bacteriostatic), this process facilitates the bacteria to develop resistance mechanism against the antibiotics. However, the phages destroy the target bacteria and phages do not act as bacteriostatic.

Population Development

Phage population will be developed based on the availability of the host, and it will multiply where the target bacteria present, and the phage population will be reduced when the target bacterial population was killed. Phage multiplication is very rapid as single phage produces $4x10^3$ progeny in an hour and increased to 410^6 after one hour.

Biological Magic Bullet

The phages have the ability to attack particular bacteria, which have the receptor for the particular phage. However, the antibiotics inhibit the growth of both target and non target bacteria.

Less Chances for Resistance Development

Narrow host range is exhibited by most of the phages; this limits the number of bacterial type, which can develop resistance. Nevertheless, this character for antibiotic is totally different from broad-spectrum activity. Phages do not develop secondary resistance, which is quite often in antibiotics.

No Cross Resistance with Antibiotics

The mechanism of bacterial control by phage and antibiotics are different, so that the antibiotic resistance mechanisms of the bacteria will not affect phage therapy. Phages can be employed to control antibiotic resistance strains of phytopathogen. Even if bacteria develop resistance to phage arises, it will not be a major concern when compared to antibiotic resistance development in bacteria. Since, the exponential growth rate of phages will overcome the bacterial growth as well as mutate at the same rate thereby new phages will arise to kill the mutated as well as resistant bacteria.

Easy Isolation

Phages for specific phytopathogen can be easily isolated from the bacterial infected parts of the plant, because the natural phage population will be present at point of bacterial infection.

Formulation and Application Stratagems

The formulation can be developed as a mixture of cocktail to have broad host ranges. There are different formulations such as liquids, creams, impregnated into solids, etc. are available.

Single Dose Potential

Single dose application of phage is advantages because the phages multiply by killing the bacterial host. So the efficacy in plant disease control can be achieved by one time application or less frequent doses.

Possible Transfer of Phages

This is the cross infection of phage from treated subjects or environment to untreated subjects. This kind of mechanisms is useful in plant diseases control.

Targeted Work of Phage

The phage contributes to kill only specific bacterial pathogens and will not affect other beneficial bacteria.

Low Cost of Production

The production process of phage is accompanied with the host and followed by purification of phages. Cost dependence on the growth of host, and purification process may be reduced by technology developments. Isolation and mass production of phages are cheaper than the development of antibiotic against particular plant disease.

Application of Bacteriophage for Plant Disease Control

Bacterial resistance due to the misuse of antibiotics has become a global issue and alternative methods are being developed that might decrease the use of antimicrobials in agricultural settings. This has led researchers to seek fresh ideas. Bacteriophage therapy is one "old" idea undergoing a renaissance, with the potential to resolve the antibiotic predicament we find ourselves in today. Bacteriophage therapy represents a novel way to control the growth of plant-based bacterial pathogens. Phages are both self-replicating and self limiting, they will only actively replicate as long as susceptible hosts are available. The potential of bacteriophage biocontrol is only just beginning to be realized. Use of bacteriophages for controlling plant diseases is an emerging field with great potential. The concern about food safety, environmental protection and sustainable agriculture necessitated the development of safer, more specific and environment-friendly pesticides. These factors, together with the expanding knowledge about phage application led to renewed interest in bacteriophage-based disease control in modern agriculture.

Prerequisites for Phage Therapy in Agriculture

The aim of phage usage is to reduce the phytopathogen infection to the level where plant defense system takes care of the remaining population [72]. Before attempting phage therapy in field level, the following prerequisites should be met:

1. Phage therapy should not be attempted before the biology of the therapeutic phage is well understood. Since the phage–host systems are extremely complicated, this prerequisite has to be faced with some common sense.
2. Phage preparations should meet all the safety requirements; the preparations should be free of bacteria and their components.
3. Phage preparations should contain infective phage particles, thus storage of the preparations should be validated.

4. The phage receptor should be known. In a bacterial population of 10^6 –10^8cfu ml^{-1}, development of spontaneous phage-resistant mutants (deficient in the receptor or with an altered receptor) is also possible.
5. Many bureaucratic and regulatory hurdles need to be cleared even when the above prerequisites are properly met.

Bacteriophages of Ralstonia Solanacearum

Ralstonia solanacearum is one of the gram negative soil born pathogen, which cause more economical loss by causing wilt disease in plants [73, 74, 75]. It has a wide host range, can affect 200 plant species belongs to more than 50 families and it shows phenotypic and genotypic diversity between the strains [74]. Based on host range it is subdivided into five races, based on physiological and biochemical characters it is divided into five biovars and based on the geographic origins it is divided into four phylotypes. Strains from Asia are named as phylotype one, two from America, three from Africa and surrounding islands in the Indian Ocean, and phylotype four from Indonesia. *R. solanacearum* can easily spread in the field conditions *via* soil, irrigation water, farm equipment and infected biological material. It can survive in soil for many years by infecting alternate host plants. Control of *R. solanacearum* with soil fumigation is less effective. Because of the fast spreading and tolerance to fumigation treatments, it creates serious problem in many field grown crops in many parts of the world.

Bacteriophages against *R. solanacearum* were successfully isolated from field soil and tobacco plants affected by *Ralsatonia* sp. Isolated phage P4282 showed effective control of *R. solanacerum* M4S strain [76]. A decade after the isolation of this phage P4282 it was found to have a bacteriolytic protein, which is encoded by a single gene present in the phage genome [77]. Lytic enzyme may be the main reason for controlling *R. solanacerum* M4S by phage P4282. Most of the isolated phages against *R. solanacearum* exhibited similar characters [76, 77, 78]. The phages belong to *Myoviridae* family with dsDNA genome of 35-39 kbp had narrow host range. The phages (φRSL1, φRSA1, φRSM1 and φRSS1) isolated by Yamada et al., [79] against *Ralsatonia* sp. had different morphology, genomic size and wide host range. φRSA1, φRSM1 and φRSS1 were sequenced to illustrate the exact mechanism in controlling *R. solanacearum* [80, 81].

φRSA1 showed lysis for 15 different biovars of *R. solanacearum* in plate assay. The titer value for the strain M4S was low (2 x 10^8 to 1 x 10^9 PFU/ml) but phage recovery was increased after addition of 1 mM EGTA (ethylene glycol tetra acetic acid). Similar results were also observed for *P. aeruginosa* phage φCTX, where addition of EGTA increased the titer value. *P. aeruginosa* phage φCTX use the lipopolysaccharide core as a receptor site on the cell surface requires Ca^{2+} ions for binding to the receptor [82]. Based on this observation it can be conclude that the lipopolysaccharide and Ca^{2+}ion may act as a receptor for the φRSA1 phage.

φRSB1 resemble like T7 bacteriophages, which can infect many strains of *E. coli* and φRSB1 also had wide host range as φRSA1. Genome sequence revealed that it has 43,079 bp of double stranded DNA with 61.7% of G+C content. Strong activity for tandem early promoters and wide specificity of phage promoters of φRSB1 were demonstrated [83]. φRSB1 can completely convert host RNAP (*RNA polymerase*) to phage RNAP within 90 min of contact was identified based on the rifampicin treatment (antibiotic which affect the bacterial RNAP).Genome sequences from ϕRSL1 revealed that the ORF181 has the genes responsible for chitinase-like bacterial lytic enzyme and ORF172 has similar sequence to the α-1,3- glucosidase of *Bacillus circulans*. It has been proved that these enzymes can hydrolyze the cell wall of *Schizophyllum commune* in combination with chitinase I [84]. Murugaiyan et al. have isolated 36 phages against *R. solanacearum* infecting tomato and chillies and 15 phages were selected based on the host diversity, lytic nature and time of lysis [85]. All nine tested hosts have susceptibility to phages PE226 and TM227. These two phages exhibited very similar physical dimension consisting of a filamentous nature and belongs to the *Inoviridae* family of bacteriophages.

One of the interesting genes in the PE226 genome is the presence of *Zonular occludens* toxin (Zot) in ORF7. The deduced amino acid sequence of ORF7 showed the highest identity to Zot from phage p12J infecting *R. pickettii* (69%) and φRSS1 infecting *R. solanacearum* (31%). PE226 may have the same mechanism of control of *Ralsatonia* sp. with p12J and φRSS1.Successful control *R. solanacearum* was reported by Fujiwara et al., [24], where they used φRSA1, φRSL1 and φRSB1 alone and in combination to control tomato wilt disease. All tested pathogens were effectively controlled by φRSL1 and it is stable at higher temperatures (37°to 50°C). The presence of φRSL1 was observed in treated plants and in soil even after 4 month of infection. Controlling *R. solanacearum* with cocktail of φRSA1, φRSB1, and φRSL1 is not suitable, because the pathogens showed resistance against φRSA1 and φRSB1 but not to φRSL1 inoculation which was confirmed in *gfp* tagged *R. solanacearum* [24]. The authors found that *gfp* labeled bacterial cells moved upward in the xylem until 72h and entire tape root after 96h resulted in wilting of hypocotyls and young leaves. The lateral root formation was reduced in pathogen alone inoculated treatment compared to the host φRSL1 inoculated plants.

Bacteriophages of *Erwinia Amylovora*

Erwinia amylovora is the enterobacterial phytopathogen, which causes the fire blight in several species of Rosaceae family. Fire blight is the most destructive disease in many of the apple and pear growing regions of the world. *E. amylovora* pathogen first infects the open blossoms during spring season and affects the vascular system of the plant [86].

Fire blight disease was first originated in North America and spreads to Europe during 1950s. Due to the ban of antibiotic streptomycin use in agriculture, the control measure of this phytopathogen is highly difficult at present. Several alternative management practices like use of antagonistic bacterial saprophytes [87], yeast [88], avirulant strain of *E. amylovora* [89], plant SAR (systemic acquired resistance) [14], construction of a transgenic plant [90], application of plant extracts and new antibiotic from symbiotic bacteria of entomopathogenic nematodes *Xenorhabdus budapestiensis* and *X. szentirmaii* [91] were also tested. In addition, a novel and promising method could be the use of bacteriophages against *E. amylovora*.

Phage isolation against *E. amylovora* was started during 1950s. In 1979, Ritchie and Klos [92], was first to report on the phage isolation from aerial part of the plant infected by *E. amylovora*. They isolated fourteen phages and divided into two groups (Group 1 and 2) based on the plague morphology. Phage Peal (h) was very effective against encapsulated *E. amylovora* and mixture of three phages Pea1, Pea116B and Pea116C were effective against *E.*

amylovora infected apple orchards [89]. The population of these three phages on the apple bloom was higher when the population of *E. amylovora* was higher and reduced the disease incidence by 37 percent. During 1998, Gill reported that the isolation of phage against *E. amylovora* was not successful from non infected plants. They isolated 9 phage from the infected plant parts and 45 from the infected rhizopshere soil. Molecular analysis of this 54 phages help to group into 6 types. Based on the bioassay using discs of immature pear fruit, 23 phages significantly reduced the population of all tested *E.amylovora* strains in pear fruit and it reduced the disease incidence by 96%.

Fifty strains of phages were isolated from apple, pear and raspberry tissues and 5 distinct phages were identified based on the genome size and restriction fragment pattern. Out of five phages φEa1 and φEa7 were related to previously described phages and three novel phages (φEa100, φEa125 and φEa116C) were identified. φEa1 is most widely distributed in the isolated area, with a genome size of 46 kb. φEa100, φEa7 and φEa125 have the genome size of 35 kb and these can be distinguished by the *EcoRI* restriction fragment pattern. φEa1, φEa7, φEa100, φEa125, and φEa116C were able to infect 39, 36, 16, 20 and 40 *E. amylovora* strains respectively isolated from apple orchards in Michigan and 8, 12, 10, 10, and 12, *E. amylovora* strains respectively from raspberry fields [93]. φEa116C was the most efficient phage, which controlled all 52 bacterial strains tested.

Control of blossom infection by *E. amylovora* is difficult because *E. amylovora* multiplies on the stigmatic surface of blossom prior to blossom infection. Phages must reach the stigmatic surface for effective control of pathogens. If *E. amylovora* specific phages are applied at initial stage of blossom formation, the growth of pathogen might be suppressed. Application of avirulant strains of *E. amylovora* has been important for the successful establishment and survival of phage population on the blossom because phage population rapidly declines in the absence of host, UV light and desiccation [93].

Fifty phages were isolated in and around the Niagara region of Southern Ontario and the Royal Botanical Garden, Hamilton, Ontario by enrichment culture using 6 *E.amylovora* strains. Majority of the phages were isolated from the soil surrounding the infected plants and only five phages were isolated from the infected aerial parts of the plant. The phages were grouped into 6 different groups based on the molecular characterization with the combination of PCR and restriction endonuclease digestions. Based on the electron microscopy study the group 1 and 2 were tailed and contractile, and group 3 to 6 were tailed and openings with thin appendages [94]. Based on morphotypes, the bacteriophages were belongs to *Caudovirales* order, in the family of *Myoviridae* and *Podoviridae*.

Four novel (L1, M7, S6, and Y2) and wide host range phages against *E. amylovora* were isolated from Switzerland[95]. L1 has the genome size of 39.3 kb featuring invariable genome ends with direct terminal repeats, which belongs to the *Podoviridae* family. Another podo phage(S6) has the same kind of direct terminal repeats with larger genome size of 74.7kb and exhibits a complex tail fiber structure. Two novel strains M7 and Y2 belong to the *Myoviridae* family has 84.7 kb and 56.6 kb genome size respectively with direct terminal repeats. Comparative genome analysis revealed that M7 phage is similar to the *E. amylovora* phage φEa21-4, whereas other 3 phages (L1, S6, and Y2) are unrelated to any other *E. amylovora* phage. The architecture of all four phage genome is typical for tailed phages, i.e., organized into functional specific gene clusters. Based on the genome sequences this four novel isolates do not have any functional genes for the lysogeny life cycle. This clearly indicates that these four phages can effectively control *E. amylovora* by the lytic cycle, which is one of the most important criteria for the phage therapy.

Born et al., [23] found that two phage cocktail concept as a potential source for biocontrol of phytopathogens. Even though the phages showed broad host range, the host specificity is more important for phage therapy. Based on the infection experiments the two phage cocktails enhanced the control of *E.amylovora* compared to the single phage treatment. The effect was higher because the two phages used different receptor sites on the target bacteria. Enhanced efficacy through the combined action of different types of phages with complementary host ranges represents a useful strategy and has been suggested for *E. amylovora*. Because of the growing interest on the phage therapy in agriculture, more study should be conducted to reveal the genomic function followed by finding the exact mechanism of pathogen control and the receptor site in the host. Muller et al., [95] sequenced four phages (φEa1h, φEa100, φEa104, and φEa116) isolated from USA and the size of their genomes was 45 to 85kb. Phage φEa1h and φEa100 belongs to *Podoviridae*, φEa104 and φEa116 to *Myoviridae* family. These phages showed wide host range for *E. amylovora* and weakly attached to *E. billingiae* and not with *E. tasmaniensis*. The *Podoviridae* phages (φEa1h and φEa100) are depend on the amylovoran capsule of *E. amylovora* but high EPS producing strain of *E. pyrifoliae* are less sensitive, because lack of receptor site on the capsules. The *Myoviridae* family phages (φEa104 and φEa116) did not depend on the capsule/EPS layer for the receptor sites; it also effectively lyses the capsule mutant strain of *E. amylovora*. This four phages showed effective control of fire blight in orchard and have more potential to use as a biocontrol agents against fire blight disease. Another recent success of phage in fire blight control was exhibited by Schwarczinger et al., [22], the *invitro* study showed inhibition of *E. amylovora* population in three different apple cultivars.

Boule et al., [96] isolated phage belongs to *Myoviridae* and *Podoviridae* family to control *E. amylovora*. φEa1337-26 and φEa2345-6 reduced the infection by 84% and 96% in detached pear blossoms with *Pantoea agglomerans* Eh21-5 bacterium as a carrier. This epiphyte not only supports the multiplication of the phages but also act as a competitor to *E. amylovora*[97]. A different strain of *P. agglomerans* is available commercially, as Bloomtime FD (Northwest Agricultural Products, Pasco, WA), and has shown to inhibit *E. amylovora* by antibiosis [98]. Another interesting feature is that these phages have compatible with *P. fluorescens* strains, which is widely used biocontrol agent in agriculture. Because of this compatible nature these phages can be combined with BlightBan® A506 (Nufarm Americas Inc., Burr Ridge IL), a registered BCA for fire blight that contains *P. fluorescens* A506 as an active ingredient. Therefore, this kind of combined biocontrol application will be more effective in control of fire blight disease. Most of the isolated phages against fire blight were belongs to *Myoviridae* and *Podoviridae* family and it showed wide host range to *E. amylovora* strains. Therefore, *Myoviridae* and *Podoviridae* family bacteriophages may have the potential to be used as biological control of fire blight disease [22].

Bacteriophages of *Xanthomonas Oryzae*

The genus *Xanthomonas* is a diverse and economically important group of bacterial phytopathogen belongs to the gamma-subdivision of the Proteobacteria. *X. oryzae* are gram negative rod shaped, capsulated and polar flagellated bacteria causes blight disease in rice plant (*Oryza sativa* L.). This bacterium enters into the plants through hydathodes (type of secretory tissue in leaves) and wounds on the leaf and spreads through wind and rain. Phage OP1 and OP2 was isolated during 1960 by Yoshimura et al., [99] had wide host range. Another *X. oryzae* bacteriophage was isolated by Kuo et al., [100], showed 6 month viability in dry filter paper at room temperature. Civerolo and Keil, [101] showed 86 to 100% reduction in bacterial spot caused by *X.oryzae* by using phageXp12, isolated by Kuo et al., [100]. Xp12 showed effective control of blight disease when applied one day before pathogen inoculation in rice plants [102]. Addition of calcium ion was found to enhance the Xp12 phage penetration into *X.oryzae* bacterium [Kuo et al., 103]. The same scientists tested the efficiency of five phages (Xp10, Xp13, Xp20, Xf and Xp12) isolated from blight infected leaf surface during 1974 against *X.oryzae* [Kuo et al., 104]. Xp12 also controlled *X. pruni*, *X. phaseoli*, *X.malvacearum*, *X.monihotis* and *X. citri*.

Lin et al., [105] isolated a filamentous phage φXo against *X. oryzae*, with a genome size of 7.6 kb. Another effective strain Xp10 was sequenced by Yuzenkova et al., [106], which was originally isolated by Kuo et al., [102]. Based on the electron microscopic view, Xp10 was grouped under *Siphoviridae* family of double stranded DNA bacterial viruses. The result was controversial to previous report, because Xp10 encodes a single subunit RNAP but all known phage which encode single-subunit RNAPs belongs to the *Podoviridae* family which has short-tail [106]. Kuo et al., [107] showed phages were resistance to wide range of mechanical stresses and withstand high temperature and resistant to treatment with DNase and trypsin. Xf phage showed tolerance character to various environmental stresses, this character supports to use these phages in field level to control *X. oryzae*. Lee et al., [108] isolated a phage (φXo411) against *X. oryzae* and small fragment of gene responsible for the lysozyme production was successfully cloned and expressed in *E. coli*. The gene fragment cloned in *E.coli* showed effective control of *X. oryzae*. This illustrates that the lysozyme is the main enzyme produced by phage φXo411 to control phytopathogens. Knowing the interaction between phage and pathogen is important in successful control of phytopathogens. Recently Chae et al., [109] isolated 34 phages from flooded rice field water against *X. oryzae* in which 29 were belongs to *Myoviridae* family. This indicates that the *Myoviridae* is dominated in that ecosystem.

Bacteriophages of Xanthomonas Campestris

X.campestris is a gram negative aerobic bacterium, causing black rot disease in plants. So far 20 different pathovars of *X. campestris* were identified. This pathogen persists in soil for more than a year and easily spread through the irrigation water. With the help of pruniphage isolated by Civerolo, [110] plant bioassay was performed in peach leaves. Results revealed that the isolated pruni phage can control the *X. campestris* but latter this pathogen developed resistance mechanism against the particular pruni phage strain upto 50% [111]. Phage resistance was developed because of the race shifts among populations of *Xanthomonas* spp. [112, 113].

During 1980 to 1990, there were many reports on copper and antibiotic resistance by *X.campestris* pv. *vesicatoria*. Control of this pathogen with a various chemical, biological, and genetic strategies had limited success [114, 115, 116]. Because of ineffective fungicide and antibiotic, phage therapy had gained more attention. Peoples started to use one of the commercially available phage preparation Agriphage™against *X. campestris* (product from USA based company), which contains mixture of four H mutant bacteriophages specific for tomato race 1 (T1) and race 3 (T3) of the bacterial spot pathogen, *X. campestris* pv. *vesicatoria*. The effect of Agriphage was evaluated during two seasons (Fall 1997 and Fall 1998) with the other bactericides (copper/mancozeb). The results showed that the disease incidence was very low for the Agriphage treatments compared to the control and other chemical bactericides. Use of this phage product increased the fruit weight by 37.8% (1997) and 23.9% (1998) compared to the plants treated with the chemical bactericides [117].

X.campestris pv.*juglandis* is a causative agent of Walnut blight. These bacteria rapidly developed the resistance to copper spray during 1990's, and the alternative chemicals showed unreliable results. Phages against *X. campestris* pv.*juglandis* were isolated from soil nearby walnut blight infected plants. Six isolated phages showed effective control of three different *X. campestris* pv.*juglandis* strains and the survivability of phage was more than 50% even after three month storage at 4°C [118].

Bacteriophages of Xanthomonas Axonopodis

X.aonopodis is anaerobic gram negative Proteobacteria consisting of an outer membrane made up of lipopolysaccharide (LPS) and a phospholipid inner membrane. They are found in leaves, stems and fruits of citrus tree, including lime, oranges and grapefruits. Kuo et al., [107], first isolated *X. aonopodis* phage Xf, this phage was originally isolated from citrus canker of orange infected fruits, and it takes only 10 h for destroying the pathogen. Based on the thermal denaturation characters and the hydroxyl apatite column elusion pattern Xf has the single stranded DNA [119]. Wakimoto [120] observed two groups of citrus phage and Goto and Starr [121] found highly specific virulent citrus phages to *X. axonopodis* pv. *citri* strains. During 2001, twenty eight strains of bacteriophages were isolated from canker affected leaves of citrus in South Korea [122]. The isolated strains showed effective multiplication in Wakimotos potato semisynthetic media at 25°C with the host at pH 6.5 [123]. *X.axonopodis* pv. *allii*, (leaf blight on onion) can be effectively controlled using bacteriophage and a plant defense activator (acibenzolar-S-methyl) under greenhouse and field conditions. Biweekly application of bacteriophage in field grown onion reduced 26 to 50% of disease incidence was better than weekly application of copper based bactericide [124]. Xf phage was unable to infect *X. axonopodis* pv. *citri* mutant strain (*pilA* gene), this results clearly indicates that the phage attachment site is present on the pili of this phytopathogen [125]. Greenhouse experiment was conducted with CP2, ΦXac2005-1, ccΦ7, ccΦ13, ΦXacm2004-4, ΦXacm2004-16, ΦX44, ΦXaacA1, which showed consistent control of citrus canker diseases [126]. Filamentous phage XacF1 isolated by Ali Ahmad et al., [127] controlled seven out of eleven *X.* axonopodis pv. *citri* strain tested. Infection of XacF1 phages found to reduce the bacterial EPS production, motility, reduced the growth rate of the bacteria and virulence.

Bacteriophages of *Pseudomonas* sp.

Pseudomonas syringae is a gram negative, rod shaped Proteobacteria with polar flagella, which can infect wide range of economically important plants, exist as over 50 different pathovars. Bacteriophage isolation against *P.syringae* was successfully done by Crosse and Garrett [128] where they isolated 23 phages from Pseudomonas mors-prunorum and P. syringae infected plants. The isolated phages showed wide host range and the biochemical characterization of phage helped to differentiate phage against the two pathogens mentioned above. Twenty years later Thomas and Leary, [129] isolated fifty six phages against *P. syringae* from sewage water and only six phages showed wide host range. Electron microscopic results revealed that the six phages had double stranded DNA with four distinct morphologies [129]. Out of eight phages isolated from sewage against *P. syringae*, five showed different restriction patterns with *EcoRI* restriction enzyme. One phage from each restriction pattern was studied for morphological differentiation under electron microscope and came up with four different morphotype phages [130]. Cuppels [131] isolated sixteen phages against *P. syringae* pv. *tomato* strains from the pathogen infected field soil and plant debris. Only four phage strains (PT1, PT18, PT20, and PT32) were shown effective control (upto 90%) against all phytopathogen tested. Phage PT1 and PT18 are having isometric head and long, striated, noncontractile tail, whereas PT20 and PT32 has isometric head and short, noncontractile tail. Minor et al., [132] isolated 20 phages against *P. syringae* pv. *syringae* by enrichment technique. Prior et al., [133] isolated elongated polyhedral head and a long tail phage belongs to *Siphoviridae* family. The phages were isolated by enrichment of *P. syringae* pv. *tomato* (*P. tomato*) pathovar. Recently φPSA1 and φPSA2 was isolated against *P. syringae* pv. *actinidiae*, the causal agent of Kiwifruit canker [134]. The authors suggested that the phage φPSA2 may be considered as a good candidate for kiwifruit disease control, while φPSA1 are more specific to *P. syringae* pv. *actinidiae* strain. *P. solanacearum*, is a plant pathogenic bacteria which cause wilt diseases in Solanaceae family plants. Tanaka et al., [76], isolated phage against this particular pathogen from diseased tobacco stem, which has polygonal head and a short tail. The same group isolated PK-101 phage against *P. solanacearum* strain K -101 from phytophathogen infected field soil [78]. *P. solanacearum* is an economically important phytopathogen in mushroom industries, which causes brown blotch disease. The virulent phage was isolated from sewage water in South Korea. Totally 21 phages were isolated and pitting test on mushroom showed complete inhibition of *P.tolassii* population [135].

Bacteriophages of other Important Phytopathogens

Streptomyces Scabies

S. scabies is a soil born plant pathogen mainly affects the tuber crops such as potato, radish, beets, turnip, carrot and parsnips. Phage φL and φS was isolated from *S. scabies* MR13 infected field, which was irrigated with sewage water. Based on the electron microscopic study these two actinophages has hexagonal head and non contractile tail. φS has long tail and φL has short tail. These two phages showed effective control of S. scabies strain

MR13 and S-63 [136]. In vivo study using phage strain φAS1 [137] showed effective control of S. scabies on potato tubers [138].

Another in vitro study with Stsc1 and Stsc3 showed effective control of disease development in *S.scabies* infected radish seedlings. These phages were originally isolated from soil around the pathogen infected plants. The morphology study revealed that both the phages has symmetric structure and icosahedral head with a long striated tail [139].

Agrobacterium Tumefaciens

A.tumefaciens is the causal agent of crown gall diseases, which affects around 140 dicot plant species. It belongs to the alpha proteobacteria, rod shaped and gram negative soil bacterium. Bacteriophage (PB1) against A. tumefaciens B6-806 was isolated by Stonier et al., [140] *and did not completely eliminated A. tumefaciens under in vitro condition, but in plate assay this phage showed effective control of the tested bacterial strains. Parsons and Beardsley,* [141] *isolated phage PS8 from tumor causing A.tumifaciens strain B6 from sunflower plant. Three phages isolated during 1970 by Boyd* et al. [142], *effectively inhibited the A. tumefaciens strains PTll, IIBNV6, and IIBNV6-C. Phage PT11 is a polyhedral, bacillary shaped phage that has a tail with a slightly knob or granule at the distal end. Phage PIIBNV6 has hexagonal shaped head indicating icosahedral particles and PIIBNV6-C with icosahedral particle with short tail. Phage PS8 isolated by Parsons and Beardsley,* [141] *was characterized during 1974* [143] *which has a hexagonal head and a flexible tail.*

Dickeya sp.

Dickeya sp. is a phytopathogen of potato, which causes tuber soft rot disease during storage and blackleg disease in the field [144, 145]. This strain was previously named as *Erwinia chrysanthemi*, gram negative, non sporulating and facultative anaerobic bacteria belong to Enterobacteriaceae. Recently a more virulent *Dickeya* type was isolated which belongs to biovar3 of *E. chrysanthemi*, tentatively named as '*Dickeya solani*' [146]. At this stage, there is no recommended chemical control measure for this phytopathogen, resulting in significant economic losses. Two phage strains (LIMEstone1 and LIMEstone2) isolated from disease infected potato field showed control of around 20 *D. solani* strains. Based on the electron microscopy the isolated phages were placed in *Myoviridae* family with icosahedrical head and a tail. The *invitro* and field level test of this two phages showed effective control of phytopathogen and increased the tuber weight [147]. This result clearly indicates that the phage therapy is a promising biocontrol method against the phytopathogens. Recently ϕD5 phage was isolated by Czajkowski et al., [148] which inhibits *D. solani* strain and also *D. dianthicola*, *D. zeae*, *D. dadantii* and *D. chrysanthemi*. Genome sequences of ϕD5 phage revealed the lytic cycle of the phage which can effectively inhibits *Dickeya* spp.[149].

Factors Influences on Phage Therapy Success

Even though, there are numerous reports showing the control of phytopathogen using phages, the efficiency in field condition is still a major problem. The efficiency of phage in plant disease control is not only depends on the susceptibility of the target phytopathogen but also the environmental factors which influences the phage survival. Field and greenhouse studies reported that the phages are inactivated when exposed to sunlight, high temperature, extreme pH and water flow[150]. There are few measures need to be taken to overcome these environmental problems, such as time of phage application, temperature, osmotic shock, survivability and compatibility with other bactericides (Figure 3).

Phages/Bacteria Ratio

Application of bacteriophages against pathogenic bacteria has been studied by passive and active approaches [152]. In passive approach, the bacteriophages are added in a sufficient amount to ensure that all target bacteria are infected and lysed in a short period. While active methods depend on the addition of small amount of phage, the replicated phages infect the remaining bacterial population. It appears that the passive treatment is more efficient than the active approach, because the time required to control bacteria is less in passive treatment. The bacteriophages/bacteria ratio is explained by the term multiplicity of infection (MOI) which refers to the number of virus that are added per cell during infection. Normally the MOI for *in vitro* and *in vivo* experiments ranges between 0.01 and 100. If the MOI is 100 means there are enough phages in the media.

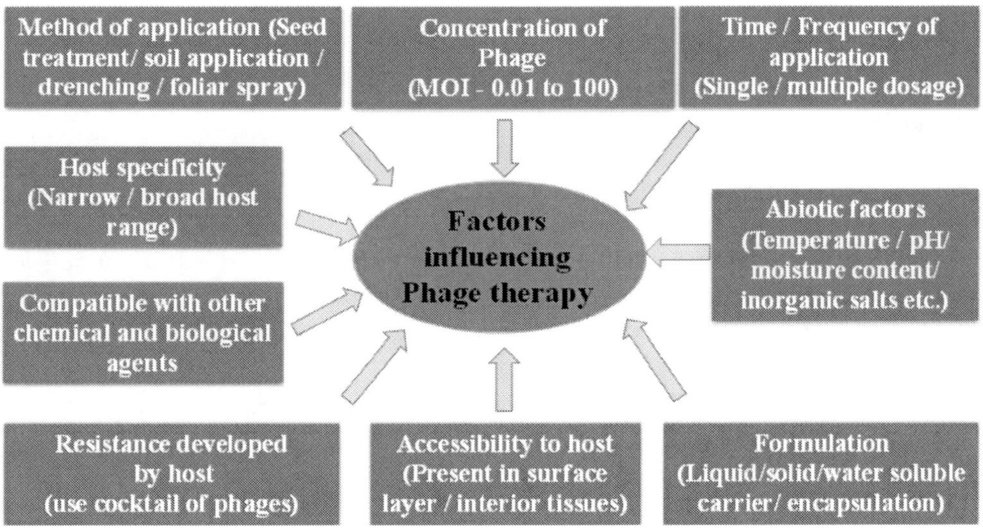

Figure 3. Factors influencing the phage therapy in agriculture (Modified from Ly-Chatain, [151]).

Time of Application

One of the important criteria to improve phage therapy is the time of application, because it plays a major role in the survivability of phage. Civerolo and Keil [101] found application of phage an hour or one day before occurrence of pathogen is effective and ineffective when applied a day after disease occurrence. This is because phage does not have the ability to enter into the intercellular spaces of plant, were the bacteria can enter and causes the disease [153]. *X. campestris* pv. *campestris* (cabbage black rot) and *X. campestris* pv. *vesicatoria* (bacterial spot of pepper) was controlled when phage was applied three days before bacterial infection and one day after infection of pathogen under greenhouse condition [154].

During field level application of phages damges due to UV rays should be considered. Flaherty et al. [117] found that the bacterial spot could be effectively controlled when phages were applied early morning rather than late morning. However, the efficacy was higher for disease control when phages were applied during evening than in the morning [155]. In general, the phage application during the minimal sunlight can be more effective by avoiding the irradiation damage. Effective control of soil borne bacterial diseases can be achieved when the phages are applied during the seedling stage. Effective control of disease was achieved, when the φRSL1 phage is applied to plants at the seedling stage. The formulation with 10^6 or 10^8 PFU/ml is more effective than other concentration of phage tested for tomato spot disease control [155]. Phages applied at the time of phytopathogen inoculation was found to be more effective in disease control whereas phage application was not effective when the phages were applied three days before or after the pathogen inoculation [156].

Nonetheless, post-treatment of *R. solanacearum* phage PE204 exhibited biocontrol activity. Tomato plants not showing initial wilt symptoms after phage treatment survived without showing further development of disease symptoms [157]. Application of phage to tomato plants at regular time intervals may control bacterial wilt development and phage stability under various conditions, suggesting that the phage application through irrigation should be considered for root disease management.

Temperature

Temperature is another important factor for better establishment, survival and control of phytopathogens [158]. Temperature higher than optimal leads to the prolonged latent period of phage [159] and at the same time low temperature leads to reduced phage multiplication. So finding the cardinal temperature for phage multiplication is very important for phage therapy success [160]. If we did not consider temperature effect on phage, then the phage therapy efficiency in field level will be a doubtful one.

Osmotic Shock

Developing phage formulation for soil borne phytopathogens with due consideration should be given for soil physicochemical properties, which affect the survivability and multiplication. pH and Ec are the two soil factor which has more influence on phage efficiency, hence, the phage formulation must be developed from the native soil will have better survival. Phages also exhibits salt tolerance, Seaman and Day [161] isolated phage from salt plain soil, where the salinity level ranges from 0.3% to 27%. Normally the phage with larger genome (340kb) may have more adoption to harsh environmental conditions. Selection of phage with more environmental stress tolerance will be more effective in controlling phytopathogens.

Phage Survival in Soil and in Phyllosphere

Survival of phage in the phyllosphere region is more challenging, because of sunlight and leaching by rainwater. Sunlight is the most destructive factor, which can damage the phage DNA by forming thymine dimer. Phage ability to survive on desiccation stress on leaf surface is another factor; desiccation tolerant phage may perform better in the phyllosphere than the non desiccation tolerant phages [156]. There were several approaches to increase the phage survival in the plant phyllosphere *viz*, application of phages during minimal sunlight, mixed with protective agents and using carrier bacteria. Phage isolated from aerial part will have better adaptation and more efficient in controlling phyllosphere diseases. Phage therapy might be more successful for soil borne pathogen control, because phages are rapidly isolated from soil; however many factors also hinders the phage activity in soil. Phage may not find the target phytopathogen in soil because phages are trapped in biofilm and absorbed by soil particles. φRSL1 showed high stability in soil even at high temperature and this phage was reisolated from the field after 4 month of application. Phages applied in the soil were found to be translocated to the lower leaf of the tomato plants and application of attenuated phytopathogens increased the phyllosphere survival of phages [162]. Finding the suitable phage which has high survival ability in soil and phyllosphere may be effective in field condition. Recently soil microcosm experiment was conducted to find out theΦRsS6 phage survivability in soil [160]. The results showed that the detectable level of phage population on the plant roots was observed even after 15 days of inoculation in the absence of the host bacterium. *Ralstonia solanacearum* phage PE204 is stable under a wide range of temperature and pH, and also stable in the presence of the surfactant Silwet L-77 [157]. The artificial soil microcosm (ASM) study showed phage survival was less stable under elevated soil temperature.

Compatibility with Chemical Bactericide

Biocontrol of phytopathogen are more effective when combining biological agent with chemical agents. However, the chemical agent should not affect the survivability and efficiency of the biological agent. Even though there are many reports showing disease control by combined use of biological and chemical agents, the compatibility of phage to chemical bactericides are very low and chemicals reduces the phage population under *invitro* condition. Phage can be successfully combined with antibiotic like streptomycin, which is used to control bacterial phytopathogens. Nevertheless, in most of the countries streptomycin was banned to use in agriculture because of antibiotic resistance pathogen enters in food chain. Gill and Abedon [37] proposed some other factors like the location of the target pathogen resides, presence of water as a medium for phage diffusion, rate of phage decay, *insitu* multiplication ability and fitness to phage resistant bacterial mutant.

Specificity

The specificity of phage to a bacterial cell can limit the effectiveness of phage use [151]. To conform that the bacteria can be lysed by the phage, the isolated bacterial strains should be tested for its susceptibility to the phage. Selected phage must have lytic cycle for the successful application of phage to control pathogenic bacteria. Polyvalent phages may act as a broad spectrum bacterial controlling agent which can infect several strains of the same bacterial species. A cocktail of phages can replace these polyvalent phages. Nevertheless, a single unique phage with broad host range has shown effective control of bacterial wilt compared to the phage cocktails [24]. Several solutions were proposed to overcome the phage specificity problem: (1) isolating the bacteria from the infection site and test the sensitivity to the group of phages, (2) selecting polyvalent phages with lytic activity (3), and prepare a cocktail of phages that may increase the broad spectrum activity of the phages.

Resistance to Phage

Similar to antibiotics, bacteria has the ability to develop resistance against phages, which may reduce the effectiveness of phage application. Bacteria resistance against phage occurs in three ways (1) Bacteria loses the receptor site on the cell surface (2) horizontal transfer of restriction modification system which degrades the nucleic acid of the injected phage (3) mutation in the bacterial gene. Nevertheless, bacterial resistance to phage is not a major problem in phage application, because the rate at which the bacteria develop resistance to phage is 10 fold lower than the antibiotics [163]. However, the problem related to resistance to phage may be overcome by using the cocktail of phage with one preparation, even if one phage fail another phage will destroy the bacteria. Garcia et al., [164] and Moineau and Fortier, [165], gave some protocols for successful isolation of phages and application. Time required for the isolation and screening of new phage is faster than the development of novel antibiotics [166].

Formulation Development and Commercialization

Formulation Development for Phages

Only scarce information is available on formulation development and commercialization of phages for agricultural use. As we know that, the formulation development is one of the most important parts in commercialization of biocontrol agent. Formulation must have some important properties like easy handling, protect the biological agent present in the formulation, increase the shelf life, the carrier should not cause environmental pollution and cost effective. Different kinds of materials were found to increase the virus tolerance to UV, rain leaching and other external factors. For example light adsorbing or reflecting dyes [167, 168], activated charcoal [167], starch and flour [169, 170], casein [171], alkaline gluten [172], and lignin [150] based formulations were developed. Only few papers were published in bacteriophage formulation development against the phytopathogens. Balogh [173] developed the formulation using pregelatinized corn flour (PCF), sucrose, casecrete NH- 400, skim milk powder and sucrose. The authors developed this phage formulation against tomato bacterial spot disease and tested in greenhouse and field condition. These formulations were found to increase the persistence of phage population in the phyllosphere region under field condition. Compared to non formulated phage population on leaf, the formulated products such as PCF, casecrete, and skim milk formulations showed 4700-, 38,500- and 100,000-fold increase in phage numbers.

In greenhouse condition these formulations showed minimal improvement in the disease control. Whereas casecrete and skim milk formulations were significantly reduced the disease incidence but PCF did not showed any significant differences with the unformulated phage products. Under field condition the above three formulations significantly reduced the disease development [155]. PCF, casecrete and skim milk reduced 11, 21 and 10% disease severity respectively compared to the un-formulated phage application [173]. Phage formulated using skim milk reduced the bacterial leaf blight of rice [109]. Proteins and sugars present in the formulations provide nutritional sources for the pathogens and helped in the breakdown of surface tension and forming aqueous layer on the leaf surface [174]. Application of phage along with low number of their host may lead to increased success rate and the persistence of phage in the environment. For example a mild pathogenic strain of *X. perforans*, which causes less virulent disease incidence, can be combined with phage formulation for effective control of wild *X. perforans* [175]. Because in the absence of host, the phage survivability on the phyllosphere will be around six days but the presence of host increases the persistence for more than two weeks [174]. Fire blight was successfully controlled using the formulations of phage and *P. agglomerans* as delivering bacteria. The phages present in the formulation can lysis *P. agglomerans* and *E. amylovora* [176]. Same kind of formulation was also used to control tobacco bacterial wilt, where phage was applied along with avirulant strains of *R.solanacearum* [76].

Chattopadhyay et al. [177] showed that the most tested surfactants, such as a nonionic Triton X-100, an anionic sodium dodecyl benzene sulfonate, a cationic hexadecyl trimethylammonium bromide, and surfactin, significantly reduced the viral activity of T-2 and ΦX-174 below 0.1% concentration. However, use of Silwet L-77 does not affect phage stability upto a concentration of at least 0.1%, nor does it affect biocontrol activity. Silwet L-77 is an organo silicone surfactant based on trisiloxane ethoxylate [157]. It has also been reported that Silwet L-77 enhanced the entry of bacterial pathogen into plant leaf tissues. Therefore, the results suggested that Silwet L-77 could be used as a supplement for phage formulation as it might enhance phage stability and phage delivery into plants.

Commercialization

Phage products have been approved and commercialized. Regularly, approvals have been granted in the USA for commercial phage products. In 2006 the FDA (Food and Drug Administration) approved the use and the preparation of bacteriophages generally recognized as safe (GRAS) as food additives for the control of the pathogenic bacterium *L. monocytogenes* in meat and poultry products. In Europe, the use of Listex™ was also approved in Switzerland for cheese making and has recently been approved for other food types. The Listex™ has also been approved for use in food processing by Food Standards Australia and NewZealand (FSANZ) in 2012. Several phage products are currently produced on a commercial scale.

Before commercialization of phage products few criteria such as the phage must have lytic cycle rather than lysogenic [178], genome sequence and structure of the phage must be analyzed and applicability of the phage against specific pathogen [179] should be tested. Omnilytics (formerly AgriPhi, Inc.) is the first company to commercialize the phage products for plant disease control and got the patent for the company AgriPhi, Inc., [180]. They developed the product with the wild type phage and H- mutant phage strains to inhibit phytopathogens. The product was developed by using the phages isolated from the plant leaves, stem, soil and water of the diseased fields. The product was a combination of two phages - wild and H mutant phage. If the pathogen develops the resistance to wild type phage then H-mutant phage will act on the particular pathogen. They conducted series of experiments in greenhouse and field level to prove the efficiency of their product and received the registration for the product from US Environmental Protection Agency (Registration # 679861). The registered product is highly host specific for *X.campestris* pv. *vesicatoria*, or *P.syringae* pv. *tomato* [181]. Only few phage products were available in the market, but in last five years numerous phytopathogenic phages were isolated and the efficient strains were sequenced and lot of successful greenhouse and field level experiments were done. So most of the basic requirements for the commercialization of phages in phytopathogen control were fulfilled now and can expect more phage products against particular phytopathogen in near future.

Summary and Conclusion

Continuous development of antibiotic and copper resistance by phytopathogens leads to major crop loss in agriculture. At the same time, chemical control creates residual effect on human health and causes environmental pollution. Considering this negative impact, biological methods of plant diseases control is gaining more attention, which is environmentally safe and not harmful to beneficial microorganism. One of the biological agents, which found environmentally safe and not harmful to beneficial bacterial populations are bacteriophages. The use of bacteriophages as antimicrobial agents controlling pathogenic bacteria has appeared as a promising new strategy and it seems that phage therapy may provide a good alternative solution to antibiotics. The abundance of phages in the environment highlights their potential use for control of pathogenic bacteria. Even though it was found in 1900's it usage in agriculture was not successful in early period because of its structure and complex growth nature. After the molecular and microscopic development, we are now able to track almost every aspects of phage life. With the complete knowledge on phage, we are now successful in using phage against phytopathogens. Considering its application in agriculture is numerous, as more than 100 phages were well characterized and effectively applied in plant disease control, but we still suffer in some aspects of phage therapy in agriculture. Most importantly, the field level success is less when compared to controlled environmental conditions. In field level, phages survival and multiplication are affected by many factors, but it is not the problem only for phage - almost every biocontrol agent will have environmental influences. The field level success of phage application may be improved by developing effective formulation, standardizing application method and time. In addition, to improve the phage application few more criteria[151]should be taken into account:

- Use high phage concentration
- Use phage to treat the bacterial infection as soon as possible
- Test the stability of phage in real environmental conditions
- Protect the phage by developing novel formulations
- Screen and develop phage cocktail, which can infect many bacterial strains
- Use polyvalent bacteriophages with broad cross-strain lytic activity or develop a phage cock tail to lyse the majority of bacterial strains and limit the development of resistance to phage

By overcoming, the problems related to phage application and following standard protocols for phage preparation may leads to the successful application of phages in field level. In near future phage may act as a most effective agent in plant disease control compared to other chemical and biological agent.

REFERENCES

[1] FAO, (2004). FAOSTAT: FAO statistical data bases. Rome, Italy (http://faostat.fao.org).

[2] Waddington, S.R., Li, X., Dixon, J., Hyman, G., de Vicente, C., (2010). Getting the focus right: production constraints for six major food crops in Asian and African farming systems. *Food Security*. 2, 27-48.

[3] Thomas, M.B., Waage, J.K., (1996). Integration of biological control and host plant resistance breeding: a scientific and literature review. Technical Center for Agriculture and Rural Cooperation of the European Union, Wageningen, The Netherlands.

[4] Alabouvette, C., Olivain, C., Steinberg, C., (2006). Biological control of plant diseases: The European situation. *Eur. J. Plant Pathol.*114, 329-341.

[5] Nicot, P.C.; Decognet, V.; Fruit, L., Bardin, M.; Trottin, Y., (2002).Combined effect of microclimate and dose of application on the efficacy of biocontrol agents for the protection of pruning wounds on tomatoes against *Botrytis Cinerea*. *Bulletin IOBC/SROP*.25,73-77.

[6] Whetzel, H.H., (1929). The terminology of plant pathology. Proc. Int. Cong. Plant Science 2, 1204-1215.

[7] Arneson, P.A., (2001). Plant Disease Epidemiology, Temporal Aspects. *The Plant Health Instructor*. DOI: 10.1094/PHI-A-2001-0524-01.

[8] Francl, L.J., (2001). The disease triangle: A plant pathological paradigm revisited. The Plant Health Instructor in Teaching Articles, Education Center. *The Plant Health Instructor*DOI: 10.1094/PHI-T-2001-0517-01.

[9] Sankarana, S., Mishraa, A., Ehsania, R., Davis, C., (2010). A review of advanced techniques for detecting plant diseases. *Comput. Electron. Agr.* 72, 1-13.

[10] Sundin, G.W., Jones, A.L., Fulbright, D.W., (1989). Copper resistance in *Pseudomonas syringae*pv. *syringae* from cherry orchards and its associated transfer *in vitro* and *in planta* with a plasmid. *Phytopathol.* 79, 861-865.
[11] Cooksey, D.A., (1990). Genetics of bactericide resistance in plant pathogenic bacteria. *Annu. Rev. Phytopathol.* 28,201-219.
[12] Voloudakis, A.E., Bender, C.L.,Cooksey, D.A., (1993). Similarity between copper resistance genes from *Xanthomonas campestris* and *Pseudomonas syringae. Appl. Environ. Microbiol.* 59,1627-1634.
[13] Manulis, S., Zutra, D., Kleitman, F., Dror, O., David, I., (1998). Distribution of streptomycin resistant strains of *Erwinia amylovora* in Israel and occurrence of blossom blight in the autumn. *Phytoparasitica*.26,223-230.
[14] Saikia, R., Sindh, T., Kumar, R., Srivastava, J., Srivastava, A.K., Singh, K., Arora, D.K., (2003). Role of salicylic acid in systemic resistance induced by *Pseudomonas fluorescens* against *Fusarium oxysporum* f. sp. *ciceri* in chickpea. *Microbiol. Res.* 158, 203-213.
[15] Romero, A.M., Kousik, C.S., Ritchie, D.F., (2001). Resistance to bacterial spot in bell pepper induced by acibenzolar-S-methyl. *Plant Dis*.85,189-194.
[16] Gent, D.H., Schwartz, H.F., (2005). Management of *Xanthomonas* leaf blight of onion with a plant activator, biological control agents, and copper bactericides. *Plant Dis*., 89,631-639.
[17] Kousik, C.S., Ritchie, D.F., (1996). Disease potential of pepper bacterial spot pathogen races that overcome the *Bs2* gene for resistance. *Phytopathol*.86,1336-1343.
[18] Gassmann, W., Dahlbeck, D., Chesnokova, O., Minsavage, G.V., Jones, J.B., Staskawicz, B.J., (2000). Molecular evolution of virulence in natural field strains of *Xanthomonas campestris* pv. *vesicatoria*. *J. Bacteriol.* 182,7053-7059.
[19] Liu, T., (1998). Biological control with tomato bacterial spot with hrp− mutants of *Xanthomonas campestris pv. vesicatoria*. MS thesis. Univ. FL, Gainesville.
[20] Byrne, J.M., Dianese, A.C., Ji, P., Campbell, H.L., Cuppels, D.A., Louws, F.J., Miller, S.A., Jones, J.B., Wilson, M., (2005). Biological control of bacterial spot of tomato under field conditions at several locations in North America. *Biol. Control.* 32, 408-418.
[21] Yim, W., Seshadri,S., Kim, K.,Lee, G.,Sa,T., (2013). Ethylene emission and PR protein synthesis in ACC deaminase producing *Methylobacterium* spp. inoculated tomato plants (*Lycopersicon esculentum* Mill.) challenged with *Ralstonia solanacearum* under greenhouse conditions. *Plant Physiol. Bioch*. 67, 95-104.
[22] Schwarczinger, I., Kiss, E., Sule, S., Toth, M., Hevesi, M., (2011). Control of fire blight by bacteriophages on apple flowers. *Actahortic*. 896, 457-462.
[23] Born,Y.L., Fieseler, J., Marazzi, R., Lurz, B., Duffy., Loessner, M.J., (2011). Novel Virulent and broad-host-range *Erwinia amylovora* bacteriophages reveal a high degree of mosaicism and a relationship to Enterobacteriaceae phages. *Appl. Environ. Microbiol.* 77, 5945-5954.
[24] Fujiwara, A., Fujisawa, M., Hamasaki, R., Kawasaki, T., Fujie, M., Yamada, T., (2011). Biocontrol of *Ralstonia solanacearum* by treatment with lytic Bacteriophages. *Appl. Environ. Microbiol.* 77, 4155-4162.
[25] Hendrix, R.W., (2002). Bacteriophages: Evolution of the majority. *Theoret. Popul. Biol*.61, 471-480.

[26] Pal, C., Macia, M., Oliver, A., Schachar, I., Buckling, A., (2007). Coevolution with viruses drives the evolution of bacterial mutation rates. *Nature* 450,1079-1081.

[27] Sau, S., Chattoraj, P., Ganguly, T., Chanda, P.K.,Mandal, N.C., (2008). Inactivation of indispensable bacterial proteins by early proteins of bacteriophages: implication in antibacterial drug discovery. *Curr. Protein Pept. Sci.* 9,284-290.

[28] Borysowski, J., Weber-Dabrowska, B.,Gorski, A., (2006). Bacteriophage endolysins: Bacteriophage Therapy as a novel class of antibacterial agents. *Exp. Biol. Med.* 231,366 - 377.

[29] d'Herelle,F., (1918). Technique de la recherche du microbe filtrant bactériophage (*Bacteriophagum intestinale*).*Compte Rendu Hebdomadaire des Séances et Mémoires de la Société de Biologie* 81, 1160-1162.

[30] Matsuzaki, S., Rashel, M., Uchiyama, J., Sakurai, S., Ujihara, T., Kuroda, M., Ikeuchi, M., Tani, T., Fujieda, M., Wakiguchi, H., Imai,S., (2005). Bacteriophage therapy: a revitalized therapy against bacterial infectious diseases. *J. Infect. Chemo. Ther.* 11,211-219.

[31] Ackermann, H.W., (1999). Tailed bacteriophages. The order *Caudovirales. Adv.Virus.Res.*51,135-201.

[32] Lwoff, A., Horne, R.W., Tournier, P., (1962). A system of viruses, *Cold Spring Harbor Symp. Quant. Biol.*27, 51-62.

[33] Ackermann, H.W., (2001). Frequency of morphological phage descriptions in the year 2000. *Arch.Virol.*146,843-857.

[34] Mindich, L., Qiao, X., Qiao, J., Onodera, S., Romantchuk, M., Hoogstraaten, D., (1999). Isolation of additional bacteriophages with genomes of segmented double-stranded RNA. *J.Bacteriol.*181, 4505-4508.

[35] Day, L., Maniloff, J., (2000). Inoviruses, In Van Regenmortel and coll., pp. 267-498.

[36] Peng, X., Blum, H., She, Q., Mallok, S., Brugger, K., Garrett, R.A., Zillig, W., Prangishvili, D., (2001). Sequences and replication of genomes of the archaeal rudiviruses SIRV1 and SIRV2: Relationships to the archaeal lipothrixvirus SIFV and some eukaryal viruses. *Virol.*291, 226-234.

[37] Gill, J.J., Abedon, T.S., (2003). Bacteriophage ecology and plants. *APS net.* doi: 10.1094/APSnetFeature-2003-1103.

[38] Twort, F.W., (1915). An investigation on the nature of ultramicroscopic viruses. Lancet II, 1241.

[39] d'Herelle, F., (1917). Sur un microbe invisible antagoniste des bacilles dysentériques. CR. *Acad.Sci.Ser.D.* 165, 373.

[40] Beckerich, A., Hauduroy, P., (1922). Le bacteriophage dans le traitment de la fievre typhoide. *C. R. Soc. Biol.*86,168.

[41] Davison, W.C., (1922). The bacteriolysant therapy of bacillary dysentery in children. Archives of Pediatrics and Adolescent Medicine. *Am.J.Dis.Child.* 23,531-534.

[42] Mallmann, W.L., Hemstreet, C.J., (1924). Isolation of an inhibitory substance from plants. *Agric. Res.* 6, 599-602.

[43] Coons, G.H., Kotila, J.E., (1925). The transmissible lytic principle (bacteriophage) in relation to plant pathogens. *Phytopathol.* 15,357-370.

[44] Kotila, J.E., Coons, G.H., (1925). Investigations on the blackleg disease of potato. Michigan Agricultural Experimental Station Technical Bulletin. 67,3-29.

[45] Massey, R.E., (1934). Studies on blackarm disease of cotton III. *Empire Cotton Growing Rev.*11,188-193.

[46] Thomas, R.C., (1935). A bacteriophage in relation to Stewart's disease of corn. *Phytopathol.*25,371-372.

[47] Okabe, N., Goto, M., (1963). Bacteriophages of plant pathogens. *Ann. Rev. Phytopathol.* 1, 397-418.

[48] Goto, M., (1992). Fundamentals of bacterial Plant Pathology, Academic Press, New York.

[49] McManus, P.S., Stockwell, V.O., Sundin, G.W., Jones, A.L., (2002). Antibiotic use in plant agriculture. *Annu. Rev. Phytopathol.* 40, 443-465.

[50] Chiou, C.S., Jones, A.L., (1991). The analysis of plasmid-mediated streptomycin resistance in *Erwinia amylovora. Phytopathol.* 81, 710-714.

[51] Loper, J.E., Henkels, M.D., Roberts, R.G., Grove, G.G., Willet, M.J., Smith, T.J., (1991). Evaluation of streptomycin, oxytetracycline, and copper resistance of *Erwinia amylovora* isolated from pear orchards in Washington State. *Plant Dis.* 75, 287-290.

[52] Fukusawa, K., Sakurai, H., Shimizu, S., Naganawa, H., Kondo, S., Kawabe, H., Mitsuhashi, S., (1980). 3"-phosphoryldihydro streptomycin produced by the inactivating enzyme of *Erwinia carotovora. J. Antibiot.* 33,122-123.

[53] Yano, H., Fujii, H., Mukoo, H., Shimura, M., Watanabe, T., Sekizawa, Y., (1978). On the enzymic inactivation of dihydrostreptomycin by *Pseudomonas lachrymans*, cucumber angular leaf spot bacterium: isolation and structural resolution of the inactivated product. *Ann. Phytopathol. Soc. Japan.* 44, 413-419.

[54] Jones, A.L., Norelli, J.L., Ehret, G., (1991a). Detection of streptomycin-resistant *Pseudomonas syingae* pv. *papulans* in Michigan apple orchards. *Plant Dis.* 75,529-531.

[55] Minsavage, G.V., Canteros, B.I., Stall, R.E., (1990). Plasmid-mediated resistance to streptomycin in *Xanthomonas campestris* pv. *vesicatoria. Phytopathol.* 80, 719-723.

[56] Bender, C.L., Cooksey, D.A., (1986). Indigenous plasmids in *Pseudomonas syringae* pv. *tomato*: Conjugative transfer and role in copper resistance. *J. Bacteriol.* 165,534-541.

[57] Andersen, G.L., Menkissoglou, O., Lindow, S.E., (1991). Occurrence and properties of copper-tolerant strains of *Pseudomonas syringae* isolated from fruit trees in California. *Phytopathol.* 81, 648-656.

[58] Marco, G.M., Stall, R.E., (1983). Control of bacterial spot of pepper initiated by strains of *Xanthomonas campestris* pv. *Vesicatoria* that differ in sensitive to copper. *Plant Dis.*67,779-781.

[59] Adaskaveg, J.E., Hine, R.B., (1985). Copper tolerance and zinc sensitivity of Mexican strains of *Xanthomonas campestris* pv. *vesicatoria*, causal agent of bacterial spot of pepper. *PlantDis.*195, 993-996.

[60] Cooksey, D.A., (1987). Characterization of a copper resistance plasmid conserved in copper-resistant strains of *Pseudomonas syringae* pv. *tomato. Appl. Environ. Microbiol.* 53, 454-456.

[61] Bender, C.L., Malvick, D.K., Conway, K.E., George, S., Cooksey, D.A., (1990). Characterization of pXv10A, a copper resistance plasmid in *Xanthomonas campestris* pv. *vesicatoria. Appl. Environ. Microbiol.* 56, 170-175.

[62] Stall, R.E., Loschke, D.C., Jones, J.B., (1986). Linkage of copper resistance and avirulence loci on a self-transmissible plasmid in *Xanthomonas campestris* pv. *vesicatoria*. *Phytopathol.* 76, 240-243.

[63] Madsen, L., Aarestrup, F.M., Olsen, J.E., (2000). Characterization of streptomycin resistance determinants in Danish isolates of *Salmonella thyphimurium*. *Ver. Microbiol.* 75, 73-82.

[64] Scholz, P., Haring,V.B., Liebold, W., Ashman, K., Bagdasarian, M., Scherzinger, E., (1989). Complete nucleotide sequence and gene organization of the broad-host range plasmid RSF1010. *Gene* 75, 271-288.

[65] Barth, P.T., Grinter, N.J., (1974).Comparison of the deoxyribonucleic acid molecular weights and homologies of plasmids conferring linked resistance to streptomycin and sulphonamides. *J. Bacteriol.* 120, 618-630.

[66] Han, H.S., Nam, H.Y., Koh, Y.J., Hur, J.S., Jung, J.S., (2003). Molecular Bases of High-Level Streptomycin Resistance in *Pseudomonas marginalis* and *Pseudomonas syringae* pv. *actinidiae*. *J. Microbiol.* 41, 16-21.

[67] Li, M.Y., (2007). Determination of *Pseudomonas syringae* pv. *tabaci* resistance to agricultural streptomycin. *Chin. Agric. Sci. Bull.* 23,328-332.

[68] Xiang, P.A., Zhou, Y.,Gao, B.D., (2003). Plasmids and streptomycin - or copper - tolerance in *Xanthomonas vesicatoria* and *X. oryzae*pv. *oryzicola* in China. *Acta. Phytopathol. Sin.* 33, 330-333.

[69] Norelli, J.L., Burr, T.J., Lo Cicero, A.M., Gilbert, M.T., Katz, B.H., (1991). Homologous streptomycin resistance gene present among diverse gram-negative bacteria in New York state apple orchards. *Appl. Environ. Microbiol.* 57, 486-491.

[70] Sundin, G.W., Bender, C.L., (1995). Expression of the *strA-strB* streptomycin resistance genes in *Pseudomonas syringae* and *Xanthomonas campestris* and characterization of IS*6100* in *X. campestris*. *Appl. Environ. Microbiol.* 61, 2891-2897.

[71] Ritchie, D.F., Dittapongpitch, V., (1991). Copper- and streptomycin- resistant strains and host differentiated races of *Xanthomonas campestris* pv. *vesicatoria* in North Carolina. *Plant Dis.* 75, 733-736.

[72] Levin,B.R., Bull, J.J., (2004). Population and evolutionary dynamics of phage therapy. *Nat. Rev. Microbiol.* 2, 166-173.

[73] Yabuuchi, E., Kosako, V., Yano, I., Hotta, H., Nishiuchi, Y., (1995). Transfer of two *Burkholderia* and an *Alcaligenes* species to *Ralstonia* gen. nov.: proposal of *Ralstonia picketii* (Ralston, Palleroni and Doudoroff 1973) comb. nov., *Ralstonia solanacearum* (Smith 1896) comb. nov. and*Ralstonia eutropha* (Davis 1969) comb. nov. *Microbiol Immunol.* 39, 897-904.

[74] Hayward, A.C., (2000).*Ralstonia solanacearum*. In: *Encyclopedia of Microbiology*, vol. 4 (Lederberg, J., Ed.), Academic Press, San Diego, C. A., pp. 32-42.

[75] Zulperi, D., Sijam, K., Mior Ahmad, Z.A., Awang, Y., Rashid, T.S., (2014). Occurrence of *Ralstonia solanacearum* Race 2 Biovar 1 Associated with Moko Disease of Banana (*Musa paradisiaca* cv. Nipah) in Malaysia. *J. Phytopathol.* 162, 697-702.

[76] Tanaka, H., Negishi, H., Maeda, H., (1990). Control of tobacco bacterial wilt by an avirulent strain of *Pseudomonas solanacearunm* M4S and its bacteriophage. *Ann. Phytopathol. Soc. Jpn.* 56, 243-246.

[77] Ozawa, H., Tanaka, H., Ichinose, Y., Shiraishi, H., Yamada, T., (2001). Bacteriophage P4282, a parasite of *Ralstonia solanacearum*, encodes a bacteriolytic protein important for lytic infection of its host. *Mol.Genet. Genom*. 265, 95-101.

[78] Toyoda, H., Kakutani, K., Ikeda, S., Goto, S., Tanaka, H., Oucm, S., (1991). Characterization of Deoxyribonucleic Acid of virulent bacteriophage and its infectivity to host bacteria, *Pseudomonas solanacearum. J.Phytopathol*. 131, 11-21.

[79] Yamada, T., Kawasaki, T., Nagata, S., Fujiwara, A., Usami, S., Fujie, M., (2007). New bacteriophages that infect the phytopathogen *Ralstonia solanacearum. Microbiol*. 153, 2630-2639.

[80] Kawasaki, T., Satsuma, H., Fujie, M., Usami, S., Yamada, T., (2007). Monitoring of phytopathogenic *Ralstonia solanacearum* cells using green fluorescent protein expressing plasmid derived from bacteriophage ϕRSS1. *J. Biosci. Bioeng*. 104, 451-456.

[81] Fujiwara, A., Kawasaki, T., Usami, S., Fujie, M., Yamada, T., (2008). Genomic characterization of *Ralstonia solanacearum* phage φRSA1 and its related prophage (φRSX) in strain GMI1000. *J. Bacteriol*. 190, 143-156.

[82] Nakayama, K., Kanaya, S., Ohnishi, M., Terawaki, Y., Hayashi, T., (1999). The complete nucleotide sequence of CTX, a cytotoxin-converting phage of *Pseudomonas aeruginosa*: implications for phage evolution and horizontal gene transfer via bacteriophages. *Mol. Microbiol*. 31, 399-419.

[83] Kawasaki, T., Shimizu, M., Satsuma, H., Fujiwara, A., Fujie, M., Usami, S., Yamada, T., (2009). Genomic characterization of *Ralstonia solanacearum* phage ϕRSB1, a T7-like wide-host-range phage. *J. Bacteriol*. 191, 422-427.

[84] Yamada, T., Satoh, S., Ishikawa, H., Fujiwara, A., Kawasaki, T., Fujie, M., Ogata, H., (2010). A jumbo phage infecting the phytopathogen *Ralstonia solanacearum* defines a new lineage of the *Myoviridae* family. *Virol*. 398, 135-147.

[85] Murugaiyan, S., Bae, J.Y., Wu, J., Lee, S.D., Um, H.Y., Choi, H.K., Chung, E., Lee, J.H., Lee, S.W., (2010). Characterization of filamentous bacteriophage PE226 infecting *Ralstonia solanacearum* strains. *J. Appl. Microbiol*. 110, 296-303.

[86] Zwet, T.V.D., Beer, S.V., (1995). Fire Blight-Its Nature, Prevention and Control. United States Department of Agriculture, Agriculture Information Bulletin 631.

[87] Stockwell, V.O., Johnson, K.B., Loper, J.E., (1998). Establishment of bacterial antagonists of *Erwinia amylovora* on pear and apple blossom as influenced by inoculum preparation. *Phytopathol*. 88, 506-513.

[88] Kunz, S., von Eitzen-Ritter, M., Schmitt, A., Haug, P., (2004). Control of fire blight during ecological fruit production [Feuerbrandbekampfung im ökologischen Obstbau], Ökoobstbau, 4, 2-7 (in German).

[89] Schnabel, E.L., Fernando, W.G.D., Meyer, M.P., Jones, A.L., Jackson, L.E., (1999). Bacteriophage of *Erwinia amylovora* and their potential for biocontrol. *Acta. Hortic*. 489, 649-654.

[90] Sule, S., Kiss, E., Kim, W.S., Geider, K., (2002). Transformation of SR1 tobacco and JTE-H apple rootstock with the EPS-depolymerase gene from *Erwinia amylovora* phage. *Acta.Hortic*. 590, 407-409.

[91] Lengyel, K., Lang, E., Fodor, A., Szállás,E., Schumann, P., Stackebrandt, E., (2005). Description of four novel species of *Xenorhabdus*, family Enterobacteriaceae: *Xenorhabdus budapestensis* sp. nov., *Xenorhabdus ehlersii* sp. nov. *Xenorhabdus*

innexi sp. nov., and *Xenorhabdus szentirmaii* sp. nov. *Syst. Appl. Microbiol.* 28, 115-122.

[92] Ritchie, D.F., Klos, E.J., (1979). Some properties of *Erwinia amylovora* bacteriophages, *Phytopathol.* 69, 1078-1083.

[93] Schnabel, E.L., Jones, A.L., (2001). Isolation and characterization of five *Erwinia amylovora* bacteriophages and assessment of phage resistance in strains of *Erwinia amylovora*, *Appl. Environ. Microbiol.* 67, 59-64.

[94] Gill, J.J., Svircev, A.M., Smith, R., Castle, A.J., (2003). Bacteriophages of *Erwinia amylovora*. *Appl. Environ. Microbiol.* 69, 2133-2138.

[95] Muller, I., Jelkmann, W., Geider, K., Lurz, R., (2011).Properties of *Erwinia amylovora* phages from North America and Germany and their possible use to control fire blight. In: *Proc. 12th Int. Workshop on Fire Blight* (P. Sobiczewski et al.Eds.) *Acta. Hort.*, 896.

[96] Boule, J., Sholberg, P.L., Lehman, S.M., Ogorman, D.T., Svircev, A.M., (2011). Isolation and characterization of eight bacteriophages infecting *Erwinia amylovora* and their potential as biological control agents in British Columbia, Canada. *Can. J. Plant Pathol.* 33, 308-317.

[97] Hattingh, M.J., Beer, S.V., Lawson, E.W., (1986). Scanning electron microscopy of apple blossoms colonized by *Erwinia amylovora* and *Erwinia herbicola.Phytopathology*. 76, 900-904.

[98] Pusey, P.L., Stockwell, V.O., Rudell, D.R., (2008). Antibiosis and acidification by *Pantoea agglomerans* strain E325 may contribute to suppression of *Erwinia amylovora*. *Phytopathol.* 98, 1136-1143.

[99] Yoshimura, S., Saito, M., Yoshino, R., Morihashi, T., (1960). Properties of a new *Xanthomonas oryzae* phage OP2. *Proc. Assoc. Plant Prot. Hokuriku.* 8, 15-20 (in Japanese).

[100] Kuo, T.T.Huang, T.C., Wu, R.Y., Chen, C.P., (1968).Phage Xp12 of *Xanthomonas oryzae* (Uyeda et Ishiyama) Dowson. *Can. J. Microbiol.* 14, 1139-1142.

[101] Civerolo, E.L., Keil, H.L., (1969). Inhibition of bacterial spot of peach foliage by *Xanthomonas pruni* bacteriophage. *Phytopathol.* 59, 1966-1967.

[102] Kuo, T.T., Yang, C.M., Chow, T.Y., Lin, Y.T., (1970). Bacterial leaf blight of rice plant II. The formation and properties of spheroplasts of *Xanthomonus oryzae*. *Bot. Bull. of Acad. Sinica*. 11, 36-45.

[103] Kuo, T.T., Chow, T.Y., Lin, Y.T., (1972). Role of calcium ion in proliferation of phage Xp12 of *Xanthomonus oryzae*. *Bot. Bull. Acad. Sinica*. 13, 104-110.

[104] Kuo, T.T., Chang, L.C., Lin, Y.T., (1974). Host specificities of five bacteriophages isolated from *Xanthomonus oryzae*. *Bot. Bull. of Acad. Sinica*. 15, 28-35.

[105] Lin, N.T., You, B.Y., Huang, C.Y., Kuo, C.W., Wen, F.S., Yang, J.S., Tseng, Y.H.G., (1994). Characterization of two novel filamentous phages of *Xanthomonas*. *J. Gen. Virol.* 75, 2543-2547.

[106] Yuzenkova, J., Nechaev, S., Berlin, J., Rogulja, D., Kuznedelov, K., Inman, R., Mushegian, A., Severinov, K., (2003). Genome of *Xanthomonas oryzae* bacteriophage Xp10: an odd T-odd phage. *J. Mol. Biol.* 330,735-748.

[107] Kuo, T.T., Huang, T.C., Chow, T.Y., (1969). A filamentous bacteriophage from *Xanthomonasoryzae*. *Virol.* 39, 548-555.

[108] Lee, C.N., Lin, J.W., Chow, T.Y., Tseng, Y.H., Weng, S.F., (2006).A novel lysozyme from *Xanthomonas oryzae* phage varphiXo411 active against *Xanthomonas* and *Stenotrophomonas*. *Protein Expr. Purif.* 50, 229-237.

[109] Chae, J.C., Hung, N.B., Yu, S.M., Lee, H.K., Lee, Y.H., (2014). Diversity of bacteriophages infecting *Xanthomonasoryzae* pv. *oryzae* in paddy fields and its potential to control bacterial leaf blight of rice. *J. Microbiol. Biotechnol.* 24, 740-747.

[110] Civerolo, E.L., (1973). Relationship of *Xanthomonas pruni* bacteriophages to bacterial spot disease in *prunus*. *Phytopathol.* 63, 1279-1284.

[111] Randhawa, P.S., Civerolo, E.L., (1986). Interaction of *Xanthomonas campestris* pv. *pruni*with pruniphage and epiphytic bacteria on detached peach leaves. *Phytopathol.* 76, 549-553.

[112] Jones, J.B.; Woltz, S.S.; Jones, J.P.; Portier, K.L., (1991b). Population dynamics of *Xanthomonas campestris* pv. *vesicatoria*on tomato leaflets treated with copper bactericides. *Phytopathol.* 81, 714-719.

[113] Jones, J.B.; Stall, R.E.; Scott, J.W.; Somodi, G.C.; Bouzar, H.; Hodge, N.C., (1995). A third tomato race of *Xanthomonas campestris* pv. *vesicatoria*. *Plant Dis.* 79, 395-398.

[114] Jones, J.B.; Jones, J.P., (1985). The effect of bactericides, tank mixing time and spray schedule on bacterial leaf spot of tomato. *Proc. Fla. State Hort. Sco.* 98, 244-247.

[115] Jones, J.B., Scott, J.W., (1986). Hypersensitive response in tomato to *Xanthomonas campestris* pv. *vesicatoria*. *Plant Dis.* 70, 337-339.

[116] Scott, J.W., Jones,J.B., (1986). Sources of resistance to bacterial spot (*Xanthomonas campestris* pv. *Vesicatoria* (Doidge) Dye) in tomato. *Hort. Sci.* 21, 304-306.

[117] Flaherty, J.E., Jones, J.B., Harbaugh, B.K., Somodi, G.C., Jackson, L.E., (2000). Control of bacterial spot on tomato in the greenhouse and field with H-mutant bacteriophages. *Hort. Science*.35, 882-884.

[118] Mcneil, D.L., Romero, S., Kandula, J., Stark, C., Stewart, A., Larsen, S., (2001). Bacteriophages: a potential biocontrol agent against walnut blight (*Xanthomonas campestris* pv. *juglandis*). *NZ. Plant Prot.* 54, 220-224.

[119] Dai, H., Chiang, K.S., Kuo, T.T., (1980).Characterization of a New Filamentous Phage Cf from *Xanthomonas citri*. *J.Gen.Virol.* 46, 277-289.

[120] Wakimoto, S., (1967). Some characteristics of citrus canker bacteria, *Xanthomonas citri* (Hasse) Dowson, and the related phages isolated from Japan. *Ann. Phytopathol. Soc. Japan*. 33, 301-310.

[121] Goto, M., Starr, M.P., (1972). Phage host relationships of *Xanthomonas citri* compared with those of other *Xanthomonus*. *Ann. Phytopatha. Soc. Japan*. 38, 226-248.

[122] Myung, I.S., Cho, Y.S., Lee, Y.H., Kwon, H.M., (2001). Phage typing and lysotype distribution of *Xanthomonas axonopodis* pv. *citri*, the causal agent of citrus bacterial canker in Korea. *Plant Pathol. J.* 17, 336-341.

[123] Myung, I.S., Nam, K.W., Chol, Y., (2002). Cultural Characteristics of *Xanthomonas axonopodis* pv. *citri*bacteriophages CP1 from Korea. *Plant Pathol. J.* 18, 333-337.

[124] Lang, J.M., Gent, D.H., Schwartz, H.F., (2007). Management of *Xanthomonas* leaf blight of onion with bacteriophages and a plant activator. *Plant Dis.* 91, 871-878.

[125] Yang, Y.C., Chou, C.P., Kuo, T.T., Lin, S.H., Yang, M.K., (2004). PilR Enhances the Sensitivity of *Xanthomonas axonopodis* pv. *citri* to the Infection of Filamentous Bacteriophage Cf. host. *Curr. Microbiol.* 48, 251-261.

[126] Balogh, B., Canteros, B.I., Stall, R.E., Jones, J.B., (2008). Control of citrus cancer and citrus bacterial spot with bacteriophages. *Plant Dis*. 92, 1048-1052.

[127] Ali Ahmad, A., Askora, A., Kawasaki, T., Fujie, M., Yamad, T., (2014). The filamentous phage XacF1 causes loss of virulence in *Xanthomonas axonopodis*pv. *citri*, the causative agent of citrus canker disease. *Front. Microbiol.* 5, 321. doi: 10.3389/fmicb.2014.00321.

[128] Crosse, J.E., Garrett, C.M.E., (1963). Studies on the bacteriophagy of *Pseudomonas morsprunorum*, pv. *syringae* and related organisms. *J. Appl. Bacteriol.* 26, 159-177.

[129] Thomas, M.D., Leary, J.V., (1983). Bacteriophages from sewage specific for fluorescent phytopathogenic *Pseudomonads*. *Phytopathol*, 73, 403-406.

[130] Nordeen, R., Morgan, K., Currier, T., (1983). Isolation and partial characterization of bacteriophages of the phytopathogen *Pseudomonas syringae*. *Appl. Environ. Microbiol.* 45, 1890-1898.

[131] Cuppels, D.A., (1983). Isolation and characterization of phages useful for identifying *Pseudomonas syringae* pv. *tomato*. *Phytopathol.* 73,1376-1381.

[132] Minor, S.M., Nordeen,R.O., Paschall, C.E., (1996). Partial characterization of bacteriophages of *Pseudomonas syringae* pv. *Tomato.J. Ark. Acad. Sci.* 50, 137-140.

[133] Prior, S.E., Andrews, A., Nordeen, R.O., (2007). Characterization of bacteriophages of *Pseudomonas syringae* pv. *tomato*. *J. Ark. Acad. Sci.* 61, 84-89.

[134] Lallo, G.,Evangelisti,M., Mancuso, F., Ferrante,P., Marcelletti, S., Tinari,A., Superti, F., Migliore, L., D'Addabbo, P., Frezza,D., Scortichini M., Thaller, M.C., (2014). Isolation and partial characterization of bacteriophages infecting *Pseudomonas syringae* pv. *actinidiae*, causal agent of kiwifruit bacterial canker. *J. Basic Microbiol.*, DOI: 10.1002/jobm.201300951.

[135] Kim, M.H., Park, S.W., Kim, Y.K., (2011). Bacteriophages of *Pseudomonas tolaasii* for the Biological Control of Brown Blotch Disease. *J. Korean Soc. Appl. Biol. Chem.* 54, 99-104.

[136] El Sayed, E.S.A., El Didamony, G., Mansour, K., (2001). Isolation and characterization of two types of actinophage infecting *Streptomyces scabies*. *Folia Microbiol.*, 46, 519-526.

[137] Williams, S.T., Wellington, E.M.H., Tipler, L.S., (1980). The taxonomic implications of the reactions of representative Nocardia strains to actinophage. *J. Gen. Microbiol.* 119, 173-178.

[138] McKenna, F., El-Tarabily, K.A., Hardy, G.E.S.T., Dell, B., (2001). Novel *in vivo* use of a polyvalent *Streptomyces* phage to disinfest *Streptomyces* scabies-infected seed potatoes. *Plant Pathol.* 50, 666-675.

[139] Goyer, C., (2005). Isolation and characterization of phages Stsc1 and Stsc3 infecting *Streptomyces scabiei* and their potential as biocontrol agents. *Can. J. Plant Pathol.* 27, 210-216.

[140] Stonier, T., McSharry, J., Speitel, T., (1967).*Agrobacterium tumefaciens* Conn IV. Bacteriophage PB21 and its inhibitory effect on tumor induction. *J. Virol.* 1, 268-273.

[141] Parsons, C.L,Beardsley, R.E., (1968). Bacteriophage activity in homogenates of crown gall tissue. *J. Virol.* 2 (6), 651.

[142] Boyd, J., Hildebrandt, A.C., Allen, O.N., (1970).Electron Microscopy of Phages for *Agrobacterium tumefaciens*. *Arch. Microbiol.* 73, 47-54.

[143] Knopf, U.C., (1974). Studies on the Bacteriophage PS-8 of *Agrobaeterium tumefaciens* (Smith and Towsend) Conn: Purification and properties. *Arch. Ges. Virusforsehg*. 46, 205-216.

[144] Perombelon, M.C.M., (2002). Potato diseases caused by soft rot *Erwinias*: an overview of pathogenesis. *Plant Pathol*. 51, 1-12.

[145] Czajkowski, R., Perombelon, M.C.M., van Veen, J.A., van der Wolf, J.M., (2011). Control of blackleg and tuber soft rot of potato caused by *Pectobacterium* and *Dickeya* species: a review. *Plant Pathol*. 60, 999-1013.

[146] Toth, I.K., van der Wolf, J.M., Saddler, G., Lojkowska, E., Helias, V., Pirhonen, M., Tsror, L., Elphinstone, J.G., (2011). Dickeya species: an emerging problem for potato production in Europe. *Plant Pathol*. 60, 385-399.

[147] Adriaenssens, E.M., Vaerenbergh, J.V.,Andenheuvel,D.V., Dunon, V., Ceyssens, P.J., Proft, M.D., Kropinski, A.M., Noben, J.P., Maes, M., Lavigne, R., (2012). T4-Related Bacteriophage LIME stone Isolates for the Control of Soft Rot on Potato Caused by '*Dickeya solani'*. *PLoS ONE*. 7, 1-10.

[148] Czajkowski, R., Ozymko, Z., Lojkowska, E., (2013). Isolation and characterization of novel soil borne lytic bacteriophages infecting *Dickeya* spp. biovar 3 ('*D. solani'*). *Plant Pathol*., doi:10.1111/ppa.12157.

[149] Czajkowski, R., Ozymko, Z., Zwirowski, S., Lojkowska, E., (2014). Complete genome sequence of a broad-host-range lytic *Dickeya* spp. bacteriophage ϕD5. *Arch. Virol*., doi:10.1007/s00705-014-2170-8.

[150] Behle, R.W., Tamez-guerra, P., Mcguire, M.R., (2003). Field activity and storage stability of *Anagrapha falcifera* nucleopolyhedrovirus (af MNPV) in spray-dried lignin-based formulations. *J. Econ. Entomol*. 96, 1066-1075.

[151] Ly-Chatain, M.H., (2014). The factors affecting effectiveness of treatment in phages therapy. *Fron. Microbiol*. 5, 1-7.

[152] Gill, J.J., (2010).Practical and theoretical considerations for the use of bacteriophages in food systems,in *Bacteriophage in the Control of Food- and Waterborne Pathogens*, Vol. 11,eds P. M. Sabourand M. W. Griffiths (Washington: ASM Press American Society for Microbiology), pp. 217-235.

[153] Civerolo, E.L., (1972). Maas Geesteranus HP, editor. Interaction between bacteria and bacteriophages on plant surfaces and in plant tissues. Proceedings of the third international conference of plant pathogenic bacteria, April 14-21 1971, Wageningen. Wageningen: Centre for Argricultural Publishing and Documentation pp. 25-37.

[154] Bergamin, F.A., Kimati, H., (1981). Estudos sobre um bacteriofago isolado de *Xanthomonas campestris*. II. Seu emprego no controle de *X. campestris* e *X. vesicatoria*. Sum.*Phytopath*. 7, 35-43.

[155] Balogh, B., Jones, J.B., Momol, M.T., Olson, S.M., Obradovic, A., (2003). Improved efficacy of newly formulated bacteriophages for management of bacterial spot on tomato. *Plant Dis*. 87, 949-954.

[156] Iriarte, F.B., Balogh, B., Momol, M.T., Smith, L.M., Wilson, M., Jones, J.B., (2007). Factors affecting survival of bacteriophage on tomato leaf surfaces. *Appl. Environ. Microbiol*. 73, 1704–1711.

[157] Bae, J.Y., Wu, J., Lee, H.J., Jo, E.J., Murugaiyan, S., Chung, E.,Lee, S.W., (2012). Biocontrol Potential of a Lytic Bacteriophage PE204 against Bacterial Wilt of Tomato. *J. Microbiol. Biotechnol*. 22, 1613-1620.

[158] Olson, M.R., Axler, R.P., Hicks, R.E. (2004). Effects of freezing and storage temperature on MS2 viability. *J. Virol. Meth.* 122,147-152.

[159] Tey, B.T., Ooi, S.T., Yong, K.C., Tan Ng, M.Y., Ling, T.C., Tan, W.S., (2009). Production of fusion M13 phage bearing the disulphide constrained peptide sequence (C-WSFFSNI-C) that interacts with hepatitis B core antigen. *J. African. Biotechnol.* 8,268-273.

[160] Lopez soto, F.S. M., Allende-Molar, R., Márquez-Zequera, I., Heredia, J.B., León-Félix, J., Valdez-Torres, J.B., Bojórquez-Ramos, C., García-Estrada, R.S., (2014). Agricultural soil supports the survival of *Ralstonia solanacearum* bacteriophage φrss6 with biocontrol potential.*J. of Exp. Biol. and Agri. Sci.* 2, 90-97.

[161] Seaman, P.F., Day, M.J., (2007). Isolation and characterization of a bacteriophage with an unusually large genome from the Great Salt Plains National Wildlife Refuge, Oklahoma, USA. *FEMS Microbiol. Ecol.* 60,1-13.

[162] Iriarte, F.B., Obradovic, A., Wernsing, M.H., Jackson, L.E., Balogh, B., Hong, J.A., Momol, M.T., Jones, J.B., Vallad, G.E., (2012). Soil-based systemic delivery and phyllosphere *in vivo* propagation of bacteriophages two possible strategies for improving bacteriophage persistence for plant disease control. *Bacteriophage.* 2(4), 215-224.

[163] Carlton, R.M., (1999). Phage therapy – past history and future prospects. *Arch. Immunol. Ther .Exp. (Warsz.)* 47, 267-274.

[164] Garcia,P., Madera,C., Martinez,B., Rodriguez,A.,EvaristoSuarez, J., (2009). Prevalence of bacteriophages infecting *Staphylococcus aureus* in dairy samples and their potential as biocontrol agents. *J. Dairy Sci.* 92, 3019-3026.doi: 10.3168/jds.2008-1744.

[165] Moineau, S., Fortier,L.C., (2009). Phage production and maintenance of stocks, including expected stock lifetimes. *Methods Mol. Biol.* 501, 203-219. doi, 10.1007/978-1-60327-164-6-19

[166] Sulakvelidze, A., Alavidze, Z., Morris, J.G.J., (2001). Bacteriophage therapy. *Antimicrob. Agents Chemother.*45, 649-659. doi: 10. 1128/AAC.45.3. 649- 659. 2001.

[167] Ignoffo, C.M., Garcia, C., Saathoff, S.G., (1997). Sunlight stability and rain-fastness of formulations of *Baculovirus heliothis*. *Environ. Entomol.* 26, 1470-1474.

[168] Guerra, P.T., McGuire, M.R., Behle, R.W., Shasha, B.S., Wong, L.J.G., (2000). Assessment of microencapsulated formulations for improved residual activity of *Bacillus thuringiensis*. *J. Econ. Entomol.* 93, 219-225.

[169] McGuire, M.R., Shasha, B.S., Lewis, L.C., Bartelt, R.J., Kinney, K., (1990). Field evaluation of granular starch formulations of *Bacillus thuringiensis* against *Ostrinia nubilalis* (Lepidoptera: Pyralidae). *J. Econ. Entomol.* 83, 2207-2210.

[170] McGuire, M.R., Shasha, B.S., Lewis, L.C., Nelsen, T.C., (1994). Residual activity of granular starch-encapsulated *Bacillus thuringiensis*. *J.Econ. Entomol.* 87, 631-637.

[171] Behle, R.W., McGuire, M.R., Shasha, B.S. (1996). Extending the residual toxicity of *Bacillus thuringiensis* with casein-based formulations. *J. Econ. Entomol.* 89, 1399-1405.

[172] Behle, R.W., McGuire, M.R., Gillespie, R.L, Shasha, B.S. (1997). Effects of alkaline gluten on the insecticidal activity of *Bacillus thuringiensis*. *J. Econ. Entomol.* 90, 354-360.

[173] Balogh, B., (2002).Strategies of improving the efficacy of bacteriophages for controlling bacterial spot of tomato. MS thesis. Univ. FL: Gainesville.

[174] Jones, J.B., Jackson, L.E., Balogh, B., Obradovic, A., Iriarte, F.B., Momol, M.T., (2007). Bacteriophages for Plant Disease Control. *Annu. Rev. Phytopathol.* 45, 245-262.

[175] Balogh, B., (2006).Characterization and use of bacteriophages associated with citrus bacterial pathogens for disease control. Ph.D thesis. Univ. FL: Gainesville.

[176] Svircev, A.M., Lehman, S.M., Kim, W., Barszcz, E., Schneider, K.E., Castle, A.J., (2006). Control of the fire blight pathogen with bacteriophages. *Proc. Int. Symp. Biol. Control Bact. Plant Dis., 1st, Seeheim / Darmstadt, Germany,* 259. Berlin: Dtsch. Bibl., CIP-Einh.aufn.

[177] Chattopadhyay,D.,Chattopadhyay,S., Lyon, W.G., Wilson, J.T., (2002). Effects of surfactants on the survival and sorption of viruses. *Environ. Sci. Technol.* 36, 4017-4024.

[178] Skurnik, M., Pajunen, M., Kiljunen, S., (2007). Biotechnological challenges of phage therapy. *Biotechnol. Lett.* 29, 995-1003.

[179] Carlton, R.M., Noordman, W.H., Biswas, B., De Meester, E.D., Loessner, M.J., (2005). Bacteriophage P100 for control of *Listeria monocytogenes* in foods: Genome sequence, bioinformatic analyses, oral toxicity study, and application. *RTP*.43, 301–312.

[180] Jackson, L.E. (1989).*U.S. Patent No. 4828999.*

[181] Jones, J.B., Iriarte, F.B., Obradovic, A., Balogh, B., Momol, M.T., Jackson, L.E., (2006). Management of bacterial spot on tomatoes with bacteriophages. *Proc. Int. Symp. Biol. Control Bact. Plant Dis., 1st, Darmstadt, Germany,* 408:154. Land-Forstwirtsch: Mitt. Biol. Bundesanst.

In: Microbes in Soil and Their Agricultural Prospects ISBN: 978-1-63482-824-6
Editors: K. Kishore Choudhary and D. Wattal Dhar

Chapter 13

BIOSURFACTANT AND ITS APPLICATIONS IN AGRICULTURE

Nisha Daxini and Kinnari Mistry*
Ashok & Rita Patel Institute of Integrated Studies
in Biotechnology & Allied Sciences
(ARIBAS), New Vallabh Vidhya Nagar – (Gujarat) India

ABSTRACT

The development in agricultural sector to meet the growing demands of human population is a matter of great concern for all countries. New incentives and policies may ensure the sustainability of agriculture and ecosystem services. It will be crucial to meet the demands of improving yields without compromising environmental integrity or public health. Use of green compounds to achieve the sustainable agriculture is the present necessity and increases day by day. Surfactants are molecules that concentrate at interfaces and decrease surface and interfacial tension. Naturally occurring surface-active compounds derived from microorganisms known as biosurfactants. They serve as green surfactant and are attracting attention in recent years because they offer several advantages over chemical surfactants, such as low toxicity, inherent good biodegradability and ecological acceptability. These compounds find applications in an extremely wide variety of agricultural processes involving emulsification, wetting, dispersing or solubilization. In agriculture, surface active compounds are also used for hydrophilization of heavy soils to obtain good wettability and to achieve even spreading of fertilizer in the soil. The present chapter highlights physiological role and uses of biosurfactants in agricultural soil and agrochemical industries.

Keywords: biosurfactant; physiological role; microbial surfactants, agricultural applications

* Corresponding Author address: 1Ashok & Rita Patel Institute of Integrated Studies in Biotechnology & Allied Sciences (ARIBAS), New Vallabh Vidhya Nagar – 388121, (Gujarat) India. E-mail: kinarimistry@aribas.edu.in

INTRODUCTION

Many prokaryotic and eukaryotic microorganisms can grow on compounds that are poorly soluble in water and often associated with the production of surface-active compounds or Biosurfactants. Biological surface-active agents are capable of reducing surface or interfacial tension between liquids, solids and gases, thereby allowing them to mix and disperse readily in water or other liquids. Biosurfactants are amphipathic molecules consisting of a hydrophilic and a hydrophobic moiety that interacts with the phase boundary in heterogeneous systems. The hydrophobic "tail" is typically a hydrocarbon chain whereas the hydrophilic r "head" may be comprised of many different varieties such as carbohydrates, amino acids phosphates or in combination. The classification of biosurfactants is mainly based on their chemical structure and microbial origin. The main classes are glycolipids, phospholipids, polymeric biosurfactants and lipopeptides (e.g., surfactin). The best known glycolipids are rhamnolipids, sophorolipids and trehalolipids. They can also be grouped into two categories; low-molecular mass compounds that lower surface and interfacial tension and high molecular mass compounds that bind tightly to surfaces [1].

Production of biosurfactants is a common feature in microorganisms, and in particularly in plant-associated bacterial species, which include many plant beneficial and plant pathogenic bacteria. Gram positive and gram negative bacteria both can produce different kinds of cyclic lipopeptides [2]. Fluorescent *Pseudomonas* species is a comman one. The synthesis of cyclic lipopeptides is non-ribosomal and catalyzed by large peptide synthetase complexes [3]. *Pseudomonas*-derived biosurfactants are involved in many important bacterial functions. The biosurfactant synthesized from *P. fluorescens* shows antifungal activity against phytopathogenic fungi *Pythium ultimum* and R. solani [4]. Amphisin, a biosurfactnat produced by *P. fluorescens* terminate the growth of pathogen and prevent the plant from infection by pathogen [5]. Biosurfactants can alter the bio-availability of exogenous compounds, as nutrients, to enhance their uptake, and of endogenous metabolites, including phenazine antibiotics, resulting in an increased biological activity. Antimicrobial activity of biosurfactants could play a key role in intraspecific competition, self-defence and pathogenesis. Additionally, biosurfactants specifically bacterial surfactants can affect plants in different ways, either protecting them from disease, or acting as a toxin in a plant-pathogen interaction.

Comparing with chemical surfactants, Biological surface active compounds have many advantages such as lower toxicity, higher biodegradability and diversity, effectiveness at extreme temperatures or pH values and widespread applicability. Additionally, their unique structures provide new properties that classical surfactants may lack. Some advantages are:

Biodegradability

Biological surfactants are easily degraded by microorganism as compared to synthetic surfactants. Research indicated that biosurfactants like rhamnolipid are biodegradable under aerobic and anaerobic conditions, whereas synthetic surfactant Triton X-100 is nonbiodegradable under anaerobic conditions and only partially biodegradable under aerobic conditions [6]. Biodegradation of rhamnolipid in two types of soil (loamy and sandy soil) was relatively low in the first two days of incubation, but sharply increased on the third day and after seven days of incubation 92% of rhamnolipid was degraded in both kinds of soil examined [7]. Currently, the main application of biosurfactant is for enhancement of oil recovery and hydrocarbon bioremediation due to their biodegradability and low critical micelle concentration (CMC) [8].

Low Toxicity

Biosurfactant demonstrate lower toxicity than the chemical-derived surfactants. Biosurfactants , derived from renewable resources are low or nontoxic, biodegradable, demonstrate excellent surface activity, possess high specificity, show effectiveness under extreme conditions, and can be reused through regeneration too as compared to synthetic surfactants [9, 10, 11], hence commercially exploited. It is also reported that biosurfactants showed higher EC 50 (effective concentration to decrease 50% of test population) values than synthetic dispersants [12].

Availability of Raw Materials

Biosurfactants can be produced from cheap raw materials which are available in huge quantities. The carbon source as a substrate may come from carbohydrates, hydrocarbons and/or lipids, which may be used separately or in combination with each other [13]. Where permitted by intended use, biosurfactants can be produced even from industrial wastes and byproducts, which remains a promising area for bulk production [14, 15].

Physical Factors

Many biosurfactants are not affected by environmental factors such as temperature, pH and ionic strength tolerances. Lichenysin produced by *Bacillus licheniformis* strain was not affected by temperature ranges up to 50°C, a pH range of 4-5- 9.0, and NaCl concentration of 50g/l and Ca concentration of 25g/l [16].

Surface and Interface Activity

Mulligan (2005) reported that a good quality surfactant can lower surface tension of water from 75 to 35mN/m and the interfacial tension water/hexadecane from 40 to 1mN/M [17]. Surfactin have the ability to reduce the surface tension of water to 25m N/M and the interfacial tension of water/hexadecane to < 1mN/M [16]. The sophorolipids from *T. bombicola* have been reported to reduce the surface tension to 33 mN/m and the interfacial tension to 5 mN/m . In general, biosurfactants are more effective and efficient and their CMC is about 10–40 times lower than that of chemical surfactants and less surfactant is required to reduce surface tension [12].

Economical and Promising Substrates

Production economy is the major bottle-neck in biosurfactants production, as in the case with most biotechnological processes. For biosurfactants, to compete with synthetic surfactants currently in the market, the cost of the production needs to be decreased. Often the amount and type of a raw material can contribute considerably to the production cost. Thus to reduce this cost, it is desirable to use low-cost raw-materials for the production of desired biosurfactants. Several studies with plant- derived oils have shown that they can act as effective and cheap raw material for biosurfactant production: for example, rapeseed oil and corn oil [18]. Table 1 shows some oil based substrates used for the production of biosurfactants.

Table 1. Oil-based substrates for production of biosurfactant

No	**Substrate**	**Biosurfactant**
1.	Rape-seed oil	Rhamnolipid
2.	Babassu oil	Sophorolipid
3.	Turkish corn oil	Sophorolipid
4.	Sunflower oil	Lipopeptide
5.	Soybean oil	Rhamnolipid
6.	Waste frying sunflower oil	Sophorolipids
7.	Waste frying soybean oil	Rhamnolipid

Mulligan and Gibbs [19] analyzed the economics of biosurfactant production. Many of their conclusions are still relevant today. The aspects included; choosing inexpensive raw materials, increasing yields and production rates, optimizing the fermentation step, reducing product recovery costs, producing biosurfactants for suitable applications.

Other Advantages

Kosaric explained about biocompatibility and digestibility of biosurfactants which allows their application in agriculture, pharmaceuticals and as functional food additives [13]. In addition, Biosurfactants occur naturally in soil, which makes them acceptable from a social and ecological point of view.

Physiological Role of Biosurfactant in Microorganisms

Biosurfactants are produced by a variety of microorganisms; they are secreted either extracellular or attached to the cells predominantly during growth on water immiscible substrates. The major physiological role or mechanism of biosurfactants is to permit microorganisms to grow on water-immiscible substrate by reducing the surface tension at the phase boundary, thus making the substrate more readily available for uptake and also metabolise in adverse environmental situation as shown in Figure 1. Ron, and Rosenberg, recommended that when the surface area becomes limiting, biomass increases arithmetically rather than exponentially, in support that emulsification is a natural process brought about by extra-cellular agents is indirect and there are certain conceptual difficulties to understand how emulsification can offer an advantage for the micro-organisms producing emulsifiers [1].

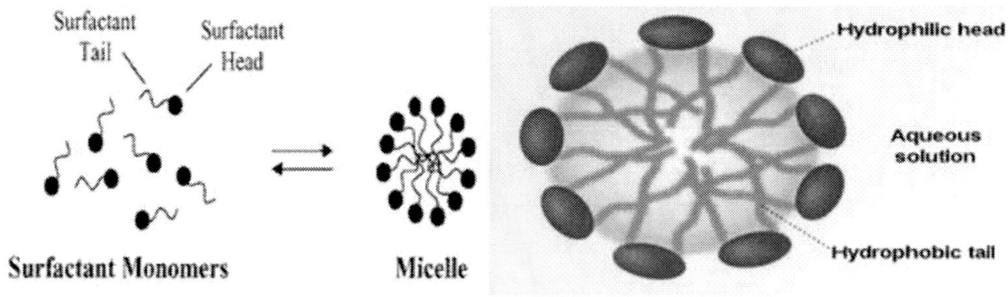

Figure 1. Mechanism of Surfactant.

Another physiological role of biosurfactants is their antimicrobial activities towards various micro-organisms. Usually, different surfactant inhibits different taxonomy. In addition biosurfactants have been shown to be involved in cell adherence which imparts greatest stability under hostile environmental conditions and virulence and in cell desorption when organisms need to find new habitats for survival.

Types of Biosurfactants

A typical biosurfactant is composed of a hydrophilic component (mainly amino acids, peptides or proteins, mono/disaccharides, polysaccharides) and a hydrophobic component (mainly saturated or unsaturated fatty acids, hydroxyl fatty acids or fatty alcohols) [20]. Biosurfactants can be classified mainly by their chemical composition and their origin [21]. Biosurfactants are structurally diverse and can have various chemical structures mainly consisting of glycolipids, lipopeptides, phospholipids, neutral lipids, polymeric surfactants and particulate biosurfactant, depending on the producing microorganism, raw matter and process conditions [22]. Biosurfactants can be grouped into categories according to their origin as plant-derived biosurfactants, animal-derived surfactants and microbially derived surfactants. This review mainly focused on microbial derived surfactants.

Microbial Derived Surfactants

Microorganisms utilize a variety of organic compounds as the source of carbon and energy for their growth. When the carbon source is an insoluble substrate like a hydrocarbon (CxHy), microorganisms facilitate their diffusion into the cell by producing a variety of substances, the biosurfactants. Surfactants derived from microorganisms are surface-active agents synthesized by bacteria, yeasts and fungi to facilitate growth on various substrates like sugars, oils, alkanes and wastes. Rhamnolipids from *Pseudomonas aeruginosa*, surfactin from *Bacillus subtilis*, emulsan from *Acinetobacter calcoaceticus* and sophorolipids from *Candida bombicola* are some examples of microbial-derived surfactants (Table 2).

Table 2. Examples of some biosurfactants produced from microorganisms

No.	Microbial species	Type of surfactant
1.	*Bacillus subtilis*	lipoprotein
2.	*Pseudomonas aeruginosa*	Glycolipid
3.	*Bacillus licheniformis*	Lipoprotein
4.	*Torulopsis bombicola*	Glycolipid
5.	*Pseudomonas* sp.	Lipoprotein
6.	*Pseudomnonas* sp	Rhamnose lipid
7.	*Pseudomonas fluorescens*	Rhamnose lipid
8.	*Arthrobacter* sp.	Glycolipid
9.	*Arthrobacter paraffineus*	Glycolipid
10.	*Torulopsis petrophilurn*	Glycolipid
11.	*Candida tropicalis*	Polysaccharide-fatty acid
12.	*Corynebacterium lepus*	Phospholipid
13.	*Candida petrophilum*	Peptidolipid
14.	*Acinetobacter calcoaceticus*2CAC	Whole cells (lipopeptide)

The increasing interest in microbial derived surfactants is based on their wide diversity in structure and function and on their low cost, which is enabled by their production from cheaper agro-based substrates and waste materials [22]. However, their high surface activity, biodegradability, low or non-toxicity, emulsifying and demulsifying ability, antimicrobial activity, and tolerance to wide ranges of pH, temperature and ionic strength make them promising for agricultural, industrial and pharmaceutical applications.

Applications of Biosurfactants in Agriculture

The market for biosurfactants was 2210 million USD in 2011 and will grow from 2011 to 2018 at a rate of 3.5% (Transparency Market Research) [23]. Surfactants are molecules that intervene in nearly every product and every aspect of human daily life. The major functions of biosurfactants include solubilization, emulsification, dispersion, wetting, foaming, and detergent capacity.

Several Biosurfactants exhibit antibacterial, antifungal and antiviral activities, which make them relevant molecules for applications in agriculture to combat with many diseases and infections. Two cyclic lipopeptides isolated from *Pseudomonas fluorescens* have also shown antifungal and anti-weed activities [24].

Also in plant protection, apart from their general use as a formulation and dispersion aid, certain surfactants are actually the active ingredient. Biosurfactants such as lippopeptides are known to have very high and specific antimicrobial activity against *Alternaria*, one of the most important phytopathogenic fungi. Few applications of biosurfactants in agriculture are highlighted in Figure 2.

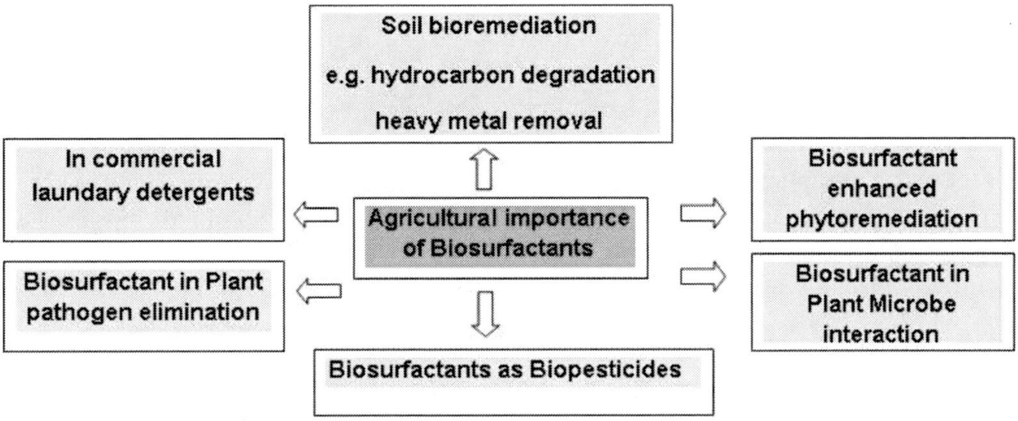

Figure 2. Applications of Biosurfactant in Agriculture.

Biosurfactant and Soil Remediation

The productivity of agriculture land is affected by presence of organic and inorganic pollutants that impart abiotic stress on the cultivated crop plant. To increase the quality of such soil contaminated by hydrocarbon and heavy metals, process of bioremediation is required. The desorption of tightly bound hydrophobic pollutants from soil can be accelerated by biosurfactants. Biosurfactants can be effectively used for hydrocarbon degradation as well as heavy metal removal [25]. Degradation of soil is dependent on the presence of hydrocarbon-degrading species, oxygen availability, hydrocarbon components, water, temperature, pH, inorganic nutrients and also physical state of hydrocarbon. Chemical surfactant or biosurfactant increases solubility and mobility of CxHy, which is necessary for microbial degradation [26]. Oberbremer, et al. reported that addition of biosurfactants such as sphorolipids, increased both the final biomass yield as well as degradation [27]. Several strains of anaerobic bacteria produce biosurfactants reduce surface tension and enhance solubilization of toxic organic chemicals. *Pseudomonas ceparia* AC 1100 produced an emulsifier that formed a stable suspension with 2,4,5-T, and also exhibited some emulsifying activity against chlorophenols. Thus, this emulsifier can be used to enhance bacterial degradation of organo chlorine compounds [28, 29]. Biosurfactants can also improve the degradation of certain chemical insecticides which are accumulated in the agricultural soil [30, 31, 32, 33]. There are several reports which suggest role of biosurfactants in improving the health of agriculture soil by the process of soil remediation. Mata-Sandoval, et al. reported pesticide biodegradation supported by surfactin type of biosurfactant [34].

The heavy metals serve as essential micronutrients and are required for various important physiological processes in plant metabolism. But it can be detrimental to plant growth at higher concentrations causing damage to plant in form of root tissue necrosis and purpling of foliage. Biosurfactants such as rhamnolipid and surfactin are known to remove heavy metals such as Ni, Cd, Mg, Mn, Ca, Ba, Li, Cu, and Zn (ions) from soil with a new method of foaming-surfactant technology [24, 10]. Thus, rather than use of harmful synthetic surfactants, biosurfactants can be the most useful for soil bioremediation [35].

Biosurfactants are said to help microbes adsorb to soil particles occupied by pollutants, thus decreasing the diffusion path length between the site of absorption and site of biouptake by the microorganisms [36]. In agriculture, surfactants are used to prevent the caking of certain fertilizer during storage and promote spreading and penetration of the toxicants in pesticides [12]. The rhamnolipid, mostly produced by the genus *Pseudomonas* is known to possess strong antimicrobial activity. Further, no adverse effects on humans or the environments are anticipated from aggregate exposure to rhamnolipid biosurfactants. Fengycins are also reported to possess antifungal activity and therefore may be employed in biocontrol of plant diseases.

Biosurfactants As Biopesticide

Conventional arthropod control strategy involves applications of broad-spectrum chemicals and pesticides, which often produce undesirable effects. Further, emergence of pesticide resistant insect populations as well as rising prices of new chemical pesticides have stimulated the search for new eco-friendly vector control tools. Biosurfactant like lipopeptides produced by several bacteria exhibit insecticidal activity against fruit fly *Drosophila melanogaster* and hence it is promising to be used as biopesticide [15].

Applications of Biosurfactants in Commercial Laundry Detergents

Surfactant is an important component used in modern day commercial laundry detergents generally chemically synthesized and exert toxicity to fresh water living organisms. Growing public awareness about the environmental hazards and risks associated with chemical surfactants has stimulated the search for ecofriendly, natural substitutes of chemical surfactants in laundry detergents. Biosurfactants such as cyclic lipopeptide are stable over a wide pH range (7.0- 12.0) and also high temperature resistant without affecting its surface-active property [36]. They showed good emulsifying capability with vegetable oils and demonstrated tremendous compatibility and stability with commercial laundry detergents favoring their inclusion in laundry detergents formulation [37].

Biosurfactants in Beneficial Plant Microbe Interaction

Biosurfactant production is a common phenomenon observed in bacteria, and particularly in plant-associated bacterial species. Many rhizosphere and plant associated microbes produce biosurfactant; these biomolecules play vital role in motility, signaling, and biofilm formation, indicating that biosurfactant governs plant-microbe interaction [38]. Bacillus subtilis produce cyclic lipopeptides stimulate plant defense responses, and may be involved in the elicitation of this induced systemic resistance phenomenon. It is reviewed that acyl homoserine lactone (AHL) molecules involved in quorum sensing are required for synthesis of antifungal compounds by the rhizobacteria. Studies also indicate that the concentration of these molecules is high in rhizosphere as compared to that to bulk soil (soil away from plant roots) suggesting the role of AHL-like molecules in rhizosphere competence [39,40]. Dusane, et al. have reported that the biosurfactant (rhamnolipid) produced by *Pseudomonas spp.* regulates the process of quorum sensing (cell to cell communication) [41]. Hence, these green surfactants are important parameters for microbes to achieve a beneficial association with the plant roots and improve the growth of the plant.

Plant Pathogen Elimination

Pathogenic microorganisms affecting plant health are a major and chronic threat to food production and ecosystem stability worldwide. Biosurfactants from microbes have antimicrobial activity against plant pathogens and therefore they are considered to a promising biocontrol molecule for achieving sustainable agriculture. Antibiotic activity of biosurfactants towards microbes could play a role in intraspecific competition, self-defence and pathogenesis. An agricultural application of chemical surfactants and biosurfactants also facilitates biocontrol mechanism of plant growth promoting microbes such as parasitism, antibiosis, competition, induced systemic resistance, and hypovirulence [42]. Additionally, surfactants are used in combination with fungus (*Myrothecium verrucaria*) to eradicate weed species which affect the land productivity and also the spread of weed species have adverse effect on biodiversity [43]. It is reported that biosurfactants are an essential component for inhibition of phytopathogens. The lipopeptide type of biosurfactants produced by *Bacillus* exhibited growth inhibition of phytopathogenic fungi such as *Fusarium* spp., *Aspergillus* spp., and *Biopolaris sorokiniana*. Such biosurfactant can be used as biocontrol agent [44]. Apart from these anti-phytopathogenic properties, addition of biosurfactant is also known to promote compositing process by providing favorable conditions for microbial growth and thus offers an additional advantage of use of these green surfactants. Thus, the biosurfactants play diverse role in plant pathogen elimination directly or indirectly and at different processes related to agriculture [45, 46].

Biosurfactant and Phytoremediation

The use of biosurfactants to enhance the apparent aqueous solubility and bioavailability of heavy metals and organic compounds in soil solution had been well documented. However, there is little information available to elucidate the rate and extent of plant uptake of heavy metals and organic compounds in the presence of biosurfactants. Biosurfactants have shown their potential for remediation of contaminated soils. Both organic and inorganic contaminants can be removed through desorption or biodegradation processes.The efficacy of phytoremediation could be increased by using surfactants such as EDTA, EDDS and LAS to increase organic and inorganic contaminants accumulation by plants. But such synthetic chelating agents at high concentrations can also be toxic to plants. Biosurfactants, such as saponin, surfactin, rhamnolipid, and sophorolipids, because of their high biodegradability, low toxicity to plants, possibility of reuse, had been used to enhance the plants uptake of heavy metals or organic compounds. Hence, the biosurfactant enhanced phytoremediation was considered to be a potential technology for organic and inorganic contaminated soils. Rhamnolipids appeared to be effective in enhancing the uptake of phenanthrene and pyrene into ryegrass within a certain range of concentrations [38, 47, 48].

Conclusion

Over the last few years, the global biosurfactants market has grown incrementally. Increasing awareness among consumers for bio-based products will give rise to new markets for the biosurfactants. Among the various bioactive compounds, biosurfactants or bioemulsifiers are attracting biomolecules due to their structural and functional diversity. The versatile properties of biosurfactant show numerous applications in various agriculture based industries. Consequently, further insight into the roles and activities of surfactants produced by bacteria could provide means to optimize the applications of biological control as an alternative crop protection strategy. Application of microbial biosurfactants in agriculture and agro-industries is anticipated to open up immense opportunities in the next few years. Additionally, due to high biodegradability, low toxicity and environmental compatibility, biosurfactants are very promising to apply to remediation technologies. Nowadays, biosurfactants enhanced phytoremediation also been reviewed and proved to be a promising technology for the removal of contaminants from polluted sites.

Acknowledgment

Authors are grateful to Charutar Vidya Mandal (CVM) Vallabh Vidyanagar, Gujarat for providing platform for this research work. We are also thankful to DR. Nilanjan Roy, director Ashok and Rita Patel Institute of Integrated Study & Research in Biotechnology and Allied Sciences (ARIBAS) New Vallabh Vidya Nagar, for providing the facilities and their valuable suggestions during our research work.

References

[1] Rosenberg, E., Ron, E.Z., (1999). High and low-molecular-mass microbial surfactants. *Appl. Microbiol. Biotechnol.* 52, 154-162.

[2] Katz, E, Demain A.L., (1977). The peptide antibiotics of Bacillus: chemistry, biogenesis and possible functions. *Bacteriol. Rev.*, 41, 449-474.

[3] Marahiel, M.A., (1997). Protein templatesfor the biosynthesis of peptide antibiotics. *Chem. Biol.* 4, 561-567.

[4] Nielsen, T.H., Thrane, C., Nielsen, M.N., Olsson, S., Sorensen, J., (2000). Viscosinamide-producing Pseudomonas fluorescens DR54 exerts a biocontrol effect on Pythium ultimum in sugar beet rhizosphere. *FEMS Microbiol. Ecol.* 1151, 1-8.

[5] Andersen, J.B., Koch, B., Nielsen, T.H., Sorensen, D., Hansen, M., Nybroe, O., Christophersen, C., Sorensen, J., Molin, S., Givskov, M., (2003). Surface motility in Pseudomonas sp. DSS73 is required for efficient biological containment of the root-pathogenic microfungi Rhizoctonia solani and Pythium ultimum. *Microbiol.* 149, 37-46.

[6] Mohan, P.K., Nakhla, G., Yanful, E.K., (2006). Biokinetics of biodegradability of surfactants under aerobic, anoxic and anaerobic conditions. *Water Res* 40: 533-540.

[7] Pei, X., Zhan, X. and Zhou, L. (2009) Effect of biosurfactant on the sorption of phenanthrene onto original and H2O2-treated soils. *J. Env. Sci.* 21 (10), 1378-1385.

[8] Hirata, Y., Ryua, M., Odaa, Y., Igarashia, K., Nagatsukaa, A., Furutaa, T. and Sugiuraa, M. (2009) Novel characteristics of sophorolipids, yeast glycolipid biosurfactants, as biodegradable low-foaming surfactants. *J. Biosci. Bioeng.* 108 (2), 142-146.

[9] Koglin, A., Doetsch, V., Bernhard, F., (2010) Molecular engineering aspects for the production of new and modified biosurfactants. *Adv. Exp. Med. Biol.* 672:158–169.

[10] Xu Q, Nakajima M, Liu Z, Shiina, T., (2011a) Biosurfactants for microbubble preparation and application. *Int. J. Mol. Sci.* 12:462–475.

[11] Lima, T.M., Procópio, L.C., Brandão, F.D., Leão, B.A., Tótola, M.R. and Borges, A.C., (2011b) Biodegradability of bacterial surfactants. *Biodegradation* 22, 585-592.

[12] Desai, J.D., Banat, I.M., (1997). Microbial production of surfactants and their commercial potential. Microbiol. *Mol. Biol. Rev.* 61, 47-64.

[13] Kosaric, N., (2001). Biosurfactants and their Applications for Soil Bioremediation. *Food Biotechnol.* 39, 295-304.

[14] Maneerat, S., (2005) Production of biosurfactants using substrates from renewable-resources. Songklanakarin *J. Sci. Technol.* 27, 675-683.

[15] McInerney, M.J., Javaheri, M., Nagle, D.P., (1990) Properties of the biosurfactant produced by Bacillus liqueniformis strain JF-2. *I. J. Microbiol. Biotechnol.*5, 95-102.

[16] Krishnaswamy, M., Subbuchettiar, G., Ravi, T.K., Panchaksharam, S., (2008). Biosurfactants properties, commercial production and application. *Curr. Sci.* 94, 736-747.

[17] Mulligan, C.N., (2005). Environmental applications for biosurfactants. *Environ. Pollut.* 133, 183-198.

[18] Oliveira, F.J.S., Vazquez, L., de Campos, N.P., de França F.P., (2009). Production of rhamnolipids by a Pseudomonas alcaligenes strain. *Process Biochem.* 44, 383-389.

[19] Mulligan, C.N. and Gibbs, B.F., (1993) Factors influencing the economics of biosurfactants. In *Biosurfactants, Production, Properties and Applications,* Kosaric, N. (ed.) Marcel Dekker Inc., New York, Chapter 13, pp. 329-372.

[20] Makkar, R.S., Cameotra, S.S., Banat, I.M., (2011). Advances in utilization of renewable substrates for biosurfactant production. *AMB Express,* 1, 5.

[21] Muthusamy, K., Gopalakrishnan, S., Ravi, T.K., Swachidambaram P., (2008). Biosurfactants properties, commercial production and application. *Curr. Sci.*, 94, 6.

[22] Syldatk, C., Hausmann, R., (2010). Microbial biosurfactants. *Eur. J. Lipid Sci. Technol.*, 112, 616-617.

[23] Sekhon, K.K., Khanna, S., Cameotra, S.S., (2012) Biosurfactant production and potential correlation with esterase activity. *Petrol and Envir.Biotech.* 3, 11-10.

[24] Van Hamme, J.D., Singh, A., Ward, O.P., (2006). Physiological aspects. part 1 in a series of papers devoted to surfactants in microbiology and biotechnology. *Biotechnol. Adv.*, 24, 604.

[25] Sun, X., Wu, L., Luo, Y., (2006). Application of organic agents in remediation of heavy metals- contaminated soil. *Ying Yong Sheng Tai Xue Bao* 17, 1123-1128.

[26] Morgan, P., Watkinson, R.J., (1989). Hydrocarbon degradation in soils and methods for soil biotreatment. CRC Crit. *Rev. Biotechnol.*, 8, 305-333.

[27] Oberbremer, A., Muller-Hurtig, R., Wagner, F., (1990). Effect of the addition of microbial surfactants on hydrocarbon degradation in a soil population in a stirred reactor. *Appl. Microbiol. Biotechnol.* 32 (4), 485-489.

[28] Awashti, N., Kumar, A., Makkar, R., Cameotra, S., (1999). Enhanced biodegradation of endosulfan, a chlorinated pesticide in presence of biosurfactant. *J. Environ. Sci. Heal.* B 34, 793-803.

[29] Zhang, C., Wang, S., Yan, Y., (2011a). Isomerization and biodegradation of beta-cypermethrin by Pseudomonas aeruginosa CH7 with biosurfactant production. *Bioresour. Technol.* 102, 7139-7146.

[30] Singh, P.B., Sharma, S., Saini, H.S., Chadha, B.S., (2009). Biosurfactant production by Pseudomonas sp. and its role in aqueous phase partitioning and biodegradation of chlorpyrifos. *Lett. Appl. Microbiol.* 49, 378-383.

[31] Sharma, S., Singh, P., Raj M., Chadha, B.S., Saini, H.S., (2009). Aqueous phase partitioning of hexachlorocyclohexane (HCH) isomers by biosurfactant produced by Pseudomonas aeruginosa WH-2. *J. Hazard. Mater.* 171, 1178-1182.

[32] Wattanaphon, H.T., Kerdsin, A., Thammacharoen, C., Sangvanich, P., Vangnai, A.S., (2008). A biosurfactant from Burkholderia cenocepacia BSP3 and its enhancement of pesticide solubilization. *J. Appl. Microbiol.* 105, 416-423.

[33] Neilson, J.W., Artiola, J.F., Maier, R.M., (2003). Characterization of lead removal from contaminated soils by non-toxic soil-washing agents. *J. Environ. Qual.* 32, 899-908.

[34] Mata-Sandoval, J.C., Karns, J., Torrents, A., (2001). Influence of rhamnolipids and Triton X-100 on the biodegradation of three pesticides in aqueous and soil slurries. *J. Agri. Food Chem.* 49, 3296-3303.

[35] Rosenberg, E., Ron, E.Z., (1998). Surface active polymers from the genus Acinetobacter. In: *Biopolymers from renewable resources* (Kaplan, D.L., Ed.), Springer, Berlin, pp. 281-291.

[36] Mukherjee, A.K., (2007). Potential Application of Cyclic lipopeptide biosurfactants produced by Bacillus subtilis strains in laundry detergent formulations. *Lett. Appl. Microbiol.* 45, 330-335.

[37] Das, K., Mukherjee, A.K., (2007). Crude petroleum-oil biodegradation efficiency of Bacillus subtilis and Pseudomonas aeruginosa strains isolated from petroleum oil contaminated soil from north-east India. *Bioresource Technol.* 98, 1339-1345.

[38] Nihorimbere, V., Marc Ongena, M., Smargiassi, M., Thonart, P., (2011). Beneficial effect of the rhizosphere microbial community for plant growth and health. *Biotechnol. Agron. Soc. Environ.* 15, 327-337.

[39] Loh, J., Pierson, E.A., Pierson, L.S., Stacey, G., Chatterjee, A., (2002). Quorum sensing in plant-associated bacteria. *Curr. Opin. Plant Biol.* 5, 1-5.

[40] Newton, J.A., Fray, R.G., (2004). Integration of environmental and hostderivedsignals with quorum sensing during plant–microbe interactions. *Cell Microbiol.* 6, 213-224.

[41] Dusane, D., Rahman, P., Zinjarde, S., Venugopalan, V., McLean, R., Weber, M., (2010). Quorum sensing; implication on rhamnolipid biosurfactant production. *Biotech. Genetic Eng. Rev.* 27, 159-184.

[42] Singh, A., Van Hamme, J.D., Ward, O.P., (2007). Surfactants in microbiology and biotechnology: part 2: application aspects. *Biotechnol. Adv.* 25, 99-121.

[43] Kapadia, S.G., Yagnik, B.N., (2013). Current Trend and Potential for Microbial Biosurfactants. *Asian J. Exp. Biol.* Sci. 4(1), 1-8.

[44] Velho, R.V., Medina, L.F., Segalin, J., Brandelli, A., (2011). Production of lipopeptides among Bacillus strains showing growth inhibition of phytopathogenic fungi. *Folia Microbiol.* (Praha) 56, 297-303.

[45] Zhang, F., Gu, W., Xu, P., Tang, S., Xie, K., Huang, X., Huang, Q., (2011b). Effects of alkyl polyglycoside (APG) on composting of agricultural wastes. *Waste Manag* 31, 1333-1338.

[46] Sachdev, D.P., Cameotra, S.S., (2013). Biosurfactants in agriculture. Appl, Microbiol, Biotechnol. 97(3), 1005-1016.

[47] Zhu, L.Z., Zhang, M., (2008). Effect of rhamnolipids on the uptake of PAHs by rye grass. *Environ.Pollut.* 156, 46-52.

[48] Banat, I., Franzetti, A., Gandolfi, I., Bestetti, G., Martinotti, M., Fracchia, L., (2010) Microbial biosurfactants production, applications and future potential. *Appl, Microbiol, Biotechnol.* 87(2), 427-44.

In: Microbes in Soil and Their Agricultural Prospects
Editors: K. Kishore Choudhary and D. Wattal Dhar
ISBN: 978-1-63482-824-6

Chapter 14

SUITABILITY OF ARBUSCULAR MYCORRHIZAL (AM) FUNGI FOR SUSTAINABLE AGRICULTURE

R. Lakshmipathy[1], V. C. Suvarna[2] and A. N. Balakrishna[2*]

[1] Post Harvest Technology Centre, Bapatla, India
[2] Department of Agricultural Microbiology, University of Agricultural Sciences, G.K.V.K., Bengaluru, India

ABSTRACT

Sustainable farming systems strive to minimize the use of synthetic pesticides and to optimize the use of alternative management strategies to control soil-borne pathogens. AM fungi represent an important component of the soil microbial biomass. They are involved in many bio-geo chemical cycles at plant soil interface. Further, they form the most dominant fungal association in agricultural cropping systems. Hence, they are very much suited to meet the concepts and needs of sustainable agriculture. AM fungi enhance uptake of P and diffusion limited nutrients like Zn and Cu. AM fungi can also enhance tolerance or resistance to root pathogens and abiotic stresses such as drought and metal toxicity. AM fungi in sustainable farming systems reduce the use of agrochemicals to control plant diseases. Bioprotection conferred to plants against many soil-borne pathogens. In addition to increasing the absorptive surface area of their host plant root systems, the hyphae of these symbiotic fungi provide an increased area for interactions with other microorganisms and an important pathway for the translocation of energy-rich plant assimilates to the soil. In this way AM fungi increase the microbial activity stimulated by the leakage and exudation of organic substances from the root. However, the association of mycorrhiza with higher plant roots is commonly observed both under natural and semi natural conditions. Hence, the concept of "rhizosphere" is replaced by "Mycorrhizosphere" to include the fungal component also. The mycorhizosphere enhances the population of N fixing bacteria or P solubilizing bacteria and may have synergistic interaction with AM fungi and thereby benefit plant development and growth. In this chapter, the role of AM fungi and the mechanisms by which it sustains the crop growth is discussed in detail.

[*] Corresponding Author: E-mail – anbalakrishna@gmail.com.

Keywords: AM fungi; abiotic stress; bio-protection; pathogens; plant-microbe interaction; sustainable agriculture

INTRODUCTION

Soil microorganisms have a significant influence on soil fertility and plant health [1]. Symbiotic mycorrhizal fungi, such as arbuscular mycorrhizal (AM) fungi form key components of the microbial populations influencing plant growth and uptake of nutrients. The arbuscular mycorrhizal (AM) fungi are the major components of rhizosphere microflora in natural ecosystems and play a significant role in nutrient cycling. They modify the structure and function of plant communities [2] and are useful indicators of ecosystem change [3]. They are associated with about 80% of the plant families in the world [4]. The fungi responsible for AM formation in plants in majority of the terrestrial ecosystems show great diversity. The conservation and efficient utilization of this diversity are of crucial importance for sustainable plant production systems [5].

The benefits of AM fungi in agricultural ecosystems are now widely known. Increase in P uptake and growth response due to AM fungi is well documented [6, 7]. The increased growth of plants inoculated with AM fungi is not only attributed to improved phosphate uptake but also better availability of other diffusion limited nutrients like Zn and Cu [8].

Sustainable agriculture is the act of farming using principles of *ecology*, the study of relationships between organisms and their environment. The term sustainable farming was coined by the Australian agricultural scientist Gordon Mc Clymont. It has been a widely used phrase ever since. Explaining sustainable agriculture can be understood by examining three goals-improving environmental health, economic profitability and producing a social benefit. One among the natural resources that supply nutrients for plant growth is soil microbiota, thus improving soil health. The AM fungi as explained in the beginning, supply nutrients, produce plant growth hormones and control root pathogens can be better exploited in sustainable agriculture. This review chapter explains about the AM fungi; its biology, classification and current knowledge concerning the interactions between arbuscular mycorrhizal fungi and other microorganisms, particularly bacteria and discusses the role of these interactions on sustainable agriculture.

ARBUSCULAR MYCORRHIZAL FUNGI (AMF) AND ITS CLASSIFICATION

The arbuscular mycorrhizal (AM) fungi belong to the phylum Glomeromycota, class Glomeromyctes. It is divided into four orders Archaeosporales, Diversisporales, Glomerales and Paraglomerales comprising of 10 families and 13 genera [9]. The order Archaeosporales (C. Walker & Schuessler c) contains three Families; Ambisporaceae (Genus: *Ambispora*), Archeosporaceae (Genera: *Archaeospora* J.B. Morton & D. Redecker Intraspora Oehl & Sieverd.) and Geosiphonaceae (Genus: *Geosiphon* (Kütz.) F. Wettst.). The Order Diversisporales C. Walker & Schuessler contains five Families; Acaulosporaceae J.B. Morton & Benny (Genera: *Acaulospora* Gerd. & Trappe emend. S.M. Berch, Kuklospora Oehl &

Sieverd.), Diversisporaceae C. Walker & Schuessler (Genera : *Diversispora* C. Walker & Schuessler and *Otospora* Oehl, J. Palenzuela & N. Ferrol), Entrophosporaceae Oehl & Sieverd.(Genus: *Entrophospora* R.N. Ames &R.W. Schneid. emend. Oehl & Sieverd), Gigasporaceae J.B. Morton & Benny (Genus: *Gigaspora* Gerd. & Trappe emend. C. Walker & F.E. Sanders and *Scutellospora* C. Walker & F.E. Sanders) and Pacisporaceae C. Walker, Blaszk., Schuessler & Schwarzott (Genus: *Pacispora* Oehl & Sieverd.). The Order Glomerales J.B. Morton & Benny consists of one Family; Glomeraceae Piroz. & Dalpé (Genus: *Glomus* Tul. & C. Tul.). The order Paraglomerales C. Walker & Schuessler contains a family Paraglomaceae J.B. Morton & D. Redecker (Genus: *Paraglomus* J.B. Morton & D. Redecker).

The fungi of the order Archaeosporales form endocytosymbioses with photoautotrophic prokaryotes (*Geosiphon pyriformis*) or produce mycorrhizae with arbuscules, with or without vesicles. Their spores are colorless and do not react in Melzer's reagent. Glomoid spores (identical to those of fungi of the genus Glomus) form singly or in clusters on or under the soil surface. Acaulosporioid spores (similar to those of members of the genus Acaulospora) develop singly in the soil. The fungi differ from other arbuscular fungi by the possession of the rRNA SSU gene signature YCTATCYKYCTGGTGAKRCG, corresponding to homologous position 691 of the *Saccharomyces cerevisiae* SSU r RNA sequence J01353, with the coloured nucleotides being specific for the taxon. The order Archaeosporales contains three families, Archaeosporaceae with the genera *Archaeospora* and *Intraspora,* Ambisporaceae with the genus *Ambispora*, and Geosiphonaceae with the genus *Geosiphon*.

Members of the order Diversisporales form mycorrhizae with arbuscules, frequently lacking vesicles with or without auxiliary cells. The spores develop either inside (entrophosporioid spores of the genera *Entrophospora* and *Kuklospora*) or laterally on the neck of a sporiferous saccule (acaulosporioid spores of the genus *Acaulospora* and *Otospora*) from a bulbous base on the sporogenous hypha (gigasporioid spores of the genera *Gigaspora* and *Scutellospora*), or blastically at the tip of a sporogenous hypha (glomoid spores of the genera Diversispora and Pacispora). They differ from other arbuscular fungi by the possession of the rRNA SSU gene sequence signature YVRRYW/1-5/NGYYYGB corresponding to the homologous position 658 of the *Saccharomyces cerevisiae* SSU rRNA sequence J01353 SSU rRNA, GTYARDYHMHYY/2-4/GRADRKKYGWCRAC, corresponding to homologous position of the *S. cerevisiae* SSU rRNA sequence position 1346 of the *S. cerevisiae* SSU rRNA sequence J01353, TTATCGGTTRAATC, corresponding to the homologous position 650 of the *S. cerevisiae* rRNA SSU sequence J01353 and ACTGAGTTMATYT corresponding to the homologous position 1481 of the *S. cerevisiae* rRNA SSU sequence J01353 with the coloured nucleotides being specific for the taxon. The order Diversisporales is represented by five families, Diversisporaceae with the genera *Diversispora* and *Otospora*, *Acaulosporaceae* with the the genera *Acaulospora* and *Kuklospora*, Entrophosporaceae with the genus *Entrophospora*, Gigasporaceae with the genera *Gigaspora* and *Scutellospora* and Pacisporaceae with the genus *Pacispora*.

The fungi of the order Glomerales usually are hypogeous, rarely epigeous. They produce mycorrhizae with arbuscules, vesicles, and spores. The pores form either blastically at the tip of a sporogenous hypha or intercalary inside them. The spores occur singly in clusters or sporocarps having a peridium. They differ from other arbuscular fungi by the possession of the rRNA SSU gene sequence signature YTRRY/2-

5/RYYARGTYGNCARCTT CTTAGAGG GACTATCGGT GTYTAACCGRTGG, corresponding to homologous position 1353 of the *S. cerevisiae* SSU rRNA sequence J J01353 with the coloured nucleotides being specific for the taxon. The order Glomerales includes one genus *Glomus*. The species of the order Paraglomerales form arbuscular mycorrhizae, rarely with vesicles. The spores are glomoid and colourless. The fungi differfrom other arbuscular fungi by the possession of the rRNA SSU gene sequence signature GCGAAGCGTCATGG CCTTAACCGGCCGT corresponding to homologous position 703 of the *S. cerevisiae* SSU rRNA sequence J01353 with the coloured nucleotides being specific for the taxon. The order Paraglomerales is represented by one family, Paraglomeraceae, containing one genus *Paraglomus*. At present, the number of described species of arbuscular fungi may come up to 2700.

IMPORTANCE AND OCCURRENCE

Improved plant growth due to inoculation of soil with AM fungi has been demonstrated especially under P deficient condition [10]. The growth improvement is mainly because of enhanced P uptake. AM fungi can also enhance tolerance or resistance to root pathogens [11] and abiotic stresses such as drought and metal toxicity [12]. There is well documented evidence that VA mycorrhiza have important effects on plant P uptake. Greater soil exploration by mycorrhizal roots as a means of increasing phosphate uptake is well established. In phosphate deficient soils immobile phosphate ions develop a phosphate depletion zone around the roots. The hyphae beyond this zone directly translocate nutrients from the soil to the root cortex [13]. Experiments with ^{32}P labeled phosphate indicate that AM fungal hyphae obtain their extra phosphate from the labile pool rather than dissolving soluble phosphate [14]. Sparingly soluble rock phosphate is better utilized by the hyphae by closer physical contact with the ions dissociating at the particle surface. AM fungi also play a role in the formation of stable soil aggregates, building up of macroporous structure of soil that allows penetration of water and air and prevents erosion [15]. Glomalin, a soil protein produced by arbuscular mycorrhizal fungi plays an important role in soil aggregation. Glomalin in soil was quantified operationally in soil as glomalin related soil protein (GRSP), which ranges as high as several mg g^{-1} soil and GRSP is highly correlated with aggregate water stability [16,17].

The increased growth of plants inoculated with AM fungi is not only attributed to improved phosphate uptake but also to better availability of other elements like Zn, Cu, K, Al, Mn, Fe *etc*. AM fungi affect the levels of plant hormones. It was illustrated that level of plant hormones like cytokinins and gibberellin-like substances will be produced by AM fungi which enhance plant growth [18]. AM fungi can tolerate a wide range of soil water regimes and also improve water relations of many plants. It is still unclear whether the observed effects are directly due to fungus itself or indirectly to some alteration in host physiology as a result of improved P nutrition. Anatomical and other physiological studies have brought out that mycorrhizal plants have increased rates of respiration, photosynthesis and increased amounts of sugars, amino acids, RNA etc. and larger and / or more number of chloroplasts, mitochondria, xylem vessels, motor cells *etc*. Changes in the root exudations and altered rhizosphere microorganisms (which also affect plant growth) may also result because of colonization of roots by AM fungi [19, 20].

Mycorrhizal colonization may also allow introduced populations of beneficial soil organisms like, *Azotobacter, Azospirillum* and phosphate solubilizing bacteria to be maintained in high numbers than around non-mycorrhizal plants and to exert synergistic effects on plant growth. It is apparent from the investigations on AM fungi-plant pathogen interactions that AM fungi can usually (though not always) deter or reduce the severity of disease caused by soil-borne pathogens. All these studies bring out that AM fungi help host plant in more than one way and that AM fungal inoculation helps plant growth.

AM fungi are said to establish a mutualistic relationship with 80% of vascular plants [21, 22, 23]. Plants that rarely form AM fungal association include Caryophyllaceae, Brassicaceae, Chenopodiaceae and Cyperaceae [24]. In addition to their widespread distribution throughout the plant kingdom arbuscular mycorrhizae are ubiquitous and occur in plants grown in arctic, temperate and tropical regions [6]. They have been reported to be associated with plants grown in sand dunes, coal mines [25] and aquatic environment [26]. It was observed variation in AM fungal diversity with the change in plant species [27]. Plants of a particular family are colonized by specific type of AM fungi and a few AM fungal genera were found only in the plants of a particular family.

The AM fungi have the widest host range and distribution of all the mycorrhizal associations. It is estimated that about 90 % of vascular plants normally establish mutualistic relationship with AM fungi. Arbuscular mycorrhizae have been observed in 1000 genera of plants representing some 200 families. There are at least 300,000 receptive hosts in the world flora [28] and there are about 150 species of AM fungi [29]. If the hosts were divided evenly among the fungi with no overlap in host range, each fungus would have more than 2500 potential partners. We know that host range overlap extensively, suggesting that some individual AM fungi may well have access to thousands of hosts [28].

AM fungi, in addition to their widespread distribution throughout the plant kingdom, are also geographically ubiquitous and occur in plants growing in arctic, temperate and tropical regions [6]. As an explanation for their remarkable widespread distribution, it was proposed that AM fungi were disseminated inter-continentally prior to continental drift [22]. The super continent Gondwanaland is thought to have begun to break apart and drift north about 125 million years ago. Fossil records of plants containing AM like structures have occurred as Trappe suggested. In general, population of Am fungi is more in cultivated soil compared to virgin soil [30]. They are mostly seen in the top 15-30 cm of soil and their numbers decrease markedly below the top 15 cm [31]. They are normally not found in depths beyond the normal root range of plants [6]. Although AM fungi are ubiquitous in soils, the patterns of distribution of individual species have not been fully understood. Studies on the distribution of species have either covered large geographical areas [30, 32] or have been limited to smaller regions [33]. The distribution of species of AM fungi varies with climatic and edaphic environment as well as with land use. For example, *Acaulospora laevis* is common in Western Australia [33] and New Zealand [30] but occurs less frequently in soils of Eastern Australia [30].*Glomus* spp. appear to have the widest distribution. *Gigaspora* and *Sclerocystis* spp. are more common in tropical soils. *Acaulospora* seems to be better adapted to soils with pH <5.0. In fact, certain AM fungi have been linked to particular kinds of soil: *Glomus mosseae* with fine textured, fertile high pH soils; *Acaulospora laevis*with coarse textured, acid soils and *Gigaspora* species with sand dune soils [28].

A fundamental problem in studies on the distribution of AM fungi lies in identifying the fungi. It relies almost entirely on spore morphology [32] which may change with the spore age [32, 36]. Description of many of the taxa has been based on single collections of uncertain age. This coupled with the possibility of finding undescribed taxa can make the identification of spores collected from field samples very difficult. Another difficulty in determining the distribution of AM fungi is that spores of all species are not equally easy to recover from soil. Soil often contains spores of more than one species of AM fungi [33]. Moreover, the same root may become colonized by more than one species of AM fungi [34, 35]. Some AM fungi develop characteristic morphological features within plant roots [34, 36]. This can allow quantitative estimates to be made on the total amounts of coarse or fine AM fungal hyphae within plant roots [37, 38]. It is also possible to assess the extent of infection formed by different species of AM fungi with in coarse hyphae occurring in the same soil [39].

Information on the distribution and frequency of occurrence of specific AM fungi in the Western Ghats area is very scarce. However, preliminary surveys performed in Western Ghats of Kerala and Karnataka emphasized the fact that *Glomus* species were dominant in the root zone soils of tree species, thus suggesting their wide distribution [40, 41]. Occurrence of *G. eternicatum* from the root zone soil of cashew for the first time in Western Ghats was reported by [42]. Scientists while studying the distribution of AM fungi in acidic soils of North Eastern India in Assam, Arunachal Pradesh, Manipur, Mizoram and Nagaland found that the hilly soils contained species of AM fungi *viz.*, *Scutellospora nigra, Sclerocystis rubriformis* and *Glomus macrocarpum* [43]. Similarly, the rhizosphere soil samples of 20 species of tree legumes in the reserve forest of Alger hills in the Eastern Ghats of Tamil Nadu were found to contain altogether 21 species of AM fungi belonging to six genera *viz. Acaulospora, Entrophospora, Gigaspora, Glomus, Sclerocystis* and Scutellospora [44]. Spores of five *Glomus* spp., 18 *Gigaspora* spp. and one *Sclerocystis* sp. in barren low vegetation soil with pH ranging from 9.0-10.4 while in the same region with pH 8-10, seven *Glomus* spp., two *Gigaspora* spp. and two *Sclerocystis* spp. were isolated [45].

Arbuscular mycorrhizal associations in ten members of Asteranceae and seven Amaranthaceae growing on infertile, red latertic soils found that all plant species examined were mycorrhizal, but the intensity of colonization varied with the plant species, the highest colonization occurred in Blumea and the lowest in Zinnia [46]. Chlamydospores of *G. mosseae* and *G. macrocarpum var. macrocarpum* were common in the rhizosphere of almost all plants. Azygospores of *Gigaspora margarita var. macrocarpum*, *Sclerocystis, clavispora, S. pakistanica, S. rubiformis* and *S. sinuosa*were also found. Spores of *Glomus, Gigaspora, Sclerocystis* and *Acaulospora* were more abundant in sandy soils as compared to loamy sands. The occurrence of AM fungi is not an exception even to deserts [47]. It was observed the association of AM fungi with 14 plant species in a magnetic mine spoil poor in available P [48]. The spores of *G. fasciculatum* and *Gigaspora giganlea* were the predominant species in those.

It was also observed that AM fungi in about 20 randomly selected trees and found that all tree spp. to be colonized by AM fungi, to varying degrees with maximum colonization in rubber and minimum in sugar apple [49]. AM colonization in 59 forest tree species belonging to 22 families in Karnataka [50]. The intensity of colonization was high in four species, moderate in 23 species and low in 32 species. Therefore, it appears that occurrence of AM fungi in different forest tree species defines ecological niches to that particular tree species, thus determining plant community composition.

ARBUSCULAR MYCORRHIZAL SYMBIOSIS

Arbuscular mycorrhizal symbiosis is the oldest (>460 million years BP) and the most widespread type of mycorrhizal association. It is estimated that 250,000 species of plants worldwide, including many arable crops are capable of forming the symbiosis [51].The plant hosts include angiosperms, gymnosperms and pteridophytes, all having true roots [52]. Approximately 160 fungal taxa of the order Glomales (Glomeromycota) have been described on the basis of their spore morphology [53], although recent molecular analyses indicate that the actual number of AM taxa may be much higher [54,55].

During the formation of AM symbiosis the fungus penetrates the root cortical cell walls and forms haustoria-like structures (arbuscules or coils) that interface with the host cytoplasm [51]. These fungal structures (especially the highly branched arbuscules) provide an increased surface area for metabolic exchanges between the plant and the fungus. Some AM fungi also produce vesicles, which are structures, believed to function as storage organs [51]. It has been estimated that in natural ecosystems plants colonized with AM fungi may invest 10–20% of the photosynthetically fixed carbon in their fungal partners [56]. Clearly, this represents a significant input of energy into the soil ecosystem and this carbon may be crucial to microorganisms associated with the mycorrhizosphere.

Arbuscular mycorrhizal fungi also interface directly with soil by producing extraradical hyphae that may extend several centimeters out into the soil [57]. Extraradical hyphae can have a total surface area of several orders of magnitude greater than that of roots alone, which increases the potential for nutrient uptake, and possibly also water uptake, although results concerning the latter are still conflicting [57, 58]. Hyphae of AM fungi have been shown to play an important role in soil stabilization through formation of soil aggregates [59]. Additionally, the extraradical hyphae are generally believed to be important to the plants for acquisition of phosphorus (P) and other mineral nutrients [60]. The extraradical mycelium of AM fungi can also enhance mobilization of organically bound nitrogen from plant litter [61].

Arbuscular mycorrhizal symbiosis can also alleviate negative effects of plant pathogens [62, 63, 64, 65] and toxic levels of metals [66]. In addition, the extraradical hyphae may interact with other soil organisms either indirectly by changing host plant physiology, including root physiology and patterns of exudation into the mycorrhizosphere or directly by physically and/or metabolically interacting with other organisms in the mycorrhizosphere.

EFFECTS OF AM FUNGI ON SOIL BACTERIA

Colonization of plant roots by AM fungi can affect bacterial communities associated with the roots in both direct and indirect ways. Direct interactions include provision of energy-rich carbon compounds derived from host assimilates, which are transported to the mycorrhizosphere via fungal hyphae, changes in pH of the mycorrhizosphere induced by the fungus, competition for nutrients and fungal exudation of other inhibitory or stimulatory compounds. Indirect interactions can also take place in the form of mycorrhiza-mediated effects on host plant growth, root exudation and soil structure.

Using a simple dilution plate technique, it was demonstrated that a single bacterial isolate significantly increased in the mycorrhizosphere of *Glomus mosseae* inoculated plants compared to the rhizosphere of non-mycorrhizal plants, even though AM colonisation never exceeded 5.5 per cent [67]. It was also demonstrated selective differences in populations of naturally occurring taxonomic and functional groups of bacteria in the rhizosphere and the rhizoplane of mycorrhizal and non-mycorrhizal sweet corn (*Zea mays*) and subterranean clover (*Trifolium subterraneum*) plants [68]. The total bacterial populations in the rhizosphere of guinea grass (*Panicum maximum*) were higher in plants colonized by *Glomus fasciculatum*, *Gigaspora margarita* and *Sclerocystis dussi* than in non-mycorrhizal plants [69]. These authors suggested that increased amino acid exudation by P-deficient mycorrhizal plants might have stimulated the growth of amino acid-requiring bacteria.

Organic compounds produced by extraradical hyphae and the hyphae themselves play a role in aggregation of soil particles [59], which could provide microsites for microbial colonization and growth. The microbial composition of such aggregates was analyzed and identified a range of bacteria, actinomycetes and algae [70]. Compartmentalized systems in which roots and hyphae were separated by fine mesh in order to investigate the qualitative and quantitative effects of AM fungi on microbial communities in the mycorrhizosphere and the stability of soil aggregates associated with it [73]. They found that total bacteria and P-solubilizing bacteria isolated from the water-stable soil-aggregate (WSA) fraction tended to be more numerous than from the unstable fraction. Increases in WSA in mycorrhizal soybean (*Glycine max*), were dependent on the mycorrhizal fungal species involved [72]. Differences in the bacterial communities were also found among the fungal species, suggesting that AM species may influence both WSA and bacterial composition.

Bacteria associated with the mycorrhizosphere and the hyphosphere of the AM fungal species were *Glomus etunicatum*, *G. intraradices* and *G. mosseae* [71]. The observed changes in the bacterial community in the hyphosphere were not due to the amount of AM mycelium per se, suggesting that qualitative effects (*e.g.*, composition of exudates) of the fungal species on the hyphosphere are more important to the composition and proliferation of rhizobacteria than the quantitative development of AM mycelia in the soil.

While, many studies have demonstrated qualitative and quantitative effects of AM fungi on bacterial communities, the underlying mechanisms are unclear. There are still few or no studies concerning the total amounts or spectrum of compounds released from AM mycelium. Further experiments employing spatial compartmentation of roots and mycelium are needed to distinguish between direct and indirect effects of mycorrhiza on bacterial populations. Although, changed bacterial community structure has been shown in association with the mycorrhizosphere, there are no direct demonstrations of compounds produced by AM fungi, which stimulate or inhibit bacteria. There is some evidence from studies using PLFA (phospholipid fatty acid) analysis and BIOLOG™ that the effect on bacterial populations is influenced by the mycorrhizal dependency of plants [74], however, the issue of culturability remains a general problem in studies on bacterial community structure.

ROLE OF AM FUNGI IN BIO-CONTROL

AM fungi may also interact with other root-associated microorganisms, such as pathogenic fungi. The possible mechanisms of interaction are the same as those mentioned in the previous section. The differential effects of a crude extract from the growth medium of the

AM fungus *Glomus intraradices* on the sporulation of two pathogenic fungi and on the growth of two bacterial species were investigated under *in vitro* conditions [75]. Conidial germination of the mycoparasitic fungus *Trichoderma harzianum* and the growth of *Pseudomonas chlororaphis* were stimulated, whereas conidial germination of the plant root pathogen *Fusarium oxysporum* was reduced, and the growth of *Clavibacter michiganensis* was unaffected. The measured effects were correlated with extract concentration and no significant influence of pH on growth or germination was detected. The authors concluded that the release of unspecified substances by the AM fungus into the growth medium was the main factor explaining the differential growth of the tested microorganisms. Bacteria isolated from 17 year old *G. mosseae* pot cultures were screened [76]. They found that many of the bacterial isolates within the different zones of the mycorrhizosphere were actively antagonistic against *in vitro* growth of the soil-borne pathogens, *Fusarium* and *Phytophthora*. Their results also suggest the possibility of integrated use of AM fungi and their associated bacteria in biological control of soil-borne pathogens.

Many authors have suggested that the ability of AM-colonized plants to better withstand an attack from root pathogens can be ascribed to an increased nutritional status in the host plant due to the presence of the AM fungus. However, there are reports which contradict this theory. In field experiments, transplanted *Glomus* sp.-inoculated and non-inoculated seedlings of the annual grass *Vulpia ciliata* into a natural population, and found that AM inoculation did not affect P concentrations in the plants [65]. However, the mycorrhiza protected the plants from the deleterious effects of *Fusarium oxysporum* infection on shoot and root growth. Apparently, the AM suppressed pathogen development in the roots. Following transplantation, comparison of root-infecting mycofloras of AM and non-AM plants revealed that AM plants had fewer naturally occurring infections of *Fusarium oxysporum* and *Embellisia chlamydospora* [65]. It was proposed that the main benefit supplied by AM fungi to *V. ciliata* is the protection from pathogenic fungi, rather than improved P uptake. This theory was also proposed by other scientists [62]. Peat based medium containing the AM fungus *G. intraradices* to test whether it could suppress the tuber dry rot (*Fusarium sambucinum*) in minitubers of potato (*Solanum tuberosum*) [62]. Minitubers grown in this medium had significantly less (20–90%) tuber dry rot. They were also able to demonstrate these effects in a high-input commercial greenhouse, despite the fact that AM colonisation was very low and no evidence of enhanced plant P nutrition could be found. In addition, it was demonstrated that the presence of *Tagetes patula* plants colonized by the AM fungus *G. intraradices* can inhibit root pathogen development in soil and thereby reduce disease severity in co-cultured non-mycorrhizal carnation (*Dianthus caryophyllus*) [77]. In other experiments, reduction in *Fusarium* populations was observed in the soil surrounding mycorrhizal tomato (*Lycopersicon esculentum*) roots, and suggested that there was a potential role for AM fungi in biocontrol of soil-borne diseases [78].

Effect of AM Fungi on N-Transforming Bacteria

The presence of AM fungi is known to enhance nodulation and N fixation by legumes. Mycorrhizal and nodule symbioses often act synergistically on infection rate, mineral nutrition and plant growth [79]. The increased P uptake conferred by the AM symbiosis is beneficial for the functioning of the nitrogenase enzyme of the bacterial symbiont, leading to increased N fixation and consequently promotion of root and mycorrhizal development [80]. Response of other N-transforming microorganisms to two different *Glomus* species was studied [79]. The occurrence of autotrophic nitrifying bacteria in pot cultures of sweet corn colonised by the AM fungi *G. mosseae* and *G. fasciculatum* was significantly higher than in non-mycorrhizal cultures, whereas ammonifying and denitrifying bacterial populations significantly decreased in pot cultures of mycorrhizal plants. Evidently, the presence of AM fungi can modify populations of N-transforming microorganisms and these interactions may affect nutrient availability in soils [79].

Although qualitative effects of AM fungi on fungal pathogens have been repeatedly demonstrated, there is still a lack of information concerning direct effects on these pathogens. There are few or no studies on interactions of AM fungi with bacterial pathogens and as in the case of fungal pathogens, it is often difficult to distinguish between direct effects on the pathogens and indirect effects brought about by the improved nutritional status of the mycorrhizal plants.

EFFECTS OF MYCORRHIZOSPHERE BACTERIA ON AM FUNGI

Mycorrhizosphere bacteria may affect AM fungi and their plant hosts through a variety of mechanisms. Some of these have been more fully studied in ectomycorrhizal fungi [81], but possibilities include effects on (1) the receptivity of the root (2) the root-fungus recognition (3) the fungal growth (4) modification of the chemistry of the rhizospheric soil and (5) the germination of the fungal propagules.

In an early study [82], it was observed that glomalean spores germinated better on agar when bacteria were present. Later, it was shown that *Glomus versiforme* germinated in non-sterile soils, but not in autoclaved or irradiated soils [83]. Germination was also enhanced following addition of kaolin or activated charcoal, indicating that the spores contained self-inhibitors, which were possibly inactivated by soil bacteria, or were immobilized by substances with a high ion exchange capacity. Similarly, surface-sterilized spores of *G. versiforme* germinated less frequently than non-surface sterilized spores [84]. Isolation of bacteria from these spores showed that several genera, including *Pseudomonas* and *Corynebacterium*, enhanced spore germination. The stimulatory effects of actinomycetes and *Streptomyces orientalis* on *Gigaspora margarita* spore germination was tested and found that amounts of volatile compounds produced by the isolates correlated well with AM spore germination [85]. Conversely, other studies using pasteurization, fumigation or sterilisation of soils have demonstrated that the presence of some soil bacteria may also inhibit spore germination [86] or AM fungi sporulation [87, 88].

Nitrogen fixing bacteria clearly have the potential to influence AM fungi. The presence of genes for N fixation has been shown in endosymbiotic *Burkholderia* sp. [89], but expression of this activity at levels significantly influencing the growth of the mycorrhizal association has yet to be demonstrated. *Rhizobium* spp. may act synergistically with AM fungi on their plant hosts. Nodulation and N fixation are commonly increased in legumes following AM colonization, probably because the mycorrhiza supplies the plant and the rhizobacteria with P, which is essential for the enzymes involved in the N fixation process.

Nitrogen fixation further promotes mycorrhizal development [81]. Some mycorrhizosphere bacteria may be able to promote mycorrhizal establishment through improved spore germination [82], but so far there are no direct demonstrations of this in the field. The colonization enhancement may also be mutual between associated microorganisms and this has been reported following dual inoculation of *Pseudomonas* sp. and *Glomus* sp., which additionally increased the growth of the host plant in an additive manner [90].

A central problem in characterizing effects of bacteria on AM fungi is the difficulty of identifying the active bacteria against a biological background of immense diversity. Clearly, mycorrhizosphere bacteria have a number of potential effects on AM fungi. Remaining challenge is to determine whether effects demonstrated under laboratory conditions can be reproduced in the field.

Effect of Agricultural Practices on AM Fungal Population and Diversity

Modern agricultural practices to enhance food production to meet the needs of increasing human population are posing problems to AM fungi. The agricultural intensification declined the AM fungal abundance and effectiveness with respect to root colonization and plant growth promotion. The important agricultural practices commonly followed are cropping pattern, crop rotation, tillage operations, organic amendments, seasons, fertilizer application, weeding, crop protection e*tc*.

Monocropping with a particular crop results in the development of a predominant AM fungus in soil. Continuous cultivation of maize in Philadelphia, USA, for three years resulted in the development of *Gigaspora gigantia* in a soil [91]. Mixed cropping is common practice in the tropics. In an experiment on the effect of mono and mixed cropping with soybean and maize on AM fungal population in soil, mixed cropping stimulated the proliferation of AM fungi compared with monocropping with maize or soybean [92]. Soybean being a legume possibly provides nitrogen to maize through AM fungi. It was found that mycorrhizal root infection of cassava increased by intercropping with legumes [93]. One reason for the higher propagule density under mixed cropping may be the more intensively rooted soil in the mixed system. Additionally, through higher plant density, nutrients are extracted faster from the soil, thereby stimulating AM fungal reproduction. Mixed cropping is a common practice in the tropics. Mixed cropping of maize and soybean was found to increase the proliferation of AM fungi compared to monocropping of maize or soybean [92]. Similar results were also observed wherein AMF infective propagules increased in ragi-groundnut-ragi crop compared to monocropping with ragi [94].

Graminaceous and leguminous crops are generally believed to increase AM fungal population, while non-mycotrophic plants decrease the population of mycorrhizal fungi [93]. Taking non-mycorrhizal hosts like mustard or leaving the land follow will reduce the propagules of AM fungi in soil [95]. In contrast, use of a crop, which is strongly mycorrhizal, will increase their numbers. An experiment carried out in a P deficient soil where in the first season, finger millet was grown in all the plots and in the second season, a mycorrhizal host (cowpea) was grown in two thirds of the plot and rest was left fallow. In the third season, cowpea was grown in all the plots. The results showed considerable reduction in the mycorrhizal propagules in the soil left fallow [95]. Growing a non-mycorrhizal host significantly reduced the native mycorrhiza but the reduction was not as bad as when the land was left fallow. A mycorrhizal host taken in the third season resulted in a slow build up of mycorrhizal population in the soil. However, at the end of the third season it did not reach the same level as in plots cropped continuously with mycorrhizal host. This suggests the reduction in mycorrhizal population caused by leaving the land fallow or growing a non-mycorrhizal host. It may take at least two cropping seasons with mycorrhizal host to rebuild the reduced mycorrhizal population to the original level. Therefore, it is best to grow a variety of crops in rotation. Further, some plants do not get colonized by AM fungi and therefore will depress populations of these fungi.

Weeds can act as kind of an instantaneous crop rotation. Since the diversity of the AM fungus community can be proportional to the associated plant community, strict and complete weed control decreases the diversity and efficacy of the indigenous community of AM fungi. Earlier studies have shown that less tillage of soil is better for the buildup of mycorrhizal population [96]. Less disturbance of soil with ridge tillage resulted in more mycorrhizal population compared with mould board ploughing. Studies have shown that less tillage of soil is better for the buildup of mycorrhizal population [96]. Disturbance of soil with ridge tillage resulted in more mycorrhizal population compared with mould board ploughing. Further, a cover crop of hairy vetch planted after harvest of winter wheat became significantly more colonized by AM fungi in non-tilled soil where hyphal networks were intact than in plots subjected to mould board tillage. Non tilled soils may have more mycorryzal spores in the top 8 cm of soil while tilled soils may have more at 8-15 cm depth. Further, a cover crop of hairy vetch planted after harvest of winter wheat became significantly more colonized by AM fungi in no tilled soil, where hyphal networks were intact than in plot subjected to mould board plough tillage [7]. The mycorrhizal hyphae in the soil act as the nutrient-absorbing organ of the mycorrhiza and the way in which new roots are colonized. Tillage disrupts both of these functions. On the other hand, seedlings grown in no tilled soils become colonized by AM fungi more rapidly and have greater phosphorus status than those grown in tilled soils.

Another practice that has negative impact on AM fungi is over winter bare fallow. This removes potential host roots from which the fungi can receive sugar during mild fall and spring weather, thereby decreasing viability and ability of the fungi to colonize the next crop. An over winter cover crop may not only be useful for the mycorrhizal fungi but will also boost the amount of AM fungi in the soil. Studies conducted in temperate countries revealed that AM fungi sporulate during summer with higher temperature and longer day length [98, 99]. Experiments conducted revealed that AM fungi sporulate during winter in the tropics [95,100]. This is probably because the optimum soil temperature for sporulation occurs during winter in the tropics while it occurs during summer in the temperate countries. The optimum temperature for sporulation by mycorrhizal fungi appears to be around 25^0 C.

In tropical soils, application of organic matter either in the form of FYM, compost or organic amendments stimulates proliferation of AM fungi [101]. This is probably because of the low organic matter content in tropical soils. Addition of organic amendments such as paddy straw, maize straw and pongamia leaf increased the mycorrhizal activity [95]. Of the three amendments studied, the addition of pongamia leaf encouraged AM fungi to the maximum, followed by maize straw. In another study, it was observed that the application of maize residue plus inorganic fertilizers to the fields over the years stimulated AM colonization [94].

For modern agriculture, fertilizer application is an essential and often the most promising method to increase crop production in infertile soils. In tropical soils, phosphorus is one of the most limiting elements for crop production. High P availability is reported to be negatively correlated with AM fungal activity [102]. Apparently, the internal P content of plants regulates AM fungal infection and reproduction [103]. Tissue P concentration is also not always a good estimate for mycorrhizal colonization, because the mycorrhizae themselves influence the factor. It is likely that P influences AM colonization by affecting concentrations of root carbohydrates or the amount of root exudates. The content of P in plants at the time of AM colonization is the best indicator to identify a soil, which provides good AM colonization [104]. The rock phosphate applied at 100 ppm P level resulted in more infective propagules of *Glomus fasciculatum* [105]. Rock phosphate encouraged proliferation of AM fungi, compared with bone meal and super phosphate [106]. The research findings show that years of P fertilization can lead to very high soil P levels. Plants that are able to absorb sufficient P via their roots alone in high nutrient soils inhibit the spread of colonization by the fungus. This reduces the flow of sugars to the fungus, which lessens the amount of AM fungi in soil. Hence, low or no P fertilization is necessary in such soils. Addition of phosphatic fertilizer decreased the AM fungal formation compared to no added P control and P supplied through organic matter. Even disturbance of soil pre-established with AM fungi affected its further establishment [107].

Application of heavy doses of nitrogen fertilizers (188 kg N per hectare per year) can have a large negative effect on AM population [108] and nitrate source of nitrogen has been shown to be inhibitory to AM development than ammonium salts [109, 110]. Regular fertilization of citrus with more than 100ppm N as a mixture of NO_3 and NH_4 retarded mycorrhizal development [109]. Among the salts, calcium nitrate, urea and calcium ammonium nitrate at different levels were compared and it was found that calcium ammonium nitrate applied at 80 ppm N level to soilrite: perlite mix substrate with negligible N produced maximum number of infective propagules of *Glomus fasciculatum,* in association with Rhodes grass [110]. Levels greater than 80 ppm N decreased the number of infective propagules. These results conclude that nitrogen content in soil could greatly influence the distribution and abundance of AM fungi. Potassium content of acidic tropical soils is generally low, and sustainable crop yields depend on K application. Cassava has a very high K demand. About 5.8 kg has to be applied for each tonne of cassava yield [111]. It was found that increasing K application levels up to 200 kg K / ha increased mycorrhizal root infections as well as tuber yield in Columbia.

Most pesticides inhibit colonization and development of AM fungi in plants [112]. Although the majority of pesticides tested adversely affect the symbiosis [113], others do not appear to damage mycorrhizal fungi. Some may even increase mycorrhizal colonization [114]. Fumigation of soil with biocides such as methyl bromide, chloropicrin, etc. effectively kills endophytes. However AM fungi can reinvade most fumigated soils within several years [114]. The systemic fungicides like thiobendazole, benomyl and triademefon are most toxic to these fungi [115]. Pentachloro nitrobenzene which is not systemic is also toxic [116]. Fungicides like Captan and Rilon applied at half the recommended level had no adverse effect on AM fungi [117]. The insecticide metasystox and aldrin differ in their activity on AM fungi, the former being less toxic than the latter whilst most nematicides exhibited or slight inhibitory effect on AM fungi. Interestingly some nematicides such as 1, 2- dibromo-3-chloropropane (DBCP) could stimulate root infection in host plants [103]. This response could be due to control of competitive and pathogenic microflora and possible stimulation of root exudates of host plants or other factors [114]. The results of the different studies were contradictory; some shows that, pesticides affect AM colonization but others favour AM fungi.

CONCLUSION

The AM fungi benefit the plants by (1) mobilizing the nutrients and water (b) producing growth promoting hormones (3) protecting the plants against pests and diseases. These characters have made them, as essential components of sustainable agriculture. Its capacity of nutrient mobilization has been very well proved but to establish it as a biocontrol agent, requires further detailed studies.

REFERENCES

[1] Gianinazzi, S., Schüepp, H., (1994). Impact of arbuscular mycorrhizas on sustainable agriculture and natural ecosystems. Birkhäuser Verlag Basel. p. 226.

[2] Douds, D.D., Millner, P., (1999) Biodiversity of arbuscular mycorrhizal fungi in agroecosystems. *Agric. Ecosyst. Environ.* 74: 77-93.

[3] McGonigle, T.P., Miller, M.H., (1996) Development of fungi below ground in association with plants growing in disturbed and undisturbed soils. *Soil Biol. Biochem.* 28, 263–269.

[4] Giovannetti, M., Sbarna, C., (1998). Self and Non self responses in hyphal tips of arbuscular mycorrhizal fungi. In: *Cell Biology of Plant and Fungal Tip Growth.* (Geitmann, A., Ed.), IOS Press, Amsterdam, The Netherlands, pp. 221-231.

[5] Giovannetti, M., Gianinnazzi-Pearson, V., (1994). Biodiversity in arbuscular mycorrhizal fungi. *Mycol. Res.* 98, 705-715.

[6] Mosse, B., Stribley, D.P., Le Tacon, F., (1981). Ecology of mycorrhizae and mycorrhizal fungi. *Adv. Microbial Ecol.* 5, 137-209.

[7] Doss, D. D., Bagyaraj, D. J., (2001) Ecosystem dynamics of mycorrhizae. In: *Innovative Approaches in Microbiology* (Maheswari D.K., Dubey, R.C., Eds.), Bishen Singh Mahendra Pal Singh Pub., Dehra Dun, India. pp.115-129

[8] Krishna, K.R., Bagyaraj, D.J., (1991). Effect of vesicular-arbuscular mycorrhiza and soluble phosphateon *Abelmoscusesculentus* (L.) Moench. *Plant Soil* 64, 209-213.

[9] Palenzuela J., Ferrol N., Boller T., Azcón-Aguilar C., Oehl, F., (2008). Otospora bareai, a new fungal species in the Glomeromycetes from a dolomitic shrubland in the Natural Park of Sierra de Baza (Granada, Spain). *Mycologia* 100, 296-305

[10] Mosse, B., (1973). Advances in the study of vesicular-arbuscular mycorrhiza. *Annu. Rev. Phytopathol.* 11,171-196.

[11] Borowicz, V.A., (2001). Do arbuscualar mycorrhizal fungi alter plant pathogen relations? *Ecol.* 82, 3057-3068.

[12] Meharg, A.A., Cairney, J.W.G., (2000). Co-evolution of mycorrhizal symbionts and their hosts to metal-contaminated environments.*Adv. Ecol. Res.* 30, 69-112.

[13] Hayman, D.S., (1983). The physiology of vesicular arbuscular endomycorrhizal symbiosis. *Can. J. Bot.* 61, 944-963.
[14] Raj. J., Bagyaraj, D.J., Manjunath, A., (1981). Influence of soil inoculation with vesicular arbuscular mycorrhiza and a phosphate dissolving bacterium on plant growth and ^{32}P-uptake. *Soil Biol. Biochem.* 13, 105-108
[15] Miller, R.M., Jastrow, J.D., (1992). The application and reclamation. In: *Mycorrhizal functioning* (Allen, M.F., Ed.), Chapman and Hall Ltd., London, England, pp. 214-229
[16] Rilling, M.C., Ramsey, P.W., Morris, S., Paul, E.A., (2003). Glomalin, an arbuscular mycorrhizal fungal soil protein, responds to land use changes, *Plant Soil.* 253, 293-299.
[17] Rilling, M.C., (2004). Arbuscular mycorrhizae, glomalin and Soil aggregation. *J. Soil. Sci.* 84, 355-363.
[18] Allen, M.F., Moore, T.S., Christensen, M., (1982). Phytohormone changes in *Boutelina gracilis* infected by vesicular arbuscular mycorrhizae. II .Altered levels of gibberillin-like substances and abscisic acid in the host plant. *Can. J. Bot.* 60, 468-471
[19] Mada, R.J., Bagyaraj, D.J., (1993). Root exudations from *Leuceana leucocephala* in relation to mycorrhizal colonization. *World J. Microbiol. Biotechnol.* 9, 342.
[20] Caroline Machado, Bagyaraj, D.J., (1995). Mycorrhization bacteria and its influence on growth of cowpea. In: *Mycorrhizae, Biofertilizers for the future* (Adholeya, A., Singh, S., Eds.), Tata Energy Research Institute Pub., New Delhi, pp. 192-196.
[21] Harley, J.L., Smith, S.E., (1983). *Mycorrhizal symbiosis.* Academic press, London, U.K.
[22] Trappe, J.M., (1977). Biogeography of hypogeous fungi: Trees, mammalsand continental drift. Abstr. 2nd Internatl. Mycol. Congr. (Bigelow, H.E., Simmons, E.G., Eds.) University of South Florida, Tampa, pp. 675.
[23] Gianinazzi, S., Gianinazzi-Pearson, V., (1986). Progress and headaches in endomycorrhiza biotechnology. *Symbiosis* 2, 139-149.
[24] Hirrel, M.C., Mehravaran, H., Gerdemann, J.W., (1978). Vesicular arbuscular mycorrhizae in Chenopodiaceae and Cruciferaceae:do they occur?. *Can. J. Bot.* 56, 2813-2817.
[25] Khan, A.G., (1978). VA mycorrhizae in plants colonizing black wastes from bituminono coal wining in Illwara ware regron of New South Wales. *New Phytol.* 81, 53-63.
[26] Bagyaraj, D.J. Manjunath, A., Patil, R.B., (1978). Occurrence of vesicules arbuscular mycorrhiza in some tropical aquatic plants. *Trans. Br. Mycol. Soc.* 72, 164-167.
[27] Blaszkowski, J., (1994). Comparative studies on the occurrence of arbuscular fungi and mycorrhizae (Glomales) in cultivated and uncultivated soils of Poland. *Acta Mycol.* 28, 93-140.
[28] Kendrick, B., Berch, S., (1985). Mycorrhizae: Applications in agriculture and forestry. In: *Comprehensive Biotechnology*, (Robinson, C. W., Ed.) Pergamon Press, Oxford, pp. 109-150.
[29] Schenk, N.C., Perez, Y., (1999). *Manual for Identification of VA Mycorrhizal Fungi.* 3rdEdition, Synergistic Publications, Gainesville, Florida, USA.
[30] Mosse, B., Bowen, G.D., (1968). The distribution of *Endogone* spores in some Australian and New Zealand soils and in the experimental field soil at Rothamsted. *Trans. Br. Mycol. Soc.* 51, 485-492.

[31] Redhead, J.F., (1977). Endotrophic mycorrhizae in Nigeria; Species of the Endogonaceae and their distribution.*Trans.Br. Mycol. Soc.* 69, 275-280.

[32] Hall, I.R.S., (1977). Species and mycorrhizal infections of New Zealand Endogonaceae. *Trans. Br. Mycol. Soc.* 68, 341-356.

[33] Abbott, L.K., Robson, A.D., (1977). The distribution and abundance of vesicularendophytes in some western Australian soils. *Aust. J. Bot.* 25, 515-522.

[34] Cooper, K.M., (1976). A field survey of mycorrhizas in New Zealand ferns. *New Zealand J. Bot.* 14, 169-181.

[35] Mosse, B., (1977). Plant growth responses to vesicular –arbuscular mycorrhiza. X. Responses of Styloxanthus and maize to inoculation in unsterile soils. *New Phytol.* 78, 277-288.

[36] Abbott, L.K., Robson, A.D., (1979). A quantitative study of the spores and anatomy of mycorrhizas formed of *Glomus*, with reference to its taxonomy. *Aust. J. Bot.* 27, 363-375.

[37] Sparling, G.P., Tinker, P.B., (1978). Mycorrhizal infection in Penning grassland. II. Effects of mycorrhizal infection on growth of some upland grasses on irradiated soil. *J. Appl. Ecol.* 15, 951-958.

[38] Powell, C.L. (1979). Spread of mycorrhizal fungi through soil. *New Zealand J. Agric. Res.* 22, 335-339.

[39] Abbott, L.K., Robson, A.D., (1981). Infectivity and effectiveness of five endomycorrhizal fungi: Competition with indigenous fungi in field soils. *Aust. J. Bot.* 32, 621-630.

[40] Muthukumar, T.K.V., Manian, S., (1993). Vesicualar arbuscular mycorrhizal fungi Western Ghats. *Indian Bot. Contractor.* 10, 79-83.

[41] Vasanthakrishna, M., Muthanna, M.B., Bagyaraj, D.J., (1994). Succession of vesicular arbuscular mycorrhizalfungi associated with *Casuarina equisetifolia L. Ann. For.* 2, 123-126.

[42] Lakshmipathy. R., Sumana, D.A., Balakrishna, A.N., Bagyaraj, D.J., Kumar, D.P., (2004). Evaluation, Grafting Success and Field Establishment of Cashew Rootstock as Influenced by VAM Fungi. *Indian J. Exp. Biol.* 42, 1132-1135.

[43] Venkataramanan, M.N., Rumi-Kotoky., Singh, H.D., Kotoky, R., (1990). Occurrence and distribution of vesicular-arbuscular mycorrhizal fungi in acid soils of North Eastern India. In: *Current trends in mycorrhizal research* (Jalali, B.L., Chand, H., Eds.) Proceedings of the National Conference on Mycorrhiza, held at Haryana Agricultural University, Hisar, India, Feb. p. 14-16.

[44] Santhaguru, K., Ponmalar, S.B.G., Karunakaran, R., (1995). Vesicular-arbuscular mycorrhizae in tree-legumes and its rhizospheric soils in Alagar hills. *Indian Forester* 121, 817-823.

[45] Seema-Bhadauria., Rekha-Yadav., Ritu-Singh., Bhadauria, S., Yadav, R., Singh, R., (1998). Survey for detection of native high pH-tolerant VAM fungi in alkaline soils. *Proc. Natl. Acad. Sci. Ind.* 68, 141-146.

[46] Vijayalakshmi, M., Rao, A. S. (1988). Vesicular-arbuscular mycorrhizal associations of some *Asteraceae* and *Amaranthaceae*. *Acta Bot Indica* 16,168-174.

[47] Bhardwaj, S., Dudeja, S.S., Khurana, A.L. (1997). Distribution of vesicular-arbuscular mycorrhizal fungi in the natural ecosystem. *Folia Microbiol.* 42, 589-594.

[48] Raman, N., Nagarajan, N., Gopinathan, S., Sambandan, K., (1993). Mycorrhizal status of plant species colonizing a magnesite mine spoil in India. *Biol. Fert. Soils* 16, 76-78.

[49] Nair, S.K., Girija, V.K. (1988). Incidence of vesicular-arbuscular mycorrhiza in certain tree crops of Kerala. *J. Plant Crops*. 16, 1, 67-68.

[50] Byra Reddy, M.S., Krishna Naik, L., Bhaskar, V., Bagyaraj, D.J., (1994). Status of Mycorrhizal association in some tropical tree sprcies. *J. Soil. Biol. Ecol.* 14, 51-57.

[51] Smith, S.E., Read, D.J., 1997, Mycorrhizal Symbiosis. Academic Press, San Diego. 605 pp.

[52] Read, D.J., Duckett, J.G., Francis, R., Ligrone, R., Russell, A., (2000). Symbiotic fungal associations in 'lower' land plants. *Philosophical Trans. Royal Soc. London Series B- Biol. Sci.* 355, 815-830.

[53] Schussler, A., Schwarzott, D., Walker, C., (2001). A new fungal phylum, the Glomeromycota: phylogeny and evolution. *Mycological Res.* 105, 1413-1421.

[54] Vandenkoornhuyse, P., Husband, R., Daniell, T.J., Watson, I.J., Duck, J.M., Fitter, A.H., Young, J.P.W., (2002). Arbuscular mycorrhizal community composition associated with two plant species in a grassland ecosystem. *Mol. Ecol.* 11, 1555-1564.

[55] Daniell, T.J., Husband, R., Fitter, A.H., Young, J.P.W., (2001). Molecular diversity of arbuscular mycorrhizal fungi colonising arable crops *FEMS Microbiol. Ecol.* 36, 203-209.

[56] Johnson, D., Leake, J.R., Ostle, N., Ineson, P., Read, D.J., (2002). *In situ* (CO_2)–C-13 pulse-labelling of upland grassland demonstrates a rapid pathway of carbon flux from arbuscular mycorrhizal mycelia to the soil. *New Phytol.* 153, 327-334.

[57] Rhodes, L.H., Gerdemann, J.W., (1975). Phosphate uptake zones of mycorrhizal and non-mycorrhizal onions. *New Phytol.* 75, 555-561.

[58] Auge, R.M., (2001). Water relations, drought and vesicular-arbuscular mycorrhizal symbiosis. *Mycorrhiza* 11, 3-42.

[59] Tisdall, J.M., Oades, J.M., (1979). Stabilization of soil aggregates by the root systems of rye grass. *Austr. J. Soil Res.* 17, 429-441.

[60] Read, D.J., Perez-Moreno, J., (2003). Mycorrhizas and nutrient cycling in ecosystems – a journey towards relevance. *New Phytol.* 157, 475-492.

[61] Hodge, A., Campbell, C.D., Fitter, A.H., (2001). An arbuscular mycorrhizal fungus accelerates decomposition and acquires nitrogen directly from organic matter. *Nature* 413, 297-299.

[62] Niemira, B.A., Hammerschmidt, R., Safir, G.R., (1996). Postharvest suppression of potato dry rot (*Fusarium sambucinum*) in prenuclear minitubers by arbuscular mycorrhizal fungal inoculum. *Amer. Potato J.* 73, 509-515.

[63] Azcon-Aguilar, C., Barea, J.M., (1996). Arbuscular mycorrhizas and biological control of soil-borne plant pathogens – An overview of the mechanisms involved. *Mycorrhiza* 6, 457-464.

[64] Arnaud, St., Hamel, C., Vimard, B., Caron, M., Fortin, J.A., (1997). Inhibition of *Fusarium oxysporum* f. sp. dianthi in the non-VAM species *Dianthus caryophyllus* by co-culture with *Tagetes patula* companion plants colonized by *Glomus intraradices*. *Can. J. Bot.* 75, 998-1005.

[65] Newsham, K.K., Fitter, A.H., Watkinson, A.R., (1995). Arbuscular mycorrhiza protect an annual grass from root pathogenic fungi in the field. *J. Ecol.* 83, 991-1000.

[66] Khan, A.G., Kuek, C., Chaudhry, T.M., Khoo, C.S., Hayes, W.J., (2000). Role of plants, mycorrhizae and phytochelators in heavy metal contaminated land remediation. *Chemosphere* 41, 197-207.

[67] Ames, R.N., Reid, C.P.P., Ingham, E.R., 1984(). Rhizosphere bacterial population responses to root colonization by a vesicular-arbuscular mycorrhizal fungus. *New Phytol.* 96, 555-563.

[68] Meyer, J.R., Linderman, R.G., (1986b). Selective influence on populations of rhizosphere or rhizoplane bacteria and actinomycetes by mycorrhizas formed by *Glomus fasciculatum*. *Soil Biol. Biochem.* 18, 191-196.

[69] Secilia, J., Bagyaraj, D.J., (1987). Bacteria and actinomycetes associated with pot cultures of vesicular arbuscular mycorrhizas. *Can. J. Microbiol.* 33, 1069-1073.

[70] Forster, S.M., Nicolson, T.H., (1981). Aggregation of sand from maritime embryo sand dune by microorganisms and higher plants. *Soil Biol. Biochem.* 13, 199-203.

[71] Andrade, G., Mihara, K.L., Linderman, R.G., Bethlenfalvay, G.J., (1997). Bacteria from rhizosphere and hyphosphere soils of different arbuscular-mycorrhizal fungi. *Plant Soil* 192, 71-79.

[72] Schreiner, R.P., Mihara, K.L., McDaniel, H., Bethlenfalvay, G.J., (1997). Mycorrhizal fungi influence plant and soil functions and interactions. *Plant Soil* 188, 199-209.

[73] Andrade, G., Mihara, K.L., Linderman, R.G., Bethlenfalvay, G.J., (1998). Soil aggregation status and rhizobacteria in the mycorrhizosphere. *Plant Soil* 202, 89-96.

[74] Söderberg, K.H., Olsson, P.A., Baath, E., (2002). Structure and activity of the bacterial community in the rhizosphere of different plant species and the effect of arbuscular mycorrhizal colonisation. *FEMS Microbiol. Ecol.* 40, 223-231.

[75] Filion, M., St-Arnaud, M., Fortin, J.A., (1999). Direct interaction between the arbuscular mycorrhizal fungus Glomus intraradices and different rhizosphere microorganisms. *New Phytol.* 141, 525-533.

[76] Citernesi, A.S., Fortuna, P., Filippi, C., Bagnoli, G., Giovannetti, M., (1996). The occurrence of antagonistic bacteria in *Glomus mosseae* pot cultures. *Agronomie,* 16, 671-677.

[77] St-Arnaud, M., Hamel, C., Vimard, B., Caron, M., Fortin, J.A., (1997). Inhibition of *Fusariumoxysporum* f.sp.dianthi in the non-VAM species *Dianthus caryophyllus* by co-culture with *Tagetes patula* companion plants colonized by *Glomus intraradices*. *Can. J. Bot. Revue Canadienne De Botanique,* 75, 998-1005.

[78] Caron, M., (1989). Potential use of mycorrhizae in control of soil-borne diseases. *Can.J. Plant Pathol. Revue Canadienne De Phytopathologie* 11, 177-179.

[79] Amora-Lazcano, E., Vazquez, M.M., Azcon, R., (1998). Response of nitrogen-transforming microorganisms to arbuscular mycorrhizal fungi. *Biol. Fert. Soil* 27, 65-70.

[80] Puppi, G., Azcón, R., Höflich G., (1994). Management of positive interactions of arbuscular mycorrhizal fungi with essential groups of soil microorganisms. In: *Impact of Arbuscular Mycorrhizas on Sustainable Agriculture and Natural Ecosystems,* (Gianinazzi, S., Schüepp, H., Eds.), Birkhäuser Verlag, Basel, Switzerland, pp. 201-215.

[81] Garbaye, J., (1994). Helper bacteria – a new dimension to the mycorrhizal symbiosis. *New Phytologist* 128, 197-210.

[82] Mosse, B., (1959). The regular germination of resting spores and some observations on the growth requirements of an Endogone sp. causing vesicular-arbuscular mycorrhiza. *Trans. Brit. Mycol. Soc.* 42, 273-286.

[83] Daniels, B.A., Trappe, J.M., (1980). Factors affecting spore germination of the vesicular-arbuscular mycorrhizal fungus, *Glomus epigaeus*. *Mycologia 72*, 457–471.

[84] Mayo, K., Davis, R.E., Motta, J., (1986). Stimulation of germination of spores of *Glomus versiforme* by spore-associated bacteria. *Mycologia* 78, 426-431

[85] Carpenter-Boggs, L., Loynachan, T.E., Stahl, P.D., (1995). Spore germination of *Gigaspora margarita* stimulated by volatiles of soil-isolated actinomycetes. *Soil Biol.Biochem.*27, 1445-1451.

[86] Tommerup, I.C., (1985). Inhibition of spore germination of vesicular-abuscular mycorrhizal fungi in soil. *Trans. Brit. Mycol. Soc.* 85, 267-278.

[87] Ross, J.P., (1980). Effect of nontreated soil on sporulation of vesicular-arbuscular mycorrhizal fungi associated with soybean. *Phytopathol.* 70, 1200-1205.

[88] Wilson, G.W.T., Hetrick, B.A.D., Kitt, D.G., (1988). Suppression of mycorrhizal growth-response of big bluestem by non-sterile soil. *Mycologia* 80, 338-343.

[89] Minerdi, D., Fani, R., Gallo, R., Boarino, A., Bonfante, P., (2001). Nitrogen fixation genes in an endosymbiotic Burkholderia strain. *Appl. Environ. Microbiol.* 67, 725-732.

[90] MacDonald, R.M., Chandler, M.R., Mosse, B., (1982). The occurrence of bacterium-like organelles in vesicular-arbuscular mycorrhizal fungi. *New Phytologist.* 90, 659-663.

[91] Schenck, N. C., Kinloch, R.A., (1980). Incidence of mycorrhizal fungi on six field crops in monoculture on a newly cleared woodland site. *Mycologia* 72, 229-443.

[92] Harinikumar, K.M., Bagyaraj, D.J., Mallesha, B.C., (1990). Effect of intercropping and organic soil amendments on native VA mycorrhiza in semi-arid tropics. *Arid Soil Res. Rehabil.* 4, 193-197.

[93] Sieverding, E., Leihner, D., (1984). Effect of herbicides on population dynamics of VA mycorrhiza with cassava. *Angew Botanic.* 58, 283-294

[94] Balakrishna, A.N., Lakshmipathy, R., Sudhir, K., Bagyaraj, D.J., (2001). Effect of long Term application of inorganic fertilizers on native VA mycorrhizal fungi and soil microbial biomass in a finger millet-maize-fallow rotation. *J. Soil Boil. Ecol.* 21, 1-6.

[95] Harinikumar, K.M., Bagyaraj, D.J., (1988). The effect of season on VA mycorrhiza of leucaena and mango in semi-arid tropic. *Arid Soil Res. Rehabil.* 7, 139-143.

[96] McGonigle, T.P., Miller, M.H., (1993). Mycorrhizal development and phosphorusadsorption in maize under conventional and reduced tillage.*Amer. J. Soil Sci. Soc.* 57, 1002-1006.

[97] Abbott, L.K., Robson, A.D. (1991). Field management of VA mycorrhizal fungi. In: *The Rhizosphere and Plant Growth.* (Kelster, D.L., Cregan, P.B., Kluwu, Norwell, Eds.), Kluwer Academic Publisher, Dordrecht, The Netherlands, pp. 355-362.

[98] Hayman, D.S., (1974). Plant growth responses to vesicular-arbuscular mycorrhiza VI.Effect of light and temperature. *New Phytol.* 73, 71-80.

[99] Furlan,V., Fortin, J.A., (1977). Effects of light intensity on the formation of vesicular-arbuscular endomycorrhizas on *Allium cepa* by *Giaspora calospora. New Phytol.* 79, 335-340.

[100] Mallesha, B.C., Bagyaraj, D.J., (1991). Season favouring sporulation of VA mycorrhizal fungi in cardamom plantations. *Soil Boil. Ecol.* 11, 75-78.

[101] Harinikumar, K.M., Bagyaraj, D.J., (1989). Effect of cropping sequence, fertilizers and farmyard manure on vesicular-arbuscular mycorrhizal fungi in different crops over three consecutive seasons. *Boil. Fert. Soils* 7, 173-175.

[102] Krishna, K.R., Bagyaraj, D. J., (1982). Effect of Vesicular-arbuscular Mycorrhiza and Soluble Phosphate on *Abelmoscus esculentus* (L.) *Moench. Plant Soil*, 64, 209-213.

[103] Menge, J.A., Steirle, D., Bagyaraj, D.J., Johnson, E.L.V., Leonard, T.T., (1978). Phosphorus concentrations in plants responsible for inhibition of mycorrhizal infection. *New Phytol.* 80, 575-578.

[104] Jasper, P.A., Robson, A.D., Abbott, L.B., (1979). Phosphorus and the formation of vesicular-arbuscular mycorrhizae on growth and metabolism of sweet orange. *New Phytol.* 90, 665-670.

[105] Sreenivasa, M.N., Bagyaraj, D.J., (1989). Suitable form and level of phosphorus for mass production of the VA mycorrhizal fungus *Glomus fasiculatum. Zbl. Mikrobiol.* 144, 34-36.

[106] Clarke, C., Mosse, B., (1981). Plant growth response to vesicular-arbuscular mycorrhiza. XII. Field inoculation responses of barley at two soil P levels. *New Phytol.* 87, 695-703.

[107] Boddington, C.L., Dodd, J.C., (2000). The effect of agricultural practices on the development of indigenous arbuscular Mycorrhizal fungi, II.Studies in experimental microcosms. *Plant Soil* 218, 145-157.

[108] Hayman, D.S., (1975). The occurrence of mycorrhiza in crops as affected by soil fertility. In: *Endomycorrhizas* (Sanders, F.E., Mosse, B., Tinker, P.S., Eds) Academic Press, London, pp. 495-509.

[109] Menge, J.A., (1984). Inoculum production. In: *VA mycorrhiza*. (Powell, C.L., Bagyaraj, D.J., Eds.) CRC Press Inc., Boca Raton, Florida, pp. 188-189.

[110] Sreenivasa, M.N., Bagyaraj, D.J., (1988). *Chloris gayana* (Rhodes grass) a better host for mass production of *Glomus fasiculatum* inoculum. *Plant Soil* 106, 109-112.

[111] Howeler, R.H., (1980). Related cultural practices for cassava. In: *Cassava Cultural Practices*. Proceedings of a workshop held in Salvador, Bahia, Brazil, 18-21 March (Webes, E.J., Toro, M.J.C., Graham Alexander, M.I.J., Eds.), CAB International, Oxon, U.K. pp. 159-169.

[112] Ocampo, J. A., Martin, J., Hayman, D.S., (1980). Influence of plant interactions of vesicular-arbuscular mycorrhizal infections. I. Host and non-host plants grown together. *New Phytol.* 84, 27-35.

[113] Menge, J.A., (1982). Effect of soil fumigants and fungicides on vesicular-arbuscular fungi. *Phytopathol.* **72**, 1125-1132.

[114] Trappe, J.M., Molina, R., Castellano, M., (1984). Reactions of mycorrhizal fungi and mycorrhiza formation to pesticides. *Annu. Rev. Phytopathol.* 22, 331-359.

[115] Nemec, S., (1980). Effects of eleven fungicides on endomycorrhizal development in sour oranges. *Can. J. Bot.*58, 522-527.

[116] Nemec, S., Tucker, D., (1983). Effects of herbicides on endomycorrhizal fungi in Florida citrus (Citrus spp.) soils. *Weed Sci.* 31, 417-431.

[117] Praveen Kumar, K.A., Bagyaraj, D.J., (1999). Mass production of arbuscular mycorrhiza as influenced by some agrochemicals. *Proc. Natl. Acad. Sci. India*, 69, 61-66

In: Microbes in Soil and Their Agricultural Prospects
Editors: K. Kishore Choudhary and D. Wattal Dhar
ISBN: 978-1-63482-824-6

Chapter 15

On Agricultural Importance of Cyanobacteria Mediated Phosphate Availability in Soil

Keshwanand Tripathi [1,2]***, Ashwani K. Rai*** [2]
and Naveen K. Sharma [3]
[1] Department of Microbiology,
Indian Agricultural Research Institute, New Delhi, India
[2] Department of Botany, Banaras Hindu University,
Varanasi (UP), India
[3] Department of Botany, Indira Gandhi National Tribal University,
Amarkantak (MP), India

Abstract

Life on earth is dependent on phosphorous (P). In many natural ecosystems, P is one of the principle macronutrients limiting the growth and development of living forms. Diminishing global pool, escalating cost and limited supply have made P a geo-strategically important mineral. In many agro-ecosystems, contrary to the common belief P is available in plenty but in forms that cannot be utilized directly by the crop plants. The limited availability of P to the crops is a serious threat to the global food security. In tropical regions, cyanobacteria is an agronomically important group of prokaryotic photoautotrophs often constitute a dominant microflora in agro-ecosystems such as paddy and sugarcane-fields. They are bestowed with efficient systems to sense *in milieu* phosphate deficiency. Moreover, they have developed different strategies to scavenge organic phosphate available in the external medium into usable inorganic forms. Apart from fulfilling their own need, the freed inorganic P becomes available to the crop plants. In this chapter, we have reviewed various strategies adopted by cyanobacteria to meet out their phosphate need under Pi-deficient conditions; and agricultural significance of cyanobacteria-mediated Pi enrichment of soil in increasing crop productivity.

Keywords: agriculture; cyanobacteria; fertilizer; phosphatases; phosphorus

INTRODUCTION

Besides climate change, ensuring global food security is a major threat to the security of nation states. The burden of increasing global population on agricultural productivity is tremendous. The first phase of green revolution has adversely affected the fertility of soil [1]. Studies implicate nutritional limitations especially, nitrogen and phosphorus and increasing salinization of soil owing to the excess irrigation, use of poor quality of irrigation water, excessive use of fertilizers and rising global temperature as factors mainly responsible for the reduction in global productivity.

Phosphorous (P) is an essential macronutrient for all the life forms [2] and stands second to nitrogen (N) as the most frequent limiting macronutrient for plant growth [3]. It limits the crop yield by more than 30% (i.e., 30 – 40%) of the world arable land [4]. Phosphorous constitutes nearly 0.2% (0.05 - 0.50%) of plants dry weight [4, 5]; and is an integral component of many key molecules such as nucleic acids, phospholipids, co-enzymes, ATP and NADPH [4].

Global population is projected to reach 9 billion by 2040. Vance [3] suggested that in order to meet the food production demands of 8 to 9 billion people an additional 40 to 20 Tg (10^{12} g) of N and P fertilizers would be required. Furthermore, the crop yield on 5.7 billion hectare (40%) of the world's arable land is limited by the availability of inorganic phosphate (Pi) [3, 4, 6). Between 1960 and 2000, the use of P-fertilizers has increased 4 – 5-folds, which is projected to increase further by 20 Tg Y^{-1} by 2030 [3, 7]. Reports also suggest that world resource of rock phosphate reserves will deplete drastically by 2050 [8, 9].

The excessive use of fertilizers has resulted in the salinization of soils thereby reducing the productivity of soil. In 1960, one ton of fertilizer produced 80 tons of cereals, which decreased to 20 tons by 1995 [4]. Thus, improved P-acquisition and efficient use of the acquired P by plants are critical for economic, environmental and humanitarian reasons [4]. Microbial populations of natural and artificial ecosystems notably, agro-ecosystems play important role in the mobilization and immobilization of phosphate. They help in releasing the immobile forms of P into the medium increasing the availability of usable phosphate for themselves as well as for plants.

Phosphorous (P): An Overview

Phosphorous (P, non-metal, group-15, period 3, atomic number - 15, atomic weight - 31.97) is a lithogenic element. It is colorless, transparent in its pure form and insoluble in water, but soluble in carbon disulfide. In nature because of its highly reactive nature, P does not occur as a free element. Typically, it is found in phosphate rocks. Annual global phosphate production is around 153,000,000 tons. The primary P mining countries include Russia, USA, Morocco, Tunisia, Tongo and Nauru.

Phosphate minerals (PO_4^{3-}) are sparingly soluble in water (≈1 μM at pH 7, 25 ^{0}C). Owing to slow geochemical cycling P often becomes a major limiting nutrient for the development of life. Inorganic phosphate molecules (Pi) used by organisms include orthophosphate, pyrophosphate, phosphate, hypophosphate, phosphine, and other condensed phosphates

(Figure 1). Under presently existing terrestrial redox conditions of the earth surface, orthophosphate is the most stable and dominant form of Pi. Other forms may be present at low concentrations but, generally not in thermodynamic equilibrium and hydrolyze or oxidize slowly to form orthophosphate. Orthophosphate is mainly found as life-mediated apartite deposit [$Ca_5(PO_4)_3$ (OH, F, Cl)], and pre-biotic whitlockite [$Ca_9MgH(PO_4)_7$] and brushite [$CaHPO_4.2H_2O$] deposits [10]. Bulk of P comes from the mining of P-containing rocks [11]. Properties such as thermodynamic instability, kinetic stability, charge and co-ordination state, and a constant oxidation state under typical redox conditions make phosphate advantageous to biological systems [10]. In cells, orthophosphate acts as a buffer keeping their internal pH constant at 7.

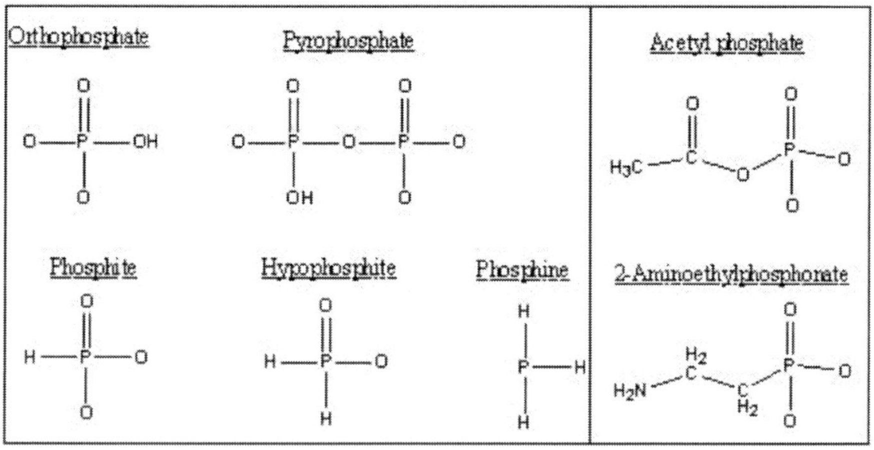

Figure 1. Chemical structure of P forms available to organisms (redrawn from Pasek 2008).

Plants acquire P as phosphate anions (PO_4^{3-}, molar mass - 95.97 g mol^{-1}) from the environment where it rapidly forms insoluble complexes with cations such as calcium, iron, and aluminum (soluble complex with sodium, potassium and ammonium). The concentration of usable orthophosphate forms such as $H_2PO_4^-$ and HPO_4^{2-} formed through desorption and solubilization in the soil solutions is on an average 1 μM, rarely exceeding 10 μM [12-14]. Their availability in the environment varies according to pH of the soil. Below pH 6.0, most of the Pi is present in monovalent $H_2PO_4^-$ form, while H_3PO_4 and HPO_4^{2-} forms are present only in minor proportions [5].

The dwindling global P pool, escalating cost and limited supply could result in a serious threat to the future global food security. More so, as despite the presence of substantial amount of total P in the ambient environment, it is largely unavailable and often becomes limiting for uptake by plants and other organisms. In nature, P is largely present in unavailable forms [5]. For example, in soil, P is present in both organic (e.g., 20 – 80 % of total soil P-content is in organic pools) and inorganic forms (the rest as mineral pool) [5]. Organic forms include the compounds with C-O-P and C-P linkages [10]. Phytic acid (an inositol hexaphosphate) is a major constituent of soil organic-P [15]. Inorganic fraction of P contains 170 mineral forms [16].

Upon application as fertilizer nearly 80% of Pi in soils becomes immobile hence, unavailable to the plants [5, 16]. This is largely due to the adsorption to soil particles and other soil contaminants (insoluble complexes with cations), precipitation and/or conversion to the organic forms by microbes [5, 16]. Processes controlling the amount of available Pi in soil solution include: dissolution – precipitation of P-bearing minerals, adsorption-desorption of phosphate on soil surfaces, and hydrolysis of organic matter, pH, ionic strength, concentration of P and other metals, and the presence of competing ions including organic acids in the environments [4, 17] (Figure 2). Apart from its low availability, rapid utilization of orthophosphate by plants and microbes also causes P deficiency in the environments.

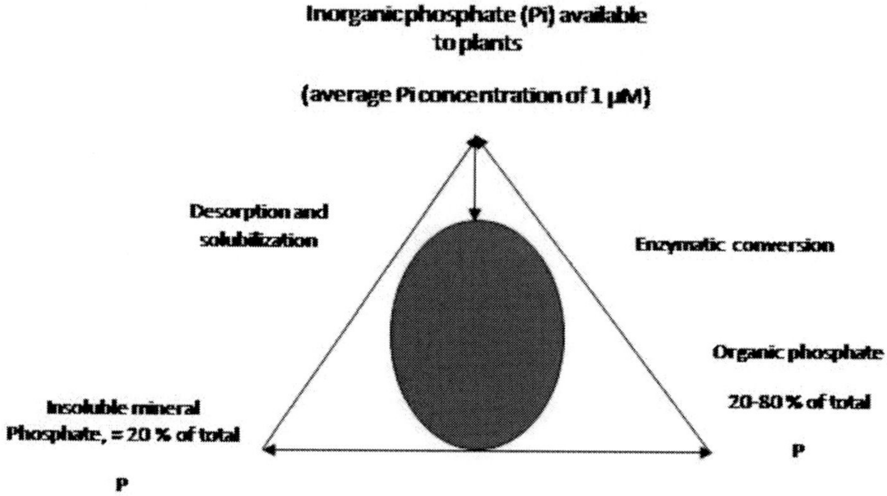

Figure 2. Status of the different P-forms in ecosystems.

In most of the natural ecosystems (e.g., soils, marine and freshwater environments) unavailability of usable form of P often limits nitrogen fixation and primary productivity [18]. Microbial diversity (i.e., phosphate solubilizing microorganisms, PSMs) plays a significant role in increasing the availability of usable phosphate ions in the medium [19]. They secrete low molecular weight extracellular organic acids (e.g., acetic, lactic, malic and 2-ketogluconic acids) in the medium that act as good chelators of divalent cations. Consequently, releasing the phosphate ions from insoluble phosphate salts into the medium [20]. In aquatic systems, phytoplankton facilitates the release of immobile P into water. Presence of free Pi often remains at very low concentrations, whereas concentration of the dissolved organic P (DOP) remains sufficiently high [21]. The DOP supports the primary production in freshwater and marine habitats [22]. The extracellular phosphatases (especially acidic and alkaline phosphatase) secreted by microbes cleave phosphate from organic monophosphate esters facilitating the release of Pi from DOP [23]. Pouring of organic wastes via agricultural runoff, industrial and domestic waste discharge (i.e., cultural eutrophication) is increasing the DOP concentration in water bodies.

Cyanobacteria, is a group of morphologically and metabolically diverse photoautitrophs bestowed with higher plants-like oxygenic photosynthesis. They had a long evolutionary history and merit amongst the most successful and oldest life forms (2.8 -3.2 Ga) to inhabit the earth [24]. Never before in the history of earth, a group of organisms had such a profound impact on the evolution of subsequent life. They enriched the earth atmosphere with oxygen, thus, had played a key role in the evolution of entirely different kind of life. Further, they play a crucial role in regulating the global biogeochemical cycles of nitrogen, carbon and oxygen.

The role of cyanobacteria in sustaining and improving the soil productivity is well known. They fix a significant quantity of atmospheric nitrogen (20 - 30 kg ha^{-1}) into the soil, add organic content, release plant growth stimulating hormones in soil, and improve water-holding capacity of the soil by releasing extracellular polysaccharides that enhance the aggregation property of the soil [2,24 and references therein]. Cyanobacteria are reported to mobilize different insoluble forms of P by synthesizing chelators and releasing organic acids. Mandal et al. [25] reported cyanobacteria-mediated increase in the phosphate availability to rice plants. Further, phosphate assimilated by cells during growth is released upon death and decomposition of cyanobacteria (i.e., organic phosphate), and is mineralized to orthophosphate consequently increasing the availability of Pi to the plants [25].

Alternatively, under P-limitation, cyanobacteria release enzyme alkaline phosphatases (APases and ADPases) into the soil, which hydrolyze the non-available organic phosphates in the soil to the utilizable inorganic forms [26, 27]. This may also increases the availability of Pi to the crops.

STRATEGIES ADOPTED BY CYANOBACTERIA AND PLANTS TO OVERCOME PI-LIMITATION

Microbes and plants have developed two broad strategies to enhance use and acquisition of phosphate (Table 1) [23, 7]. These strategies are genetically controlled hence subject to genetic improvement. The strategies adopted by cyanobacteria to deal with P-deprivation involve accessing the stored intracellular polyphosphate bodies, increasing Pi-uptake rate and induction of the synthesis and release of extracellular phosphatases for scavenging of Pi moiety from a variety of organic molecules present in the surroundings [28, 29]. During their exponential growth phase, cyanobacteria store phosphate in the form of polyphosphate (PolyP), a linear polymer of phosphate units linked through high-energy phosphoanhydride bonds, granules at high level [30]. PolyP bodies are degraded under P-deficiency through the activity of exophosphatase (Ppx).

Likewise, under low P availability, plants either use cellular P in conservative manner or, enhance the acquisition of Pi from the environment. In later case, there is increased production and secretion of phosphatases, exudation of organic acids and enhanced expression of Pi transporters [4]. Ae et al. [31] reported exudation of piscidic acid (a phenolic) from pigeon pea (*Cajanus cajan*) roots that increased the availability of Pi (from Fe-PO_4) in the soil. The acquisition of phosphate has also been improved by enhanced synthesis and release of citrate and malate [32]. The amount of carbon exuded in these two carbon rich acids ranges from 10% to >25% of the total dry weight of plant. Photosynthetic CO_2-fixation provides bulk (65%) of the carbon exuded. While, dark CO_2-fixation via enhanced activities of phosphoenol pyruvate carboxylase, PEPC; malate dehydrogenase, MDH; and citrate synthase, CS provided the rest [33, 34]. The type and concentration of organic acids produced by organisms varied according to the source of available phosphates [35]. Koyama et al. [36] developed carrot (*Daucus carota*) cell lines with overexpressed mitochondrial citrate synthase that grew well on Al-PO_4^- medium. Similarly, overexpression of a bacterial citrate synthase with ^{35}S promoter of cauliflower mosaic virus (CaMV^{35}S) resulted in the increased secretion of citrate into rhizosphere [37].

Table 2. Strategies adopted by plants against P-deficiency (Vance 2001)

Strategy	Adaptation	References
Enhanced acquisition or uptake	Enhanced expression of PO_4^- transporters, biosynthesis and release of acid/alkaline phosphatases, and organic acids (e.g., citrate, malate etc.)	Gilbert et al., (1998) Raghothama (1999) Gilroy and Jones (2000)
Conserved use	Internal remobilization, decreased growth rate, more growth per unit P (p-use efficiency); metabolic flexibility; modified C and N metabolism that by-pass P-requiring steps, alternative respiratory pathways, modification in membrane structure.	Schachtman et al., (1998) Raghothama (1999) Uhde-Stone et al., (2003 a,b)

Scavenging of Inorganic Phosphate from Organic P-Sources

Bulk of organic phosphorus present in environments is enzymatically broken down to usable inorganic form. Cyanobacteria use a diverse range of organic substrates for obtaining Pi [38]. Phosphatases are a group of enzyme produced by plants and microbes in response to Pi-starvation [39]. The phosphatases that are involved in Pi mobilization from organic sources include - RNAase [40, 41, 42], extracellular cyclic nucleotide phophodiesterase (from nucleotides) [43] and apyrases (from extracellular ATP) [44].

Acid/alkaline phosphatase is other group of phosphatases synthesized and secreted by plants and microbes to solubilize organically bound P. In acidic conditions, acid phosphatases hydrolyze monoester organic phosphate in pH-dependent manner. In outside medium (extracellular), acid phosphatses exhibit broad substrate specificity, whereas they are highly substrate specific inside the cells (intracellular) [45, 46]. A family of 29 genes with conserved domains of purple acid phosphatases has been identified in the *Arabidopsis* genome [47].

Cyanobacteria are generally found in alkaline conditions and respond to Pi-limitation by synthesizing extracellular alkaline phosphatase (APase), which are secreted out to scavenge Pi from dissolved organic sources [48, 49, 50, 51].

ALKALINE PHOSPHATASE

Alkaline Phosphatases (APase and ADPase) are ubiquitous in bacteria, cyanobacteria and higher plants. An APase is functionally defined as a phosphatase that works (non-specific hydrolysis of phosphomonoesters) at alkaline pH. The best-characterized alkaline phosphatase is APase; E.C.3.6.1.1 inorganic diphosphatase. Since the enzymes appear in a response to P-limitation, their appearance is regarded as an indication of P limitation [52-53]. Regulation of APase by P supply has been well studied in various organisms [54-58]. In *Synechococcus* sp. PCC 7942 growing under P-limiting conditions, APase activity increased with subsequent increase (50-fold increase in V_{max}) in Pi transport into the cells [59-62]. The dynamics of initial response of APase activity to Pi- limitation as well as the final levels of APase activity at minimum cellular P- quota vary with species [63]. In freshwater *Scenedesmus quadricauda* and *Asterionella formosa*, cellular P content increased rapidly after P addition but, APase activity remained high for several days [63]. Three types of APases have been reported in cyanobacteria namely, PhoA, PhoX, and PhoD. They differ in homology and metal requirements for their activity [64]. A functional APase may include several different gene products [65-66].

PhoA

PhoA, an atypical large alkaline phosphatase has been reported from different species including humans [67-68]. It is a member of super-family of enzymes [69]. Its structure consists of a dimer with active site containing two Zn^{+2} ions and one Mg^{+2}, and a conserved serine residue, which is phosphorylated during catalysis [70]. In *Synechococcus* sp. PCC 7942 PhoA (M_r 145kDA) is located in the periplasmic space as soluble enzyme [48] whereas, it is membrane bound in Gram positive *Enterococcus faecalis* [71]. PhoA is secreted via the 'sec' protein secretion pathway [72-73]. In cyanobacteria, PhoA transcription is induced under P-starvation (i.e., P-repressible), and is regulated by a two-component SphS/SphR environmental sensor/response regulator signal transduction pathway [74]. Wagner et al. [51] reported a second alkaline phosphatase (61.3 kDa) from cyanobacterium *Synechococcus* sp. PCC 7942 encoded by *PhoV*, which is P-irrepressible.

PhoX

PhoX is more widely distributed in marine bacteria and cyanobacteria (e.g., *Trichodesmium* sp.) compared to that of PhoA [75]. It is a monomeric enzyme activated by calcium, magnesium and copper, and has no homology with PhoA, [76-80]. PhoX is exported via the twin arginine transport (Tat) system [81].

PhoX sequences contain a conserved calcium-binding site in gluconolactonases and senescence marker proteins (SMP-30) [79]. The PhoX calcium-binding motif aligns with a part of Ca^{+2} catalytic sites present in the central cavity of a six-bladed b-propeller structure [79]. The aspartate residue coordinating with the putative catalytic Ca^{+2} is conserved in all PhoX sequences. At the active site, oxygen atoms from amino acid residues donate other ligands of the catalytic calcium. However, in diisopropylfluoro phosphatase, a histidine is involved in diisopropylfluorophosphate hydrolysis [82]. Proline preceding the coordinated aspartate is also conserved in all PhoX sequences except *Enterobacter sakazakii* and plasmid of *Nostoc* sp. PCC 7120.

PhoX produces phosphatase activity with *p*NPP and other natural phosphatase substrates at pH 7.5 and high protein load [83]. The analysis of PhoX with 89 natural phosphate substrates revealed high phosphatase activity towards 79 substrates including all major classes of known phosphorylated metabolites (nucleotides, carbohydrates, amino acids and organic acids). Except for phytic acid and phosphatidic acid, PhoX dephosphorylated all metabolites with C-O-P bonds. Amongst the N-P bond substrates, phosphoarginine and imidodiphosphate were hydrolyzed, but no activity was observed with phosphoramidite substrates (AMP-ramidate, UMP-morpholidate, GMP-morpholidate etc). Likewise, no activity was reported for five phosphono-substrates containing C-P bond.

Pho D

Marine microbes also contain another family of alkaline phosphatase ie., PhoD, which hydrolyzes both mono- (APase) and diester organic phosphates (APDase) [26, 64]. PhoD is secreted into extracellular medium via 'Tat' pathway, and shows Ca^{2+} dependent phosphatase activity. 'Tat' pathway recognizes PhoD by their N-terminal twin-arginine signal peptides containing the Tat consensus (SRRXFLX) motif [26, 84]. In $PhoD_{BS}$ (PhoD of *Bacillus subtilis*), TatC acts as a specificity determinant. The inactivation of the *tatCd* completely inhibits the secretion of PhoD while, inactivation of the second *tatC* gene (*tatCy*) has no effect on the secretion of PhoD. This indicates the existence of multiple 'Tat' pathways in a single bacterial cell with separate substrate specificity. The *tatA*/*tatC* gene pair (*tatAd*/*Cd*) is localized downstream from *phoD*, and is co-regulated with the expression of *phoD*.

Role of Cyanobcateria in Enhancing the Pi-Availability to Crop Plants

Use of intensive agricultural technologies especially, excessive use of synthetic fertilizers, has not only increased the cost of produces but also resulted in decrease in the soil fertility. As a response, much attention has been paid to the development of sustainable agricultural practices. Both in natural and agro-ecosystems, microbes are present in close proximity (structural and functional symbioses) to plants (as ecto- and endosymbionts). Where, they play crucial role in the overall growth and development of vegetative and reproductive phases of plants. Cyanobacteria are principle microbial flora of many agricultural fields, supporting many agronomically important functions.

An important component of sustainable agriculture is to enhance the productivity of crops using their natural adoptive potentials, with a minimal disturbance to the environment [85]. A promising strategy to sustainable agriculture could be the use of environment friendly microbes as a substitute to hazardous agrochemicals [86]. The information gathered from microbial ecology need to be translated into agricultural biotechnologies.

In agro-ecosystems P is present in excess (not less) but in forms not available to the crops. Crops take up only 15–30% of applied P fertilizer in the year of its application [87]. The rest becomes immobilized and add up to the already existing P pool. Improving the efficiency of P (fertilizer) use for crop growth requires enhanced availability of Pi (Pi-mobilization), improved uptake of phosphate by plants from soil (P-acquisition efficiency) and improved productivity per unit P taken up (P-use efficiency) [refer 88,89 for more information]. Veneklass et al. [89] have provided a detailed account on the P-use efficiency of various crops plants and the way it could be achieved in case of low P availability. The optimal distribution and redistribution of P in plants allows maximum growth and biomass allocation to harvestable plant parts. It is possible to improve the P distribution within plants by increasing remobilization from tissues that no longer needed by plant (e.g., senescing leaves) and reduced partitioning of P to developing grains. This would prolong and enhance the productive use of P in photosynthesis, and will have nutritional and environmental benefits [89].

Studies indicate that cyanobacteria are capable of enriching Pi content in soil solution. However, majority of researches on the ability of soil/freshwater cyanobacteria to release Pi from the externally present organic P forms have exclusively been carried out in laboratories with isolated cyanobacterial systems [see 90 for other relevant works]. *In vivo* studies conducted in agricultural fields to assess the potential of cyanobacteria either present as natural population or used as biofertilizers to ameliorate soil with Pi are lacking. There exists a complex cyanobacteria-plant system with many other microbial forms. Therefore, extrapolation of isolated lab data to field will be erroneous and unrealistic. For this, we need quantitative estimates of Pi availability in soil, its co-relation to growth and development of native cyanobacterial communities (growth phases and time series analysis), and partitioning of freed Pi amongst cyanobacteria and associate microbes and crop plants. Moreover, developmental stages of crop plants and cyanobacteria are asynchronous differing in time. Therefore, it is possible that Pi scavenged by cyanobacteria as per their own need may not be utilized by crop plant at that time. Besides, phosphate transporters of crop plants and cyanobacteria are most likely to vary in their uptake efficiency (i.e., different Km value), which may lead to misappropriate acquisition of phosphate from the medium. This may cause P-deficiency for slow growing plants or vice-versa. Moreover, both plant and cyanobacteria produce their own sets phosphatases. Therefore, it is difficult to assign whether increase in soil Pi level is either due to plant or cyanobacterial phosphatases. Conclusively, more authentic and focused data are needed to establish the fact that use of cyanobacteria will be helpful in improving phosphate availability to crop plants in complex agro-ecosystems.

REFERENCES

[1] Tilman, D., Fargione, J., Wolff, B., D'Antonio, C., Dobson, A., Howarth, R., Schindler, D., Schlesinger, W.H., Daniel Simberloff, D., Swackhamer, D. (2001). Forecasting agriculturally driven global environmental change. *Science* 292, 281-284.

[2] Rai, A.K., Sharma, N.K. (2006). Phosphate metabolism in the cyanobacterium *Anabaena doliolum* under salt stress. *Curr Microbiol,* 52, 6-12.

[3] Vance, C.P. (2001). Symbiotic nitrogen fixation and phosphorus acquisition: plant nutrition in a world of declining renewable resources. *Pl Physiol,* 127, 390-397.

[4] Vance, C.P., Uhde-Stone, C., Allan, D.L. (2003). Phosphorus acquisition and use: critical adaptation by plants securing a non-renewable resource. *New Phytol*, 157, 423-447.

[5] Schachtman, D.P., Reid, R.J., Ayling, S.M. (1998). Phosphorus uptake by plants: from soil to cell. *Pl Physiol,* 116, 447-453.

[6] Batjes, N.H. (1997). A world data set of derived properties by FAO- UNESCO soil unit for global management. *Soil Use Managt,* 13, 9-16.

[7] Lajtha, K., Harrison, A.F. (1995). Strategies of phosphorus acquisition and conservation by plant species and communities. In: *Phosphorus in the Global Environment* (Tiessen, H., Ed.), Chichester, UK: John Wiley and Sons; 1995, pp. 140-147.

[8] Cordell, D., Jan-Olof, D., Stuart, W. (2009). The story of phosphorus: Global food security and food for thought. *Global Environ Change,* 19, 292-305.

[9] United States Geological Survey (2009). *Mineral Commodity Summaries.* U.S. Geological Survey.

[10] Pasek, M.A. (2008). Rethinking early Earth phosphorus geochemistry. *Proc Natl Am Soc* USA, 105, 853-858.

[11] Dery, P., Anderson, B. (2007). *Peak phosphorus.* Energy Bulletin. Enerybullentin.net Node 33164.

[12] Bieleski, R.L. (1973). Phosphate pools, phosphate transport and phosphate availability. *Annu Rev Pl Physiol,* 24, 225-252.

[13] Healey, F.P. (1982). Phosphate. In: *The Biology of Cyanobacteria* (Carr, N.G., Whitton, B.A., Eds.), London: Blackwell Scientific Publishers; 1982, pp. 105-124.

[14] Raghothama, K.G. (1999) Phosphate acquisition. *Annu Rev Pl Physiol Pl Mol Biol,* 50, 665-693.

[15] Richardson, L.L. (1994). Algal accessory pigment detection using AVIRIS image-derived spectral radiance data. *J Geophys Res* 100, 1565- 1578.

[16] Holford, I.C.R. (1997). Soil phosphorus: Its measurement, and its uptake by plants. *Aust J Soil. Res.* 35, 227-239.

[17] Hinsinger, P. (2001). Bioavailability of soil inorganic P in the rhizosphere as affected by root-induced chemical changes: A review. *Pl Soil,* 237, 173-195.

[18] Sanudo-Wilhelmy, S.A., Kustka, A., Capone, D.G., Hutchins, D., Gobler, C., Yang, M., Lwiza, K., Burns, J., Raven, J.A., Carpenter, E.J. (2001). Phosphorus limitation of N_2 fixation in the central Atlantic Ocean. *Nature* 411, 66-69.

[19] Yandigeri, M.S., Yadav, A.K., Meena, K.K., Pabbi, S. (2010). Effect of mineral phosphates on growth and nitrogen fixation of diazotrophic cyanobacteria *Anabaena variabilis* and *Westiellopsis prolifica*. *Antonie van Leeuwenhoek* 97, 297-306.

[20] Bolan, N.S., Naidu, R., Mahimairaja, S., Baskaran, S. (1994). Influence of low molecular-weight organic acids on the solubilization of phosphates. *Biol Fert Soils,* 18, 311-319.

[21] Nausch, M., Nausch, G., Wasmund, N. (2004). Phosphorus dynamics during the transition from nitrogen to phosphate limitation in the central Baltic Sea. *Mar Ecol Prog Ser* 266, 15-25.

[22] Cotner, J.B., Wetzel, R.G. (1992). Uptake of dissolved inorganic and organic phosphorus compounds by phytoplankton and bacterioplankton. *Limnol. Oceanogr.* 37, 232-243.

[23] Rodriguez, A.A., Stella, A.M., Storni, M.M., Zulpa, G., Zaccaro, M.C. (2006). Effects of cyanobacterial extracellular products and gibberellic acid on salinity tolerance in *Oryza sativa* (L). *Saline Syst,* 2-7.

[24] Sharma, N.K., Tiwari, S.P., Tripathi, K.N., Rai, A.K. (2011). Sustainability and cyanobacteria (blue- green algae): facts and challenges. *J Appl Phycol*, 23, 1059-1081.

[25] Mandal, B., Vlek, P.L.G., Mandal, L.N. (1999). Beneficial effects of blue green algae and *Azolla*, excluding supplying nitrogen, on wet land rice fields: A review. *Biol Fert Soils,* 28, 329-342.

[26] Kageyama, H., Tripathi, K.N., Rai, A.K., Cha-um, S., Waditee-Sirisattha, R., Takabe, T. (2011). An Alkaline Phosphatase/Phosphodiesterase, PhoD, Induced by Salt Stress and Secreted out of the cells of *Aphanothece halophytica,* a halotolerant cyanobacterium. *Appl Environ Microbiol,* 77, 5178-5183.

[27] Tripathi, K., Sharma, N.K., Kageyama, H., Takabe, T., Rai, A.K. (2013b). Physiological, biochemical and molecular responses of the halophilic cyanobacterium *Aphanothece halophytica* to Pi-deficiency. *Eur J Phycol*, 48, 461-473.

[28] Romans, K.M., Carpenter, E. (1994). Buoyancy regulation in the colonial diazotrophic cyanobacterium *Trichodesmium tenue*: ultrastructure and storage of carbohydrate, polyphosphate and nitrogen. *J Phycol*, 30, 935-942.

[29] Bhaya, D., Schwarz, R., Grossman, A. (2000). Molecular responses to environmental stress In: *The Ecology of Cyanobacteria: Their Diversity in Time and Space* (Whitton, B.A., Potts, M., Eds.), Dordrect: Kluwar; 2000, pp. 397-442.

[30] Kornberg, A., Rao, N.N., Ault-Riche, D. (1999). Inorganic polyphosphate: a molecule of many functions. *Annu Rev Biochem,* 68, 89-125.

[31] Ae, N., Arihara, J., Okada, K., Yoshihara, T., Johansen, C. (1990). Phosphorus uptake by pigeon pea and its role in cropping systems of Indian subcontinent. *Science* 248, 477- 480.

[32] Marschner, H. (1995). *Mineral Nutrition in Plants,* 2nd ed. San Diego, CA: Academic Press.

[33] Johnson, J.F., Allan, D.L., Vance, C.P., Weiblem, G. (1996a). Root carbon dioxide fixation by phosphorus-deficient *Lupinus albus* contribution to organic acid exudation by protoid roots. *Pl Physiol,* 112, 31-41.

[34] Johnson, J.F., Vance, C.P., Allan, D.L. (1996b). Phosphorus deficiency in *Lupinus albus* altered lateral root development and enhanced expression of phosphoenolpyruvate carboxylase. *Pl Physiol,* 112, 31-41.

[35] Scervino, J.M., Mesa, M.P., Mónica, I.D., Recchi, M., Moreno, N.S., Godeas, A. (2010). Soil fungal isolates produce different organic acid patterns involved in phosphate salt solubilization. *Biol Fert Soils,* 46, 755-763.

[36] Koyama, H., Takita, E., Kawamura, A., Hara, T., Shibata, D. (1999). Over expression of mitochondrial citrate synthase gene improves the growth of carrot cells in Al-phosphate medium. *Plant Cell Physiol,* 40, 482-488.

[37] López-Bucio, J., de la Vega, O.M., Guevara-García, A., Herrera-Estrella, L. (2000). Enhanced phosphorus uptake in transgenic tobacco plants that overproduce citrate. *Nat Biotechnol,* 18, 450-453.

[38] Whitton, B.A., Grainger, S.L.J., Hawley, G.R.W., Simon, J.W. (1991). Cell-bound and extracellular phosphatase activities of cyanobacterial isolates. *Microbial Ecol*, 21, 85-98.

[39] Duff, S.M.G., Sarath, G., Plaxton, W.C. (1994). The role of acid phosphatase in plant phosphorus metabolism. *Physiologia Plantarum* 90, 791-800.

[40] Bariola, P.A., Howard, C.J., Taylor, C.P., Verburg, M.T., Jaglan, V.D., Green, P.J. (1994). The *Arabidopsis* ribonuclease gene RNS1 is tightly controlled in response to phosphate limitation. *Plant J,* 6, 673-685.

[41] Green, P.J. (1994). The ribonuclease of higher plants, *Annu Rev Plant Physiol Pl Mol Biol,* 45, 421-445.

[42] Kock, M., Loffier, A., Abel, S., Glund, K. (1995). Structural and regulatory properties of a family of phosphate starvation induced ribonuclease from tomato. *Pl Mol Biol* 27, 477-485.

[43] Abel, S., Nurnberger, T., Ahnert, V., Krauss, G., Glund, K. (2000). Induction of an extracellular cyclic nucleotide phosphodiesterase as an accessory ribonucleolytic activity during phosphate starvation of cultured tomato cells. *Pl Physiol,* 122, 543-552.

[44] Thomas, C., Sun, Y., Naus, K., Lloyd, A., Roux, S. (1999). Apyrase function in plant phosphate nutrition and mobilizes phosphate from extracellular ATP. *Pl Physiol,* 119, 543-551.

[45] Duff, S.M.G., Lefebvre, D.D., Plaxton, W.C. (1991). Phosphate starvation response in plant cells: *de novo* synthesis and degradation of acid phosphatases. *Proc Natl Acad Sci USA,* 88, 9538-9542.

[46] Raghothama, K.G., Karthikeyan, A.S. (2005) Phosphate acquisition. *Pl Soil,* 274, 37-49.

[47] Li, D., Zhu, H., Liu, K., Liu, X., Leggewie, G., Udvardi, M., Wang, D. (2002). Purple acid phosphatases of *Arabidopsis thaliana*. *J Biol Chem,* 277, 27772-27781.

[48] Ray, J.M., Bhaya, D., Block, M.A., Grossman, A.R. (1991). Isolation, transcription, and inactivation of the gene for an atypical alkaline phosphatase of *Synechococcus* sp. strain PCC 7942. *J Bacteriol,* 173, 4297-4309.

[49] Tandeau de Marsac, N., Houmard, J. (1993). Adaptation of cyanobacteria to environmental stimuli: new steps towards molecular mechanisms. *FEMS Microbiol Rev* 104, 119-190.

[50] Grossman, A.R., Bhaya, D., Collier, J.L. (1994). Specific and general responses of cyanobacteria to macro-nutrient deprivation. In: *Phosphate Metabolism and Cellular Regulation in Microorganisms* (Torriani, G.A., Yagil, B.E., Silver, S., Eds.), Washington DC, USA: ASM Press, 1994, pp. 112-118.

[51] Wagner, K.U., Mosepohl, B., Pistorious, E.K. (1995). The cyanobacterium *Synechococcus* sp. strain PCC7942 contains a second alkaline phosphatase encoded by PhoV. *Microbiology* 141, 3049-3058

[52] Beardall, J., Berman, T., Heraud, P., Kadiri, M.O., Light, B.R., Patterson, G., Roberts, S., Sulzberger, B., Sahan, E., Uehlinger, U., Woods, B.A. (2001). Comparison of methods for detection of phosphate limitation in microalgae. *Aquat Sci,* 63, 107-121.

[53] Gouvea, S.P., Melendez, C., Carberry, M.J., Bullerjahn, G.S., Wilhelm, S.W., Langen, T.A., Twiss, M.R. (2006). Assessment of phosphorus-microbe interactions in Lake Ontario by multiple techniques. *J Gt Lakes Res*, 32, 455-70.

[54] Lee, T.M. (2000). Phosphate starvation induction of acid phosphatase in *Ulva lactuca* L. (Ulvales, Chlorophyta). *Bot Bull Acad Sin,* 41, 19-25.
[55] Hernandez, I., Niell, F.X., Whitton, B.A. (2002). Phosphatase activity of benthic marine algae. *J Appl Phycol,* 14, 475-487.
[56] Rengefors, K., Ruttenberg, K.C., Haupert, C.L., Taylor, C., Howes, B.L., Anderson, D.M. (2003). Experimental investigation of taxon-specific response of alkaline phosphatase activity in natural freshwater phytoplankton. *Limnol Oceanogr,* 48, 1167-1175.
[57] Strojsova, A., Vrba, J., Nedoma, N., Komarkova, J., Znachor, P. (2003). Seasonal study of extracellular phosphatase expression in the phytoplankton of a eutrophic reservoir. *Eur J Phycol,* 38, 295-306.
[58] Strojsova, A., Vrba, J., Nedoma, J., Simek, K. (2005). Extracellular phosphatase activity of freshwater phytoplankton exposed to different in situ phosphorus concentrations. *Mar Freshw Res,* 56, 417-24.
[59] Grillo, J.F., Gibson, J. (1979). Regulation of phosphate accumulation in the unicellular cyanobacterium *Synechococcus*. *J Bacteriol,* 140, 508-517.
[60] Aiba, H., Nagaya, M., Mizuno, T. (1993). Sensor and regulator proteins from the cyanobacterium *Synechococcus* species PCC 7942 that belong to the bacterial signal-transduction protein families: implication in the adaptive response to phosphate limitation. *Mol Microbiol,* 8, 81-91.
[61] Aiba, H., Mizuno, T. (1994). A novel gene whose expression is regulated by the response-regulator, SphR in response to phosphate limitation in *Synechococcus* species PCC 7942. *Mol Microbiol,* 13, 25-34.
[62] Nagaya, M., Aiba, H., Mizuno, T. (1994). The sphR product, a two-component system response regulator protein, regulates phosphate assimilation in *Synechococcus* sp. strain PCC 7942 by binding to two sites upstream from the phoA promoter. *J Bacteriol,* 176, 2210-2215.
[63] Litchman, E., Nguyen, B.L.V. (2008). Alkaline phosphataes activity as a function of internal phosphorous concentration in freshwater phytoplankton. *J Phycol,* 44, 1379-1383.
[64] Luo, H., Benner, R., Long, R.A., Hu, J. (2009). Subcellular localization of marine bacterial alkaline phosphatases. *Proc Natl Acad Sci USA,* 106, 21219-21223.
[65] Dyhrman, S.T., Chappell, P.D., Haley, S.T., Moffett, J.W., Orchard, E.D., Waterbury, J.B., Webb, E.A. (2006). Phosphonate utilization by the globally important marine diazotroph *Trichodesmium*. *Nature* 439, 68-71.
[66] Xu, Y., Wahlund, T.M., Feng, L., Shaked, Y., Morel, F.M.M. (2006). A novel alkaline phosphatase in the coccolithophore *Emiliania huxleyi* (Prymnesiophyceae) and its regulation by phosphorus. *J Phycol,* 42, 835-844.
[67] Narisawa, S., Huang, L., Iwasaki, A., Hasegawa, H., Alpers, D.H., Millán, J.L. (2003). Accelerated fat absorption in intestinal alkaline phosphatase knockout mice. *Mol Cell Biol,* 23, 7525-7530.
[68] Song, L. (2006). A soluble form of phosphatase in *Saccharomyces cerevisiae* capable of converting farnesyl diphosphate into E, E-farnesol. *Appl Biochem Biotechnol* 128, 149-158.
[69] Galperin, M.Y., Jedrzejas, M.J. (2001). Conserved core structure and active site residue in alkaline phosphatase suferfamily enzymes. *Proteins* 45, 318-324.

[70] Coleman, J.E. (1992). Structure and mechanism of alkaline phosphatase. *Annu Rev Biophys Biomol Struct,* 21, 441-483.

[71] Ghosh, R., Ghosh, A., Ghosh, B.K. (1997). Properties of the membrane bound alkaline phosphatase from glucose and lactate grown cells of *Bacillus subtilis* SB 15. *J Biol Chem,* 252, 6813-6822.

[72] Angelini, S., Moreno, R., Gouffi, K., Santini, C., Yamagishi, A., Berenguer, J., Wu, L. (2001). Export of *Thermus thermophilus* alkaline phosphatase via the twin-arginine translocation pathway in *Escherichia coli. FEBS Lett,* 506, 103-107.

[73] Rosch, J., Caparon, M. (2004). A microdomain for protein secretion in Gram-positive bacteria. *Science* 304, 1513-1515.

[74] Suzuki, S., Ferjani, A., Suzuki, I., Murata, N. (2004). The SphS-SphR two component system us the exclusive sensor for the induction of gene expression in response to phosphate limitation in *Synechocystis*. *J Biol Chem*, 279, 13234-13240.

[75] Sebastian, M., Ammmerman, J.W. (2009). The alkaline phosphatase PhoX is more widely distributed in marine bacteria than the classical PhoA. *ISME J,* 3, 563-572.

[76] Boulanger, R.R. Jr, Kantrowitz, E.R. (2003). Characterization of a monomeric Escherichia coli alkaline phosphatase formed upon a single amino acid substitution. *J Biol Chem*, 278, 23497-23501.

[77] Majumdar, A., Ghatak, A., Ghosh, R.K. (2005). Identification of the gene for the monomeric alkaline phosphatase of Vibrio cholerae serogroup O1 strain. *Gene* 344, 251-258.

[78] Wang, J., Stieglitz, K.A., Kantrowitz, E.R. (2005). Metal specificity is correlated with two crucial active site residues in *Escherichia coli* alkaline phosphatase. *Biochemistry* 44, 8378-8386.

[79] Wu, J.R., Shien, J.H., Shieh, H.K., Hu, C.C., Gong, S.R., Chen, L.Y., Chang, P.C. (2007). Cloning of the gene and characterization of the enzymatic properties of the monomeric alkaline phosphatase (PhoX) from *Pasturella multocidastrain* X-73. *FEMS Microbiol Lett,* 267, 113-120.

[80] Kathuria, S., Martiny, A.C. (2011). Prevalence of a calcium-based alkaline phosphatase associated with the marine cyanobacterium *Prochlococcus* and other ocean bacteria. *Environ Microbiol,* 13, 74-83.

[81] Monds, R.D., Newell, P.D., Schwartzman, J.A., O'Toole, G.A. (2006). Conservation of the Pho regulon in *Pseudomonas fluorescens* Pf0-1. *Appl Environ Microbiol* 72, 1910-1924.

[82] Scharff, E.I., Koepke, J., Fritzsch, G., Lücke, C., Rüterjans, H. (2001) Crystal structure of diisopropylfluorophosphatase from *Loligo vulgaris*. *Structure* 9, 493–502.

[83] Kuznetsova, E., Proudfoot, M., Sanders, S.A., Reinking, J., Savchenko, A., Arrowsmith, C.H., Edwards, A.M., Yakunin, A.F. (2005). Enzyme genomics: Application of general enzymatic screens to discover new enzymes. *FEMS Microbiol Rev,* 29, 263-279.

[84] Pop, O., Martin, U., Abel, C., Muller, J.P. (2002). The twin-arginine signal peptide of PhoD and the TatAd/Cd proteins of *Bacillus subtilis* form an autonomous Tat translocation system. *J Biol Chem,* 277, 3268-3273.

[85] Noble, A.D., Ruaysoongnern, S. (2010). The nature of sustainable agriculture. In: *Soil Microbiology and Sustainable Crop Production* (Dixon, R., Tilston, E., Eds.), Berlin, Heidelberg, Germany: Springer Science and Business Media B.V.: 2010, pp. 1-25.

[86] Tikhonovich, I.A., Provorov, N.A. (2011). Microbiology is the basis of sustainble agriculture: An opinion. *Ann Appl Biol* 159, 155-168.

[87] Syers, J.K., Johnston, A.E., Curtin, D. (2008). Efficiency of soil and fertilizer phosphorus use. Reconciling changing concepts of soil phosphorus behavior with agronomic information. FAO Fertilizer and Plant Nutrition Bulletin No.18. Rome, Italy: FAO.

[88] Richardson, A.E., Lynch, J.P., Ryan, P.R., Delhaize, E., Andrew Smith, F., Sally E. Smith, S.E., Harvey, P.R., Ryan, M.H., Veneklaas, E.J., Lambers, H., Oberson, A., Culvenor, R.A., Simpson, R.J. (2011). Plant and microbial strategies to improve the phosphorus efficiency of agriculture. *Pl Soil,* 349, 121-156.

[89] Veneklass, E.J., Lambers, H., Bragg, J., Finnegan P.M., Lovelock, C.E., Plaxton, W.C., Price, C.A., Scheible, W.R., Shane, M.W., White, P.J., Raven, J.A. (2012). Opportunities for imroving phosphorous-use efficiency in plants. *New Phytol,*195, 306-320.

[90] Tripathi, K., Sharma, N.K., Rai, V., Rai, A.K. (2013a). Low cellular P-quota and poor metabolic adaptations of the freshwater cyanobacterium *Anabaena fertilissima* Rao during Pi-limitation. *Antonie van Leeuwenhoek* 103, 277-291

[91] Gilbert, G.A., Vance, C.P., Allan, D.L. (1998). Regulation of white lupin metabolism by phosphorus availability In: *Phosphorus in Plant Biology: Regulatory Roles in Molecular, Cellular, Organismic and Ecosystem Processes* (Lynch, J.P., Deikman, J., Eds.), Rockville, MD: American Society Plant Physiology; 1998, pp. 157-167.

[92] Gilroy, S., Jones, D.L. (2000). Through form to function: root hair development and nutrient uptake. *Trends Pl Sci,* 5, 56-60.

[93] Raghothama, K.G. (2000). Phosphate transport and signaling. *Curr Opinion Pl Biol,* 3, 182-187.

[94] Schwarz, R., Forchhmmer, K. (2005). Acclimation of unicellular cyanobacteria to macronutrient deficiency: emergence of a complex network of cellular responses. *Microbiology* 151, 2503-2514.

[95] Uhde-Stone, C., Gilbert, C., Johnson, J.M.F., Litjens, R., Zinn, K.E., Temple, S.J., Vance, C.P., Allan, D.L. (2003a). Adaptation of white lupin to phosphorus deficiency involves enhanced expression of genes related to organic acid metabolism. *Pl Soil*, 248, 99-116.

[96] Uhde-Stone, C., Zinn, K.E., Ramirez-Yanez, M., Li, A., Vance, C.P., Allan, D.L. (2003b). Nylon filter arrays reveal differential gene expression in protoid roots of white lupin in response to P deficiency. *Pl Physiol,* 131, 1064-1079.

In: Microbes in Soil and Their Agricultural Prospects
Editors: K. Kishore Choudhary and D. Wattal Dhar
ISBN: 978-1-63482-824-6

Chapter 16

MOLECULAR TOOLS IN THE STUDY OF SOIL MICROBIAL DIVERSITY: WITH AN EMPHASIS ON PHOSPHATE SOLUBILIZING MICROORGANISMS

Seema B. Sharma*

Department of Earth and Environmental Science,
KSKV Kachchh University, Mundra Road, Gujarat, India

ABSTRACT

Agriculture is the backbone of human existence. It is important for ensuring food security, alleviating poverty and conserving the vital natural resources. As the world population continues to increase, the demands placed upon agriculture to supply food will be one of the greatest challenges facing the agrarian community. In order to meet this challenge, a great deal of effort focusing on the soil biological system and the agro-ecosystem as a whole is needed. Soil quality is an issue that needs to be included in discussions of agricultural sustainability. Of the various plant nutrients, even though P is abundant in soil, its availability is limited in plants due to fixation by other soil elements such as insoluble phosphates of iron, aluminium, and calcium. As a result of this, the plant available P fraction and the concentration in the soil solution may be insufficient to satisfy plant requirements. Chemical P fertilizers are cost intensive and pose environmental hazards. Providing P to plants through biological means is an eco-friendly and viable alternative. Among heterogeneously distributed soil microflora, a group of microorganisms commonly referred to as phosphate-solubilizing microorganisms (PSM) have been found active in conversion of insoluble P to soluble forms and making it accessible to plants. This chapter focuses on various molecular techniques that have been developed in the recent past to explore these PSM, which cannot be explored by conventional culture-dependent approach. Genetic intervention in the functioning of various genes and enzymes involved in microbial-mediated P solubilisation have been discussed.

* Corresponding Author Department of Earth and Environmental Science, KSKV Kachchh University, Mundra Road, Bhuj Pin-Code 370 001, Gujarat, India, Email: seemabhargavsharma@gmail.com

Keywords: microbial diversity, soil, agriculture, phosphate solubilising microorganisms, fertilizers

INTRODUCTION

Soil is a key natural resource interacting with above ground plant and animal communities contributing to the success of sustainable agriculture. Soil quality can be defined as the soil's capacity to function in a desired manner such as to produce healthy crops, resist erosion, and minimize environmental impacts. It consists of the chemical, physical, and biological components of a soil and their interactions [1]. The major emphasis in soil quality investigations has been on the use of chemical and physical attributes of soil although the biological portion is equally significant. However, these two soil features are only part of what may impart to a soil its essence or characteristics. Therefore, awareness to investigate microbial parameters to identify soil quality has been increasing day by day. The biological component of the soil is responsible for soil humus formation, cycling of nutrients, soil tilth and structure and a myriad of other functions. It is imperative that we increase our understanding of soil microbiology as a part of managing sustainable agricultural systems [2]. The concept of soil quality needs to be further defined, and parameters involved in those changes catalogued so that the characteristics or essence of that soil can be determined and tracked.

MICROBES: UBIQUITOUS JANITORS OF EARTH

Microorganisms have been integral to the history and function of life on Earth. They have played central roles in Earth's climatic, geological, geochemical, and biological evolution [3]. The prokaryotic life emerged about 3.8 billion years ago; about 2 billion years before eukaryotic life arose and have provided conditions on the planet that have made it habitable for all other species [4]. Microorganisms represent by far the richest repertoire of molecular and chemical diversity in nature. Microbes are ubiquitous janitors of earth. Even though microorganisms are rarely conspicuous in natural environments, it is estimated that about half the biomass on earth is microbial. Furthermore, microbial life is widely distributed: where there is life on earth, there is microbial life. They are found in virtually every imaginable ecological niche on Earth, from the tropics to the Arctic and Antarctica, from underground mines and oil fields to the stratosphere and the top of great mountains, from deserts to the dead sea, from above-ground hot springs to underwater hydrothermal vents [3]. The microbial world is the largest unexplored reservoir of biodiversity on the Earth. Microbial diversity in soil ecosystems exceeds, by far, that of eukaryotic organisms. One gram of soil may harbour up to 10 billion microorganisms of possibly thousands of different species [5]. In order to best exploit microorganisms, we need to know what is there and what we can use. Studies of many organisms may yield potential products. Microorganisms represent by far the richest repertoire of molecular and chemical diversity in nature. They underlie basic ecosystem processes such as biogeochemical cycles and food chains, as well as maintain vital and often elegant relationships between themselves and higher organisms. Microbes provide the fundamental underpinning of all ecosystems.

PHOSPHORUS DEFICIENCY IN SOILS: THE REAL BOTTLENECK OF AGRO-ECOSYSTEMS

In the soil various macro (Nitrogen, Phosphorus and Potassium) and micro nutrients are required for the growth of plants. Phosphorus is the most important key element in the nutrition of plants, next to nitrogen (N). It plays an important role in virtually all major metabolic processes in plant including photosynthesis, energy transfer, signal transduction, macromolecular biosynthesis and respiration [6] and nitrogen fixation in legumes [7]. Although P is abundant in soils in both inorganic and organic forms, it is a major limiting factor for plant growth as it is in an unavailable form for root uptake. Inorganic P occurs in soil, mostly in insoluble mineral complexes, some of them appearing after frequent application of chemical fertilizers. These insoluble, precipitated forms cannot be absorbed by plants [8]. Organic matter is also an important reservoir of immobilized P that accounts for 20–80% of P in soils [9]. Only 0.1 % of the total P exists in a soluble form available for plant uptake [10] because of its fixation into an unavailable form due to P fixation. The term P fixation is used to describe reactions that remove available phosphate from the soil solution into the soil solid phase [11]. There are two types of reactions (a) phosphate sorption on the surface of soil minerals and (b) phosphate precipitation by free Al^{3+} and Fe^{3+} in the soil solution [12]. The soils that exhibit highest P fixation capacity occupy 1,018 million hectares (ha) in the tropics [13]. It is for this reason that soil P becomes fixed and available P levels have to be supplemented on most agricultural soils by adding chemical P fertilizers, which not only represent a major cost of agricultural production but also impose adverse environmental impacts on overall soil health and degradation of terrestrial, freshwater and marine resources [14]. Thus, increased P levels have been identified as a main factor for eutrophication of surface waters that may lead to algal blooms [15]. The repeated and injudicious applications of chemical P fertilizers, leads to the loss of soil fertility [16] by disturbing microbial diversity, and consequently reducing crop yield [17].

Moreover the efficiency of applied P fertilizers in chemical form rarely exceeds 30% due to its fixation, either in the form of iron/aluminium phosphate in acidic soils [18] or in the form of calcium phosphate in neutral to alkaline soils [19]. It has been suggested that the accumulated P in agricultural soils would be sufficient to sustain maximum crop yields worldwide for about 100 years if it were available [20]. A major characteristic of P biogeochemistry is that only 1% of the total soil P (400–4,000 kg P/ ha in the top 30 cm) is incorporated into living plant biomass during each growing season (10–30 kg P/ha), reflecting its low availability for plant uptake [21]. Furthermore P is a finite resource and based on its current rate of use, it has been estimated that the worlds known reserves of high quality rock P may be depleted within the current century [22]. Beyond this time the production of P based fertilizers will require the processing of lower grade rock at significantly higher cost [23].The realization of all these potential problems associated with chemical P fertilizers together with the enormous cost involved in their manufacture, has led to the search for environmental compatible and economically feasible alternative strategies for improving crop production in low or P-deficient soils. The use of microbial inoculants (biofertilizers) possessing P-solubilizing activities in agricultural soils is considered as an environmental-friendly alternative to further applications of chemical based P fertilizers [17].

MICROBIALLY MEDIATED P SOLUBILIZATION

Microorganisms are an integral component of the soil P cycle and are important for the transfer of P between different pools of soil P. Phosphate Solubilzing Microorganisms (PSM) through various mechanisms of solubilization and mineralisation are able to convert inorganic and organic soil P respectively into the bioavailable form facilitating uptake by plant roots (Figure 1).

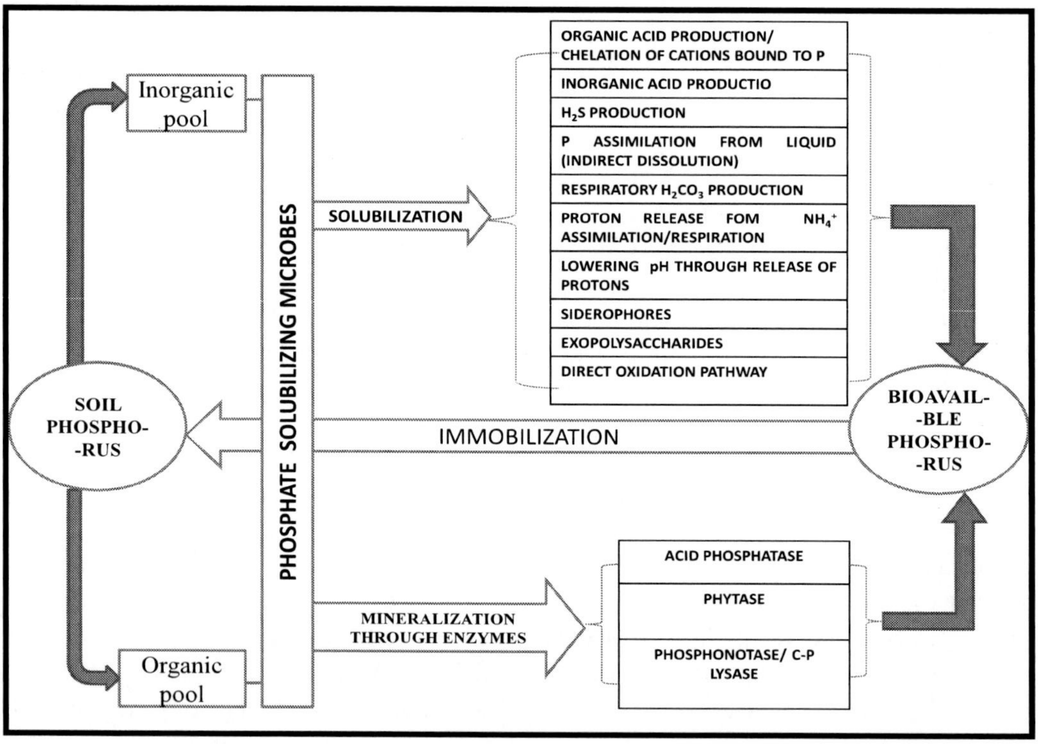

Figure 1.Schematic representation of mechanism of soil P solubilization/mineralization and immobilization by PSM (From Sharma et al. 2013 [17]).

It is important to determine the actual mechanism of P solubilisation by PSM for optimal utilization of these microorganisms in varied field conditions. Hence it is imperative to better understand the plant-soil-microbial P cycle with the aim of reducing reliance on chemical P fertilizers. This has led to increased interest in the harnessing of microorganisms to support P cycling in agro-ecosystems. Microbial mediated P management in the form of biofertilizers is an ecofriedly and cost effective approach for sustainable development of agricultural crop [17].

Molecular Tools in PSM Study

Looking at the possible avenues which can open up with exploring these environmental friendly microorganisms, it is necessary to study the composition and dynamics of these microbial populations to reach a better understanding of soil PSM diversity and P uptake by plants.

The organisms involved in phosphorus (P) cycling in soils are highly varied, and microorganisms probably play the most important role. However, more than 99% of soil microorganisms have not been cultured successfully [24].

Table 1. Summary Of Molecular Tools Used in PSM Study

Sr. No.	TECHNIQUE	ADVANTAGE	DISADVANTAGE
1	LMW RNA (Low molecular weight RNA profiling)	straightforward; no in vitro amplification step necessary; active organisms can be detected	Rapid degradation of RNA; Limited phylogenetic information and length variation of the LMW RNA
2	DGGE(denaturant gradient gel electrophoresis)/ TGGE(Temperature gradient gel electrophoresis)	Differences among samples can be detected at species or strain levels	Resolution is frequently not very good, especially in communities with higher diversity
3	SSCP (single-strand conformation polymorphism)	Differences among samples can be detected at species or strain levels	problems of reproducibility
4	RFLP(restriction fragment length polymorphism)/ARDRA(amplified ribosomal DNA restriction analysis)	Differences among samples can be detected at genus or higher levels	Sequence information is frequently too short for classification
5	T-RFLP/ (terminal RFLP)	high resolution; intra-lane markers; direct quantification of fragments	no phylogenetic information obtained; expensive equipment
6	RAPD(randomly amplified polymorphic DNA)	no special primer design required	no phylogenetic information obtained; problems of reproducibility
7	qRT-PCR (quantitative real-time polymerase chain reaction)	Quantification of specific genes	Specific primer set is required
8	Immunocapture using BrdU (5-bromo-2-deoxyuridine)	Active organisms can be analyzed. Immunoprecipitation can be applied	Dynamics of BrdU in soil are not well understood
9	FISH (fluorescence in situ hybridization)	Localization of specific microbes or genes can be visualized	amplification of signals is required for functional genes
10	ELF (enzyme-labeled fluorescence)-97 phosphate	localization of phosphatase	Application is specific
11	phosphate-reporter bacteria	P availability for bacteria can be visualized	Application is specific
12	Microarray	Whole aspects of specific microbes or mRNAs can be detected	Very expensive
13	Next generation sequencer	Metagenomic and metatranscriptomic data are Available	amplification of signals is required for functional genes

(Data inputs from Muyzer G (1999) [85] and Wasaki J and Maruyama H (2011) [41]).

Therefore, culture-independent methods are required to study the function and ecology of microbes involved in P cycling in soils. Molecular approaches for such culture-independent methods have been developed in the recent past, and each technique has its own advantages and disadvantages (Table 1). The study of populations of microorganisms, which share the common characteristic of phosphate solubilization, has great complexity because they belong to very diverse groups sometimes not closely related under a phylogenetic point of view. Therefore good techniques are needed to perform the analysis and identification of PSM populations. The molecular techniques based on nucleic acid composition like LMW RNA profiling and PCR based techniques, are excellent tools for this purpose, as they are precise, reproducible and not dependent on culture media composition or growth phase of microorganisms [25]. The starting point for all molecular approaches is the extraction of environmental DNA from the soil. Numerous methods for soil DNA extraction have been reported and several kits for soil DNA extraction are commercially provided, e.g., from Epicentre Biotechnologies (Madison, WI, USA), MO Bio Laboratories (Carlsbad, CA, USA), Nippongene (Toyama, Japan). Fungal DNA can be extracted by employing the CTAB (cetyltrimethylammonium bromide; a detergent for solubilizing membranes) method [26].

In the case of arbuscularmycorrhizal (AM) fungi, spores can be collected from soils by sieving. This method enables DNA extraction specifically from spores of AM fungi. But the yield of DNA is sometimes too low (in the nanogram order) for direct analysis hence it has to be amplified using techniques such as polymerase chain reaction (PCR)-dependent molecular tools and metagenomics, which is a culture independent technique.

DNA Based Techniques

There are numerous PCR-based techniques that can be applied to the study of biodiversity. Among them, some are based on sequencing of reaction products but techniques of this kind are more useful in taxonomy than in studies of large populations, as sequencing is still too complex a method to be applied to many strains at the same time. For this reason, other techniques have been developed in order to obtain the same results without the need of gene sequencing. Such techniques are known as DGGE and TGGE which are based on amplification of a G + C hyper-variable content zone which allows the separation of bands of the same size (generally in 16S rDNA) based on their G + C content obtained by means of variation of temperature during electrophoresis in polyacrylamide gels through a time or spatial gradient [27]. Denaturant gradient gel electrophoresis (DGGE) is frequently applied for comparing the microbial communities of various environments [28]. DNA fragments amplified from environmental DNA by PCR are separated on polyacrylamide gels containing gradients of denaturants. Differences of sequences (i.e., GC contents) result in different denaturing points, which show as bands on the denaturant gradient gel. The band patterns obtained by PCR-DGGE are used to compare samples by multivariate analyses such as principal component analysis or cluster analysis. The merit of the DGGE method is the ease with which microbial communities of different samples can be compared without DNA sequence determination. The only technical requirement for DGGE is electrophoresis equipment. DGGE can visualize differences at the line or species levels, because bands can be differentiated even if the sequences compared differ merely in one base. By contrast, DGGE is unsuitable for the detection of differences at the genus or any higher levels. When the diversity of the sample is high, the bands separated in the DGGE are poorly resolved. Marschner et al., (2004) [29] applied PCR-DGGE of 16S rDNA to investigate microbial community structures of the rhizospheres of chickpea, canola, and Sudan grass as affected by P fertilization. Similarly, Wasaki et al., (2005) [30] and Weisskopf et al., (2005) [31]studied the community structures of rhizobacteria associated with white lupin plants, in order to clarify the effects of root exudates on the diversity and community structure of rhizobacteria in cluster roots formed by P-starved white lupin plants. Their results suggested that the diversity and community structure were strongly influenced by root exudates, such as citrate and flavonoids. Temperature gradient gel electrophoresis (TGGE) resembles DGGE [32] in principle but differs in having a temperature rather than a denaturant gradient. Another fingerprinting technique, used in PSM ecology is single-stranded-conformation polymorphism (SSCP). In this technique DNA fragments, such as PCR products obtained with primers specific for the 16S rRNA gene, are denatured and directly electrophoresed on a non-denaturing gel. Separation is based on differences in the folded conformation of single-stranded DNA, which influences the electrophoretic mobility [33]. Sometimes small primers and low annealing temperatures are used; these lead to random amplifications within the microbial genome thus obtaining Randomly Amplified Polymorphic DNA RAPD profiles [34]. Intra-specific variations have been described in the RAPD both in bacteria and in fungi and therefore this technique is of great use in biodiversity studies of PSM [25].

Other techniques are based on the amplification of certain molecules followed by digestion with several restriction enzymes, obtaining RFLP (Restriction fragment length polymorphism) profiles which can be mathematically analysed to establish clusters [35, 36]. While DGGE and TGGE have been applied to prokaryotes, for the time being the RFLP profiling has also been widely applied to eukaryotes and therefore can be a very useful technique in biodiversity studies [36]. RFLP visualizes microbial communities as patterns of restriction fragment length. RFLP targeted to SSU rRNA is designated as ARDRA (amplified ribosomal DNA restriction analysis [37]. ARDRA has been used to characterize cultured isolates of potential biofertilizer*s* [38]. Because SSU rRNA is relatively short (>1,500 bp), its fragments are generally digested by a restriction enzyme that recognizes four bases. The restriction fragments are separated by electrophoresis. RFLP is a useful method for detecting relatively substantial differences (higher than genus level), but it cannot resolve small differences between the sequences compared. Specific bands can be isolated from gels and sequenced, although the sequence length is frequently too short for species identification. T-RFLP (terminal RFLP) using a primer labeled with a fluorescent dye at the 5 0 - terminus is a useful technique for the detection of specific fragments. The fluorescence can be detected with high resolution by DNA sequencers or other analytic equipment. George et al. (2009) [39] used T-RFLP to examine the effect of extracellular release of phytase from roots of transgenic plants on microbial community structures in the rhizosphere. They demonstrated that the expression of phytase in transgenic plants had little or no impact on the microbial community structure as compared with control plant lines.

Yet another procedure has been described [40] for obtaining DNA profiles that are effective in biodiversity and taxonomy. Since the variations observed are established at subspecies level within the same species. This method is based on the use of the two universal primers used to make a complete amplification of the 16S rDNA increasing its concentration about ten times and applying an annealing temperature of 50-55 °C. The sequencing of 16s r

DNA of the strain is carried out using primers. The online programme BLAST is used to find out the related sequence with known taxonomic information in the databank at NCBI website to accurately identify the strain.

RNA Based Tools

DNA work is relatively easy to apply on environmental samples, but it delivers information not only about active, but also about inactive microbes. Environmental RNA-targeted molecular approaches are required to target the major active players or key genes in the environment. However, RNA extraction from soil is still difficult because of the instability of RNA molecules, and only a few RNA-based studies have succeeded in identifying microorganisms involved in P cycling [41] and the identification of the different isolates. Certain molecules of RNA have been proposed as molecular fingerprints of microorganisms, which may be applied to the study of microbial populations [42]. These molecules have been named low molecular weight RNA (LMW RNA) and they include the 5S ribosomal RNA(r RNA) and transfer RNA(tRNA) in prokaryotes (both in archaea and bacteria). In eukaryotes, besides these molecules, the 5.8S rRNA forms part of the LMW RNA profiles. This differentiating characteristic between prokaryotes and eukaryotes is very useful when analyzing isolates from complex populations, since in the eukaryotes there is always one molecule more (5.8S rRNA) which, besides, has different sizes for each eukaryotic genus, when a voltage gradient electrophoresis is used for separating these molecules, when they have similar sizes [43]. The results obtained up to now have allowed researchers to establish that LMW RNA molecules separated by staircase electrophoresis are molecular images of each microbial species both in the case of eukaryotes and prokaryotes [44, 45].

According to the results obtained so far, each prokaryote genus has a characteristic 5S rRNA zone and each eukaryote genus has a different combined zone of 5S and 5.8S rRNA. On the other hand tRNA profiles, both class 1 and class 2, are different in each species of prokaryotes and eukaryotes belonging either to the same or different genus. It has also been demonstrated that LMW RNA profiles of all the strains belonging to the same species are identical [45]. Hence LMW RNA profiles are an excellent technique for analyzing population diversity as they can be applied to a large number of strains in a quick, easy and precise way without intra-specific variations [25]. Small subunit (SSU) rRNA and internal transcribed spacer (ITS) regions are frequently analyzed for the molecular characterization of uncultured microbes. These sequences are easily amplified by PCR. Amplified fragments are compared with respect to their length, restriction fragment patterns, and substantial differences between sequences [41].

Monteiro et al., (2009) [38] analyzed partial 16S rRNA sequences for 83 bacterial colonies, which were isolated from the rhizosphere of vetiver (*Chrysopogonzizanioides*) as potential biofertilizers. It was revealed that potentially phosphate solubilizing bacteria coexisted in the vetiverrhizosphere that were closely related to *Acinetobacter, Burkholderia, Chryseobacterium, Dyella, Enterobacter, Klebsiella,* and *Pseudomonas.*

Molecular Tools to Study the FunctionalGenes in PSM

Quantitative Real Time-PCR (qRT-PCR) for environmental DNA determines the quantity of microorganisms harboring the targeted gene. When genus-specific primers for SSU rRNA are used, the results of the technique can quantify the dominance of the genus. In the case of RNA-based methods, the amount of mRNA of the target gene can be determined. The method is useful for functional genes that are directly involved in P cycling such as phosphatases, phytases, and nucleases. Melting curves of amplified fragments produced by most of the commercial equipments provide information on the diversity of the amplified fragments. qRT-PCR was applied to evaluate interactions of arbuscularmycorrhizal fungi. Specific primers for ITS1 and rRNAs of *Glomusmosseae* and *G. intraradices* were designed for qRT-PCR [46]. The distribution of microbial species or functional groups of interest can be visualized by FISH (fluorescence in situ hybridization). A specific primer designed for SSU rRNA or a functional gene, and containing a fluorescent dye, is required for FISH. It was applied for the detection of potential polyphosphate-accumulating bacteria in biological phosphorus removal plants [47]. In a study by Bashan et al., (2010) [48] three plant growth-promoting bacteria (PGPB; *Bacillus pumilus* ES4, *B. pumilus* RIZO1, and *Azospirillum brasilense* Cd) were tested for their ability to enhance plant growth and development of the native Sonoran Desert shrub quailbush (*Atriplexlentiformis*) and FISH analysis showed that the preferred site of colonization was the root tips and root elongation area.

An analog of thymidine, 5-bromo-2-deoxyuridine (BrdU), can be used for labeling active microorganisms in the environment. BrdU-antibodies are then applied to detect the microorganisms containing BrdU. Artursson et al., (2005) [49] employed BrdUimmunocapture in combination with T-RFLP. They found distinct changes in active bacterial community compositions related to *G. mosseae* inoculation, treatment with an antifungal compound, and plant type. The dominant bacterial species that were activated as a result of *G. mosseae* inoculation were found to be mostly uncultured bacteria and *Paenibacillus species*. Phosphate-reporter bacteria have been developed and used for studies on P cycling. In particular, de Weger et al., (1994) [50] developed phosphate-reporter strains of *Pseudomonas putida WCS358* in which the production of b-glucosidase was regulated by the promoter of a gene responsive to P starvation. Kragelund et al., (1997) [51] used two phosphate-reporter strains of *P. fluorescens DF57* harboring luciferase regulated by the promoter of a gene responsive to P starvation. It can be a useful tool for studying the plant–microbe interactions involved in P cycling, because the *Pseudomonas spp.* are frequently isolated as potentially plant growth-promoting rhizobacteria. Enzyme-labeled fluorescence (ELF)-97 phosphate can be used for histochemical localization of phosphatases. This substrate provides fluorescent precipitates after hydrolysis by phosphatases [52]. Although this technique has still not been used in PSM.

Novel Technologies in PSM Study

During the last ten years various novel technologies such as omics have been developed. The term ‘Omics’ refers to the comprehensive analysis of biological systems. This new technology is developing constantly and quickly. Recent advances in bioinformatics and high-throughput technologies such as microarray analysis are bringing about a revolution in our understanding of cell biology and the molecular mechanisms underlying PSM activities [53, 54, 55]. Gene expression profiling has enabled the measurement of thousands of genes in a single RNA sample. There are a variety of microarray platforms that have been developed to accomplish this and the basic idea for each is simple: a glass slide or membrane is spotted or "arrayed" with DNA fragments or oligonucleotides that represent specific gene coding regions. Purified RNA is then fluorescently- or radioactively labeled and hybridized to the slide/membrane. In some cases, hybridization is done simultaneously with reference RNA to facilitate comparison of data across multiple experiments. The data may be analyzed by a variety of statistical algorithms. Microbes involved in P cycling could be detected by the microarray technique if a microarray targeted for soil microbes is developed. He et al., (2007) [56] developed a microarray designated the "GeoChip" containing 24,243 oligonucleotide probes and covering >10,000 genes in >150 functional groups involved in N, C, S, and P cycling, metal reduction and resistance, and organic contaminant degradation. Their array was successfully used for tracking the dynamics of metal-reducing bacteria and associated communities. To study uncultured microorganisms, environmental DNA can be recovered and sequenced. This approach is called "metagenomics." It is expected that metagenomics will provide the information required for understanding the whole micro-flora in the soil, the functions of the organisms involved (including P cycling), and the isolation of beneficial genes from uncultured microorganisms. Recently, Badri et al., (2009) [57] reported the first study on the microbial community in the rhizosphere using a pyrosequencer, one of the next-generation DNA sequencers.

Molecular approaches and recent omics analyses have could prove to be powerful tools for clarifying processes and factors involved in microbial P cycling. Adequate molecular approaches have to be selected that provide meaningful complements to conventional approaches for future investigations into this research area.

Genetic Engineering of PSM

Although knowledge of the genetics of phosphate solubilization is still scanty, and the studies at the molecular level in order to understand how precisely the PSM brings out the solubilization of insoluble P are inconclusive [58]. However, some genes involved in mineral and organic phosphate solubilization have been isolated and characterized. Initial achievements in the manipulation of these genes through genetic engineering and molecular biotechnology followed by their expression in selected rhizobacterial strains open a promising perspective for obtaining PSM strains with enhanced phosphate solubilizing capacity, and thus, a more effective use of these microbes as agricultural inoculants.

Introduction or over-expression of genes involved in soil phosphate solubilization (both organic and inorganic) in natural rhizosphere bacteria is a very attractive approach for improving the capacity of microorganisms to work as inoculants. Insertion of phosphate-solubilizing genes into microorganisms that do not have this capability may avoid the current need of mixing two populations of bacteria, when used as inoculants (nitrogen fixers and phosphate-solubilizers [59].

There are several advantages of developing genetically-modified PSM over transgenic plants for improving plant performance: (1) With current technologies, it is far easier to modify a bacterium than complex higher organisms, (2) Several plant growth-promoting traits can be combined in a single organism, and (3) Instead of engineering crop by crop, a single, engineered inoculant can be used for several crops, especially when using a nonspecifc genus like *Azospirillum* [58]. Some barriers should be overcome first to achieve successful gene insertions using this approach, such as the dissimilarity of metabolic machinery and different regulating mechanism between the donor and recipient strains. Despite the diffculties, significant progress has been made in obtaining genetically engineered microorganisms for agricultural use [60].

Genetic Intervention in Organic P Solubilization (OPS) Mechanism

As mineralization of organic p occurs through various enzyme mediated mechanisms by the PSM, Interest in these enzymes has increased during the last decade because of their potential biotechnological applications. Macaskie et al., (1997) [61] reported on the successful use of Class A NSAPs (non-specific acid phosphatases) as tools for environmental bioremediation of uranium-bearing wastewater, and Bonthrone et al., (1996) [62] on heavy metal biomineralization (particularly Ni^{2+}). A new biotechnological application for NSAPs would be to transfer and express these genes in PSM to obtain improved phosphate solubilizing strains using recombinant DNA technology. Several acid phosphatase genes from Gram-negative bacteria have been isolated and characterized [63].These cloned genes represent an important source of material for the genetic transfer of this trait to PGPB strains. Some of them code for acid phosphatase enzymes that are capable of performing well in soil. For example, the acpA gene isolated from *Francisella tularensis* expresses an acid phosphatase with optimum action at pH 6, with a wide range of substrate specificity [64]. Among rhizobacteria, a gene from *Burkholderia cepacia* that facilitates phosphatase activity was isolated [65]. This gene codes for an outer membrane protein that enhances synthesis in the absence of soluble phosphates in the medium, and could be involved in P transport to the cell. Besides, cloning of two nonspecific periplasmic acid phosphatase genes (napD and napE) from *Rhizobium (Sinorhizobium) meliloti* was accomplished [66, 67]. The napA phosphatase gene from the soil bacterium *Morganella morganii* was transferred to *Burkholderia cepacia IS-16*, a strain used as a biofertilizer, using the broad-host range vector pRK293 [68]. An increase in extracellular phosphatase activity of the recombinant strain was achieved.

Another attractive application of these enzymes that is not currently exploited is solubilization of soil organic phosphorus through phytate degradation. The growth and phosphorus nutrition of Arabidopsis plants supplied with phytate was improved significantly when they were genetically transformed with the phytase gene (phyA) from *Aspergillus niger* [69]. Thermally stable phytase genes (phy) from Bacillus sp. DS11 [70] and from*B. subtilis VTT E-68013* [71] has been cloned. Acid phosphatase/phytase genes from *E. coli* (appA and appA2 genes) have also been isolated and characterized [72]. The bi-functionality of these enzymes makes them attractive for solubilization of organic P in soil.

Genetic Intervention in Mineral P Solubilization (MPS) Mechanism

Various forms of inorganic P occur in soils. In neutral to alkaline soils calcium phosphate is the dominanat form. In the case of calcium phosphates, a significant body of evidence has been developed to show that Gram negative bacteria exhibiting superior mineral phosphate solubilizing (MPS) capabilities utilize the direct oxidase pathway [73]. This pathway (also called nonphosphorylating oxidation) produces gluconic acid (GA) and 2-ketogluconic acid directly in the periplasmic space. These strong organic acids can dissolve poorly soluble calcium phosphates such as hydroxyapatite and rock phosphate ore (e.g., fluroapatite). Therefore, the conservation of the direct oxidation pathway in rhizobacteria may, at least in part, result from the mutualistic advantage provided by the MPS trait. This provides with both a unifying metabolic strategy and a set of biochemical and genetic probes with which to systematically identify and evaluate the role of a specific subpopulation of rhizosphere bacteria in P cycling. GA biosynthesis is carried out by the glucose dehydrogenase (GDH) enzyme and the co-factor, pyrroloquinolinequinone (PQQ). Some genes involved in MPS in different species have been isolated. Goldstein and Liu (1987) [74] were the first to clone a gene involved in MPS from the Gram negative bacteria *Erwinia herbicola*. Expression of this gene allowed production of GA in *E. coli* HB101 and conferred the ability to solubilize hydroxyapatite. *E. coli* can synthesize GDH, but not PQQ, thus it does not produce GA. The cloned 1.8 kb locus encodes a protein similar to the gene III product of a pqq synthesis gene complex from *Acinetobacter calcoaceticus*, and to pqqE of *Klebsiell aneumonie* [75]. These authors suggested that the *E .herbicola* DNA fragment functions as a PQQ synthase gene, and that probably, some *E. coli* strains contain some cryptic PQQ synthase genes that could be complemented by this single open reading frame (ORF) isolated by them.

In addition, the demonstrated efficacy of the direct oxidation pathway for the dissolution of fluroapatites has provided a potential strategy for large-scale bioprocessing of rock phosphate ores. In terms of future research, successful molecular cloning of direct oxidation pathway genes provides the tools with which to study relationships between the population dynamics of MPS+ bacteria in the rhizosphere and Pi in the soil solution. Available probes include several apo-glucose dehydrogenase genes, PQQ biosynthesis genes, and gluconate dehydrogenase gene. Such genetic probes have been used to demonstrate the presence of unique populations of Gram-negative MPS+ rhizobacteria in two alkaline desert soil environments where the levels of poorly soluble calcium phosphates are extremely high but Pi is undetectable in bulk soil extracts. Both MPS+ populations were capable of high levels of direct oxidation of glucose, and the presence of the quinoprotein glucose dehydrogenase was confirmed by both enzymatic and molecular biology assays. In one case, a unique rhizobacterial population of *Enterobacter cloacae* expressed the direct oxidation pathway only in the presence of compounds washed from the root of the host plant, Helianthus sp. providing preliminary evidence for mutualism in this highly alkaline soil (pH 10) [76]. Chromosomal insertion of these genes under appropriate promoters is another interesting approach [77]. Rodriguez et al., (2000) [65] carried out a genetic construction using the broad host range vector pKT230 and plasmid pMCG898, which encode the *Erwinia herbicola*pyrroloquinolinequinone (PQQ) synthase, a gene involved in mps. The final construct was transformed and expressed in *E. coli MC1061*, and the recombinant plasmids were transferred to *Burkholderiacepacia IS-16* and *Pseudomonas sp. PSS* recipient cells by conjugation.

Clones containing recombinant plasmids produced higher clearing halos in plates with insoluble P as the unique P source, in comparison to those strains without plasmids, demonstrating the heterologous expression of the E. herbicola gene in the recipient strains. This genetic manipulation allowed the increase in mps ability of both strains, enhancing their potential as growth promoters of agricultural crops. In addition, the subcloning of the gene encoding the PhoC acid phosphatase from *Morganella morganii* (phoC gene), in a vector that permits stable chromosomal integration of this gene in plant growth-promoting bacteria, has been reported [78]. Another important rhizosphere-competent bacteria (*Pseudomonas spp.*) can form gluconic acid through the oxidative metabolism and overexpression of PQQ synthase and glucose dehydrogenase (GDH) genes make them functionally a better PS organism. Interestingly, in addition to its role in P-solubilization, PQQ also plays an important role in beneficial traits, such as antifungal activity and induced systemic resistance (ISR) of *Enterobacter intermedium*, possibly by acting as a cofactor for several enzymes including GDH [79]. In yet another approach, the mps genes can directly be transferred into the target bacteria by over-/underexpression of genes followed by the selection of transformants with mps ability. Such an approach has been used to obtain mps genes from Synechosystis PCC 6803 in *E. coli* [80]. However, it remains to be seen if this will also be effective in other bacteria. Recently a novel molecular marker for studying phylogeny and diversity of phosphate-solubilizing pseudomonads ie the pyrroloquinolinequinone biosynthetic gene pqqC has been discovered that can be used complementary to housekeepinggenes [81].

Genetic engineering could also help in increasing the survival of the inoculant strains by incorporating the abilities to utilize certain nutrients better than the rest of the microbial populations [82]. Development of novel techniques and access to new technologies will be important. Recent developments in microbial community analysis that do not rely on cultural procedures will provide better understanding of how microorganisms interact in complex environments. For example, procedures are now available for detection and direct visualization of specific microorganisms in the rhizosphere [83]. Nevertheless these examples highlight that new opportunities for direct manipulation of microorganisms and/or plants do exist. However, it is also important to recognize that any approach to increase the mobilization of soil P through genetic modification of soil microorganisms or plants will also need to satisfy a range of community and environmental issues. Although the initial achievements in the manipulation of these genes opens a promising perspective for obtaining PSM strains with enhanced P solubilizing capacity but the release of genetically modified organisms is controversial and raises some ethical questions. While some countries encourage it, others prohibit the use of the technology. However, studies carried out so far have shown that following appropriate regulations, genetically modified microorganisms can be applied safely in agriculture [60, 84].

The molecular engineering of these microbes has also provided a new insight into the promotion of crops in P-deficient soils. In this context, novel, genetically engineered and soil- and region specific PSM(s) and technologies have to be developed, pilot tested and transferred to farmers in a relatively short time in order to improve plant P-nutrition and agro-ecosystem sustainability.

Conclusion

Phosphorus is an important limiting factor in agriculture production, and looking at the hazardous effects of chemical P fertilizers, microbial intervention of PSM seems to be an effective way to solve the phosphorus availability in soil. However P-solubilization in soil is much more difficult to study than solubilization of P in broth culture. The crops respond differently to the inoculation of PSMs and are dependent on several factors such as the soil temperature, moisture, pH, salinity, source of insoluble P, method of inoculation, the energy sources and the strain of microorganism used. Hence study of PSM activity in correlation with these factors has to be done extensively before PSM can be used as a biofertilizer with promising results. The successful implementation of this approach has already been demonstrated in the fields by various workers, to a limited extent. However the large scale use of this technology would benefit from additional studies, particularly those directed towards understanding how the interaction between soil and microbial system might be facilitated.

The promise of exploiting soil microorganisms to increase mobilization of soil P remains. Whether or not this will be achieved through better management of soil microbial communities, by development of more effective microbial inoculants, through the genetic manipulation of specific organisms, or with a combination of these approaches is not known. What is clear though is that soil microorganisms play an important role in the mobilization of soil P and that detailed understanding of their contribution to the cycling of P in soil-plant systems is required for the development of sustainable agriculture and our movement from a green revolution to an evergreen revolution can be accomplished.

Despite their different ecological niches and multiple functional properties, P-solubilizing microorganisms have yet to fulfill their promise as commercial bio-inoculants. Current developments in our understanding of the functional diversity, rhizosphere colonizing ability, mode of actions and judicious application are likely to facilitate their use as reliable components in the management of sustainable agricultural systems. Although significant studies related to PSM and their role in sustainable agriculture have been done over the last few decades, the required technique remains in its infancy. Nevertheless with an awareness of the limitations of existing methods, a reassessment can be expected, so that the use of PSM as potential biofertilizers in different soil conditions becomes a reality

Enhancement in the use of PSM is one of the newly emerging options for meeting agricultural challenges imposed by the still-growing demand for food. Thus, more than ever, obtaining high yields is the main challenge for agriculture. In addition, in recent years both producers and consumers have increasingly focused on the health and quality of foods, as well as on their organoleptic and nutritional properties. Hence, this biotechnology is also likely to ensure conservation of our environments. However, before PSM can contribute to such benefits, scientists must learn more about them and explore ways and means for their better utilization in the farmers' fields. Future research should focus on managing plant–microbe interactions, particularly with respect to their mode of actions and adaptability to conditions under extreme environments for the benefit of plants.

Furthermore, scientists need to address certain issues, like how to improve the efficacy of biofertilizers, what should be an ideal and universal delivery system, how to stabilize these microbes in soil systems, and how nutritional and root exudation aspects could be controlled in order to get maximum benefits from PSM application. Biotechnological and molecular approaches could possibly develop more understanding about PSM mode of actions that could lead to more successful plant–microbe interaction. Efforts should also be directed towards the use of PSM to reduce pesticide applications. In brief, PSM biotechnology provides an excellent opportunity to develop environment-friendly phosphorus biofertilizer to be used as supplements and/or alternatives to chemical fertilizers.

REFERENCES

[1] Doran, J.W., Zeiss, M.R., (2000). Soil health and sustainability: managing the biotic component of soil quality. *Appl. Soil Ecol.* 15, 3-11.

[2] Killham, K., (1994). Soil ecology. Cambridge University Press, Cambridge, United Kingdom.

[3] Xu, J., (2006). Microbial ecology in the age of genomics and metagenomics: concepts, tools, and recent advances. *Mol. Ecol.* 15, 1713-1731.

[4] Das, S., Lyla, P.S., Khan, S.A., (2006). Marine microbial diversity and ecology: Importance and future perspectives. *Curr. Sci.* 90, 1325-1335.

[5] Daniel, R., (2004). The soil metagenome – a rich resource for the discovery of novel natural products. *Curr. Opin. Biotechnol.* 15, 199-204.

[6] Khan, M.S., Zaidi, A., Ahemad, M., Oves, M., Wani, P.A., (2010). Plant growth promotion by phosphate solubilizing fungi – current perspective. *Arch. Agron. Soil. Sci.* 56, 73-98.

[7] Saber, K., Nahla, L.D., Chedly, A., (2005). Effect of P on nodule formation and N fixation in bean. *Agron. Sustain. Dev.* 25, 389-393.

[8] Rengel, Z., Marschner, P., (2005). Nutrient availability and management in the rhizosphere: exploiting genotypic differences. *New Phytol.* 168, 305-312.

[9] Richardson, A.E., (1994). Soil microorganisms and phosphorus availability. In: *Soil biota: management in sustainable farming systems,* (Pankhurst, C.E., Doubeand, B.M., Gupta, V.V.S.R., Eds.). CSIRO, Victoria, Australia, pp 50-62.

[10] Zhou, K., Binkley, D., Doxtader, K.G., (1992). A new method for estimating gross phosphorus mineralization and immobilization rates in soils. *Plant Soil* 147, 243-250.

[11] Barber, S.A., (1995). Soil nutrient bioavailability. A mechanistic approach, Wiley, New York.

[12] Havlin, J., Beaton, J., Tisdale, S.L., Nelson, W., (1999). Soil fertility and fertilizers. An introduction to nutrient management. Prentice Hall, Upper Saddle River, NJ.

[13] Sanchez, P., Logan, T., (1992). Myths and science about the chemistry and fertility of soils in the tropics. In: *Myths and science of soils of the tropics*, (Lal, R., Sanchez, P., Eds.), Soil Science Society of America, Madison, WI, pp. 35-46.

[14] Tilman, D., Fargione, J., Wolff, B., D'Antonio, C., Dobson, A., Howarth, R., Schindler, D., Schlesinger, W.H., Simberloff, D., Wackhamer, D., (2001). Forecasting agriculturally driven global environmental change. *Science* 292, 281-284.

[15] Schindler, D.W., Hecky, R.E., Findlay, D.L., Stainton, M.P., Parker, B.R., Paterson, M.J., Beaty, K.G., Lyng, M., Kasian, S.E.M., (2008). Eutrophication of lakes cannot be

controlled by reducing nitrogen input: results of a 37-year whole-ecosystem experiment. *Proc. Natl. Acad. Sci. USA* 105, 11254-11258.

[16] Gyaneshwar, P., Naresh, K.G., Parekh, L.J., Poole, P.S., (2002). Role of soil microorganisms in improving P nutrition of plants. *Plant Soil* 245, 83-93.

[17] Sharma, S.B., Sayyed, R.Z., Trivedi, M.H., Thivakaran, G.A., (2013). Phosphate solubilising microbes: sustainable approach for managing phosphorus deficiency in agricultural soils. *Springer Plus*. 2013;2:587. doi:10.1186/2193-1801-2-587.

[18] Norrish, K., Rosser, H., (1983). Mineral phosphate. In: *Soils: an Australian viewpoint.* Academic Press, Melbourne, CSIRO/London, UK, Australia, pp. 335-361.

[19] Lindsay, W.L., Vlek, P.L.G., Chien, S.H., (1989). Phosphate minerals. In: *Minerals in soil environment, 2nd Edn.* (Dixon, J.B., Weed, S.B., Eds.), Soil Science Society of America, Madison, WI, USA, pp. 1089-1130.

[20] Khan, A.A., Jilani, G., Akhtar, M.S., Naqvi, S.M.S., Rasheed, M., (2009). Phosphorus Solubilizing Bacteria: Occurrence, Mechanisms and their Role in Crop Production. *J. Agric. Biol. Sci.* 1(1), 48-58.

[21] Blake, L., Mercik, S., Koerschens, M., Moskal, S., Poulton, P.R., Goulding, K.W.T., Weigel, A., Powlson, D.S., (2000). Phosphorus content in soil, uptake by plants and balance in three European long-term field experiments. *Nutr. Cycl. Agroecosyst.* 56, 263-275.

[22] Cordell, D., Drangert, J.O., White, S., (2009). The story of phosphorus: global food security and food for thought. *Glob. Environ. Chang.* 19, 292-305.

[23] Isherwood, K.F., (2000). Mineral Fertilizer Use and the Environment. International Fertilizer Industry Association/United Nations Environment Programme, Paris.

[24] Torsvik, V., Ovreas, L., (2002). Microbial diversity and function in soil: from genes to ecosystems. *Curr. Opin. Microbiol.* 5, 240-245.

[25] Peix, A., Velazquez, E., Martýnez-Molina, E., (2007). Molecular methods for biodiversity analysis of phosphate solubilizing microorganisms (PSM). In: *First international meeting on microbial phosphate solubilization*, (Velazquez, E., Rodrguez-Barrueco, C., Eds.), Springer, Berlin, pp. 97-100.

[26] Rogers, S.O., Bendich, A.J., (1985). Extraction of DNA from milligram amounts of fresh herbarium and minifield plant tissues. *Plant Mol. Biol.* 5, 69-76.

[27] Muyzer, G., (1999). DGGE/TGGE a method for identifying genes from natural ecosystems. *Curr. Opin. Microbiol.* 2, 317-322.

[28] Heuer, H., Krsek, M., Baker, P., Smalla, K., Wellington, E.M.H., (1997). Analysis of actinomycete communities by specific amplification of genes encoding 16S rRNA and gel-electrophoretic separation in denaturing gradients. *Appl. Environ. Microbiol.* 63, 3233-3241.

[29] Marschner, P., Crowley, D., Yang, C.H., (2004). Development of specific rhizosphere bacterial communities in relation to plant species, nutrition and soil type. *Plant Soil* 261, 199-208.

[30] Wasaki, J., Rothe, A., Kania, A., Neumann, G., R€omheld, V., Shinano, T., Osaki, M., Kandeler, E., (2005). Root exudation, P acquisition and microbial diversity in the rhizosphere of Lupinusalbus as affected by P supply and atmospheric CO2 concentration. *J. Environ. Qual.* 34, 2157-2166.

[31] Weisskopf, L., Fromin, N., Tomasi, N., Arango, M., Martinoia, E., (2005). Secretion activity of white lupin's cluster roots influences bacterial abundance, function and community structure. *Plant Soil* 268, 181-194.

[32] Henco, K., Heibey, M., (1990). Quantitative PCR – the determination of template copy numbers by temperature- gradient gel-electrophoresis (TGGE). *Nucleic Acid Res*. 18, 6733-6734.

[33] Lee, D-H., Zo, Y-G., Kim, S-J., (1996). Non-radioactive method to study genetic profiles of natural bacterial communities by single-stranded-conformation polymorphism. *Appl. Environ. Microbiol*. 62, 3112-3120.

[34] Di Cello, F., Bevivino, A., Chiarini, L., Fani, R., Paffetti, D., Tabacchioni, S., Dalmastri, C., (1997). Biodiversity of a Burkholderiacepacia population isolated from the maize rhizosphere at different plant growth stages. *Appl. Environ. Microbiol*. 63, 4485-4493.

[35] Marsh, T.L., (1999). Terminal restriction fragment length polymorphism (T-RFLP): an emerging method for characterizing diversity among homologous populations of amplification products. *Curr. Opin. Microbiol*. 2, 323-327.

[36] Osborn, A.M., Moore, E.R.B., Timmis, K.N., (2000). An evaluation of terminal-restriction fragment length polymorphism (T-RFLP) analysis for the study of microbial community structure and dynamics. *Environ. Microbiol*. 2, 39-50.

[37] Massol-Deya, A.A., Odelson, D.A., Hickey, R.F., Tiedje, J.M., (1995). Bacterial community fingerprinting of amplified 16S and 16-23S ribosomal DNA gene sequences and restriction endonuclease analysis (ARDRA). In: *Molecular microbial ecology,* (Kermans, A.D.L., van Elsas, J.D., Bruijin, F.J., Eds.), Kluwer, Dordrecht, pp. 1-8.

[38] Monteiro, J.M., Vollu, R.E., Coelho, M.R.R., Alciano, C.S., Blank, A.F., Seldin, L., (2009). Comparison of the bacterial community and characterization of plant growth-promoting rhizobacteria from different genotypes of Chrysopogonzizanioides (L.) Roberty (Vetiver) rhizospheres. *J. Microbiol*. 47, 363-370.

[39] George, T.S., Richardson, A.E., Li, S.S., Gregory, P.J., Daniell, T.J., (2009). Extracellular rerelase of a heterologous phytase from roots of transgenic plants: does manipulation of rhizosphere biochemistry impact microbial community structure? *FEMS Microbiol. Ecol*. 70, 433-445.

[40] Rivas, R., Velazquez, E., Valverde, A., Mateos, P.F., Martınez-Molina, E., (2001). A two primers random amplified polymorphic DNA procedure to obtain polymerase chain reaction fingerprints of bacterial species. *Electrophoresis* 22, 1086-1089.

[41] Wasaki, J., Maruyama, H., (2011). Molecular Approaches to the Study of Biological Phosphorus Cycling. In: *Phosphorus in Action, Soil Biology 26* (B€unemann, E.K. et al. Eds.), Springer-Verlag Berlin Heidelberg.

[42] Hofle, M.G., (1988). Identification of bacteria by low molecular weight RNA profiles: a new chemotaxonomic approach. *J. Microbiol. Meth*. 8, 235-248.

[43] Cruz-Sanchez, J.M., Velazquez, E., Mateos, P., Martınez-Molina, E., (1997). Enhancement of resolution of low molecular weight RNA profiles by staircase electrophoresis. *Electrophoresis* 18, 1909-1911.

[44] Velazquez, E., Igual, J.M., Willems, A., Fernandez, M.P., Munoz, E., Mateos, P.F., Abril, A., Toro, N., Normand, P., Cervantes, E., Gillis, M., Martı´nez-Molina, E., (2001a). Description of Mesorhizobiumchacoense sp. nov.thatnodulatesProsopis alba in the Chaco Arido region (Argentina). *Int. J. Syst. Evol. Microbiol*. 51, 1011-1021.

[45] Velazquez, E., Trujillo, M.E., Peix, A., Palomo, J.L., Garcıa-Benavides, P., Mateos, P.F., Ventosa, A., Martı´nez-Molina, E., (2001b). Stable low molecular weight RNA analyzed by staircase electrophoresis, a molecular signature for both prokaryotic and eukaryotic microorganisms. *Syst. Appl. Microbiol.* 24, 490-499.

[46] Alkan, N., Gadkar, V., Yarden, O., Kapulnik, Y., (2006). Analysis of quantitative interactions between two species of arbuscularmycorrhiz fungi, Glomusmosseae and G. intraradices, by real-time PCR. *Appl. Environ. Microbiol.* 72, 4192-4199.

[47] Kong, Y.H., Nielsen, J.L., Nielsen, P.H., (2005). Identity and ecophysiology of uncultured actinobacterial polyphosphate-accumulating organisms in full-scale enhanced biological phosphorus removal plants. *Appl. Environ. Microbiol.* 71, 4076-4085.

[48] De-Bashan, L.E., Hernandez, J.-P., Bashan, Y., Maier, R.M., (2010). *Bacillus pumilus* ES4*:* Candidate plant growth-promoting bacterium to enhance establishment of plants in mine tailings. *Environ. Experiment. Bot.* 69, 343-352.

[49] Artursson, V., Finlay, R.D., Jansson, J.K., (2005). Combined bromodeoxyuridineimmunocapture and terminal-restriction fragment length polymorphism analysis highlights differences in the active soil bacterial metagenome due to Glomusmosseae inoculation or plant species. *Environ. Microbiol.* 7, 1952-1966.

[50] De Weger, L.A., Dekkers, L.C., van der Bij, A.J., Lugtenberg, B.J.J., (1994). Use of phosphate-reporter bacteria to study phosphate limitation in the rhizosphere and in bulk soil. *Mol. Plant Microb. Interact.* 7, 32-38.

[51] Kragelund, L., Hosbond, C., Nybroe, O., (1997). Distribution of metabolic activity and phosphate starvation response of lux-tagged Pseudomonas fluorescens reporter bacteria in the barley rhizosphere. *Appl. Environ. Microbiol.* 63, 4920-4928.

[52] Wasaki, J., Kojima, S., Maruyama, H., Haase, S., Osaki, M., Kandeler, E., (2008). Localization of acid phosphatase activities in roots of white lupin plants grown under phosphorus-deficient conditions. *Soil Sci. Plant Nutr.* 54, 95-102.

[53] Uhde-Stone, C., Zinn, K.E., Ramirez-Ya´n˜ez, M., Li, A., Vance, C.P., Allan, D.L., (2003). Nylon filter arrays reveal differential gene expression in proteoid roots of white lupin in response to phosphorus deficiency. *Plant Physiol.* 131,1064-1079.

[54] Misson J, Raghothama KG, Jain A, Jouhet J, Block MA, Bligny R, Ortet P, Creff A, Somerville, S., Rolland, N., Doumas, P., Nacry, P., Herrerra-Estrella, L., Nussaume, L., Thibaud, M-C., (2005). A genome-wide transcriptional analysis using Arabidopsis thaliana Affymetrix gene chips determined plant responses to phosphate deprivation. *Proc. Natl. Acad. Sci. USA* 102, 11934-11939.

[55] Wasaki, J., Yonetani, R., Kuroda, S., Shinano, T., Yazaki, J., Fujii, F., Shimbo, K., Yamamoto, K., Sakata, K., Sasaki, T., Kishimoto, N., Kikuchi, S., Yamagishi, M., Osaki, M., (2003). Transcriptomic analysis of metabolic changes by phosphorus stress in rice plant roots. *Plant Cell Environ.* 26, 1515-1523.

[56] He, Z., Gentry, T.J., Schadt, C.W., Wu, L., Liebich, J., Chong, S.C., Huang, Z., Wu, W., Gu, B., Jardine, P., Criddle, C., Zhou, J., (2007). GeoChip: a comprehensive microarray for investigating biogeochemical, ecological and environmental processes. *ISME J.* 1, 67-77.

[57] Badri, D.V., Quintana, N., El Kassis, E.G., Kim, H.K., Choi, Y.H., Sugiyama, A., Verpoorte, R., Martinoia, E., Manter, D.K., Vivanco, J.M., (2009). An ABC transporter

mutation alters root exudation of phytochemicals that provoke an overhaul of natural soil microbiota. *Plant Physiol.* 151, 2006-2017.

[58] Rodriguez, H., Fraga, R., Gonzalez, T., Bashan, Y., (2006). Genetics of phosphate solubilization and its potential applications for improving plant growth-promoting bacteria. *Plant Soil* 287, 15-21.

[59] Bashan, Y., Moreno, M., Troyo, E., (2000). Growth promotion of the seawater-irrigated oil seed halophyte Salicorniabigelovii inoculated with mangrove rhizosphere bacteria and halotolerantAzospirillum spp. *Biol. Fertil. Soils* 32, 265-272.

[60] Armarger, N. (2002). Genetically modified bacteria in agriculture. Biochimie 84, 1061-1072.

[61] Macaskie, L.E., Yong, P., Doyle, T.C., Roig, M.G., Dı´az, M., Manzano, T., (1997). Bioremediation of uranium-bearing wastewater: biochemical and chemical factors affecting bioprocess application. Biotechnol. Bioeng. 53, 100-109.

[62] Bonthrone, K.M., Baskanova, G., Lin, F., Macaskie, L.E., (1996). Bioaccumulation of nickel by intercalation into polycrystalline hydrogen uranyl phosphate deposited via an enzymatic mechanism. *Nat. Biotechnol.* 14, 635-638.

[63] Rossolini, G.M., Shipa, S., Riccio, M.L., Berlutti, F., Macaskie, L.E., Thaller, M.C., (1998). Bacterial non-specific acid phosphatases: physiology, evolution, and use as tools in microbial biotechnology. *Cell Mol. Life Sci.* 54, 833-850.

[64] Reilly, T.J., Baron, G.S., Nano, F., Kuhlenschmidt, M.S., (1996). Characterization and sequencing of a respiratory burstinhibiting acid phosphatase from Francisellatularensis. *J. Biol. Chem.* 271, 10973-10983.

[65] Rodrıguez, H., Rossolini, G.M., Gonzalez, T., Jiping, L., Glick, B.R., (2000). Isolation of a gene from Burkholderiacepacia IS-16 encoding a protein that facilitates phosphatase activity. *Curr. Microbiol.* 40, 362-366.

[66] Deng, S., Summers, M.L., Kahn, M.L., McDermontt, T.R., (1998). Cloning and characterization of a Rhizobium. Meliloti nonspecific acid phosphatase. *Arch. Microbiol.* 170, 18-26.

[67] Deng, S., Elkins, J.G., Da, L.H., Botero, L.M., McDermott, T.R., (2001). Cloning and characterization of a second acid phosphatase from Sinorhizobiummeliloti strain 104A14. *Arch. Microbiol.* 176, 255-263.

[68] Fraga, R., Rodrıguez, H., Gonzalez, T., (2001). Transfer of the gene encoding the Nap A acid phosphatase from Morganellamorganii to a Burkholderiacepacia strain. *Acta. Biotechnol.* 21, 359-369.

[69] Richardson, A.E., Hadobas, P.A., Hayes, J.E., (2001). Extracellular secretion of Aspergillusphytase from Arabidopsis roots enables plants to obtain phosphorous from phytate. *Plant J.* 25, 641-649.

[70] Kim, Y.O., Lee, J.K., Kim, H.K., Yu, J.H., Oh, T.K., (1998). Cloning of the thermostablephytase gene (phy) from Bacillus sp. DS11 and its overexpression in Escherichia coli. FEMS Microbiol. Lett. 162, 185-191.

[71] Kerovuo, J., Lauraeus, M., Nurminen, P., Kalkinen, N., Apajalahti, J., (1998). Isolation, characterization, molecular gene cloning, and sequencing of a novel phytase from Bacillus subtilis. *Appl. Environ. Microbiol.* 64, 2079-2085.

[72] Golovan, S., Wang, G., Zhang, J., Forsberg, C.W., (2000). Characterization and overproduction of the Escherichia coli appA encoded bifunctional enzyme that exhibits both phytase and acid phosphatase activities. *Can. J. Microbiol.* 46, 59-71.

[73] Goldstein, A.H., (1996). Involvement of the quinoprotein glucose dehydrogenase in the solubilization of exogenous phosphates by Gram-negative bacteria. In: *Phosphate in Microorganisms: Cellular and Molecular Biology,* (Torriani-Gorini, A., Yagil, E., Silver, S., Eds.), ASM Press, Washington, DC, pp. 197-203.

[74] Goldstein, A.H., Liu, S.T., (1987). Molecular cloning and regulation of a mineral phosphate solubilizing gene from Erwiniaherbicola. *Biotechnology* 5, 72-74.

[75] Liu, S.T., Lee, L.Y., Taj, C.Y., Hung, C.H., Chang, Y.S., Wolfrang, J.H., Rogers, R., Goldstein, A.H., (1992). Cloning of an Erwiniaherbicola gene necessary for gluconic acid production and enhanced mineral phosphate solubilization in Escherichia coli HB101: nucleotide sequence and probable involvement in biosynthesis of the coenzyme Pyrroloquinoline Quinone. *J. Bacteriol.* 174, 5814-5819.

[76] Goldstein, A.H., Braverman, K.E., Osorio, N., (1999). Evidence for mutualism between a plant growing in a phosphatelimited desert environment and a mineral phosphate solubilizing (MPS) rhizobacteria. *FEMS Microbiol. Ecol.* 30, 295-300.

[77] Rodriguez, H., Fraga, R., (1999). Phosphate solubilizing bacteria and their role in plant growth promotion. *Biotechnol. Adv.* 17, 319-339.

[78] Fraga-Vidal, R., Rodriguez, H.M., De Villegas, T.G., (2003). Vector for chromosomal integration of the phoC gene in plant growth-promoting bacteria. In: *First international meeting on microbial phosphate solubilization*, (Velazquez, E., Rodrguez-Barrueco, Eds.), Springer, Berlin, pp. 239-244.

[79] Han, S.H., Kim, C.H., Lee, J.H., Park, J.Y., Cho, S.M., Park, S.K., Kim, K.Y., Krishnan, H.B., Kim, Y.C., (2008). Inactivation of pqq genes of Enterobacterintermedium 60-2G reduces antifungal activity and induction of systemic resistance. *FEMS Microbiol. Lett.* 282, 140-146.

[80] Gyaneshwar, P., Naresh, K., Parekh, L.J., (1998). Cloning of mineral phosphate solubilizing genes from Synechocystis PCC 6803. *Curr. Sci.* 74, 1097-1099.

[81] Meyer, J.B., Frapolli, M., Keel, C., Maurhofer, M., (2011). A novel molecular marker for studying phylogeny and diversity of phosphate-solubilizing pseudomonads: the pyrroloquinolinequinone biosynthetic gene *pqqC,Appl. Environ. Microbiol.* doi:10.1128/AEM.05434-11.

[82] Glick, B.R., Bashan, Y., (1997). Genetic manipulation of plant growth promoting bacteria to enhance biocontrol of phytopathogens. *Biotechnol. Adv.* 15, 353-378.

[83] Watt, M., Hugenholtz, P., White, R., Vinall, K., (2006). Numbers and locations of native bacteria on .eld grown wheat roots quanti.ed by Fuorescence in situ hybridization (FISH). *Environ. Microbiol.* 8, 871-884.

[84] Morrissey, J.P., Walsh, O., Donnell, U.F., Moenne-Loccoz, Y.O., Gara, F., (2002). Exploitation of genetically modi.ed inoculants for industrial ecology applications. *Antonie van Leeuwenhoek* 81, 599-606.

[85] Muyzer, G., (1999). Genetic fingerprinting of microbial communities - present status andfuture perspectives in MicrobialBiosystems: New Fronti*ers* Proceedings of the 8th International Symposium on Microbial Ecology. Bell CR, Brylinsky M, Johnson-Green P (ed) Atlantic Canada Society for Microbial Ecology, Halifax, Canada.

In: Microbes in Soil and Their Agricultural Prospects
Editors: K. Kishore Choudhary and D. Wattal Dhar
ISBN: 978-1-63482-824-6

Chapter 17

PHOSPHATE SOLUBILIZING BACTERIA AS IMPENDING BIOCONTROL AND BIOFERTILIZATION AGENTS

Kinjal Parikh, Anamika Jha* and Nisha Daxini
Ashok and Rita Patel Institute of integrated Study and Research in Biotechnology and Allied Sciences,
New Vallabh Vidyanagar, Anand (Gujarat), India

ABSTRACT

Soil nutrient content under conventional farming system is supplemented by high doses of fertilizers. But during last couple of decades, the use of chemical fertilizers and pesticides has caused tremendous harm to the environment. Thus, a balanced and responsible use of organic agriculture is to be implemented in order to make the agriculture sustainable. The principles of biofertilizer outline the similar concept where the soil health and biodiversity is built up to sustain the plant growth in longer term. A group of bacterial population closely associated with the rhizosphere region of soil is known as plant growth promoting rhizobacteria (PGPR) and is mainly involved in enhancing plant growth by several activities. Such plant growth promotion may be achieved via one or more mechanisms such as biocontrol, phosphate solubilization, nitrogen fixation, production of plant growth regulators etc. Amongst several PGPR, the Phosphate solubilizing bacteria (PSB) play a vital role in crop protection, plant growth promotion and in the improvement of overall soil health.

PSB strains have been studied for decades for their plant growth-promoting effects through effective suppression of the soil borne plant diseases. The understanding of types of pathways triggered and induced by the pathogen and PSB during phytopathogen attack is helpful in developing the disease control methods. The overall impact of such processes on the pathogen elimination needs to be understood to find efficient bacterial cultures with excellent biocontrol and growth promoting qualities. Such strains can be used to develop the effective biofertilizers for enhancing crop nutrition in different agro-

* Corresponding Author address: Ashok and Rita Patel Institute of integrated Study and Research in Biotechnology and Allied Sciences, New Vallabh Vidyanagar, Anand- 388121 (Gujarat), INDIA. E-mail - alka_mishra2k@yahoo.com

ecological climates. This chapter emphasizes on the mechanism behind plant-pathogen interaction and their possible role in biocontrol of disease and biofertilization of the soil.

Keywords: biocontrol, biofertilizer, ISR, phenolics, PGPR, phosphate solubilizing bacteria

INTRODUCTION

Microbes are the integral components of the biosphere and are accountable for various important functions in soil ecosystem. A group of bacterial population closely associated with rhizosphere region of soil is known as plant growth promoting rhizobacteria (PGPR) and is mainly involved in enhancing plant growth by several activities. PGPR functions as biocontrol, phosphate solubilization, nitrogen fixation, production of plant growth regulators etc. and contributes in soil fertility and productivity [1, 2, 3, 4, 5, 6]. Plant growth promoting rhizobacteria isolated from rhizosphere of wheat have variously been studied for phosphate solubilizing ability [7], ability to use 1-amino cyclopropan-1-carboxylate (ACC) as sole nitrogen source [1] and production of auxin [2], siderophore [3], salicylic acid [4], chitinase [5] and hydrogen cyanide [6, 7]. ACC acts as the precursor of ethylene and substrate for ACC deaminase secreted by the PGPR and thus the organism reduces stress in plants by lowering ethylene levels. Siderophore, chitinase and HCN helps in eliminating the phytopathogen via chelation activity, breakage of fungal chitinous cell wall and blocking the cytochrome oxidase pathway [3, 6]. Auxins are the major plant growth regulators produced by these microbial communities and have several physiological functions such as cell division and enlargement, stimulation of nitrogen fixation and biosynthesis of various metabolites to name few [2].

Plant growth may be enhanced either directly by supplying nutrients, vitamins or hormones or indirectly by controlling the plant diseases. PGPR strains have been studied for decades for their plant growth-promoting effects through direct fertilization effects and alternately; the effective suppression of soil borne plant diseases [9]. Phosphate solubilizing bacteria (PSB)and fungi (PSF) belong to the group of PGPR. The examples include *Pseudomonas putida*, *Burkholderia cepacia, Enterobacter* sp., *Pantoea agglomerans, Aeromonas vaga, Azotobacter* spp., *Bacillus subtilis*, *P. fluorescens*, *Rhizobium* spp., *Aspergillus* spp., *Gliocladium* spp., *Trichoderma* spp.

These species have been studied extensively for their plant growth promotion activities [10, 11, 12, 13]. Phosphate solubilizing bacteria (PSB) residing in soil can be developed as efficient biofertilizers as they support plant growth by solubilizing organic and inorganic unavailable forms of phosphates into readily available inorganic phosphates [12].

Another significant contribution of PSB is the suppression of soil borne diseases [9]. PSB suppresses soil borne disease by stimulating the mechanism of siderophore-mediated competition for iron [3], antibiosis [14], production of lytic enzymes [5], and induced systemic resistance (ISR) [15]. Induced systemic resistance (ISR) is a plant-mediated resistance mechanism exhibited by some of the PGPR strains that reduces disease caused by foliar pathogens [15].

Such resistance is effective against a broad spectrum of pathogens e.g., *Brevibacterium iodinum, Bacillus pumilus, Kluyvera cryocrescens, Enterobacter ludwigii* and *pseudomonas putida* [15]. For example, in *Arabidopsis*, ISR triggered by the root-colonizing bacterial strain

Pseudomonas fluorescens WCS417r has been found effective against several pathogens including the fungal root pathogen *Fusarium oxysporum* [16].

Realizing the plant growth promotion potential of PSB, they are being used as biofertilization and biocontrol agents for various crops [7, 17]. The relationship between biocontrol activities, plant growth promotion mechanisms and Phosphorus cycle need to be understood for the development of effective biofertilizer in totality. Thus all such events with reference to PSB-plant-pathogen interaction have been highlighted here in this chapter.

PHOSPHORUS AND ITS IMPORTANCE

Phosphorus (P) is second essential macronutrients next to nitrogen which is most commonly limiting the growth of crops and is applied to soil in the form of phosphatic fertilizers. It is an integral part of the cellular activities of plant and has a defined role in life supporting metabolism such as cell division, development, photosynthesis, breakdown of sugar and release of energy, nutrient transport within the plant and transfer of genetic characteristics from one generation to another generation and regulation of metabolic pathways [18, 19]. Phosphorus forms the structural component of several macromolecules such as the nucleic acids, DNA and RNA. It is also associated with root growth, its health and early maturity of plants. Several deficiency symptoms appear on the affected plant and plants grow slowly, become weak and stunted that may be dark green with older leaves showing a purple pigmentation. Since P is fairly mobile within the plant, deficiency symptoms initially occur in older tissue. P deficiency hampers various processes associated with energy storage and transfer. It affects root growth and bud development. Poor seed development and poor fruit quality and size may also result. The deficiency symptoms can mask other nutrient deficiencies such as nitrogen and potassium [18, 19].

Phosphorus Cycling and Availability

Phosphorus is not found free in nature and almost always occurs in fully oxidized state as phosphate. The element is widely distributed as phosphate in soils, rocks, oceans, and living cells and in many other materials (Figure 1). Phosphorus cycle is mainly sedimentary cycle as no appreciable amounts of gaseous phosphorus compounds are involved, thereby, restricting it to the lithosphere and hydrosphere only. The most important aspects of the phosphorus cycle are microbial mineralization, solubilization, immobilization and chemical fixation of phosphorus in soil.

Seventy five percent of the total P comes from sedimentary, marine phosphate rock deposits, 15-20% from igneous rocks and weathered deposits and 1-2% are biogenic (from birds and guano accumulation). More than 200 different phosphate minerals are known but only those in the apatite group occur in sufficient abundance and concentration to serve as commercial sources of the element. These are distributed in rocks in different parts of the globe.

The commonest apatite deposits mainly consist of fluorapatite [$Ca_{10}(PO_4)_6F_2$], chlorapatite [$Ca_{10}(PO_4)_6Cl_2$] and hydroxyapatite [$Ca_{10}(PO_4)_6(OH)_2$] [22].

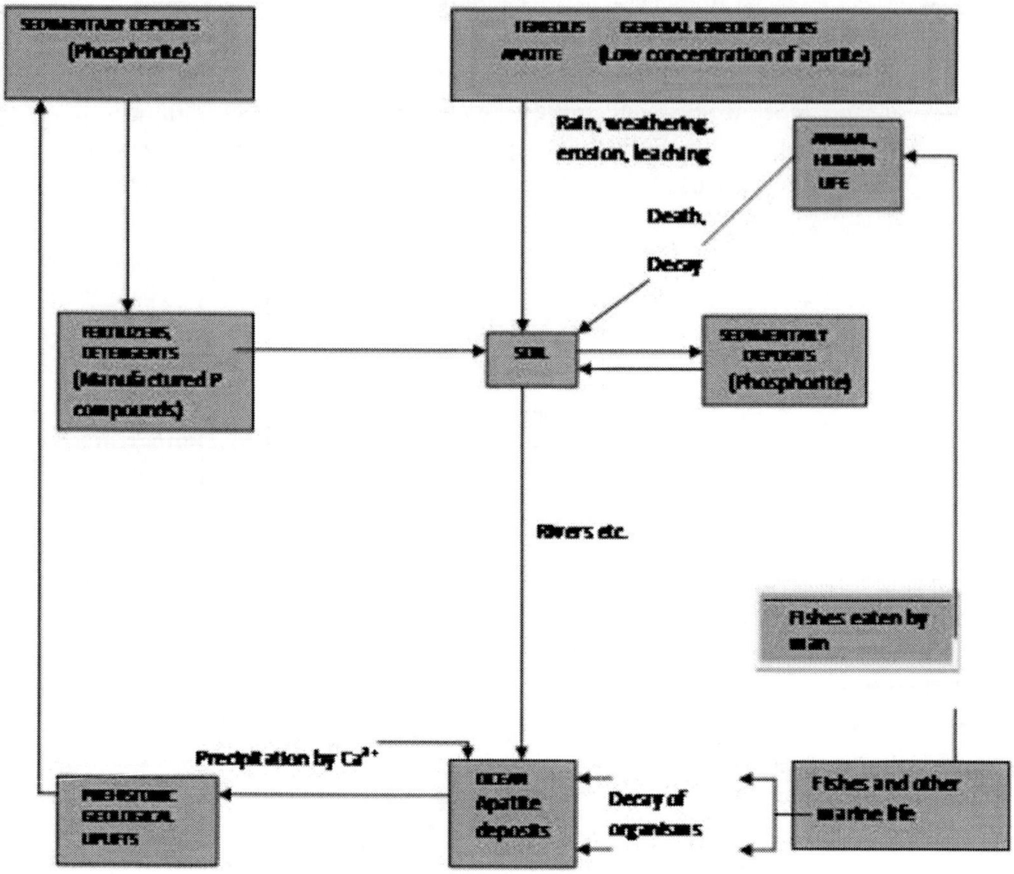

Figure 1. Natural and artificial cycles of phosphorus.

Although phosphorus is one of the most important macronutrients of all living organisms, many processes of the P cycle are still obscure. Several P forms might be taken up by the cell, but by far the greatest part is absorbed in the forms of $H_2PO_4^-$ and HPO_4^{2-}. Besides fertilization, this fraction is mainly increased by 2 different pathways:

1. The enzymatic decomposition of organic P compounds (e.g., phosphomonoesters, phosphodiesters including phospholipids, nucleic acids and phosphotriesters etc.) and inorganic P compounds (e.g., pyro and metaphosphates).
2. The non-enzymatic solubilization of different rock phosphates and inorganic phosphorus sources.

The availability of soluble Pi in many natural ecosystems is extremely low. Plants acquire P from soil solution as phosphate anions. Several soil characteristics affect the process and availability of phosphorus. The most important factors controlling the availability of P to plant roots are its concentration in the soil solution and the P-buffer capacity of the soil. The texture, amount, and type of soil strongly influence the adsorption of phosphorus. For example, finer texture soils, i.e., those with high clay content, will adsorb more phosphorus, due in part to the greater surface area availability. Additionally, phosphate anions are extremely reactive and may be immobilized through precipitation with cations such as Ca^{2+}, Mg^{2+}, Fe^{3+} and Al^{3+}, depending on the pH of soil [20, 21, 22, 23]. In acidic soil, ferric and aluminium phosphates are predominant whereas, in alkaline soil it is present as calcium and magnesium phosphate (dicalcium, tricalcium and rock phosphate). These complex forms of P are highly insoluble and unavailable to plants. As a result, the amount available to plants is usually a small proportion of this total P [24]. Further, there is considerable influence from temperature and moisture content which must be taken into consideration when understanding the maintenance and management of phosphorus availability.

The concentration of soluble P in soil is usually very low, normally at levels of 1 ppm or less than 1 ppm [25] and is not sufficient to fulfill the demand of P requirement for plants. To overcome the problem of low P availability, the phosphate is generally supplemented through chemical fertilizers for optimal crop yield. However, a large portion of soluble inorganic phosphate applied to the soil as chemical fertilizer is immobilized rapidly and becomes unavailable to the plants [26]. Thus, a phosphate load develops ignorantly and to subside the P load problem an efficient alternative is required to maintain the P availability in the soil for proper plant growth.

PSB and Plant Growth Promoting Mechanisms

Phosphorus is one of the major plant nutrients limiting plant growth. The average soil is rich in phosphorus as it contains about 0.05% (w/w) phosphorus [27] but only one tenth of this is available to plants. Approximately 95–99% is present in the form of insoluble phosphates and hence cannot be utilized by the plants due to its poor solubility and chemical fixation in the soil [28] causing low efficiency of soluble P fertilizers.

To increase the availability of phosphorus for plants, large amounts of chemical fertilizer is used on a regular basis. Unfortunately, soon after its application, significant proportion is quickly transferred to the insoluble forms. Many soil microorganisms are involved in a range of processes that affect phosphate transformation and thus influence the subsequent availability of phosphate to plant roots [29]. Free living phosphate solubilizing microorganisms (PSM) are always present in soils. As observed with other soil microbes the number of phosphate solubilizing bacteria is more important in the rhizosphere than in non rhizosphere soil [30], and the number of PSB is more important than that of fungi [31]. It is explained by the fact that PSB being rhizosphere competent, establish themselves in large numbers as compared with P solubilizing fungus.

Phosphate solubilizing bacteria are being used as biofertilizer since 1950s [32, 33] and marketed in the name of 'Phosphobacteria', 'IARI Microphos culture' etc. Phosphate solubilizing fluorescent *Pseudomonads* isolated from the groundnut rhizosphere was used as microbial P fertilizer and it was observed that the groundnut germination and grain yield was enhanced by 30% and 77% respectively [34]. Release of P by PSB from insoluble and fixed / adsorbed forms is an important aspect regarding P availability in soils. *Pseudomonas striata* and *Bacillus polymyxa* solubilized 156 and 116 mg P L-1, respectively [35].There is strong evidence that soil bacteria are capable of transforming soil P to the forms available to plant. Inorganic forms of soil P are solubilized by PSMs through organic acids production that dissolve P minerals and/or chelate cationic partners of the P ions i.e., $PO4^{3-}$ directly, releasing

P into solution [36]. Microbial biomass assimilates soluble P, and prevents it from adsorption or fixation [37].

Phosphate solubilization mechanisms are different for inorganic and organic forms. Inorganic P is solubilized mainly via organic acid production by PSB whereas, the organic forms by the phosphatase enzymes produced by PSB. The chief phosphate solubilizing mechanisms are:

a. Mineralization

Several reports have suggested the ability of different PSB species to solubilize insoluble inorganic phosphate compounds, such as tricalcium phosphate, dicalcium phosphate, hydroxyapatite, and rock phosphate [26]. In two thirds of all arable soils, the pH is above 7.0, so that most mineral P is in the form of poorly soluble calcium phosphates (CaPs). Microorganisms must assimilate P via membrane transport, so dissolution of CaPs to Pi ($H_2P0_4^-$) is considered essential in the global P cycle. Evaluation of samples from soils throughout the world has shown that, in general, the direct oxidation pathway provides the biochemical basis for highly efficacious phosphate solubilization in Gram negative bacteria via diffusion of the strong organic acids produced in the periplasm into the adjacent environment. Mineral phosphate solubilizing (mps) bacteria may be used for industrial bioprocessing of rock phosphate ore (a substituted fluorapatite) or even for direct inoculation of soils as a 'Biofertilizer' analogous to nitrogen-fixing bacteria [26].

b. Organic acid production

According to this theory, the process of phosphate solubilization by PSB is due to the production of organic acids which is accompanied by the acidification of the medium [38]. A decrease in the pH of the filtrate from the initial value of 7.0 to a final value of 2.0 was recorded by many workers [28, 39, 40, 41]. The analysis of culture filtrates of PSBs has shown the presence of number of organic acids such as malic, glyoxalic, succinic, fumaric, tartaric, alpha keto butyric, oxalic, citric, 2-ketogluconic and gluconic acid [42, 43, 44, 45]. The amount and type of the organic acid produced varied with the microorganism. The organic acids released in the culture filtrates react with the insoluble phosphate. The amount of soluble phosphate released depends on the strength and type of acid. The resulting pH change and reduction potential are thought to be responsible for dissolution of phosphate in the culture medium.

c. Enzymatic degradation

Organic P may constitute 4-90 % of the total soil P. Almost half of the microorganisms in soil and plant roots possess P mineralization potential under the action of phosphatases [46]. Organic P is converted to inorganic forms in presence of Alkaline and acid phosphatases [47]. Principal mechanism for mineralization of soil organic P is the production of acid phosphatases [48]. Release of organic anions, and production of siderophores and acid phosphatase or alkaline phosphatase [47] enzymes by plant roots / microbes [49] hydrolyze the soil organic P or split P from organic residues. The largest portion of extracellular soil phosphatases is derived from the microbial population [50]. *Enterobacter agglomerans* solubilizes hydroxyapatite and hydrolyze the organic P [51]. Mixed cultures of PSMs (*Bacillus, Streptomyces, Pseudomonas* etc.) are most effective in mineralization of organic phosphate [52]. Besides phosphate solubilization, some of the PSB promote plant growth by augmenting nitrogen content through symbiotic nitrogen fixation and also secretion of other growth promoting substances. The chief growth enhancing mechanisms among them are:

Nitrogen Fixation

In PGPR traits, some rhizobacteria can promote plant growth indirectly by affecting symbiotic N_2 fixation, nodulation, or nodule occupancy. Atmospheric nitrogen must be processed, or "fixed", to be used by plants. Some fixation occurs in lightning strikes, but most fixations are done by free-living or symbiotic bacteria. These bacteria have the Nitrogenase enzyme that combines gaseous nitrogen with hydrogen to produce ammonia, which is hydrolyzed to ammonium compounds or converted by the bacteria to organic compounds [53]. Most biological nitrogen fixation occurs by the activity of Mo-nitrogenase, found in a wide variety of bacteria, cyanobacteria and some *Archaea*. Some nitrogen fixing bacteria, such as *Rhizobium*, live in the root nodules of legumes (such as peas or beans). Here they form a mutualistic relationship with the plant, producing ammonia in exchange for carbohydrates. Nutrient-poor soils can be planted with legumes to enrich them with nitrogen.

Phosphate solubilizers have been applied for cotton and wheat fields either alone or in combination with nitrogen fixers and were found to be highly beneficial. In general, inoculation with PSB and AM fungus together resulted in higher growth and yield of wheat plants than when these organisms were used alone [54]. These results are in concurrence with the findings of Piccini and Azcon who reported significant increase in grain yields in alfalfa plants [55]. Far et al. reported that the PGPR and rhizobium bacteria significantly increased the number of grain per pod of soybean plant. The results showed that PGPR and *Rhizobium* bacteria had a positive significant effect on the number of grain per pod. Maximum number of grain per pod was obtained from inoculation with *Azospirillum* + *Rhizobium* + *Pseudomonas* treatments by 10.96 grain and the number of grain per plant increased by 15% as compared with control treatment [56].

In an investigation carried out by Zaidi and Khan, a synergistic interaction between PSM (*Pseudomonas striata* and *Penicillium*) and *Azotobacter chroococcum*, was seen which allowed better use of poorly soluble P sources (rock P) and enhanced dry-matter accumulation, seed yield, and P uptake by wheat plants [54]. Kennedy et al. [13] reported that inoculation with *Azotobacter* could increase grain yield of rice plant by 0.9 $kg.ha^{-1}$. Inoculation with *Azotobacter* also increased nitrogen accumulation by 15 $kg.ha^{-1}$ due to biological nitrogen fixation. Interestingly, the inoculation with *Azotobacter* in wheat plant provided 50% of the required nitrogen of plant in form of urea in greenhouse conditions [57]. The application of PSM and nitrogen fixing organism together could reduce P application by 50% without any significant reduction of grain yield in corn *Zea mays* [58]. A study had been carried out under green house conditions to explore the effects of combined inoculation of *Rhizobium* and phosphate solubilizing *P. striata* or *B. polymyxa* with or without added fertilizers on chickpea yield and nutritional contents [53]. Whereas, inoculation with *Rhizobium* alone was found to increase nodulation, addition of phosphate solubilizers increased the phosphorus content of the soil as well. Combined inoculation increased the nodulation and available phosphorus of the soil coupled with improved grain yield and phosphorus and nitrogen uptake by the plants.

PRODUCTION OF PLANT GROWTH REGULATORS

Plant growth regulators participate in the growth and development of cells, tissues, organs, and in fact the entire plant. These compounds are active in plants in very minute amounts and their synthesis is extremely regulated. Plants not only produce phytohormones but also, numerous plant associated bacteria both beneficial and harmful, produce one or more of these substances [59, 60]. A list of phytohormones produced by several PGPR strains and their influence on plant growth has been depicted in table 1. Among the PGPR species Azospirillum is well known for its ability to excrete phytohormone such as gibberellins, cytokinins [61] and auxins [60]. Many studies suggest the involvement of indole-3-acetic acid (IAA) produced by Azospirillum in morphological and physiological changes of the inoculated plant roots [62]. Therefore, the level of auxin appears to limit growth rate in many growing tissues supporting the idea that growth is controlled by changes in level of auxin and such mechanisms can be active simultaneously or independently at different stages of plant growth. It has been observed that IAA production by PGPR is influenced by several factors such as variation of species and strains, growth stage, culture conditions and substrate availability. Several Pseudomonas sp. along with Rhizobium sp., Bacillus sp., Enterobacter sp., Micrococcus sp., have shown remarkable production of IAA, cytokinin and Gibberellin [60].

Table 1. Plant growth promoting rhizobacteria (PGPR) mediated plant growth via secretion of various phytohormones [84]

Phytohormones	PGPR	Host
IAA	*Aeromonas veronii*	Rice
	Agrobacterium sp	Lettuce
	Alcaligenes piechaudii	Lettuce
	Azospirillum brasilense	Wheat
	Comamonas acidovorans	Lettuce
	Enterobacter cloacae	Rice
	Enterobacter sp.	Sugarcane
	Rhizobium leguminosarum	Radish
Cytokinin	*Paenibacillus polymyxa*	Wheat
	Pseudomonas fluorescens	Soybean, Pine
	Rhizobium leguminosarum	Rape & lettuce
Gibberellin	*Bacillus* sp.	Alder
ACC deaminase	*Alcaligenes* sp.	Rape
	Bacillus pumilus	Rape
	Pseudomonas cepacia	Soybean
	Pseudomonas putida	Mung bean
	Pseudomonas sp	Rape
	Variovorax paradoxus	Rape

PGP Activity of PSB in Mung Bean

Mung bean (*Vigna radiata*, synonym *Phaseolus aureus*) is an important legume crop of Asia and India accounting for around 70% of world production. Its production is seriously limited by P availability because many cultivable soils in India are deficient in available P. Management of P fertilization through chemical methods usually affects the cost of production and also soil health. By contrast, resident phosphate solubilizing bacteria (PSB) and their introduction into the rhizosphere of crops increase not only the availability of phosphorus from insoluble sources of phosphate, but also the efficiency of phosphate fertilizers such as superphosphate and rock phosphate.

Seed inoculation of mung bean with or without tricalcium phosphate (TCP) was performed by Jha et al. [12] to study the effect of single and dual bacterial inoculations with *P. fluorescens, B. cepacia* and *A. vaga* in pot trials having sterilized sandy loam soil, and was found to enhance the growth and yield of Mungbean plants [6]. Mung bean seeds inoculation with different inoculants *(Rhizobium*, PGPR and PSB) alone or in combination, significantly increased the nodulation and grain yield over uninoculated control. The combined inoculation of mungbean seeds with *Rhizobium* + PGPR + PSB gave significantly highest number of nodules/plant (21.0), dry weight of nodules/plant (87.66 mg) and grain yield (12.94 q/ha). It was at par with *Rhizobium* + PGPR with grain yield of 12.14 q/ha [63]. Mung bean seed inoculation with PSM such as *Bacillus subtilis* (TT0), *Pseudomonas striata,* Streptomycin resistant mutant of *P. striata* (M-20) and *Aspergillus niger* (TT10) showed a positive effect on rhizosphere population of phosphate solubilizers. The data on number and weight of nodules, root biomass, straw and grain yield, phosphorous and nitrogen uptake were recorded and shown significant increase in shoot and root biomass due to additional availability of phosphorous to the plants [28].

PSB and Biocontrol Potential

Plant growth promoting rhizobacteria enhance plant growth by various fertilization and disease suppression activities (Figure 2). They induce resistance in different plant species against the infection of fungal [64], bacterial [65] and viral [66] pathogens.

Phosphate solubilizing bacteria (PSB) play a vital role in crop protection, growth promotion and in the improvement of soil health.

Plenty of literature has established the role of PSB in disease suppression and Crop Improvement [10, 67]. Minaxi and Saxena have confirmed the biocontrol potential of *Pseudomonas aeruginosa* and *Burkholderia cepacia* strains isolated from semi arid region of Rajasthan in Mung beans (*Vigna radiata*) [67]. The primary mechanism of biocontrol by PSB involves the production of antibiotics such as phenazine-1-carboxyclic acid, 2, 4-diacetyl phloroglucinol, oomycin, pyoluteorin and pyrrolnitrin [68]. The antibiotics pertain to polyketides, heterocyclic nitrogenous compounds and lipopeptides which have broad-spectrum action against several plant pathogens, affecting crop plants. Though antibiotics play a vital role in disease management, their role in biocontrol is questioned due to constraints of antibiotic production under natural environmental conditions [69]. Environmental and other factors that suppress the antimicrobial action of antibiotics have to be studied to exploit the potential of antibiotics of PSB in crop protection [70].

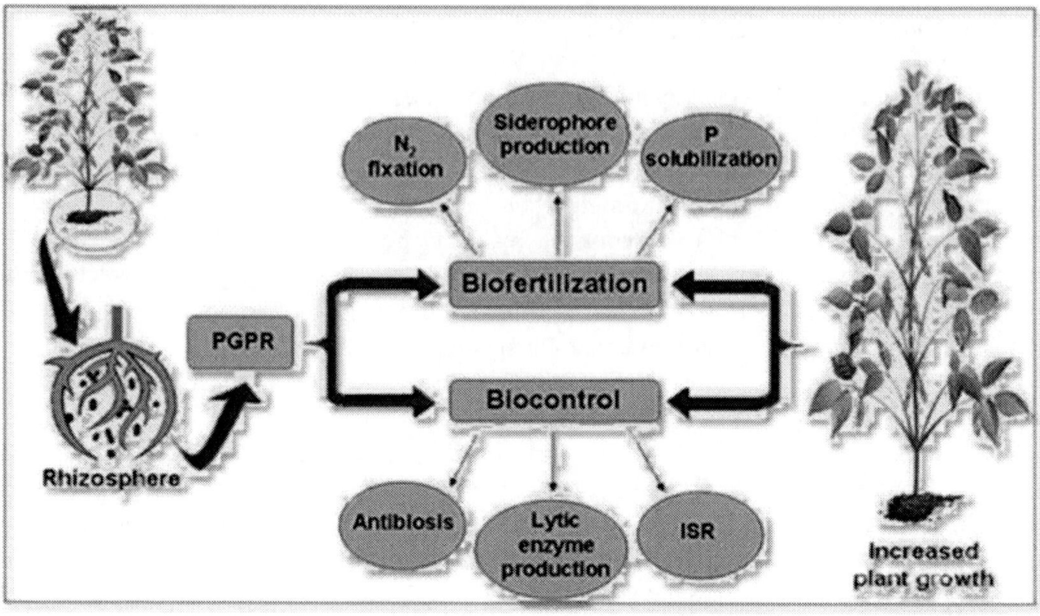

Figure 2. Diagramming representation of plant-PGPR (PSB) interactions and the important mechanisms known for pant growth promotion PSB.

Induced Systemic Resistance – The Biocontrol Weapon

Induced resistance may be defined as a physiological "state of enhanced defensive capacity" elicited in response to specific environmental stimuli and consequently the plant's innate defenses are potentiated against subsequent biotic challenges. In addition to direct antipathogenic action, PSB also serve as determinants in triggering induced systemic resistance (ISR) in the plant system. Certain bacteria trigger a phenomenon known as ISR phenotypically similar to systemic acquired resistance (SAR).

SAR develops when plants successfully activate their defense mechanism in response to primary infection by a pathogen, notably when the latter induces a hypersensitive reaction through which it becomes limited in a local necrotic lesion of brown, desiccated tissue. As SAR, ISR is effective against different types of pathogens but differs from SAR in that the inducing PGPB does not cause visible symptoms on the host plant. Manifestation of ISR is dependent on the combination of host plant and bacterial strain. Most reports of PGPB-mediated ISR involve free-living rhizobacterial strains, but endophytic bacteria have also been observed to have ISR activity. For example, ISR was triggered by *P. fluorescens* EP1 against red rot caused by *Colletotrichum falactum* on sugarcane, *Burkholderia phytofirmans* PsIN against *Botrytis cinereria* on grapevine and *Verticllium dahliae* on tomato, *P. denitrificans* 1-15 and *P. putida* 5-48 against *Ceratocystis fagacearum* on oak [71].

Determination of ISR

The ability to act as bioprotectants via ISR has been demonstrated for both rhizobacteria and bacterial endophytes, and considerable progress has been made in elucidating the mechanisms of plant-PGPR-pathogen interaction. Several bacterial traits (i.e., flagellation and production of siderophores and lipopolysaccharides) have been proposed to trigger ISR, but there is no compelling evidence for an overall ISR signal produced by bacteria. It has recently been reported that volatile organic compounds may play a key role in this process. For example, volatiles secreted by *B. subtilis* GBO3 and *B. amyloquefaciens* IN937a were able to activate an ISR pathway in *Arabidopsis* seedlings challenged with the soft-rot pathogen *Erwinia carotovora* subsp. *Carotovora* [71]. A major distinction often drawn between ISR and SAR is the dependence of the latter on the accumulation of salicylic acid (SA). Some PGPR do trigger an SA-dependent signaling pathway by producing nanogram amounts of SA in the rhizosphere. However, the majority of PGPR that activate ISR appear to do so via a SA-independent pathway involving jasmonate and ethylene signals. ISR is associated with an increase in sensitivity to these hormones rather than an increase in their production, which might lead to the activation of a partially different set of defense genes.

ISR and Defense Pathways

There has been a large body of literature describing potential uses of plant associated bacteria as agents stimulating plant growth and managing soil and plant health, but an efficient strain in terms of both fertilization and biocontrol potential has been still lacking. PSB inoculant application for the growth of mung bean and protecting it from the attack of devastating pathogen, *Sclerotium rolfsii* is reported by few workers [12, 17] and still search for such microbes is in offing by several researchers.

The signaling pathway controlling rhizobacteria mediated induction of systemic resistance (ISR) in plant clearly differs from pathogen-induced systemic acquired resistance (SAR) in that it is not associated with the accumulation of salicylic acid and induction of PRs before pathogen invasion, and is one of the mechanisms by which rhizobacteria, specially fluorescent *pseudomonads*, can suppress diseases [72]. The onset of ISR is thought to result from the perception of one or more ISR- eliciting compounds (produced by rhizobacteria) at the plant root surface. Upon binding by a receptor, transduction of plant-produced and mediated signals would lead to the state of ISR. ISR is mediated through production of several defense compounds in plants such as phenolics, flavonoids, PAL (Phenylalanine Ammonia Lyase) and other secondary metabolites. Enzymes like peroxidase (PO), polyphenol oxidase (PPO) catalyzes the formation of lignin and phenlalanine ammonia lyase (PAL) which is involved in synthesis of phytoalexins and phenolics. Other defense enzymes include PR proteins such as β-1,3-glucanases and chitinases which degrade the fungal cell wall and cause lysis of fungal cell. Chitin and glucan oligomers released during degradation of fungal cell wall act as elicitors that elicit various defense mechanisms in plants. Inhibition of the pathogen is also done by antimicrobial compounds such as antibiotics and HCN, degradation of pathogen germination factors or pathogenicity factors, and parasitism [73].

Parikh and Jha carried out an investigation to test the biocontrol potential of a phosphate solubilizing bacterial strain, LK11 isolated from mungbean rhizosphere, in Gujarat, India [17]. The most important contributing factor towards increased mungbean growth by reducing fungal (*S. rolfsii)* attack was the enhanced production of antifungal compounds (Figure 3) like PAL (47 mM^{-1} ml^{-1} mg^{-1} of tissue), phenolics (90.2 μg/ml/mg) and flavonoids (184.2 mg/ml/g). PAL is the first enzyme in phenylpropanoid metabolism and is involved in the synthesis of plant defense molecules phytoalexins and phenolics. Ramamoorthy et al. noted maximum induction of PAL activity in *P. ftuorescens* treated tomato roots after 4 days of inoculation with pathogen, *F. oxysporun* (45 nM/ml/mg of tissue) [74].

Several compounds produced by PSB such as antibiotics [75] and iron-regulated metabolites such as pseudobactin siderophore [76, 77], N-alkylated benzylamine derivative [78], Salicylic acid [79] and pyochelin siderophore [80] have been widely investigated and their substantial role in ISR phenomenon have been proved in many cases. Recently, Choudhary and Johri have established the mechanisms and role of *Bacillus* species in inducing ISR and elaborated the entire pathway of regulation mechanisms [81].

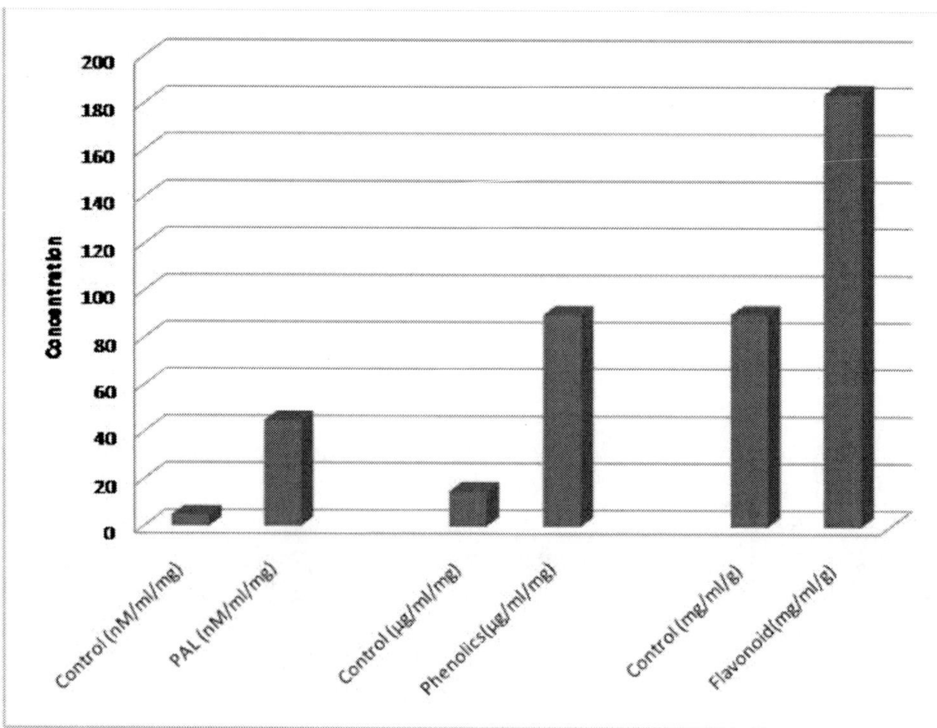

Figure 3. Production of antifungal compounds by mung bean during infection with S rolfsii in presence of a phosphate solubilizing bacterial strain LK11 (Parikh and Jha, 2012.).

Evidences on reduction in disease severity of various crops like tomato, bell pepper, muskmelon, watermelon, sugarbeet, tobacco, Arabidopsis species, cucumber, loblolly pine, and tropical crops is available. The bacterial strains that have shown significant response in respective crops are *B. amyloliquefaciens, B. subtilis, B. pasteurii, B. cereus, B. pumilus, B. mycoides*, and *B. sphaericus* in both green house and field conditions [82].

ROLE OF CELL WALL DEGRADING ENZYMES IN ANTAGONISM

Cell wall degrading enzymes also play important role in antagonistic activity of *Pseudomonas*. Chitin and glucan are cell-wall components of many pathogenic fungi and depolymerization of cell-wall by the combined activity of chitinases and glucanases could kill fungi *in vitro*. Production of lytic enzymes, chitinase and β-1, 3-glucanase by several PGPR strains is considered as a major antagonistic property of this strain. It has been reported that chitinase and β-1, 3-glucanase can function in defense against many fungal Pathogens [11]. These lytic enzymes have hydrolytic action and degrade the cell-wall of many pathogenic fungi. Moreover, these two enzymes act synergistically in the partial degradation of fungal cell-walls. It was also suggested that combinations of the two enzymes could strongly inhibit growth of many fungi, including those that could not be inhibited by chitinase or β-1, 3-glucanase alone. Chitinase and β-1,3-glucanase have not only the potential to hydrolyse cell components like chitin and β-1,3-glucan respectively, but they also release elicitors from the walls of fungi, which in turn stimulate various defense responses in plants [7, 11, 83].

CONCLUSION AND FUTURE PROSPECTS

Existence and functioning of microbes in soil for plant growth is an established fact. Due to certain erroneous practices, the soil microbiota is experiencing a threat and their number is diminishing in the ecosystem. In order to get the optimum crop growth and yield, the microbes need to be introduced in such environments in sustainable form such as biofertilizers. Fungal diseases are responsible for significant loss of crops. However, understanding the role of microbial biocontrol agents in plant growth promotion activities and plant-pathogen interaction can help in controlling the disease and increasing the crop yield.

This chapter has provided an insight into such intricate processes and application of such potential bacterial strains in the field as biofertilizers and biocontrol agents. Still some questions remain unanswered, such as the variable rhizosphere effect in different field locations and search for the strains suitable for multilocational fields. The soil conditions need to be mimicked 'in vitro' during isolation, screening and application. Further, the compatibility and abundance of introduced bacteria should be checked in soil time to time for optimum crop yields.

ACKNOWLEDGMENTS

Financial support from Gujarat State Biotechnology Mission, Dept. of Science and Technology, New Delhi is greatly acknowledged.

REFERENCES

[1] Nasciento, F., Brigido, C., Alho, L., Glick, B.R.,Oliveira, S., (2012). Enhanced chickpea growth promotion ability of a *Mesorhizobium* strain expressing an exogenous ACC deaminase gene. *Plant Soil.* 353, 221-230.

[2] Oves, M., Khan M.S., Zaidi, A., (2013). Chromium reducing and plant growth promoting novel strain *Pseudomonas aeruginosa* OSG41 enhance chickpea growth in chromium aminded soils. *Eur. J. Soil Biol.* 56, 72-83.

[3] Rajkumar, M., Ae, N., Prasad M.N.V., Freitas, H., (2010). Potential of siderophore-producing bacteria for improving heavy metal phytoextraction. *Trends Biotech.* 28, 142-149.

[4] Shrivastava, R., Shalini, M., (2009). Antifungal activity of *Pseudomonas fluorescens* against pathogenic fungi. *Electronic J. Env. Agric. Food chem.* 7, 2881-2889.

[5] Karthikeyan, M., Radhika, K., Mathiyazhagan, S. Bhaskaran, R., Samiyappan, R., Velazhahan, R., (2006). Induction of phenolics and defence-related enzymes in coconut (*Cocos nucifera* L.) roots treated with biocontrol agents. *Brazilian J. Pl. Physiol.* 18, 367-377.

[6] Devi, K.K., Seth, N., Kothamasi, S., Kothamasi, D., (2007). Hydrogen cyanide producing rhizobacteria kill subterranean termite *Odontotermes obesus* (Rambur) by cyanide poisoning under *in Vitro* Conditions. *Curr. Microbiol.* 54, 74–7.

[7] Anamika, Saxena, J., Sharma, V., (2007). Isolation of tri-calcium phosphate solubilizing strains from semi-arid agricultural fields of Rajasthan, India. *J. Pure Appl. Microbiol.* 1, 269-280.

[8] Soltani, A.A., Asadi-Rahmani, H., Khavazi, K., Omidvari, M., Dahaji, P.A., Mirhoseyni, H., (2010). *J. Agric. Sci.* 2, 106-115.

[9] Minaxi, Saxena, J., (2010). Disease Suppression and Crop Improvement in Moong Beans (*Vigna radiata*) through *Pseudomonas* and *Burkholderia* Strains Isolated from Semi arid Region of Rajasthan. *Biological Control.* 55, 799–810.

[10] Rajappan, K., Ramarej, B., (1999). Evaluation of fungal and bacterial antagonists against *Fusarium moniliforme* causing wilt of cauliflower. *Annals Pl. Protect. Sci.* 7, 205-207.

[11] Kumar, A., Prakash, A., Johri, B.N., (2011). *Bacillus* as PGPR in Crop Ecosystem In: *Bacteria in Agrobiology: Crop Ecosystems* (Maheshwari, D.K., ed.). DOI 10.1007/978-3-642-18357-7/2.

[12] Jha, A., Sharma, D., Saxena, J., (2012). Effect of single and dual phosphate solubilizing bacterial strain inoculations on overall growth of mung bean plants. *Archives Agronom. Soil Sci.* 58, 967-981.

[13] Saxena J., Minaxi, Jha, A., (2013). Impact of a phosphate solubilizing bacterium and an arbuscular mycorrhizal fungus (*Glomus etunicatum*) on growth, yield and P concentration in wheat plants. *Clean: soil, air water* DOI:10.1002/clen.201300492.

[14] Krishnan, H.B., Kang, B.R., Krishnan, A.H., Kil, Kim, K.Y., Kim, Y.C., (2007). *Rhizobium meliloti* USDA9032 engineered to produce *a* phenazine antibiotic inhibits the growth of fungal pathogens but is impaired in symbiotic performance. *Appl. Env. Microbiol.* 73, 327-330.

[15] Son, J., Sumayo, M., Hwang, Y., Kim B. Ghim, S., (2014). Screening of plant growth-promoting rhizobacteria as elicitor of systemic resistance against gray leaf spot disease in pepper. *Appl. Soil Ecol.* 73, 1-8.

[16] Bakker, P.A.H.M., Pieterse, C.M.J., Van Loon, L.C., (2007). Induced systemic resistance by fluorescent *Pseudomonas spp. Phytopathology* 97, 239-243.

[17] Parikh, K., Jha, A., (2012). Biocontrol features in an indigenous bacterial strain isolated from agricultural soil of Gujarat, India. *J. Soil Sci. Pl. Nutr.* 12, 249-256.

[18] Wang, X., Wang, Y., Tian, J., Lim, B.L., Yan, X. and Liao, H., (2009). Overexpressing AtPAP15 enhances phosphorous efficiency in soybean. *Plant Physiol.* 151, 233-240.

[19] Khan, M.S., Zaidi, A., Wani, P.A., Oves, M., (2009). Functional diversity among plant growth-promoting rhizobacteria. In: *Microbial strategies for crop improvement* (Khan, M.S., Zaidi, A., Musarrat, J., (eds,). Springer, Berlin, pp. 105-132.

[20] Paul, E. A., Clark, F. E., (1988). *Soil Microbiol. Biochem.* San Diego, CA: Academic press.

[21] Dalal, R. C., (1977). Soil organic phosphorus. *Adv. Agronom.* 29, 83-117.

[22] Harley J.L., Smith S. E., (1983). *Mycorrhizal symbiosis*. London, New York: Academic Press, pp. 483.

[23] Jha, A., Saxena, J.,Sharma, V., (2013). An Investigation on Phosphate Solubilization Potential of Agricultural Soil Bacteria as Affected by Different Phosphorus Sources, Temperature, Salt and pH. *Comm. Soil Sci. Pl. Analysis*. 44, 2443-2458.

[24] Rodriguez, E., Han, Y., Lei, X. G., (1999). Cloning, sequencing and expression of an *Escherichia.coli* acid phosphatase/phytase gene (appA2) isolated from pig colon. *Res. Comm. Biochem. Biophy.* 257, 117-123.

[25] Goldstein, A. H., (1994). Involvement of the quinoprotein glucose dehydrogenase in the solubilization of exogenous mineral phosphate by gram negative bacteria. *In*: *Phosphate in Microorganisms: Cellular and Molecular Biology*. (Torrianigorni, A.E., Yagil, S., Silver Eds.), ASM Press, Washington, DC. pp. 197-203.

[26] Goldstein, A. H., (1986). Bacterial solubalization of mineral phosphates: historical perspective and future prospects. *Am. J. Alternative Agric.* 1, 51-57.

[27] Barber, S.A., (1984). Soil nutrient availability–a mechanistic approach. New York, USA, *Wiley Interscience*. pp. 398.

[28] Gaind, S., Gaur, A.C., (1991). Thermotolerant phosphate solubilizing microorganisms and their interaction in mung bean. *Plant Soil* 133, 141-149.

[29] Richardson, A. E., (2001). Hadobas, P. A., Hayes, J. E., O'Hara, C. P., Simpson, R.J., Utilization of phosphorus by pasture plants supplied with myo-inositol hexaphosphate is enhanced by the presence of soil microorganisms. *Plant Soil* 229, 47-56.

[30] Kucey, R.M.N., Jenzen, H., Leggett, M., (1989). Microbially mediated increases in plant available phosphorus. *Advances Agron.* 42, 199-228.

[31] Kucey, R. M. N., (1983). Phosphate solubilizing bacteria and fungi in various cultivated and virgin Alberta Soil. *Can. J. Soil Sci.* 63: 671-678.

[32] Howell, C.R., Stipanovic, R.D., (1979). Control of *Rhizoctonia solani* on cotton seedlings with *Pseudomonas fluorescens* and with an antibiotic produced by the bacterium. *Phytopathology* 69, 480-482.

[33] Kudashev, I.S., (1956). The effect of phosphobacteria on the yield and protein content in grains of Autumm wheat maize and soybean. *Doki. Akad. Skh. Nauk.* 8, 20-23.

[34] Krasilinikov, N.A., (1957). On the role of soil micro-organism in plant nutrition. *Microbiologiya* 559, 659–672.

[35] Shweta,, B., Maheshwari, D., Dubey, Arora, D., Bajpai V. K., Kang, S.C., (2008). Beneficial Effects of Fluorescent Pseudomonads on Seed Germination, Growth Promotion, and Suppression of Charcoal Rot in Groundnut (*Arachis hypogea* L.) *J. Microbiol. Biotechnol.*. 18, 1578–1583.

[36] Rodriguez, H., Fraga, R., (1999). Phosphate solubilizing bacteria and their role in plant growth promotion. *Biotechnol. Adv.* 17, 319-339.

[37] He, Z. L., Bian, W., Zhu, J., (2002). Screening and identification of microorganisms capable of utilizing phosphate adsorbed by goethite. *Comm. Soil Sci. Pl. Analysis* 33, 647-663.

[38] Khan, K.S., Joergensen, R.G., (2009). Changes in microbial biomass and P fractions in biogenic household waste compost amended with inorganic P fertilizers. *Biores. Technol.* 100, 303-309.

[39] Puente, M. E., Bashan, Y., Li, C.Y., Lebsky, V. K., (2004). Microbial populations and activities in the rhizoplane of rock weathering desert plants root colonization and weathering of igneous rocks. *Pl. Biol.* 6, 629-642.

[40] Gaur, A. C., Sachar, S. (1980). Effect of rock phosphate and glucose concentration on phosphate solubilization by *Aspergillus awamori*. *Curr. Sci.* 49, 553-554.

[41] Gaur, A. C., (1990). Phosphate solubilizing microorganisms as biofertilizers. *Omega Scientific Publishers*, New Delhi, India, pp. 40-44.

[42] Ilmer, P., Schinner, F., (1992). Solubilization of inorganic phosphates by microorganisms isolated from forest soils. *Soil Biochem.* 24, 389-395.

[43] Lapeyrie, F., Ranger, J., Vareller, D., (1991). Phosphate solubilizing activity of ectomycorrhizal fungi *in vitro*. *Can. J. Bot.* 69, 342-346.

[44] Illmer, P., Schinner, F., (1995). Solubilization of inorganic phosphates by microorganisms isolated from forest soil. *Soil Biol. Biochem.* 24, 389-395.

[45] Kim, K.Y., Jordan, D., McDonald, G.A., (1997). Solubilization of hydroxyapatite by *Enterobacter agglomerans* and cloned *Escherichia coli* in culture medium. *Biol. Fert. Soil.*24, 347-352.

[46] Fasim, F., Ahmad, N., Parsons, R., Gadd, G. M., (2002). Solubilization of Zinc salts by a bacterium isolated from the air environment of a tannery. *FEMS Microbiol. Lett.* 213, 1-6.

[47] Tarafdar, J. C., Claassen, N. (1988). Organic phosphorus compounds as a phosphorus source for higher plants through the activity of phosphatases produced by plant roots and microorganisms. *Biol. Fert. Soil* 5, 308-312

[48] Beech, I.B., Paiva, M. Cause, M., Coutinhl, C., (2001). Enzymatic Activities Within Biofilms of Sulphate-Reducing Bacteria. In: *Biofilm Community Interactions: Chance or Necessity*? (Gilbert, P.G., Allison, D., Melanie Brading, M. Verran, J., Walker, J., Eds), *BioLine Cardiff*, pp. 231-239.

[49] Hilda, R., Fraga, R., (2000). Phosphate solubilizing bacteria and their role in plant growth promotion. *Biotechnol. Adv.* 17, 319-359.

[50] Yadaf, R. S., Tarafdar, J. C., (2001).Influence of organic and inorganic phosphorus supply on the maximum secretion of acid phosphatase by plants. *Biol. Fert. Soil.* 34, 140-143.

[51] Dodor, D.E., Tabatabai, M.A., (2003). Effect of cropping systems of phosphatases in soils. *J. Pl. Nutr. Soil Sci.* 166, 7-13.

[52] Kim, K.Y., Jordan, D., McDonald, G.A., (1998). Enterobacter agglomerans, phosphate solubilizing bacteria, and microbial activity in soil: effect of carbon sources. *Soil Biol. Biochem.* 30, 995-1003.

[53] Molla, M.A.Z., Choudhary, A.A., Islam, A., Hoque, S., (1984). Microbial mineralization of organic phosphate in soil. *Plant Soil* 78, 393-399.

[54] Algawadi, A.R., Gaur, A.C., (1988). Associative effect of Rhizobium and phosphate solubilizing bacteria on the yield and nutrient uptake of chickpea. *Plant Soil* 105, 241-246.

[55] Zaidi, A., Khan, M.S., (2005). Interactive effect of rhizospheric microorganisms on growth, yield and nutrient uptake of wheat. *J. Pl. Nutr. Soil Sci.* 28, 2079-2092.

[56] Piccini, D., Azcon, R., (1987). Effect of phosphate solubilizing bacteria and vesicular-arbuscular mycorrhizal fungi on the utilization of Bayovar rockphosphate by alfalfa plants using a sand vermiculite medium. *Plant Soil* 50, 45-50.

[57] Far, S.A., Shoholi, R., Ghodrati, G.H., (2008). "Evaluate soybean response to different doses and strains of the bacterium *Bradyrhizobium japonicum* in the northern region of Khuzestane bacterium. *Agronomie* 4, 81-92.

[58] Kennedy, I. R., Choudhury, A. T. A. M., Kecskes, M.L., (2004). Nonsymbiotic bacterial diazotrophs in crop-farming systems. Can their potential for plant growth promotion be better exploited"? *Soil Biol. Biochem.* 36, 1229-1244.

[59] Yazdani, M., Bahmanyar, M.A., Pirdashti, H., Esmaili, M.A., (2009). Effect of Phosphate Solubilization Microorganisms (PSM) and Plant Growth Promoting Rhizobacteria (PGPR) on yield and yield components of Corn (*Zea mays L). World Acad. Sci. Eng. Technol.* 49, 90-92.

[60] Dobbelaere, S., Vanderleyden, J., Okon, Y., (2003). Plant growth-promoting effects of diazotrophs in the rhizosphere. *Critical Rev. Pl. Sci.* 22, 107-149.

[61] Vessey, J.K., (2003). Plant growth promoting rhizobacteria as biofertilizers. *Plant Soil* 255, 571–586.

[62] Tien, T.M., Gaskins, M.H., Hubbell, D.H., (1979). Plant growth substances produced by *Azospirillum brasilense* and their effect on the growth of pearl millet (*Pennisetum americanum* L.). *Appl. Env. Microbiol.* 37, 1016–1024.

[63] Harari, A., Kigel, J., Okon, Y., (1988). Involvement of I AA in the interaction between *Azospirillum brasilense* and *Panicum miliaceum* roots. *Plant Soil* 110, 275-282.

[64] Howell, C.R., Stipanovic, R.D., (1979). Control of *Rhizoctonia solani* on cotton seedlings with *Pseudomonas fluorescens* and with an antibiotic produced by the bacterium. *Phytopathology* 69, 480-482.

[65] Park, K.S,. Kloepper, J.W., (2000). Activation of PR-1a promoter by rhizobacteria which induce systemic resistance in tobacco against *Pseudomonas syringae* pv. tabaci. *Biological Control* 18, 2-9.

[66] Maurhofer, M., Hase, C., Meuwly, P., Métraux, J.P., Défago, G., (1994). Induction of systemic resistance of tobacco to tobacco necrosis virus by the root-colonizing *Pseudomonas fluorescens* strain CHA0: Influence of the *gac*A gene and of pyoverdine production. *Phytopathology* 84, 139-146.

[67] Minaxi, Saxena, J., (2010a). Characterization of *Pseudomonas aeruginosa* RM-61 3 as a Potential Bio control Agent. *Mycopathology* 170, 181–193.

[68] Gamard, P., (1997). Novel butyrolactones with antifungal activity produced by *Pseudomonas aureofaciens* strain 63-28. *J. Antibiotics* 50, 742-749.

[69] Lee, J. Y., Moon, S. S., Hwang, B. K., (2003). Isolation and Antifungal and Antioomycete activities of Aerugine produced by *Pseudomonas fluorescens* Strain MM-B16. *Appl. Env. Microbiol.* 69, 2023-2021.

[70] Hiratani, T., Yamaguchi, H. (1993). Biological properties of aureobasidin A, a cyclic depsipeptide antifungal antibiotic. *J. Antibiotics*. 46, 1414-1420.

[71] Van Loon, L.C., (2000). Systemic induced resistance. In: *Mechanism of resistance to plant diseases*. (Slusarenko, A.J., Fraser, R.S.S., van Loon, L.C., Eds.), *Kluwer Dordrecht*, pp. 521-574.

[72] Pieterse, C.M.J., Van Wees, S.C.M., Hoffland, E., Van Pelt, J.A., VanLoon, L.C., (1996). Systemic resistance *Arabidopsis* induced by biocontrol bacteria is independent of salicylic acid accumulation and pathogenesis-related gene expression. *Plant Cell* 8, 1225-1237.

[73] Haas, D., Keel, C., (2003). Regulation of antibiotic production in root-colonizing *Pseudomonas* spp. and relevance for biological control of plant disease. *Ann. Rev. Phytopathol.* 41, 117-53.

[74] Ramamoorthy, V., Raguchander, T., Samiyappan, R., (2002). Induction of defense-related proteins in tomato roots treated with *Pseudomonas fluorescens* Pf1 and *Fusarium oxysporum* f. sp. Lycopersici. *Plant Soil* 239, 55–68.

[75] Weller, D.M., (1988). Biological control of soilborne plant pathogens in the rhizosphere with bacteria. *Ann. Rev. Phytopathol.* 26, 379-407.

[76] Meziane, H., Van der Sluis, I., Van Loon, M., Bakker, P.A.H. M., (2005). Determinants of *Pseudomonas putida* WCS358 involved in inducing systemic resistance in plants. *Mol. Pl. Pathol.* 6, 177-185.

[77] Ran, L.X., Li, Z.N., Wu, G.J., Van Loon, L.C., Bakker, P.A.H.M. (2005). Induction of systemic resistance against bacterial wilt in *Eucalyptus urophylla* by *fluorescent Pseudomonas spp*. *Eur. J. Pl. Pathol* .113, 59-70.

[78] Ongena, M., Jourdan, E., Schafer, M., Kech, C., Budzikiewicz, H., Luxen, A., Thonart, P. (2005). Isolation of an N-alkylated benzylamine derivative from *Pseudomonas putida* BTP1 as elicitor of induced systemic resistance in bean. *Mol. Plant-Microbe interact.* 18, 562-569.

[79] De Meyer, G., Capieau, K., Audenaert, K., Buchala, A., Metraux, J. P., Höfte, M., (1999). Nanogram amounts of salicylic acid produced by the rhizobacterium *Pseudomonas aeruginosa* 7NSK2 activate the systemic acquired resistance pathway in bean. *Mol. Plant-Microbe Interact.* 12, 450–458.

[80] Audenaert, K., Pattery, T., Cornelis, P., Höfte, M., (2002). Induction of systemic resistance to Botrytis cinerea in tomato by *Pseudomonas aeruginosa* 7NSK2: Role of salicylic acid, pyochelin, and pyocyanin. *Mol. Plant-Microbe Interact.* 15, 1147-1156.

[81] Choudhary, D.K., Johri, B.N., (2008). Interactions of *Bacillus* spp. and plants – with special reference to induced systemic resistance (ISR). *Microbiol. Res.* 164, 493–513.

[82] Kloepper, J.W., Ryu, C.M., Zhang, S., (2004). Induced systemic resistance and promotion of plant growth by *Bacillus* spp. *Phytopathology* 94, 1259-1266.

[83] Pandey, A., Trivedi, P., Kumar, B., Palni, L.M.S., (2006). Characterization of a phosphate solubilizing and antagonistic strain of *Pseudomonas putida* isolated from a sub-alpine location in the Indian central Himalaya. *Curr. Microbiol.* 53, 102-107.

[84] Khan, M.S., Zaidi, A., Ahmad, E., (2014). "Mechanism of phosphate solubilization and physiological functions of phosphate-solubilizing microorganisms" In: *Phosphate Solubilizing Microorganisms: Principles and Application of Microphos Technnology* (Khan, M.S., Zaidi, A., Musarrat, J., Eds.), Springer international publishing, Switzerland, pp. 31-62.

In: Microbes in Soil and Their Agricultural Prospects
Editors: K. Kishore Choudhary and D. Wattal Dhar
ISBN: 978-1-63482-824-6

Chapter 18

POTENTIAL APPLICATION OF *TRICHODERMA* SP. IN BIOCONTROL OF SOIL BORNE DISEASE

Shalini Singh*, Mohd. Aamir and R.S. Upadhyay
Centre of Advanced in Botany, Banaras Hindu University, Varanasi, India

ABSTRACT

Trichoderma sp., a genus under Deuteromycotina, Hyphomycetes, Phialasporace, Hyphales, Dematiaceae has gained immense importance since last few decades. Different mechanisms have been suggested as being responsible for their biocontrol activity, which include competition for space and nutrients, secretion of chitinolytic enzymes, antibiosis, mycoparasitism and production of inhibitory compounds. Trichoderma is a potent biocontrol agent for plant disease management especially the soil born. *Trichoderma* is one of the possible approaches to suppress disease without appreciable environmental hazards. In this chapter, future prospects of *Trichoderma* spp., abiotic stress tolerant *Trichoderma* spp., chemical fungicides tolerant *Trichoderma* spp. capable of effectively controlling the soil borne disease, their mode of action and mechanisms will be discussed.

Keywords: abiotic stress, antagonism, antibiotic, rhizosphere, transgenic

INTRODUCTION

Indiscriminate and injudicious use of chemical fungicideshas resulted in environmental pollution, ecological imbalances and cause health hazards. Hence, it is necessary to look for alternative disease management practices, which include the use of eco-friendly biological control agents (BCAs) and pathogen-resistant crop cultivars. Biocontrol agents are widely regarded by public as "natural" and non-threatening products and gained immense importance as a substitute for chemical fungicides and pesticides [47]. The unique ability of *Trichoderma*

---

* E-mail: shalini222003@yahoo.co.in

is due to the production of extracellular lytic enzymes, non-volatile and volatile toxic metabolites, high competitive saprophytic ability, high proliferation rate, etc. as well as induction of systemic resistance in host plants. *Trichoderma* spp. produce various kinds of secondary metabolites including antibacterial and antifungal antibiotics [48]. Some strains of *Trichoderma* like *Trichoderma harzianum*, *T. atroviride*, *T. viride*, *T. virens* and *T. koningii* have the ability to inhibit soil pathogen growth and thus improve the overall health of the plant.

Ordish [37] reported that "Biological control is much neglected, not because it does not work, but because not enough research is done on it". Snyder [50] stated that "The opportunities for playing one soil organisms against another to man's advantage are there and only wait man's cleverness dealing with antagonists". *Trichoderma* spp. is one possible approach to suppress diseases without appreciable environmental hazards. In recent years there has been a growing interest in biological control of pests and diseases as a strategy for integrated pest management (IPM) of crop cultures. Species of the genus *Trichoderma* have been used as antagonists for the control of some of the most important phytopathogenic fungi (*Fusarium oxysporum*, *Rhizoctonia solani*, *Botrytis cinerea*, *Sclerotinia sclerotiorum*). The biocontrol activity of *Trichoderma* is of immense importance not only to agriculture and its crops but also the environment as it does not accumulate in the food chain and thus does no harm to the plants, animals and humans [34]. The genes and gene products involved in the biocontrol mechanism of *Trichoderma* provide a vast array of research to the scientists in biotechnology and bioinformatics as well. Formulations of the most promising strains of *Trichoderma* have being applied to seeds, plantlets and soil to control plant root pathogens such as *Pythium, Rhizoctonia, Fusarium, Cylindrocladium* and *Thielaviopsis*, mostly in nurseries and greenhouses in developed countries [24]. Various strains of most ongoing research is aimed at developing new techniques or strains that can prove ecologically beneficial to the agriculture. It is necessary to increase knowledge about the effects of more enzymes and metabolites of *Trichoderma* with the aim of extending its spectrum of action as a biocidal agent.

MECHANISM OF CONTROL BY *TRICHODERMA*

Classically, three principal mechanisms of action of *Trichoderma* spp. have been recognized – mycoparasitism (parasitism of one fungus by another fungus), antibiosis (production of antimicrobial metabolites, and thus inhibiting other fungi) and the universal phenomenon of competition for food, space or oxygen. Activation of each mechanism implies the production of specific metabolites, such as plant growth factors, hydrolytic enzymes, siderophores, antibiotics, and permeases. *Trichoderma* spp. are highly efficient producer of many extra cellular enzymes like cellulases, chitinases, glucanases, proteases etc. They are being exploited in variety of ways like source of cellulases (used in foods and textiles and also in poultry feed) and chitinases (generating disease resistant transgenic), in plant disease control (through their anti-fungal and anti-nematode activity and in plant defense induction), improvement of plant growth, straw/compost decomposition and suppression of some of the weeds.

Mycoparasitism and Antibiotic (Toxin) Production

The biocontrol mechanism of *Trichoderma* spp. is a complex process mediated by the secretion of extracellular enzymes, such as chitinase, β- glucanase and proteinases, as well as secondary metabolites. Although some antibiotics may be the major factor for the biocontrol activity of certain strain, this may not be the case of others. The genome sequence analysis of two recognized biocontrol species – Trichoderma atroviride and Trichoderma virens – provide us a better understanding of how mycoparasitism arose in a common Trichoderma. The presence of fungal prey and the availability of root-derived nutrients may have been major attractors for the ancestors of Trichoderma to establish themselves in the rhizosphere and facilitate the evolution of positive interactions with plants [12]. Weindling [55] described biological control by T. lignorum of citrus seedling disease, incited by *Rhizoctonia solani* to mycoparasitism. Shalini et al., [44] discovered that the mechanism and mode of action of the *Trichoderma* against *Rhizoctonia solani* is through coiling around pathogen hyphae, penetration, and subsequent dissolution of the host cytoplasm (Figure 1). Weindling [56] reported that a strain of *T. lignorum* produced a lethal principle that was excreted into the surrounding medium, allowing parasitic activity by the biological control. Weindling [57] characterized the "lethal principle", demonstrated that it was toxic to both *R. solani* and *Sclerotinia Americana*, and named it gliotoxin. Subsequently, it has been demonstrated that the fungus that produced gliotoxin was not *T. lignorum* but *Glocladium virens* [54] a species that has recently been named *Trichoderma virens*. Howell and Stipanovic [20] isolated and described a new antibiotic, gliovirin from *Gliocladium Virens* that was strongly inhibitory to *Pythium ultimum* and a *Phytophthora* species, but not to *R. solani*, *Thielaviopsis basicola*, *Phymatotrichum omnivorum*, *Rhizopus arrhizus* or *Verticillium dahlia*. Based on the biosynthesis of either gliotoxin (Q) or gliovirin (P), Howell et al., [19] classified *T. virens* strains into two groups; "Q" groups were effective against *R. solani* and "P" groups against *Pythium ultimum*. Later, using gliotoxin or gliovirin-deficient mutants, gliotoxin was implicated to be involved in biocontrol of *Pythium* damping-off [58], but not for *R. solani* in cotton [20]. Wu et al., [59] extracted the two new antibiotic, trichodin A and trichodin B, together with the known compound, pyridoxatin from marine fungus *Trichoderma* sp. which were found effective against Staphylococcus epidermidis. Endochitinase (42kDa), chitobiosidase (40-kDa) and N-acetyl-b-D-glucosaminidase (73-kDa) from *T. atroviride* strain and *T. virens* strain were reported to have a inhibitory effect on the several fungal pathogens, viz., *Botrytis cinerea*, *Fusarium* spp., *Alternaria* spp., *Ustilago avenae*, *Uncinula necator* and virtually on all fungi containing chitin in their cell-wall [42]. Mutants of *Trichoderma harzianum* with altered antibiotic production were isolated using ultraviolet light mutagenesis and these mutants produce the isonitrile antibiotic, which was found effective against *R. solani* and *P. ultimum* [8]. Howell [22] used ultraviolet light irradiation to produce mutant strains of *T. virens* that were unable to parasitize *R. solani.* When compared for bio-control efficacy against *R. solani* incited cotton seedling diasease, the mycoparasitic deficient mutants were just as effective as the parent strains. In other studies [21], a mutant of *T. virens* deficient for both mycoparasitism and gliotoxin biosynthesis (G65) still retained biocontrol efficacy equal to that of the parent strain (G6) against both *P.ultimum* and *R. solani*. Scanning electron microscopy and fluorescence microscopy showed that *T. harzianum* and *T. hamatum* were mycoparasites of both *Sclerotium rolfsii* and *Rhizoctonia solani*. The antagonist attached to the pathogen and secreted glucanase and chitinase enzymes that act through the cell wall. *T. harzianum* isolate 1051 produced a novel protease which was biologically active against *Crinipellis perniciosa*, the causal agent of witches' broom. On the other hand three novel strains of Trichoderma (two T. harzianum and one T. atroviride) from wild mushroom showed biocontrol activity against Sclerotium delphinii infecting cultivated cotton seedlings. T. harzianum strain CICR-G, isolated as a mycoparasite on a tree-pathogenic Ganoderma sp. exhibited the highest disease suppression ability [36].

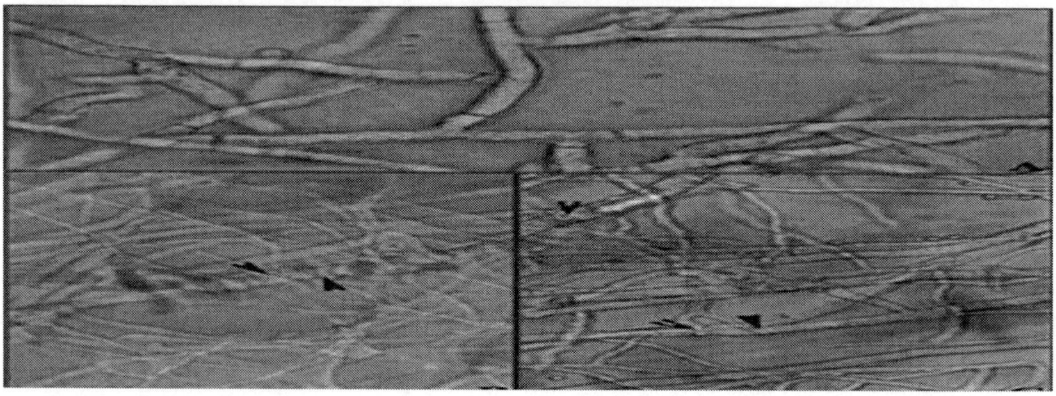

Figure 1. Mode of the action of the biocontrol agent. a. Mycelium of *R. solani* (40X); b. Coiling of the large hyphae of *R. solani* by *Trichoderma* sp. (40X); c. Initial interaction between the *Trichoderma* and *R. solani* (attachment of the *Trichoderma* hyphae by tips on the large hyphae of *R. solani*) (40X).

Inactivation of the Pathogen's Enzymes

Inactivation of the pathogen's enzymes is another biocontrol mechanism by *Trichoderma* species. Enzymes of *Botrytis cinerea*, viz., pectinases, cutinase, glucanase and chitinase were suppressed through the action of *Trichoderma* species secreted protease on plant surfaces. The *in vitro* inhibitory ability of *Trichoderma harzianum* on the phytopathogen *Alternaria alternata* was investigated in the presence of growth regulators. *A. alternata*, a pathogenic fungus secretes endopolygalacturonase (endo-PG) and pectate lyase (PL), which are responsible for the hydrolysis of pectic components of the plant cell wall. The presence of *T. harzianum* decreased endo- PG secretion by *A. alternata* to 50 per cent, and this inhibitory effect was independent of the presence of growth regulators [41]. Metcalf and Wilson [32] described the colonization of onion roots, infected with *Sclerotium cepivorum* by *T. koningii*. Hyphae of the bio-control agent penetrated into infected epidermal and cortical tissue of the root to destroy the hyphae of the pathogen, with little or damage to uninfected plant tissue. The authors ascribed this biocontrol phenomenon to production of endo and exochitinase by *T. koningii*. Baek et al., [2] disrupted or overexpressed the gene coding for chitinase (cht42) in *T. virens* (Gv298). Chitinase (ech42) activity in *T. harzianum* also showed reduced bio-control activity against *Botrytis cinerea* on bean leaves. Bolar et al., [5] demonstrated enhanced resistance to apple scab incited by *Venturia inequalis* in transgenic apple plants that has been transformed with gene encoding both endo and exochitinases from *T. atroviride*. The crop *Mentha arvensis* L. is severly affected by *S. rolfsii* and the cell walls of *S. rolfsii* composed of β-1,3 glucan and chitin. *Trichoderma* spp are reported to release active lytic enzymes, that can digest these cell wall components of *S. rolfsii* [40]. Another interesting concept related to the enzyme biosynthesis as a mechanism in the bio-control process is that advanced by Elad and Kapat [13] who suggested that bio-control of *B. cinerea* by *T. harzianum* might be due to the actions of *T. harzianum* produced proteases that inactivate the hydrolytic enzymes produced *B.cinerea* on bean leaves. Protease production by *T. harzianum* has also been associated with bio-control of the root knot nematode *Meloidogyne javanica* on tomato plants. Sharo et al., [46] showed that tomato plants treated with the bio-control agent (T203) and grown in nematode infested soil exhibited a drastic reduction in root galling when compared with the control. Migheli et al., [33] showed that transformants of *T. longibrachiatum* overexpressing a gene encoding 1,4 endoglucanase were slightly more effective in the bio-control of *P. ultimum* on cumber than the wild type. Smitha et al., [49] concluded that the protease enzyme produced by *Aspergillus niger* and *Aspergillus flavus* significantly altered the hydrolytic enzyme secretion and decreased the biocontrol efficiency of *T. viride* against these species.

Competition and Rhizosphere Competence

If mycoparasitism and antibiosis are not the principal mechanisms in the bio-control process, what is? One mechanism that has gained adherents in recent years is that of competition through rhizosphere competence. Rhizosphere competence is important because a biocontrol agent cannot compete for space and nutrients if it is unable to grow in the rhizosphere. For a Trichoderma species to be rhizosphere-competent, the fungus must colonize the rhizosphere to a depth greater than 2 cm from the seed [1] or proliferate to a concentration that exceeds the initial population coated on the seed [4]. *Trichoderma* species are generally considered to be aggressive competitors, grow very fast and rapidly colonize substrates to exclude pathogens such as *Fusarium* spp. [38]. Rhizosphere competence, following seed treatment is an important strategy to create a zone of protection against pathogens [23]. *Trichoderma* species, either added to the soil or applied as seed treatments, grow readily along with the developing root system of the treated plants. Soil treatments with *T. harzianum* spores suppressed infestations of *Fusarium oxysporum f.* sp. *vasinfectum* and *F. oxysporum* f. sp. *melonis*. Competition was a proposed mechanism, although it was not proven to be the main activity. *Trichoderma* species either added to the soil or applies as seed treatments, grow readily along with the developing root system of the treated plant [21]. The difficulty in viewing competition through rhizosphere competence as a major mechanism in biological control is that strains of *T. koningii* that are the excellent root colonizers exhibit little or no biological activity against *R. solani* on cotton seedlings. One concept that is associated with competition and rhizosphere competence, the placement of endogenous fungi on the root surface [16], can be difficult to demonstrate. *Trichoderma* species are often able to suppress the growth of endogenous fungi on an agar medium and therefore mask their presence. A rhizosphere-competent *Trichoderma polysporum* and a rhizosphere-competent *Trichoderma viride* are particularly effective biocontrol agents for plant and/or tree diseases associated with various fungi e.g., *Pythium* spp., *Sclerotium*, spp., *Rhizoctonia solani* and basidiomycetes such as *Chondrostereumm purpureum*. The rhizosphere competence of the Trichoderma atroviride isolate C52 was studied using UP-PCR band profile and the isolate was introduced into the Sclerotium cepivorum-infested soil. The use of a UP-PCR band profile also proved valuable in distinguishing T. atroviride C52 when Trichoderma isolates were recovered from the field trial site [31].

Induced Resistance

In recent years, interest in the ability of beneficial microorganisms to induce resistance in plants has grown, particularly with respect to their use as environmentally safe controllers of plant diseases. In soils, plant disease suppression by Trichoderma spp. used as biocontrol agents has been widely documented: it is considered to be a multifaceted process that requires the synergistic contribution of several mechanisms, which may include activation of the plant defense system. Specific strains of fungi in the genus *Trichoderma* colonize and penetrate plant root tissues and initiate a series of morphological and biochemical changes in the plant, considered to be part of the plant defense response, which subsequently leads to induced systemic resistance (Figure 2).

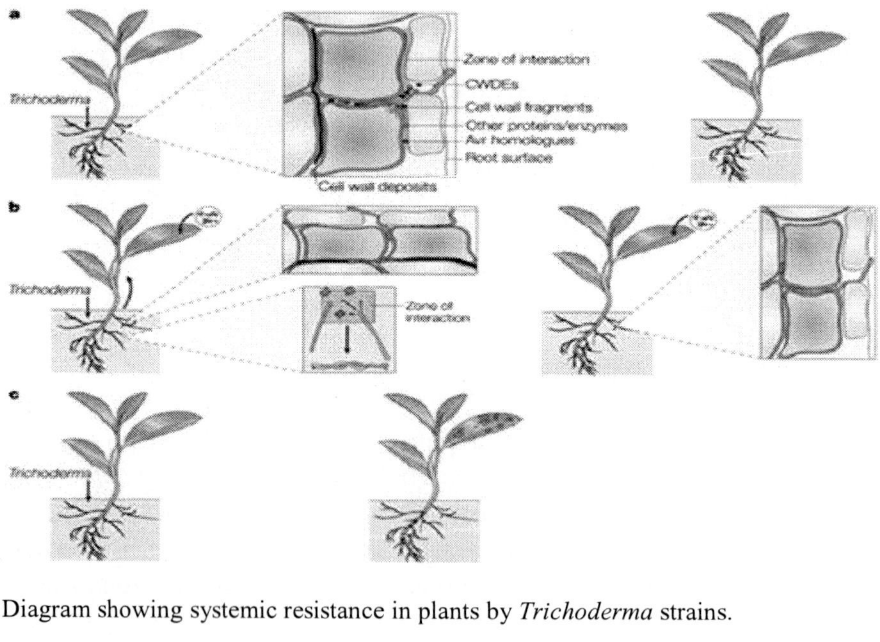

Figure 2. Diagram showing systemic resistance in plants by *Trichoderma* strains.

Root colonization by *Trichoderma* spp. also frequently enhances root growth and development, crop productivity, resistance to abiotic stresses and the uptake and use of nutrients [17]. The activation of plant defense systems in association of roots treated with *T. harzianum* strain T 203 exhibited higher activities of chitinase, ß-1, 3-glucanase, cellulase and peroxidase when compared to an untreated control 72 hours post inoculation [45]. Yedidia et al., [61] demonstrated that inoculating roots of 7 days old cucumber seedlings in an aseptic hydroponic system with *T. harzianum* (T203) spores to a final concentration of 10.5/mL initiated plant defense responses in both the roots and leaves of treated plants. They also demonstrated that hyphae of the bio-control fungus penetrated the epidermis and upper cortex of the cucumber root. The plant response was marked by an increase in peroxidase activity, an increase in chitinase activity, and the deposition of callose enriched wall appositions on the inner surface of cell walls. Later, Yedidia et al., [21] showed that inoculation of cucumber roots with *T. harzianum* (T203) induced an array of pathogenesis related proteins, including a number of hydrolytic enzymes. In another study, Howell et al., [21] demonstrated that seed treatment of cotton with bio-control preparations of *T. virens* to cotton seedling radicles induced synthesis of much higher concentrations of the terpenoids deoxyhemigossypol, hemigossypol and gossypol in developing roots than those found in untreated controls. Biocontrol activity against *R. solani* was highly correlated with induction of terpenoid synthesis in cotton roots by *Trichoderma* species, even among strains of *T. virens* that were deficient for mycoparasitism and antibiotic production. In addition to terpenoid synthesis, treatment of cotton roots with *T. virens* also induced significantly higher levels of peroxidase activity than that found in control roots. Li et al., [27] demonstrated that Trichokonins from *Trichoderma pseudokoningii* induce resistance in Chinese cabbage against Pcc infection through the activation of salicylic acid signaling pathway, which imply the potential of *Trichoderma* and peptaibols in controlling plant disease caused by Gram-negative bacteria.

PLANT GROWTH PROMOTION AND TOLERANCE TO ABIOTIC STRESS

Trichoderma spp. have beneficial effects on plant growth and enhance resistance to both biotic and abiotic stresses. Trichoderma abilities to alleviate abiotic stresses are known, although specific knowledge of mechanisms controlling multiple plant stress factors is still lacking. Mastouri et al., [30] reported that Trichoderma harzianum T22 treated seeds germinated faster than untreated seeds whether the stress condition such as osmotic, salt, or suboptimal temperatures are applied. T22 also improves tolerance to water deficit of tomato seedlings by enhancing the antioxidant defense mechanism and higher activity of ascorbate and glutathione-recycling enzymes. Trichoderma also increases root development and crop yield, the proliferation of secondary roots, the foliar area and can solubilize several plant nutrients. Colonization of cucumber roots by T. asperellum has been shown to enhance the availability of P and Fe to plants, with significant increases in dry weight, shoot length and leaf area [62]. Trichoderma spp. also produce auxins that are able to stimulate plant growth and root development [7]. An auxin-like effect has also been observed in etiolated pea stems treated with harzianolide and 6-pentyl-α-pyrone, the major secondary metabolites produced by different Trichoderma strains [53]. T. virens also induces higher photosynthetic rates and systemic increases in the uptake of CO_2 in leaves of the maize plant [52]. Mastouri et al., [29] reported that the treatment of tomato seeds with T. harzianum accelerates seed germination, increases seedling vigour and ameliorates water, osmotic, salinity, chilling and heat stresses by inducing physiological protection in plants against oxidative damage. On the other hand salt stress tolerance was observed in tobacco plants overexpressing the T. harzianum chit33 and chit42 chitinase genes [9]. T. harzianum hsp70 gene in Arabidopsis induce resistance to high temperatures, high salinity and drought without loss of vigour and growth or developmental alterations [35]. T. harzianum Thkel1 gene, encoding a kelch-repeat protein involved in the modulation of glucosidase activity that enhanced seed germination and plant tolerance to salt and osmotic stresses when it was expressed in Arabidopsis [18], and transgenic tobacco plants expressing a T. virens glutathione transferase to improve their remediation and xenobiotic degradation potential [11].

Transgenic Plants Expressing *Trichoderma* Genes

Considerable progress has been made in identification and cloning of genes involved in plant defense responses. With the aid of plant molecular biology and biotechnology, a large number of antifungal proteins and peptides have been isolated and assessed through *in vitro* bioassays. Disease resistance in transgenic plants has been improved, for the first time, by the insertion of a gene from a biocontrol fungus. Transgenic plants have been produced with genes involved in pathways in order to evaluate the effects in enhancing disease resistance. Prior to the application of genetic engineering techniques, genes were selected on the basis of *in vitro* bioassay. Development and effectiveness of specific gene(s) in disease response pathway were assessed and depending on such assessment potential molecules were identified for their utility in producing transgenic tolerant plants (Figure 3).

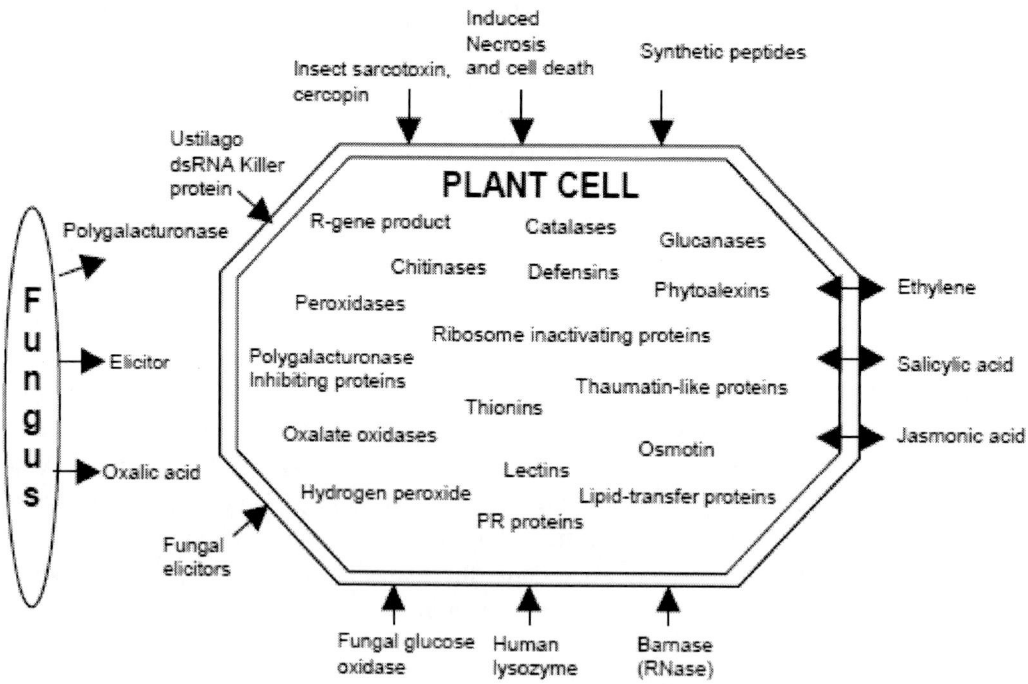

Figure 3. Overall view of various gene product having potential antifungal activity. Gene products that are secreted from fungi (Simplified from Punja, 2001).

Broglie et al., [6] constitutively expressed bean chitinase in tobacco and *Brassica napus* to enhance resistance towards *Rhizoctonia solani*. Among the PR proteins hydrolytic enzymes (chitinase and glucanase), osmotins, thionins and defensins are specially important. Genetic engineering has the advantage of incorporating resistant genes from any species to improve disease resistance genetically [14]. The gene encoding a strongly antifungal endochitinase from the mycoparasitic fungus *Trichoderma harzianum* was transferred to tobacco and potato and it was found that transgenic lines were highly tolerant or completely resistant to the foliar pathogens *Alternaria alternata*, *A. solani*, *Botrytis cinerea*, and the soilborne pathogen *Rhizoctonia solani* [28]. The lytic enzyme CHIT36 is secreted by Trichoderma harzianum that exhibits antifungal activity against Alternaria radicina and Botrytis cinerea (Baranski 2008). Shah et al., [43] developed the transgenic Nicotiana tabacum and Lycopersicon esculentum that expressed an endochitinase (ech42) gene from biocontrol fungus Trichoderma virens using Agrobacterium-mediated genetic transformation. Four genes viz., chitinase gene from bio-control gent viz., Trichoderma harzianum and Metarhizium anisopliae, polygalacturonase inhibitor protein (PGIP) a hydrolytic enzyme involved in fungal inhibition from chilli and the homologous chit42 gene from T. virens was able to enhance resistance against R. solani when it was expressed in rice. The multiple expression of rice transgenes encoding two chitinases (ech42 and nag70) and one β-1,3-glucanase (gluc78) of T. atroviride showed resistance to R. solani and Magnaporthe grisea in rice [26]. Transgenic cotton plants expressing the T. virens endochitinase gene Tv-ech1 showed significant resistance to A. alternata and R. solani and [25]. Co-expression of endo- (ech42) and exo- (nag70) chitinases of T. atroviride in apple has been correlated with increased resistance to V. inaequalis [15]. On the other hand RNA Sequence provides insights into the mechanisms of gene expression involved in mycoparasitism of T. harzianum against S.sclerotiorum and facilitate improvement of the annotation of gene models in the draft T. harzianum genome [51].

FUTURE CHALLENGES

Trichoderma strains have become established in the plant rhizosphere and evolved as intercellular root colonizers. Biological control of plant pathogens by *Trichoderma* strains has been considered a more natural and environmentally acceptable alternative to the existing chemical treatment methods. Trichoderma genomes have revealed mycotrophy and mycoparasitism as ancestral lifestyles and as a result, they stimulate plant growth and defences against pathogens. Trichoderma also produces the phytohormones ET and IAA, which play roles in interconnecting plant development and defence responses. The expression of Trichoderma genes in plants has beneficial results, mainly in the control of plant diseases and resistance to adverse environmental conditions. There is a need for more studies aimed at gaining insight into the signalling transduction pathways, related to defence mechanism, resulting from Trichoderma–plant interactions in the presence of pathogens and/or different types of abiotic stress.

ACKNOWLEDGMENTS

The authors wish to thank the Centre of Advanced in Botany, Banaras Hindu University, Varanasi – 221005, India for the financial assistance.

REFERENCES

[1] Ahmad, J.S., Baker, R., (1987). Rhizosphere competence of Trichoderma harzianum. *Phytopathology* 77, 182–9.

[2] Baek, J. M., Howell, C. R., Kenerley, C. M., (1999). The role of an extracellular chitinase from *Trichoderma virens* Gv29-8 in the biocontrol of *Rhizoctonia solani. Curr. Genet.* 35, 41-50.

[3] Baranski, R., Klocke, E., Nothnagel, T., (2008). Chitinase CHIT36 from *Trichoderma harzianum* enhances resistance of transgenic carrot to fungal pathogens. *J Phytopathol* 156, 513–521.

[4] Beagle-Ristaino, J.E., Papavizas, G.C., (1985). Survival and proliferation of propagules of Trichoderma spp. and Gliocladium virens in soil and in plant rhizospheres. *Phytopathology* 75, 729–32.

[5] Bolar, J.P., Norelli, J.L., Wong, K.W., Hayes, C.k., Harman, G.E, Aldwinckle, H.S., (2000). Expression of endochitinase from Trichoderma harzianum in transgenic apple increases resistance to apple scab and reduces vigor. *Phytopathology* 90, 72–77.

[6] Broglie, K., Holliday, C., Cressman, Biddle, Knowlton, Mauvals, and Broglie, (1991). Transgenic plants with enhanced resistance to the fungal pathogen *Rhizoctonia solani.* Science 254, 1194–1197.

[7] Contreras-Cornejo, H. A., Macias-Rodriguez, L., Cortes-Penagos, C., Lopez-Bucio, J., (2009). *Trichoderma virens,* a plant beneficial fungus, enhances biomass production and promotes lateral root growth through an auxin-dependent mechanism in Arabidopsis. *Plant Physiol* 149, 1579–1592.

[8] Cook, R.J., Granados, R.R., (1991). Biological control: making it work. In: *Agricultural Biotechnology at the Crossroads* (Macdonald, M.J., Ed.). National Agricultural Biotechnology Council., Ithaca, NY, pp. 213- 227.

[9] Dana, M. M., Pintor-Toro, J. A., Cubero, B., (2006). Transgenic tobacco plants overexpressing chitinases of fungal origin show enhanced resistance to biotic and abiotic stress agents. *Plant Physiol* 142, 722–730.

[10] De Marco, J.L., Valadares Inglis, M.C., Felix, C.R., (2003). Production of hydrolytic enzymes by *Trichoderma* isolates with antagonistic activity against *Crinipellis perniciosa* the causal agent of witches' broom of cocoa. *Brazilian Journal of Microbiology* 34, 33-38.

[11] Dixit, P., Mukherjee, P. K., Sherkhane, P. D., Kale, S. P., Eapen, S., (2011). Enhanced tolerance and remediation of anthracene by transgenic tobacco plants expressing a fungal glutathione transferase gene. *J Hazard Mater* 192, 270–276.

[12] Druzhinina, I.S., Zelazowska, M.K., Ismaiel, A., Jaklitsch, W., Mullaw, T., Samuels, G.J., Kubicek, C.P., (2012). Molecular phylogeny and species delimitation in the section *Longibrachiatum* of *Trichoderma. Fungal Genetics and Biology* 49(5): 358–368.

[13] Elad, Y., Kapat, A., (1999). The role of Trichoderma harzianum protease in the biocontrol of Botrytis cinerea. *Eur. J. Plant Pathol.* 105, 177-189.

[14] Erik A. B., (2001) Quest for antimicrobial genes for engineering disease resistant crops. *Trends in Plant Sci.* 6: 89-91.

[15] Faize, M., Malnoy, M., Dupuis, F., Chevalier, M., Parisi, L., Chevreau, E., (2003). Chitinases of *Trichoderma atroviride* induce scab resistance and some metabolic changes in two cultivars of apple. *Phytopathology* 93, 1496–1504.

[16] Harman, G. E., (2000). Myths and dogmas of biocontrol. Changes in perceptions derived from research on *Trichoderma harzianum* T22. *Plant Dis.* 84, 377–393.

[17] Harman, G. E., Howell, C. R., Viterbo, A., Chet, I. and Lorito, M., 2004. *Trichoderma* species- opportunistic, avirulent plant symbionts, A review. Nature Reviews Microbiology 2, 43-56.

[18] Hermosa, R., Botella, L., Keck, E., Jimenez, J. A., Montero-Barrientos, M., Arbona, V., Gomez-Cadenas, A., Monte, E., Nicolas, C., (2011). The overexpression in *Arabidopsis thaliana* of a *Trichoderma harzianum* gene that modulates glucosidase activity, and enhances tolerance to salt and osmotic stresses. *J Plant Physiol.* 168, 1295–1302.

[19] Howell, C. R., Stipanovic, R. D., Lumsden, R. D., (1993). Antibiotic production by strains of Gliocladium virens and its relation to the biocontrol of cotton seedling diseases. *Biocontrol. Sci. Tech.* 3, 435–441.

[20] Howell, C. R., Stipanovic, R. D., **(**1995**).** Mechanisms in the biocontrol of *Rhizoctonia solani*-induced cotton seedling disease by *Gliocladium virens:* antibiosis. *Phytopathology* 85, 469–472.

[21] Howell, C.R., Hanson, L.E., Stipanovic, R.D., Puckhaber, L.S., (2000). Induction of terpenoid synthesis in cotton roots and control of *Rhizoctonia solani* by seed treatment with *Trichoderma virens*. *Phytopathology* 90, 248-252.

[22] Howell, C. R., (2002). Cotton seedling preemergence damping-off incited by *Rhizopus oryzae* and *Pythium* spp. and its biological control with *Trichoderma* spp. *Phytopathology* 92, 177–180.

[23] Howell, C. R., (2003). Mechanisms employed by *Trichoderma* species in the biological control of plant diseases: The History and evolution of current concepts. Plant Dis.: 87, 4-10.

[24] Jorge, L., (2014). *Trichoderma* strains as biocontrol agents. *Adv Genet Eng*. 3(1), 110-115.

[25] Kumar, V., Parkhi, V., Kenerley, C. M., Rathore, K. S., (2009). Defense-related gene expression and enzyme activities in transgenic cotton plants expressing an endochitinase gene from *Trichoderma virens* in response to interaction with *Rhizoctonia solani. Planta* 230, 277–291.

[26] Liu, M., Sun, Z. X., Zhu, J., Xu, T., Harman, G. E., Lorito, M., (2004). Enhancing rice resistance to fungal pathogens by transformation with cell wall degrading enzyme genes from *Trichoderma atroviride. J Zhejiang Univ Sci* 5, 133–136.

[27] Li, H.Y., Luo,Y., Zhang, X.S., Shi, W.L., Gong, Z.T., Shi, M., Chen, L.L., Chen, X.L., Zhang, Y.Z., Song, X.Y., (2014). Trichokonins from *Trichoderma pseudokoningii* SMF2 induce resistance against Gram-negative *Pectobacterium carotovorum* subsp. *carotovorum* in Chinese cabbage. FEMS Microbiology Letters 354 (1), 75-82.

[28] Lorito, M., Woo, S. L., García, I., Colucci, G., Harman, G. E., Pintor-Toro, J. A., Filippone, E., Muccifora, S., Lawrence, C. B., (1998). Genes from mycoparasitic fungi as a source for improving plant resistance to fungal pathogens. *Proc Natl Acad Sci* 95, 7860–7865.

[29] Mastouri, F., Bjorkman, T., Harman, G. E., (2010). Seed treatment with *Trichoderma* harzianum alleviates biotic, abiotic, and physiological stresses in germinating seeds and seedlings. Phytopathology 100, 1213–1221.

[30] Mastouri, F., Bjorkman, T., Harman, G. E., (2012). *Trichoderma harzianum* enhances antioxidant defense of tomato seedlings and resistance to water deficit. *Mol. Plant Microbe. Interact.* 25, 1264–1271.

[31] McLean, K.L., Swaminathan, J., Frampton, C.M., Hunt, J.S., Ridgway, H.J., Stewart, A.,. (2005). Effect of formulation on the rhizosphere competence and biocontrol ability of *Trichoderma atroviride* C52. *Plant Pathol.* 54, 212–218.

[32] Metcalf, D.A., Wilson, C.R., (2001). The process of antagonism of *Sclerotium cepivorum* in white rot affected onion roots by *Trichoderma koningii. Plant Pathology* 50, 249–257.

[33] Migheli, Q., Gonzalez-Candelas, L., Dealessi, L., Camponogara, A., Ramon-Vidal, D., (1998). Transformants of Trichoderma longibrachiatum overexpressing the β-1,4-endoglucanas gene egl1 show enhanced biocontrol of Pythium ultimum on cucumber. *Phytopathology* 88, 673–677.

[34] Monte, E., Llobell, A., (2003). Trichoderma in organic agriculture. *Proc. V World Avocado Congress*, pp. 725–733.

[35] Montero-Barrientos, M., Hermosa, R., Cardoza, R. E., Gutierrez, S., Nicolas, C., Monte, E., (2010). Transgenic expression of the *Trichoderma harzianum* hsp70 gene increases Arabidopsis resistance to heat and other abiotic stresses. *J Plant Physiol.* 167, 659–665.

[36] Mukherjee, A.K., Kumar, A.S., Kranthi, S., Mukherjee, P.K., (2014). Biocontrol potential of three novel *Trichoderma* strains: isolation, evaluation and formulation. *Biotech.* 4(3), 275–281.

[37] Ordish, G. (1967). Biological Methods in Plant Pest Control, Constable and Company Ltd. London, pp. 242.

[38] Papavizas, G. C., (1985). *Trichoderma* and *Gliocladium*: Biology, ecology and for biocontrol. *Ann. Rev. Phytopathol.* 23, 23-534.

[39] Punja, Z.K., (2001) Genetic engineering of plants to enhance resistance to fungal pathogens –a review of progress and future prospects. Can. J. Plant Pathol. 23: 216-235.

[40] Ratnakumari, Y.R., Nagamani, A., Bhramaramba, R., kumar, S., Kumar, U.C., Shaik, M., (2011). Non-volatile and volatile metabolites of antagonistic *Trichoderma* against collar rot pathogen of *Mentha Arvensis*. *Int. J. Pharm. Biomed. Res.* 2(2), 56-58.

[41] Roco, A., Perez, L.M., (2005). *In Vitro* biocontrol analysis of *Trichoderma harzianum* on *Alternaria alternata* in the presence of growth regulators. *Plant Biotechnology* 4 (2), 68-73.

[42] Schirmbock, M., Lorito, M., Wang, Y.L., Hayes, C.K., Arisan-Atac, I., Scala, F., Harman, G.E., Kubiceck, C.P., (1994). Parallel formation and synergism of hydrolytic enzymes and peptaibol antibiotics, molecular mechanisms involved in the antagonistic action of *Trichoderma harzianum* against phytopathogenic fungi. *Appl. Environ. Microbiol.* 60: 4364-4370.

[43] Shah, J. M., Raghupathy, V., Veluthambi, K., (2009). Enhanced sheath blight resistance in transgenic rice expressing an endochitinase gene from *Trichoderma virens. Biotechnol Lett* 31, 239–244.

[44] Shalini, Narayan K.P., Lata, Kotasthane A.S., (2006). Genetic relateness among *Trichoderma* isolates inhibiting a pathogenic fungi *Rhizoctonia solani. African J. Biotech.* 5, 580-584.

[45] Sharma, R., Katoch, M., Govindappa, N., Srivastava, P.S., Sastry, K.N., Qazi, G.N., (2012). Evaluation of the catalase promoter for expressing the alkaline xylanase gene *(*alx*)* in *Aspergillus niger*. *FEMS Microbiology Letters*, 327, 33-40.

[46] Sharon, E., Bar-Eyal, M., Chet, I., Herrera-Estrella, A., Kleifeld, O., Spiegel, Y., (2001). Biological control of the root-knot nematode *Meloidogyne javanica* by *Trichoderma harzianum*. *Phytopathology* 91, 687-693.

[47] Singh, A., Shahid, M., Srivastava, M., Pandey, S., Sharma, A., (2014). Optimal Physical Parameters for Growth of *Trichoderma* species at varying pH, temperature and agitation, *Virol Mycol* 3, 127-132.

[48] Sivasithamparam, K., Ghisalberti, E.L., (1998). Secondary Metabolism in Trichoderma and Gliocladium. In: Trichoderma and Gliocladium-Basic Biology, Taxonomy and Genetics (Kubicek, C.P., Harman, G.E., Eds.), Taylor and Francis Ltd., London, pp. 139–191.

[49] Smitha,C., Finosh, G.T., Rajesh, R., Abraham, P.K., (2014). Induction of hydrolytic enzymes of phytopathogenic fungi in response to *Trichoderma viride* influence biocontrol activity. *Int. J. Curr. Microbiol. App. Sci.* 3(9), 1207-1217.

[50] Snyder, W. C., (1960). Antagonism as a plant disease control principle, In: *Biological and Chemical Control of Plant and Animal Pests*, (Reitz, L. P., Ed.), Amer. Assoc. Adv. Sci., Washington, D.C., pp. 127–136.

[51] Steindorff, A.S., Ramada, M.H.S., Coelho, A.S.G., Miller, R.N.G., Pappas, G.J., Ulhoa, C.J., Noronha, E.F., (2014). Identification of mycoparasitism-related genes against the phytopathogen Sclerotinia sclerotiorum through transcriptome and expression profile analysis in Trichoderma harzianum . BMC Genomics **15,** 204.

[52] Vargas, W. A., Mandawe, J. C., Kenerley, C. M., (2009). Plant-derived sucrose is a key element in the symbiotic association between *Trichoderma virens* and maize plants. *Plant Physiol* 151, 792–808.

[53] Vinale, F., Sivasithamparam, K., Ghisalberti, E. L., Marra, R., Barbetti, M. J., Li, H., Woo, S. L., Lorito, M., (2008). A novel role for *Trichoderma* secondary metabolites in the interactions with plants. *Physiol Mol Plant Pathol* 72, 80–86.

[54] Webster, J., Lomas, N., (1964). Does *Trichoderma viride* produce gliotoxin and viridinl. *Trans. Brit. Mycol. Soc*. 47, 535-540.

[55] Weindling, R., (1932). *Trichoderma lignorum* as a parasite of other soil fungi. *Phytopathology* 22, 837-845.

[56] Weindling, R., (1934). Studies on lethal principle effective in the parasitic action of *Trichoderma lignorum* on *Rhizoctonia solani* and other soil fungi. *Phytopathology* 24, 1153–1179.

[57] Weindling, R., (1941**)**. Experimental consideration of the mold toxins of Gliocladium and Trichoderma. *Phytopathology* 31, 991–1003.

[58] Wilhite, S. E., Lumsden, R. D., Straney, D. C., **(**1994**).** Mutational analysis of gliotoxin production by the biocontrol fungus *Gliocladium virens* in relation to suppression of *Pythium* damping-off. *Phytopathology* 84, 816–821.

[59] Wu, Q., Bai, L., Liu, W., Li, Y., Lu, C., Li, Y., Fu, K., Yu, C., Chen, J., (2013) Construction of a *Streptomyces lydicus A01* transformant with a chit42 gene from

Trichoderma harzianum P1 and evaluation of its biocontrol activity against *Botrytis cinerea*. *J Microbiol*. 51, 166–173.

[60] Yedidia, I., Benhamou, N. Chet, I., (1999). Induction of defense responses in cucumber plants (Cucumissativus L.) by the biocontrol agent *Trichoderma harzianum*. *Appl. Environ. Microbiol.* 65, 1061–1070.

[61] Yedidia, I., Benhamou, N., Kapulnik, Y. Chet, I., (2000). Induction and accumulation of PR protein activity during early stages of root colonization by the mycoparasite *Trichoderma harzianum* strain T-203. *Plant Physiol. Biochem.* 38, 863–873.

[62] Yedidia I, Srivastva AK, Kapulnik Y, Chet I (2001) Effect of *Trichoderma harzianum* on microelement concentrations and increased growth of cucumber plants. *Plant Soil* 235, 235–242.

In: Microbes in Soil and Their Agricultural Prospects
Editors: K. Kishore Choudhary and D. Wattal Dhar
ISBN: 978-1-63482-824-6

Chapter 19

INVOLVEMENT OF BACTERIAL STRAINS IN METAL EXTRACTABILITY FROM FLY ASH ISOLATED FROM RHIZOSPHERIC ZONE OF FERN *AMPELOPTERIS PROLIFERA* GROWING ON FLY ASH DUMPING SITES

Alka Kumari*
Biodiversity Division, CSIR-Institute of Himalayan Bioresource technology, Palampur, India

ABSTRACT

In present investigation, 5 bacterial strains were isolated from the rhizospheric zone of fern *Ampelopteris prolifera*. The plant was growing naturally in fly ash dumping sites of coal-based Thermal Power Plant Corporation (NTPC), which is located at Kanti of Muzaffarpur district, Bihar, India. All the strains were aerobic, showed positive result with indole production and were able to grow in MacConkey agar. However, three strains were gram positive and two were gram negative. These strains were inoculated separately in the fly ash with additional source of carbon to test their ability to increase the bioavailability or immobilization of toxic metals like Cd, Cr and As. It was observed that most of the bacterial strains either enhanced the mobility of Fe, Cr and As or immobilized Cu and Cd. However, there were few exceptions where $IHBT_5$ increased bioavailability of Cu. On the other hand, $IHBT_1$ and $IHBT_3$ enhanced bioavailability of Cu and As by inducing the immobility. The results also indicated that the mobility/immobility of trace metals from the exchangeable fractions was the specific function of bacterial strains depending upon the several edaphic and environmental factors. Based on the extractability of metals from fly ash, a consortium of high performer bacterial strains will be further used to enhance the phytoextraction of metals from fly ash by metal accumulating ferns. On the other hand, bacterial strains responsible for immobilization of metals may be for arresting their leaching to water bodies.

* Corresponding Author address: Biodiversity Division, CSIR-Institute of Himalayan Bioresource technology, Palampur-176 061 (H.P.), India. Email: kumarialkasanjay@gmail.com; alka@ihbt.res.in

Keywords: heavy metals, bioavailability, immobilization, bacteria, DTPA

INTRODUCTION

Coal is the major source of energy in India as more than 70 % energy is generated by coal-based thermal power generating stations. As a result, around 90 million ton fly ash (FA) produced during 1995 which is likely to exceed 140 million ton during 2020 [1]. Since, Indian coal contains larger quantity of ash (around 40 %) so, FA has been the main coal combustion residue and its management has been considered a very important element in terms of environmental perspective.

Chemically, FA contains almost all essential micronutrient elements that are used in normal plant metabolism [2] But, moreover FA contains a number of toxic trace elements also like Si, Al, Fe, Mn, B, Cd, Pb, As, Hg [3], which gradually leach out to contaminate the surface and ground water reservoirs and posing a threat to the receiving environment as well as health risks to millions of people also [1]. Although FA is being used in various sectors like brick and tiles manufacturing, cement industries and other construction activities including land filling and query restoration but even then a large quantity of the ash remains for eco-friendly management. Besides, the use of FA as a good soil amender to boost up crop production has also been reported [4]. Therefore, management of fly ash in agriculture provides a feasible alternative for its safe disposal to improve a healthy and aesthetic environment and enhance the crop productivity. Utilization of FA to improve agriculture productivity would not only be a solution to the problem but might also decrease the use of inorganic non-nitrogenous fertilizers. But the presence of toxic trace elements is the only disadvantage in agriculture sector, which may cause some chronic toxicity through food chain [5].

Therefore, revegetation of the FA dykes/landfills by metal tolerant plants is the only cost-effective and eco-friendly green technology solution. It also serves the purpose of stabilization and provides a pleasant landscape [6, 7]. In this technology, microbes may be employed, as they are known to play various functions in metal transformations. Generally metal transformations consist of two broad categories: redox conversions of inorganic forms and conversions from inorganic to organic forms through methylation and demethylation process. They utilize metals as terminal electron acceptors in anaerobic respiration [8, 9], while they derive energy from metal oxidation for their growth in aerobic respiration [10, 11]. In addition, the microbes possess reduction mechanisms, which are not coupled to respiration, but impart metal resistance. Aerobic and anaerobic reduction of Cr (VI) to Cr (III) [9], Se (VI) to elemental Se [12], and Hg (II) to Hg (0) [13] are widely known detoxification mechanisms in the bacteria. In redox conversion and methylation reaction, acidophilic iron and sulfur oxidizing bacteria are reported to leach very high concentration of As, Cd, Cu, Ni and Zn from contaminated soils (14, 15, a-16, b-17). Reversely, metals can be precipitated as insoluble sulfides indirectly by the metabolic activity of sulfate reducing bacteria (12, a-16, b -17).

Although in addition to, several flowering plants and some aquatic plants many ferns have been reported as hyperaccumulator of noxious metals and metalloids in which Chinese brake fern (*Pteris vittata* L.) a potential hyperaccumulator of arsenic (a-18, b-19) play a leading role in field of phytoremediation. Recently the fern *Adiatum capillus-veneris* has been reported as comparable hyper-accumulator of As to *P. vittata* [20] and comparative account of chromium accumulation in three ferns have also been reported [21].

Moreover, some works have also been done on the use of fern species in phytoremediation of heavy metals from FA and revegetating FA landfills by the same [7, 22, 23] but only a few reports are available on the microbe-assisted remediation of metals from wastes to check surface and ground water metal contamination. Xu and Ting [24] used *Aspergillus niger* – a fungus for bioleaching of metals from incinerator ash of municipal solid waste [25], reported immobilization of heavy metals by sorption and *in situ* bioprecipitation processes.

However, bacterial strains isolated from rhizospheric zones enhance the remediation of contaminated soils has been reported by Amora-Lazcano et al., [26] and bacterial strains isolated from rhizosphere of *Typha latifolia* play significant role in metal mobilization/immobilization (a-16, b-17) but no work on microbe associated metal detoxification in any fern species has been reported till date. Hence, this study was planned to measure the ability of fly ash tolerant bacterial strains isolated from rhizospheric zone of *Ampelopteris prolifera* in solublization and immobilization of metals in order to develop a microbe-assisted phytoremediation technology in ferns.

MATERIALS AND METHODS

Fly ash (FA) samples were collected from the rhizospheric zones of fern *Amelopteris prolifera* Copel. growing naturally on the FA disposal sites of NTPC coal-based Thermal Power Plant, located at Kanti, district Muzaffarpur, Bihar, India.

Physico-Chemical Analysis of Fly Ash

Electrical conductivity and pH of FA was determined by Orion electrical conductivity meter and Orion pH meter. Metal concentration of FA was determined using Atomic Absorption Spectrophotometer (Perkin Elmer 2380-AAS) after the digestion of FA samples following the method Kumari et al. [23].

Bacterial Isolation from Fly Ash

Bacterial isolation was carried out from the FA of rhizospheric zone of A. prolifera following the serial dilution method and using nutrient agar (NA composition (1 L) : 10 g peptone, 10 g beef extract, 5 g sodium chloride and 12 g agar) plates. Plates were incubated at 37 ^{0}C for 24 h and the bacterial colonies were counted to find out the CFU. This value was calculated as 2.27 x 10^{8} bacteria/g of FA.

The isolated colonies were picked up by a sterilized loop and then used to make a series of parallel non overlapping streaks on the surface of the solidified agar plates. After 24 h of incubation, the isolated pure colonies developed on the agar plates. Isolated bacterial strains from FA were named for convenience before identification as: $IHBT_1$, $IHBT_2$, $IHBT_3$, $IHBT_4$, and $IHBT_5$.

Experimental Setup

All the bacterial strains were inoculated separately in nutrient broth (NB composition (1 L): 5 g peptic digest of animal tissue, 1.5 g yeast extract, 1.5 g beef extract and 5 g sodium chloride) in glass conical flask and then incubated at 37 ^{0}C in an incubation shaker (180 rpm) for 24 h.

Then after, 500 ml inoculums (CFU: 3.2 x 10^{12} – 9.65 x 10^{13} bacteria/ml, pH 7.4) was added to 1 kg air-dried fly ash placed in earthen pots (triplet for each bacterial strain) to study their growth. However, nutrient broth without bacteria was added to control pots to simulate the conditions. Fly ash in pots was kept moist by adding 300 ml double distilled water (DDW) in each pot on alternate days and homogenized by using sterile spatula to facilitate the growth of bacteria. Bacterial growth was examined through serial dilution at different time intervals to know the time required for the optimum growth of bacteria. CFU values of all bacterial strains showed their growth at 7, 15 and 22 days of inoculation in fly ash as reflected in Table 1. However, the CFU value ranged from 4.57 x 10^8 to 10.6 x 10^{10} bacteria/g at Zero days incubation. For the extraction of metals from FA, a synthetic chelater like diethylene triamine penta acetic acid (DTPA) was used as per standard procedure (27, a-16, b-17).

Metal Extraction

Since DTPA has the potential to strongly chelate Fe, Cu, Mn and Zn from the exchangeable fraction, it is normally used as a metal extractant from the alkaline soils and FA [27, 28].

The amount of metals extracted by the DTPA gives an idea of the pool size of available metals to plants. Therefore, after 7, 15, and 22 days of inoculation, a synthetic chelater DTPA was used for the extraction of metals from the fly ash incubated with all the bacterial strains separately.

The pH of the fly ash with addition of DTPA was measured to be 7.35. DTPA extractable fraction was obtained by mechanically shaking of 10 g fly ash for 2 h with 40 ml of 0.5 M DTPA, 0.01 M $CaCl_2$ and 0.1 M TEA buffered at pH 7.3 [27] and then filtered through 44 Whatman. Using an Atomic Absorption Spectrophotometer (Perkin Elmer 2380-AAS), metal content was determined in the DTPA extractions.

Table 1. Physico-chemical characteristics of Fly Ash

Physical characteristics	Value
Fly ash pH	8.1±0.1
Electrical conductivity (in mhos cm^{-1})	3.87±0.02
Cation exchange capacity	1.73±0.01
WHC (water holding capacity)	46.18±0.2

Physical characteristics	Value
Metals (μg^{-1} dw)	
Fe	4958±137
Cu	44±0.51
Zn	78±0.24
Ni	169±5.6
Cr	43±0.04
Cd	6.9±0.01
Pb	31±0.03
As	9.8±0.02

RESULTS

Fly Ash Composition

Fly ash, a residue of coal combustion is primarily made up of oxides of Al and Si, but also enriched with several other essential (Zn, Fe, Cu, Mn, B, and Mo) and non essential metals (Ni, Cr, Cd, Pb, Al, As, Si). Its particle size ranges between 10 and 150 lm, pH 7.5 and the electrical conductivity is very high (387 lS/cm).

With negligible organic carbon, it can not support bacterial growth. Hence, nutrient broth was added as an additional source of organic carbon to enhance bacterial augmentation.

Growth of Bacteria

CFU values of bacterial strains inoculated in the fly ash indicated that the bacteria multiplied very fast using the additional source of carbon in the form of NB during first

7 days of incubation and then additional carbon was perhaps depleted. Hence the multiplication of bacterial cells was arrested as reflected by the CFU values beyond 7 days of incubation (Table 1).

Table 2. CFU/g of different fly ash tolerant bacterial strains

Bacterial strain	0 days	7 days	15 days	22 days
$IHBT_1$	$7.27.10^{10}$	$14.4 \cdot 10^{10}$	$94.2 \cdot 10^{10}$	$47.7 \cdot 10^{10}$
$IHBT_2$	$10.6.10^{10}$	$37.9 \cdot 10^{8}$	$88.5 \cdot 10^{8}$	$35.2 \cdot 10^{10}$
$IHBT_3$	$6.35.10^{10}$	$27.7 \cdot 10^{10}$	$54.8 \cdot 10^{10}$	$21.2 \cdot 10^{10}$
$IHBT_4$	$7.89.10^{10}$	$38.6 \cdot 10^{10}$	$49.7 \cdot 10^{12}$	$30.5 \cdot 10^{12}$
$IHBT_5$	$6.75.10^{10}$	$25.5 \cdot 10^{10}$	$41.5 \cdot 10^{10}$	$16.5 \cdot 10^{10}$
Control	$4.57.10^{8}$	$8.65.10^{8}$	$11.6.10^{8}$	$9.95.10^{8}$

Biochemical Characterization

The biochemical characteristics of the bacterial strains isolated from the FA dumps indicated that out of 5 strains, three were found gram positive and remaining two strains were gram negative. All of them were aerobic bacteria. As none of the bacterial strain could grow in Pseudomonas agar, this ruled out the possibility of presence of Pseudomonas species in the bacterial strains isolated from the fly ash.

Metal Mobility/Immobility

When different bacteria strains isolated from the FA were re-inoculated in the FA in high inoculums separately, they presented variable trends of solublization and immobilization with incubation period. The data on the metal mobility/immobility by the different bacterial strains have been shown in Tables 3–7.

As far as the Cu immobility is concerned, most of the bacterial strain showed immobilization but there was significant metal mobility and bioavailability found in $IHBT_5$ bacterial strain with the incubation period for 7 days with respect to control and further increased, when the incubation period was increased to 15 and 22 days. However, all other bacterial strain showed decreased rate of immobilization of Cu except $IHBT_4$. The extent of Cu immobilization varied from 25% to 31% in 15 days incubation and from 12 % to 38 % in 22 days incubation period. On the other hand, bacterial strain $IHBT_5$ induced Cu mobility by 27% in 15 days and 45% in 30 days incubation period (Table 3).

In case of Fe extractability from the FA with the incubation of bacterial strains has been shown in Table 4. It was noted that all the bacterial strains induced more Fe bioavailability than toxic heavy metals. However, $IHBT_1$ and $IHBT_4$ immobilized Fe in the FA at 15 d incubation period but mobilized at 22 d incubation period. Two bacterial strains $IHBT_2$ and $IHBT_3$ showed an increasing trend of metal bioavailability with the increasing incubation period, but $IHBT_1$, $IHBT_3$ and $IHBT_5$ indicated maximum Fe extractability at 7 days incubation period and then there was a gradual decline in $IHBT_1$ and $IHBT_5$. As compared to control, Fe mobility was enhanced between 22% and 115% at 7 days, 26% to 97% at 15 days and between 12% and 87% at 22 days incubation period.

Thus, these strains showed variable trends of iron metal mobility in the FA with the varied incubation period (Table 4).

Table 5 indicates the mobility of Cd in the FA induced by the different bacterial strains. The solubility of Cd varied remarkably depending upon the bacterial strains incubated in FA for metal extractability. It was observed that most of the bacterial strains could decrease Cd metal bioavailability significantly after 15 days and 22 days of incubation period except $IHBT_1$ and $IHBT_5$, which indicated Cd immobilization to little extent only. Two bacterial strains $IHBT_1$ and $IHBT_5$ showed significantly high Cd metal bioavailability at 22 days incubation period. The metal mobility after 15 days incubation period was only 13 % and after 22 days between 37% and 54% depending upon the specific bacterial strain inoculated in FA.

As far as Cr mobility is concerned, most of the bacterial strains enhanced its bioavailability during their incubation in the FA. It is evident from Table 6, that all the bacterial strains having high bioavailability at 7 d incubation period and further reduced at 15 d incubation period except $IHBT_1$ and $IHBT_4$. Result also showed that Cr mobility was between 45% and 375% at 7 days, between 32% and 225% at 15 days and between 22% and 217% at 22 days incubation period except $IHBT_3$. It was also observed that Cr metal showed high percentage of bioavailability in comparison to the other toxic trace elements.

The mobility and immobility of the metalloid As showed reverse trend with incubation period for most of the bacterial strains except $IHBT_1$ and $IHBT_4$ at 15 d. All the strains immobilized As between 35% to 276% at 7 d, around 45% to 125% at 15 d and in between 35% to 78% at 22 d of incubation period. $IHBT_3$ and $IHBT_5$ indicated comparatively negative metal extractability at 22 days incubation period. Like other metals, As mobility was also influenced by the incubation of different bacterial strains. In this case, all the bacterial strains without any exception enhanced immobilization of As in the FA, as evident from Table 7.

Table 3. Mobility/immobility of Cu (μg/g dw) in fly ash influenced by different fly ash tolerant bacterial strains

Bacterial strain	Metal concentration (1g/g dw) incubation period			Metal bioavailability (%) incubation period		
	7 days	15 days	22 days	7 days	15 days	22 days
$IHBT_1$	2.655	2.725	2.978	-35.57	-31.75	-26.87
$IHBT_2$	2.826	3.323	3.826	-32.24	-25.56	-12.34
$IHBT_3$	1.867	1.989	2.265	-47.67	-38.57	-30.45
$IHBT_4$	3.756	2.987	2.558	-11.35	-29.85	-38.25
$IHBT_5$	1.678	2.245	2.879	-8.97	-27.79	-45.24
Control	4.125	4.289	4.446			

* Average of three replicates.

ANOVA table for Cu

Source	df	SS	MS	f
Between days	5	23.65	4.35	1148.9
Between strains	2	3.15	1.38	356.5
Error	10	5.75	0.57	145.5
Total	35	0.14	0.002	

Table 4. Mobility/immobility of Fe (μg/g dw) in fly ash influenced by different fly ash tolerant bacterial strains

Bacterial strain	Metal concentration (lg/g dw) incubation period			Metal bioavailability (%) incubation period		
	7 days	15 days	22 days	7 days	15 days	22 days
$IHBT_1$	6.754	7.727	9.532	68.56	26.25	34.65
$IHBT_2$	5.826	8.327	11.265	22.25	52.55	48.36
$IHBT_3$	7.867	9.965	12.165	94.67	78.57	62.42
$IHBT_4$	3.759	4.995	8.257	-11.24	-2.86	12.25
$IHBT_5$	7.678	11.325	13.578	115.27	97.89	87.24
Control	4.027	5.175	6.789			

* Average of three replicates.

ANOVA table for Fe

Source	df	SS	MS	f
Between days	5	221.65	41.39	889.2
Between strains	2	125.15	61.35	1326.5
Error	10	9.75	0.927	
Total	35	1.515	0.004	

Table 5. Mobility/immobility of Cd (μg/g dw) in fly ash influenced by different fly ash tolerant bacterial strains

Bacterial strain	Metal concentration (lg/g dw) incubation period			Metal bioavailability (%) incubation period		
	7 days	15 days	22 days	7 days	15 days	22 days
$IHBT_1$	0.065	0.095	1.098	0.002	13.87	546.81
$IHBT_2$	0.086	0.056	0.030	31.54	-9.85	-79.34
$IHBT_3$	0.031	0.026	0.475	-60.65	-78.56	-70.58
$IHBT_4$	0.076	0.047	0.251	6.35	-89.85	-35.23
$IHBT_5$	0.058	0.045	0.795	-23.27	-41.59	375.25
Control	0.065	0.079	0.165			

* Average of three replicates.

ANOVA table for Cd

Source	df	SS	MS	F
Between days	5	2.05	1.05	489.97
Between strains	2	1.15	0.24	612.51
Error	10	2.75	0.25	
Total	35	0.01	0.00	

Table 6. Mobility/immobility of Cr (μg/g dw) in fly ash influenced by different fly ash tolerant bacterial strains

Bacterial strain	Metal concentration (lg/g dw) incubation period			Metal bioavailability (%) incubation period		
	7 days	15 days	22 days	7 days	15 days	22 days
$IHBT_1$	0.355	0.525	0.774	85.67	97.74	106.86
$IHBT_2$	0.606	0.765	0.996	232.26	225.57	217.35
$IHBT_3$	0.469	0.358	0.215	174.45	32.52	-110.45
$IHBT_4$	0.254	0.387	0.518	45.36	55.15	50.27
Bacterial strain	Metal concentration (lg/g dw) incubation period			Metal bioavailability (%) incubation period		
$IHBT_5$	0.778	0.543	0.417	375.57	128.39	22.24
Control	0.165	4.289	4.446			

* Average of three replicates.

ANOVA table for Cr

Source	df	SS	MS	F
Between days	5	0.1265	0.053	149.9
Between strains	2	1.87	0.384	1076.5
Error	10	0.0011	0.000	
Total	35			

Table 7. Mobility/immobility of As (μg/g dw) in fly ash influenced by different fly ash tolerant bacterial strains

Bacterial strain	Metal concentration (lg/g dw) incubation period			Metal bioavailability (%) incubation period		
	7 days	15 days	22 days	7 days	15 days	22 days
$IHBT_1$	0.035	0055	0.067	35.05	51.55	35.85
$IHBT_2$	0.056	0.075	0.088	145.21	115.56	72.34
$IHBT_3$	0.086	0059	0.035	247.65	45.56	-42.43
$IHBT_4$	0.045	0.087	0.095	85.32	125.82	78.25
$IHBT_5$	0.088	0.065	0.046	276.56	79.75	-21.24
Control	0.021	0032	0.050			

* Average of three replicates.

ANOVA table for As

Source	df	SS	MS	F
Between days	5	0.0075	0.00185	76.025
Between strains	2	0.00057	0.00030	65.245
Error	10	0.014580	0.00150	
Total	35	0.00085	0.000	

DISCUSSION

Metal contamination of water bodies like rivers, ponds and lakes due to disposal of metal-loaded industrial and sewage wastes is a serious environmental concern today. To overcome this problem, bioremediation which involves microbes, plants and animals is considered as an alternative technology which is promising, cost-effective and eco-friendly in present scenario. No doubt microbes, being ubiquitous in nature, are being used for a long time to degrade pesticides, sludge and other xenobiotic compounds like PCB, PAH, TNT etc. but their application in bioremediation of metals from industrial wastes is a new dimension. Being integral component of biogeochemical cycle, they can be used to either solubilize the toxic metals, thereby increasing their bioavailability or immobilize them to check their migration to water reservoirs [29, 30, 31]. When the bacterial strains, isolated from *A. prolifera* growing naturally on FA dumps, were augmented in the FA and metals were extracted by DTPA after different incubation periods, it was observed that most of the bacterial strains induced the bioavailability of Fe in FA, but immobilized Cu and Cd on other hand.

However, $IHBT_5$ induced more mobility of Cu in contrast to other three strains. Similarly, $IHBT_2$ and $IHBT_3$ caused high mobilization of Fe at 15 days while other 3 strains decreased the bioavailability of Fe in comparison to 7 d. As against other strains, $IHBT_3$ and $IHBT_5$ induced Cd immobilization. This shows that metal mobility/ immobility is the specific function of the bacterial strains in both aerobic and anaerobic conditions. No doubt, solublization and immobilization of metals are also governed by several edaphic and environmental factors. Besides, the extent of metal solubilization and immobilization also varied significantly among the bacterial strains and with incubation periods. Among 5 bacterial strains, better performers for more than 50% metal mobility were $IHBT_1$ and $IHBT_4$ for toxic metals like Cr and As and For $IHBT_2$ and $IHBT_5$ for Fe and Cu respectively. The bacterial strains $IHBT_1$ and $IHBT_5$ showed extraordinary mobility and percentage of bioavailability for Cd metal at 22 d incubation period.

In our studies, most of the FA tolerant bacteria induced the bioavailability of Fe which may be linked to biologically mediated reduction processes. Soil bacteria have been shown to exude organic compounds which stimulate bioavailability and thereby facilitate root absorption of various metal ions, Fe [32]. Moreover, Caccavo et al. [33] isolated Geobacter sulfurrenducens from hydrocarbon contaminated ditch which was the first bacterium described to couple the oxidation of hydrogen (or acetate) to Fe (III) reduction. Besides, mobilization and immobilization of metals are also governed by various physico-chemical characteristics, more particularly pH of the contaminated sites. However, in FA which has shown pH around neutrality, metal mobility/immobility can be attributed to bacterial actions. Increased organic matter also enhances both soluble and exchangeable metal levels in soil [34]. Cd occur in soil primarily in soluble precipitates (PO2_ 4, CO2_3, and hydroxyl-oxide) and hence unavailable to plants, but they are made available to plants by the microbial action. Fly ash has also high affinity to adsorb these elements. In our studies, most of the bacterial strains enhanced the bioavailability of Fe and Cr but immobilized Cd and Cu. While metal mobilization can be achieved by autotrophic and heterotrophic leaching, chelation by microbial metabolites and siderophores and methylation which can result in volatilization, immobilization is attributed to sorption to cell components, intracellular sequestration or precipitation as insoluble organic and inorganic compounds e.g., oxalates [35, 36], sulphides (a-37, b-38) or phosphates [39]. Thus bacteria involve various metabolic processes to enhance the solubilization and immobilization of heavy metals in the soils or industrial wastes, as metals serve as electron donors or acceptors in anaerobic and aerobic respirations.

CONCLUSION

The results of the present study conclude that the rhizospheric bacteria may enhance the phytoremediation ability of the associated fern species. These bacterial strains contribute significantly to the high metal accumulation capability of ferns .Since the extraction of metal contaminants from industrial wastes by physical and chemical methods is very cumbersome and cost-intensive, the consortium of such kind of bacterial strains may be exploited to either enhance the bioavailability of toxic metals to be removed in the phytoextraction process or immobilize them to check their leachability to water reservoirs.

ACKNOWLEDGMENTS

Authors are thankful to Dr. P S Ahuja, Director, CSIR-IHBT, Palampur for motivation of collaborating research to exchange required research facilities and Dr. K. C. Gupta then Director, CSIR-NBRI and Dr. S. N. Singh Head, Ecotoxicology and Bioremediation lab, CSIR-NBRI Lucknow for extending required laboratory facilities. Alka Kumari is thankful to DST for providing financial support for this work under WOS-A scheme (SR/LS/WOS-117/2008).

REFERENCES

[1] Kalra, N., Jain, M.C., Joshi, H.C., Choudhary, R., Harit, R.C., Vasta, V.K., Sharma, S.K., Kumar, V., (1998). Fly ash as a soil conditioner and fertilizer. *Biores. Technol.*, 64, 163-167.

[2] Klein, T., Nusing, R.M., Pfeilschifter, J., Ullrich, V., (1995). Selective inhibition of Cyclooxygenase. *Biochem. Pharmacol.*, 48, 1605–1610.

[3] Mehra, A., Ferago, M. E., Banerjee D.K., (1998). Impact of fly ash from coal-fired station in Delhi with particular reference to metal contamination. *Environ. Monitor. Assessment*. 50, 15-35.

[4] Jala, S., Goyal, D., (2006). Fly ash as a soil ameliorant for improving crop production - a review. *Biores. Technol.,* 97, 1136-1147.

[5] Basu, M., Pande M., Bhadoria, P.B.S., Mahapatra, S.C., (2009). Potential fly-ash Utilization in agriculture: A global review. *Prog. Nat. Sci.,* 19, 1173– 1186.

[6] Rai, U.N., Pandey, K., Sinha, S., Sinha, A., Saxena R., Gupta, D.K., (2004). Revegetating fly ash landfills with Prosopis julifora L.: impact of different amendments andRhizobium inoculation. *Environ. Inter.,* 30, 293-300.

[7] Kumari, A . Tripathi K.P., (2009). Management of metal contaminated wastelands with fern species, *Indian Forester,* 135, 117-225.

[8] Niggemyer, A., Spring, S., Stackebrandt, E., Rosenzweig, R.F., (2001). Isolation and characterization of a novel As(V)-reducing bacterium: implications for arsenic mobilization and the genus Desulfitobacterium. *Appl. Environ. Microbiol.,*. 67, 5568-5580.

[9] Quilntana, M., Curutchet, G., Donati, E., (2001). Factors affecting chromium (VI) reduction by Thiobacillus ferrooxidans. *Biochem. Eng. J.,* 9, 11–15.

[10] Tebo, B.M., Ghiorse, W.C., van Waasbergen, L.G., Siering, P.L., Caspi, R., (1997). Bacterially-mediated mineral formation: insights into manganese (II) oxidation from Molecular, genetic and biochemical studies. *Rev. Mineral.,* 35, 225–266.

[11] Santini, J.M., Sly, L.I., Schnagl, R.D., Macy, J.M., (2000). A new chemolitoautotrophic Arsenite-oxidizing bacterium isolated from a gold-mine: phylogenetic, physiological, and Preliminary biochemical studies. *Appl. Environ. Microbiol.,* 66, 92–97.

[12] Lloyd, J.R., Mabbett, A.N., Williams, D.R., Macaskie, L.E., (2001). Metal reduction by sulphate-reducing bacteria: physiological diversity and metal specificity. *Hydrometallurgy,* 59, 327–337.

[13] Brim, H., McFarlan, S.C., Fredrickson, J.K., Minton, K.W., Zhai, M., Wackett, L.P., Daly, M.J., (2000). Engineering Deinococcus radiodurans for metal remediation in radioactive mixed waste environments. *Nat. Biotechnol.,* 18, 85–90.

[14] Seidel, H., Ondruschka, J., Morgenstern, P., Wennrich, R., Hoffmann, P., (2000). Bioleaching of heavy metal contaminated sediments by indigenous Thiobacillus spp.: metal Solubilization and sulfur oxidation in the presence of surfactants. *Appl. Microbiol.Biotechnol.,* 54, 854–857.

[15] Groudev, S.N., Spasova, I.I., Georgiev, P.S., (2001). *In situ* bioremediation of soils contaminated with radioactive elements and toxic heavy metals. *Int. J. Miner. Process,* 62, 301–308.

[16] Tiwari, S., Kumari, B., Singh, S.N., (2008 a). Evaluation of metal mobility/immobility in Fly ash induced by bacterial strains isolated from rhizospheric zone of Typha latifolia growing on fly ash dumps. *Biores. Technol.*, 99, 1305-1310.

[17] Tiwari, S., Kumari, B., Singh, S.N., (2008 b). Microbe-induced changes in metal extractability from fly ash. *Chemosphere,* 71, 1284-94.

[18] Ma, L.Q., Komar, K.M., Tu, C., Zhang, W. Cai, Y., (2001 a). A fern that hyper accumulates Arsenic. *Nature,* 409, 579.

[19] Ma, L.Q., Komar, K.M., Tu, C., Zhang, W., Cai, Y., (2001 b). A fern that hyper-accumulates Arsenic-addendum. *Nature,* 411, 438.

[20] Singh N., Raj, A., Khare P.B., Tripathi R.D., Jamil, S., (2010). Arsenic accumulation pattern in 12 Indian ferns and assessing the potential of Adiantum capillus-veneris, in comparison to Pteris vittata, as arsenic hyperaccumulator. *Biores. Technol.,* 101, 8960–8968.

[21] Kumari A, Pakade Y.B., Chand, P., Lal, B., (2011). Assessment of bioaccumulation of heavy metals by *Pteris vittata* L. growing in the vicinity of fly ash. *International J of Phytoremediation,* 13: 779-787.

[22] Kumari, A., Pandey V.C., Rai U.N., (2013). Feasibility of fern Thelypteris dentata for revegetation of coal fly ash landfills, *Journal of Geochemical Exploration,* 128, 147–152.

[23] Kumari, A., Pakade, Y.B., Chand, P., Prasad, M.N.V., Lal, B., (2014). Comparative account of chromium accumulation in three ferns under hydroponic system *J. of Scientific & Industrial Research (JSIR),* 73 (8), 553-558.

[24] Xu, T.J., Ting, Y.P., (2004). Optimization on bioleaching of incinerator fly ash by *Aspergillus niger* use of central composite design. *Enzyme Microbial. Technol.,* 35, 444–454.

[25] Roy, S.V., Vanbroekhoven, K., Dejonghe, W., Diels, L., (2006). Immobilization of heavy metals in the saturated zone by sorption and *in situ* bioprecipitation processes. *Hydrometallurgy,* 83, 195–203.

[26] Amora-Lazcano E., Guerrero-Zúñiga , L.A., Rodriguez-Tovar, A., Rodriguez-Dorantes, A., Vasquez-Murrieta1, M.S. (2010). Rhizospheric plant-microbe interactions that enhance the remediation of contaminated soils. In Current research, Technology and Education in Applied Microbiology and Microbial Biotechnology A. Mendez Vilas (Ed.) published by FORMATEX.

[27] Lindsay, Norvell, (1978). Development of DTPA soil test for Zn, Mn and Cu. *Soil Sci. Am. J.,* 42, 421–428.

[28] Gupta, A.K., Sinha, S., (2006). Role of Brassica Juncea (L.) Czern. (var. Vaibhav) in the phytoextraction of Ni from soil amended with fly ash: selection of extractant for metal. *J. Hazard Mater.,* 136 (2), 371–378.

[29] Wainwright, M., Gadd, G.M., (1997). Industrial Pollutants. In: *The Mycota*, Volume V: *Environmental and Microbial Relationships.* Wicklow, D.T., Soderstrom, B. (Eds.), published by Springer-Verlag, Berlin, p. 85–97.

[30] White, C., Sayer, J.A., Gadd, G.M., 1997. Microbial solubilization and immobilization of toxic metals: key biogeochemical processes for treatment of contamination. *FEMS Microbiol. Rev.,* 20, 503–516.

[31] Gadd, G.M., Bridge, T.A.M., Gharieb, M.M., Sayer, J.A., White, C., (2001). Microbial processes for solublization of metals and metalloids and their potential for environmental bioremediation. *Ind. Environ. Biotechnol.*, 55–80.

[32] Bural, G.I., Dixon, D.G., Glick, B.R., (2000). Plant growth-promoting bacteria that decrease heavy metal toxicity in plants. *Can. J. Microbiol.*, 46, 237–245.

[33] Caccavo, F.J., Lonergan, D.J., Lovley, D.R., Davis, M., Stolz, J.F., McInerney, M.J., (1994). *Geobacter sulfurreducens* sp. nov., a hydrogen and acetate-oxidizing assimilatory metal-reducing microorganism. *Appl. Environ. Microbiol.*, 60, 3752–3779.

[34] Yoo, M.S., James, B.R., (2002). Zinc extractability as a function of pH in organic waste- contaminated soils. *Soil Sci.* 167, 246–259.

[35] Sayer, J.A., Gadd, G.M., (1997). Solubilization and transformation of insoluble inorganic metal compounds to insoluble metal oxalates by *Aspergillus niger. Mycol. Res.,* 101, 653–661.

[36] Gharieb, M. M., Sayer, J.A., Gadd, G.M., (1998). Solubilization of natural gypsum (CaSO4.2H2O) and the formation of calcium oxalate by *Aspergillus niger* and *Serpula himantioides. Mycol. Res.,* 102, 825–830.

[37] White, C., Gadd, G.M., (1996a). Mixed sulphate-reducing bacterial cultures for bioprecipitation of toxic metals: factorial and response-surface analysis of the effects of dilution rate, sulphate and substrate concentration. *Microbiology,* 142, 2197–2205.

[38] White, C., Gadd, G.M., (1996b). A comparison of carbon/energy and complex nitrogen sources for bacterial sulphate-eduction: potential applications to bioprecipitation of toxic metals as sulphides. *J. Ind. Microbiol.*, 17, 116–123.

[39] Young, P., Macaskie, L.E., (1995). Removal of tetravalent actinide thorium from solution by a biocatalytic system. *J. Chem. Technol. Biotechnol.*, 64, 87–95.

In: Microbes in Soil and Their Agricultural Prospects
Editors: K. Kishore Choudhary and D. Wattal Dhar
ISBN: 978-1-63482-824-6

Chapter 20

INFLUENCE OF MICROBIAL POPULATIONS ON BIOMASS C, N AND P UNDER ARECA CATECHU L. BASED TRADITIONAL HOMESTEAD GARDEN OF NORTHEAST INDIA

Sumpam Tangjang*[1*]*, A. Arunachalam*[2]*,
***K. Arunachalam*[3] *and A. K. Shukla*[4]

[1]Department of Botany, Rajiv Gandhi University, Itanagar, Arunachal Pradesh, India
[2]Division of Natural Resources Management, Indian Council of Agricultural Research, Krishi Anusandhan Bhavan II, Pusa, New Delhi, India
[3]School of Environment & Natural Resources, Doon University, Kedarpur, Dehra Dun, Uttarakhand, India
[4]Department of Botany, Indira Gandhi National tribal University, India

ABSTRACT

We investigated the seasonal and depthwise variations in bacterial and fungal populations (and their influence on microbial biomass C, N and P under *Areca catechu* (L.) based traditional homestead garden. The systems were managed by two ethnic communities' ('Kalitas' and 'Nyishis'). *Areca catechu* palm was abundantly found in Harmutty which also recorded greater species diversity. Bacterial population was highest during spring and that of fungi during autumn. Altogether, 24 soil micro-fungal forms were recorded from the three sites. *Aspergillus* and *Penicillium* were the abundant genera in the sites. Microbial biomass and activities are sensitive indicators of management effects. Across the land-use, the value of microbial biomass C ranged between 47.5-1167.6μg g-1 and was highest in Harmutty than the sites in Arunachal Pradesh. Microbial biomass C, N and P were highest during rainy season and lowest during winter in all the stands. It was observed that the type of plant species composition, plant residues and

[*] Corresponding Author address: Department of Botany, [1]Department of Botany, Rajiv Gandhi University, Rono Hills-791 112, Itanagar, Arunachal Pradesh, India. Email: sumpam@gmail.com

organic matter and soil mineral nutrients altered the microbial populations as well as their species composition. Nonetheless, the low soil microbial biomass in the sites Arunachal Pradesh could be a reflection of the low vegetation abundance.

Keywords: ethnic communities; homestead garden; microbial biomass; Northeast India

INTRODUCTION

Soil represents a favourable habitat for microorganisms and is inhabited by a wide range of microorganisms [1]. Nonetheless, agricultural landscapes exhibit a high degree of spatial variability, including variation in soil physicochemical characteristics and agroecosystem management [2]. This can affect the activity and composition of the soil biota [3, 4]. Most species of soil bacteria and fungi play pivotal roles in recycling of organic compounds [5] and also influence aboveground ecosystems by contributing to plant nutrition, plant health, soil structure and soil fertility [6]. The numbers and kinds of microorganisms present in soil depend on many environmental factors: amount and type of nutrients available, moisture, degree of aeration, pH, temperature etc. [7].

The quantity and quality of soil organic matter and carbon and nitrogen inputs are the overriding controls on soil microbial biomass and activity [8]. Soil microbial biomass is an agent of transformation of added and native organic matter and acts as a labile reservoir of plant nutrients [9, 10]. It is also used as an early indicator of changes in soil properties resulting from soil management practices in agricultural ecosystems [11]. The flow of C, N, and P through microbial biomass has been identified as the most useful property for assessing soil health in ecosystems [12] and is considered as an important early indicator of changes that may occur in the long term with regard to soil fertility and constitutes an important source and sink of nutrients [13] and a driving force behind soil organic matter transformations. Because of its importance in the functioning of different ecosystems, dynamics of microbial biomass and its role in plant nutrition under different ecosystem conditions has assumed greater significance [14]. Soil microbiological properties such as microbial biomass may be used as comparisons of soils under different managements systems [15], and that high levels of microbial activity are fundamental in maintaining soil quality [16].

An integrated approach to the seasonal changes in microbial diversity and their influence on soil microbial biomass C, N and P soil nutrient conservation in traditional homestead gardens in a humid tropical environment in the north-eastern India is deficient, although such a challenge may give an insight into the microbial population dynamics as influenced by soil organic matter and nutrient build up. Therefore, the present chapter aimed at understanding the impact of microbial diversity and microbial C, N and P dynamics in traditional homestead agroforestry in and around Arunachal Pradesh, northeast India.

MATERIALS AND METHODS

The study was conducted during 2003 January to 2004 November in traditional homestead gardens of three villages (namely, Doimukh and Nirjuli (126m asl) of Papum Pare district in Arunachal Pradesh (27°60′ N latitude & 94°21′ E longitude) and Harmutty village (120m asl) of North Lakhimpur district of Assam (26°46′ N latitude & 93°50′ E longitude) bordering Arnnachal Pradesh. The 'Kalitas' dominated Harmutty village while Nirjuli and Doimukh villages are inhabited by the 'Nyishis', one of the major tribes of Arunachal Pradesh. Three systems were selected in each site. The average area of the homegarden plots varied between 200-400 m^2 in Nirjuli and Doimukh villages and 320-490 m^2 in Harmutty village. The sites are characterized by a climate with most rainfall occurring during summer months (May-July) with relatively a little or none during the winter months.

Total annual rainfall is typically 1100-1600 mm in all the sites, 90% of which falls in the rainy season (June-August) and daytime temperatures average from minimum of 12° C to a maximum of 37° C. Nirjuli and Doimukh sites are composed of newer alluvium (newer terrace deposits) represented by valley fill deposits comprising of sediments, while Harmutty area is composed of alluvium belonging to Pleistocene and recent times [17]. Over all, soil was sandy loam in all sites and slightly acidic (Table 1).

The soil samples from three depths (viz. 0-10, 10-20 and 20-30 cm) were sampled during January, March, May and November months representing winter, spring, rainy and autumn seasons respectively. Composite soil samples of ten corers (6.5 cm inner diameter) were collected from each plot aseptically in sterilized plastic bags to prevent moisture loss and were transported to the laboratory for the isolation of bacteria and fungi within 24 h. The remaining soil were sieved through a 2 mm mesh-screen and divided into two parts. One part was used in field moist condition to determine soil pH, moisture content, texture, bulk density, ammonium-N, nitrate-N, available-P and microbial C, N and P and the other part was air-dried for the determination of water holding capacity (WHC), organic carbon and total N following standard procedures given in Anderson and Ingram [18].

Isolations of Bacterial and Fungal Population

Soil bacterial population was estimated by Waksman's [19] method using nutrient agar medium at 10^5 dilutions. Fungal population was estimated by dilution plate method [20] using Martin's Rose Bengal agar medium at 10^3 dilution in water. The inoculated Petri-dishes were incubated at 30±1°C for 24 h and 25±1°C for 5 days for bacteria and fungi respectively. To calculate the bacteria and fungi populations, the resultant colonies after incubation were counted using a digital colony counter and expressed colony forming units (CFUs) g^{-1} dry weight of soil. Representative isolates of fungi were identified under microscope with the help of standard manuals [21, 22].

Microbial C was estimated following chloroform fumigation incubation (CFI) procedure given by Jenkinson and Powlson [23] as modified by Srivastava and Singh [24]. Microbial N and P were extracted following chloroform fumigation extraction (CFE) procedures using 0.5 M $K2SO_4$ and $NaHCO_3$ respectively [18]. Correction factors used in microbial C, N and P were: Microbial C/0.45 [25], Microbial N/0.54 [26] and Microbial P/0.40 [27].

Table 1. Soil properties in the traditional agroforestry systems

Sites										
Soil properties	Harmutty			Nirjuli			Doimukh			LSD (p=0.05)
	Soil depth (cm)									
	0-10	10-20	20-30	0-10	10-20	20-30	0-10	10-20	20-30	
Texture	SL			LS			LS			
WHC (%)	43.2	31.09	33.41	45.50	40.27	36.91	47.40	41.38	36.30	0.99
BD (g/cm^3)	0.58	1.10	1.15	1.13	1.21	1.27	0.38	0.41	0.58	0.24
pH	5.57	5.68	5.82	5.47	5.51	5.42	5.60	5.51	5.85	0.24
Organic C (%)	1.44	1.34	1.27	1.56	1.40	1.31	1.54	1.42	1.41	0.05
Total N (%)	0.33	0.27	0.20	0.28	0.25	0.22	0.28	0.25	0.23	0.05
C/N (%)	4.36	4.96	6.35	5.57	5.60	5.95	5.50	5.68	6.13	0.45
Available P ($\mu g\ g^{-1}$)	3.09	1.97	1.28	6.16	5.06	4.45	5.10	4.44	3.26	0.17

Values in parentheses are % contribution to respective soil nutrients. SL, sandy loam; LS, Loamy sand; WHC, Water holding capacity; BD, Bulk density

The values presented are the means of five replicated determinations and have been expressed on oven-dry weight basis (24 h at 105 °C). The data were analysed using ANOVA to test the significance level of variations in soil physico-chemical and microbial properties under different systems.

RESULTS AND DISCUSSION

Soil is a multifaceted ecosystem in a condition of dynamic symmetry, delimited by physico-chemical parameters that hold enormous numbers of diverse living organisms [28]. The soil of the three study sites was sandy loam to loamy sand soil and slightly acidic in nature ranging from 5.51 to 5.82 (Table 1). Quantitatively, both bacterial and fungal CFUs were greater in the surface soil (0-10 cm) layer as compared to other depths in all the seasons except the rainy season where the counts were greater in the subsurface soil (10-20 cm) layer (Table 2).

Higher microbial counts in the surface soil (0-10 cm) layer may be because this layer is usually provided with high organic matter content which in the presence of adequate moisture supply is acted upon by microorganisms to decompose the complex organic residues into simpler forms while reduction in microbial population in the lower depths may be attributed to fewer amounts of minerals, low oxygen content and increased carbon-dioxide concentration [29, 30]. On the other hand, higher CFUs in the subsurface soil (10-20 cm) layer during rainy season corroborates those of Mishra [31] who pointed out that during rainy months, this layer occasionally harbours more fungal populations caused by temperature and moisture regimes than the topsoil layer.

They also reported that the dynamics of fungal species was greatly influenced by the changes in canopy cover and soil nutrient status. It may also be due to percolation and leaching of the organic matter to this layer.

Among the sites, Harmutty village (Assam) recorded the higher bacterial and fungal CFUs as compared to the sites in Arunachal Pradesh populations were more in Harmutty site which may probably be due to the dense growth of plants and greater availability of nutrients on account of greater accumulation of litter [7], patterns of rhizodeposition from the fine roots [32] and may also be due to spreading of other biodegradable domestic wastes into the system by the local farmers [30]. Mishra and Sharma [33] further suggested that plant species growing on the soil also exert very important influence on the population and species composition of the soil fungi.

However, although there were differences in the averages total bacterial and fungal CFUs of the different sampling locations, these differences were not statistically significant. The non-significance of the differences in the fungal and bacterial counts of different samples, irrespective of sampling locations supports the finding of Amir and Pineau [34].

Seasonally, bacterial CFUs were more during rainy season (Table 2) while fungal CFUs were more during autumn in all the sites. Nevertheless, bacterial counts were generally higher than those of fungi irrespective of the sites in all seasons which may be attributed to their faster multiplication rate [30]. The predominance of bacteria over fungi observed throughout the sampling time has been reported by other workers [35].

Table 2. Seasonal variation in microbial population in three traditional agroforestry systems at three depths

Sites										
Microbial Population	Time	Harmutty			Nirjuli			Doimukh		
		0-10	10-20	20-30	0-10	10-20	20-30	0-10	10-20	20-30
Bacterial Population	Jan' 03	105.33	98.33	43.66	101.33	87.66	51.33	82.67	53.33	21.67
	Mar' 04	177.41	134.00	99.72	143.70	119.00	58.20	141.16	109.52	78.06
	May' 04	172.67	186.33	96.67	139.67	152.67	78.00	127.33	141.70	59.00
	Nov' 04	135.67	128.33	61.67	119.69	85.33	71.00	86.20	34.67	21.67
Fungal Population	Jan' 03	29.63	24.57	16.19	27.57	18.30	15.87	22.00	20.00	17.00
	Mar' 04	87.30	68.52	63.12	61.02	56.87	34.69	71.12	37.14	34.52
	May' 04	34.07	37.30	20.20	21.03	32.11	15.13	22.07	31.12	21.03
	Nov' 04	118.00	105.00	39.00	112.00	91.07	29.09	101.00	86.34	43.10

Table 3. Microfungi isolated from three soil depths in the three traditional agroforestry systems

Sites									
Species	Harmutty			Nirjuli			Doimukh		
	0-10	10-20	20-30	0-10	10-20	20-30	0-10	10-20	20-30
Absidia van Tieghem	+	+	-	+	-	-	-	-	-
A. glauca Hagem.	-	+	-	-	-	-	+	-	-
A. spinosa Lendner	+	-	+	-	-	-	-	-	-
Aspergillus Mich. ex. Fr.	+	-	+	+	+	-	+	+	+
A. clavatus Desm.	-	-	-	+	-	-	-	+	-
A. flavus Linker Gray.	-	+	-	-	-	+	-	-	-
A. fumigatus Fres.	-	-	-	-	-	-	+	-	-
A. niger van Tieghem.	+	-	-	+	-	-	-	-	-
Curvularia Boedijn.	-	+	-	-	-	-	-	-	+
Geotrichum candidum Linker Leman.	+	-	-	+	+	+	-	+	-
Mucor Mich. ex. St.-Am.	+	-	+	-	-	-	-	-	+
M. hiemalis Wehmer.	-	-	-	+	-	+	-	+	-
M. mucedo Mich. ex. St.-Am.	-	+	-	-	-	-	+	-	-

Sites									
Species	Harmutty			Nirjuli			Doimukh		
	0-10	10-20	20-30	0-10	10-20	20-30	0-10	10-20	20-30
Penicillium Link. ex. Fr.	+	-	+	+	-	-	-	+	+
P. chrysogenum Thom.	+	+	-	s-	+	-	-	-	+
P. claviforme Bain.	-	-	-	+	-	-	+	-	-
P. expansum Link. ex. Gray.	-	-	+	-	-	-	-	+	-
P. frequentaus Westling.	-	+	-	-	-	+	-	-	-
P. sacculum Dale.	+	-	-	-	-	-	+	-	-
P. spinulosa Thom.	-	-	-	+	-	-	-	+	-
Rhizopus Ehrenb.	+	-	-	-	+	-	-	-	-
R. stolonifer Exrenb.ex. Link.	-	-	-	+	-	-	-	+	-
Trichoderma Pers. ex. Fr.	+	-	-	-	-	-	-	-	-
T. viride Pers. ex. Fr.	-	-	-		+	-	-	-	-

The higher bacterial and fungal counts observed during rainy and autumn season respectively may probably be due to prevailing favorable moisture and temperature conditions. Nonetheless, minimum microbial count during winter may presumably be due to existing low temperature and greater physiological water stress which are otherwise critical for the growth and activity of microbes [36].

Soil microbial communities are arguably the most diverse communities ranging from myriad of invisible microbes to the more familiar macro-fauna that plays a critical role for the maintenance of the sustainable ecosystems. Microorganisms are found in prominent amount with great species diversity in the soil of the earth [1]. Altogether, 24 forms of micro-fungi belonging to eight genera were isolated in the three sites (Table 3). The relative proportion of different fungal taxonomic groups was almost identical in all the three sites indicating their wide ecological amplitudes [36]. *Penicullium* with 7 species and *Aspergillus* with 5 species were the most abundant group of species while some species were restricted to a particular site. For instance, *Aspergillus fumigatus*, *Trichoderma viride* was exclusively found in Nirjuli, while *Trichoderma* sp. was isolated only in Harmutty site.

The dominance of the genus *Penicillium* and *Aspergillus* in the present study sites may be due to their greater spore production and dispersal rate [36] and partly due to their resistance over extreme environmental conditions [37]. Many workers have also recorded a correlation between fungal species composition and the species composition of the aboveground vegetations [38].

Microbial Biomass C, N and P

The importance of microorganisms in ecosystem functioning has led to an increased interest in the determination of soil microbial biomass. Soil microbial biomass is a potential source of plant nutrients, and a higher level of soil microbial biomass is an indicator of soil fertility.

The values of microbial C (47.52-1167.64μg g^{-1}, Table 4) in the present study fell in the lower end of the range (61-2000 μg g-1) reported for the various temperate and tropical forest soils [39], but was comparable to a Japanese cropland (1076μg g^{-1}, Marumoto [40], subtropical humid forest regrowths (203.74-1087.70 μg g^{-1}, Maithani et al. [7] and jhum fields (801.4-14319 μg g^{-1}, Arunachalam and Pandey, [41].

Variation in microbial biomass has been attributed to differences in the quantity and quality of the present and past substrate properties [42], total dissolved organic C and [43] and labile soil organic matter [44]. However, Zeller et al. [45] reported site factors as greater determinants of soil microbial C than management factors in sub-alpine meadows. In this study, although seasonal fluctuations in soil microbial C differed among the three sites, they did follow similar patterns under different management within each site. High value of microbial C in Harmutty may be partly attributed to greater plant density and diversity and partly to the dense fibrous and coral root system of *Areca catechu* [46] in the site that may have favoured the growth of microbial population and accumulation of larger microbial biomass. However, occasional lower values may be due to extreme rainfall in the area leading to greater run-off and leaching of the top soil. Further, decrease in microbial C during rainy months might be attributed to loss of microbial propagules.

Table 4. Seasonal variations in soil microbial C ($\mu g\ g^{-1}$)

Sites									
Season	Harmutty			Nirjuli			Doimukh		
	Soil depth (cm)								
	0-10	10-20	20-30	0-10	10-20	20-30	0-10	10-20	20-30
Microbial C									
Jan'3	358.16 (±1.3)	320.32 (±3.0)	58.96 (±5.0)	343.96 (±11.41)	295.68 (±0.7)	95.92 (±3.7)	259.60 (±0.01)	225.28 (±5.4)	47.52 ±1.7)
Mar'4	1167.64 (±33.3)	548.88 (±4.8)	137.73 (±1.7)	826.53 (±0.3)	792.08 (±0.1)	482.16 (±8.0)	1102.04 (±1.7)	860.98 (±3.7)	516.57 (±1.4)
May'4	182.37 (±5.3)	213.44 (±4.0)	166.28 (±3.0)	161.62 (±0.2)	309.04 (±3.7)	142.88 (±10.3)	285.33 (±3.3)	380.06 (±0.03)	150.52 (±0.4)
Nov'4	474.32 (±15.0)	429.44 (±1.7)	71.28 (±3.3)	420.80 (±23.7)	390.72 (±6.3)	119.68 (±2.7)	305.36 (±5.3)	261.36 (±0.3)	117.92 (±3.3)
Microbial N									
Jan'3	20.74 (±0.03)	19.01 (±0.05)	9.50 (±0.01)	19.88 (±0.03)	8.64 (±0.02)	5.18 (±1.07)	19.88 (±0.06)	12.97 (±0.02)	9.50 (±0.05))
Mar'4	26.42 (±0.03)	12.22 (±0.08)	4.14 (±0.04)	12.86 (±0.05)	8.23 (±0.67)	2.68 (±0.01)	15.93 (±0.07)	7.96 (±0.02)	4.25 (±0.08)
May'4	32.32 (±0.03)	86.19 (±0.07)	29.63 (±3.33)	61.64 (±0.02)	74.71 (±3.25)	49.27 (±0.10)	21.67 (±0.03)	24.22 (±0.03)	9.51 (±0.06)
Nov'4	25.06 (±0.04)	18.15 (±0.01)	19.37 (±0.03)	23.33 (±0.01)	19.87 (±0.15)	8.63 (±0.07)	20.74 (±0.03)	12.10 (±0.01)	12.10 (±0.08)
Microbial P									
Jan'3	6.52 (±3.03)	4.05 (±0.02)	2.90 (±0.03)	5.15 (±6.66)	4.08 (±0.22)	2.57 (±0.03)	7.86 (±0.04)	3.63 (±0.02)	1.90 (±0.07)
Mar'4	4.00 (±2.03)	3.71 (±3.08)	1.33 (±3.04)	19.02 (±3.05)	7.86 (±0.02)	4.05 (±0.02)	14.00 (±0.03)	6.92 (±0.09)	3.17 (±0.01)
May'4	2.79 (±0.03)	3.34 (±0.07)	2.76 (±0.05)	3.35 (±0.02)	3.42 (±3.39)	2.47 (±0.07)	1.88 (±0.03)	2.59 (±0.08)	1.18 (±0.12)
Nov'4	6.51 (±0.09)	4.05 (±2.54)	1.81 (±0.03)	4.78 (±0.01)	3.68 (±0.03)	2.02 (±0.17)	7.66 (±0.13)	4.04 (±0.03)	4.06 (±0.04)

In the present study, microbial N and P had similar spatial variations (Table 4). The microbial N values (2.68-86.19 μg g^{-1}) was lower than those reported in several croplands (9-25μg g^{-1}, Brookes et al., [26], coniferous forest soils (52-125 μg g^{-1}, Martikainen and Palojarvi [47], soils of broadleaved deciduous and evergreen forests (132-240 μg g^{-1} and 42-242 μg g^{-1} respectively, Diaz-Ravina et al. [48] and 226.52-547.22 μg g^{-1} from plantation forest by Upadhyaya and Arunachalam [49]. Nevertheless, in agricultural soils, several authors [50] observed significant annual variations in soil microbial biomass. Microbial P varied between 1.18 and 7.66 μg g^{-1} (Table 4) which was comparable to those reported by Arunachalam and Arunachalam [51] from secondary bamboo forest (1.5-9 μg g^{-1}) in Arunachal Pradesh, but was far lower than the reported range of 8.43- 19.25 μg g^{-1} from jhum fields and plantation forests [49] and from forest ecosystem (17-35 μg g^{-1}) in the central Himalaya region [24]. Plants supply organic materials as energy sources for microbial growth, so the low soil microbial biomass could be a reflection of the low vegetation abundance.

CONCLUSION

From the present study, it can be concluded that overall bacterial and fungal CFUs is influenced by vegetation, density and species composition. However, the role of macro- and micro-climatic seasonality and soil nutrient status cannot be completely ruled out. It is also understood that litter accumulating in these traditional homestead garden serves as a substrate that is converted into soil organic matter by the microbial biomass that are furthermore important and would play a vital role in soil nutrient management within the system. Amongst the study sites, litter quality was comparatively good in Harmutty. Thus, relatively greater accumulation of litter and fine roots in Harmutty site might have favored the growth of microbial population and accumulation of microbial biomass.

However, the role of soil properties such as clay content and soil organic C in soil microbial biomass dynamics cannot be completely ruled out. In this context, temporal changes in soil moisture, temperature and carbon input from crop roots, rhizosphere products (i.e., root exudates, mucilage, sloughed cells, etc.), and crop residues may have a large effect on soil microbial biomass and its activity. This in turn, affect the ability of soil to supply nutrients to plants through soil organic matter turnover, particularly in the production systems.

ACKNOWLEDGMENTS

We thank the villagers of Harmutty, Nirjuli and Doimukh for their cooperation during the field study.

The authors are also thankful to Mr. Moharam Ingti, Laboratory Assistant for helping in the laboratory studies. This study is part of a research project funded by the Indian Council of Agricultural Research (ICAR), New Delhi. The first authors also thank the Council of Scientific and Industrial Research (CSIR), New Delhi for awarding Junior Research Fellowship.

REFERENCES

[1] Ishaq, F., Khan, A., (2011). Isolation, Identification and Comparative Study of Fungal and Bacterial Strains Found in Organic and Inorganic Soils of Different Agricultural Fields. *Recent Research in Science and Technology*, 3(11), 30-36.

[2] Vasseur, C., Joannon, A., Aviron, S., Burel, F., Meynard, J.-M., Baudry, J., (2013). The cropping systems mosaic: how does the hidden heterogeneity of agricultural landscapes drive arthropod populations? *Agr. Ecosyst. Environ.* 166, 3-14.

[3] Acosta-Martínez, V., Acosta-Mercado, D., Sotomayor-Ramírez, D., Cruz- Rodríguez, L., (2008). Microbial communities and enzymatic activities under different management in semiarid soils. *Appl. Soil Ecol.* 38, 249-260.

[4] Schipanski, M.E., Drinkwater, L.E., (2012). Nitrogen fixation in annual and perennial legume-grass mixtures across a fertility gradient. *Plant Soil* 357, 147-159.

[5] Wall, D.H., Virginia, R.A., (1999). Controls on soil biodiversity insights from extreme environments. *Appl. Soil Ecol.* 13, 137–150.

[6] O' Donnell AG, Seasman M, Macrae A, Waite I, Davies JT (2001). Plants and Fertilizers as drivers of change in microbial community structure and function in soil. *Plant Soil* (232), 135 – 145.

[7] Maithani, K., Tripathi, R.S., Arunachalam, A., Pandey, H.N., (1996). Seasonal dynamics of microbial biomass C, N and P during regrowth of a disturbed subtropical humid forest in north-east India. *Appl. Soil Ecol.* 4, 31-37.

[8] Kallenbach, C., Grandy, A.S., (2011). Controls over soil microbial biomass responses to carbon amendments in agricultural systems: A meta-analysis. *Agr. Ecosyst. Environ.* 144, 241–252

[9] Jenkinson, D.S., Ladd, J.N., (1981). Microbial biomass in soil: measurement and turnover. In: Soil Biochemistry, (Paul, E.A., Ladd, J.N., eds.) 5, pp. 415- 417. Book chapter page number publisher etc.

[10] Logah, V., Ewusi - Mensah, N., Tetteh, E.N., Quansah, C., Danso, I., (2013). Seasonal variations of soil microbial biomass under different nutrient management and cropping systems on a Ferric Acrisol in Ghana. *J. Trop. Agri.* 51(1-2), 98-104.

[11] Jordan, D., Kremer, R.J., Bergfield, W.A., Kim, K.Y., Cacnio, V.N., (1995). Evaluation of microbial methods as potential indicators of soil quality in historical agricultural fields. *Biol. Fert. Soils* 19, 297 – 302.

[12] Sparling, G.P., Shepherd, T.G., Kettles, H.A., (1991). Changes in soil organic C, microbial C and aggregate stability under continuous maize and cereal cropping, and after restoration to pasture in soil from the Manawater region. *New Zeal. Soil Till. Res.* 24, 225–241.

[13] Moussa S, Van Rensburg, L., Kellner, K., Bationo, A., (2007). Soil microbial biomass in semi-arid communal sandy rangelands in the western bophirima district, South Africa. *Appl. Ecol. Env.* Res. 5(1), 43-56

[14] Azam, F., Farooq, S., Lodhi, A., (2003). Microbial biomass in agricultural soils-determination, synthesis, dynamics and role in plant nutrition. *Pak. J. Biol. Sci.* 6 (7), 629-639.

[15] Kieft, T.L., (1994), Grazing and plant canopy effects on semi-arid soil microbial biomass and respiration. *Biol. Fert. Soils* 18, 155-182

[16] Garcia, C., Hernandez, T., Roldan, A., Martin, A., (2002). Effect of plant cover decline on chemical and microbiological parameters under Mediterranean climate. *Soil Biol. Biochem.* 34, 635-642

[17] Kumar, G., (1997). Geology of Arunachal Pradesh Geological Society of India, Bangalore.

[18] Anderson, J.M., Ingram, J.S.I., (1993). Tropical Soil Biology and Fertility A Handbook of Methods, 2nd Edition, CAB. International, Vallingford, U.K.

[19] Waksman, S.A. (1952) Soil Microbiology, John Wiley and Sons, New York.

[20] Johnson, L.F., Curl, E.A., (1972). Methods for the Research on Ecology for Soil-Borne Plant Pathogens. Burgess Publishing Co., Minneapolis, Minnesota.

[21] Barnett, H.L., Hunter, B.B., (1972). Illustrated genera of imperfect fungi. Burgess Publishing Company. Minneapolis. Page number

[22] Domsch, K.H, Gems, W., Anderson, T.H., (1980). Compendium of soil fungi Vol.2, Academic Press A Subsidiary of Harcourt Brace Jovanovich, Publishers.

[23] Jenkinson, D.S., Powlson, D.S., (1976). The effect of biocidal treatments on metabolism in soil-V. A method for measuring soil biomass. Soil Biol. Biochem. 8, 209-213.

[24] Srivastava, S.C., Singh, J.S., (1988). Carbon and phosphorus in the soil biomass of some tropical soils of India. *Soil Biol. Biochem.* 20, 743-747.

[25] Jenkinson, D.S., (1988). Determination of microbial biomass C and N in soil. In: *Advances in Nitrogen Cycling in Agricultural Ecosystems*, (Wilson, J.R., Ed.), CAB International, Wallingford, pp.368-386.

[26] Brookes, P.C., Kragt, J.F., Powlson, D.S., Jenkinson, D.S., (1985). Chloroform fumigation and release of soil nitrogen: The Effect of fumigation time and temperature. *Soil Biol. Biochem.* 17, 831-835.

[27] Brookes, P.C., Polwson, D.S., Jenkinson, D.S., (1984). Phosphorus in the soil microbial biomass. *Soil Biol. Biochem.* 16, 169-175.

[28] Nandhini, B., Josephine, R.M., (2013). A study on bacterial and fungal diversity in potted soil. *Int J. Curr. Microbiol. Appl. Sci.* 2, 1-5

[29] Shukla, A.K., Tiwari, B.K., Mishra, R.R., (1989). Temporal and depthwise distribution of microbes, enzyme activities and soil respiration in potato field soil under different agricultural systems in northeastern hill region of India. *Revue a Ecologie et de Biologic du sol* 26, 249-265.

[30] Tangjang, S., Arunachalam, K., Arunachalam, A., and Shukla, A.K., (2008). Microbial population dynamics of soil under traditional agroforestry system in northeast India. Res. J. Soil Biol. 1(1), 1-7.

[31] Mishra, R.R., (1965). Seasonal distribution of fungi in four different grass consociations of Varanasi (India). Trop. Ecol. 6, 133-140.

[32] Arunachalam, A., Pandey, H.N., Tripathi, R.S., Maithani, K., (1996). Fine root decomposition and nutrient mineralization patterns in a subtropical humid forest following tree cutting. *Forest Ecol. Manage.* 86, 114-150.

[33] Mishra, R.R., Sharma, G.D., (1977). Ecology of soil fungi. Population variation in relation to varying cover vegetation and soil factors *Sydowia Annales Mycologici Ser II* Vol. XXX, Heft 1-6, pp. 134-140. Check reference book title editors publisher etc.

[34] Amir, H., Pineau, R., (1998). Influence of plants and cropping on microbiological characteris tics of some new Caledonian Ultramafic soils. *Aust. J. Soil Res.* 36 (3), 457 – 470.

[35] Okoh, L.A., Badejo, M.A., Nathaniesl, I.T., Tian, G., (1999). Studies on the bacteria, fungi and springtails (collembola) of an agroforestry arboretum in Nigeria. *Pedobiologia* 43, 18-27.

[36] Arunachalam, K., Arunachalam, A., Tripathi, R.S., Pandey, H.N., (1997). Dynamics of microbial population during the aggradation phase of a selectively logged subtropical humid forest in north-east India. *Trop. Ecol.* 38, 333-341.

[37] Schimel, J., (1995). Ecosystem consequences of microbial diversity and community structure Ecological Studies Vol. 113. In: book title (Chapin and Korner Eds. Full details of editors), Springer-Verlag, Berlin, Heidelberg, pp. 239-254.

[38] Morall. R.A.A., Vanterpool, T.C., (1968). The soil microfungi of upland bereal forest at Candle Lake Saskatchewari. *Mycologia* 60, 642-654.

[39] Diaz-Ravina, M., Acea, M. J., Carballas, T., (1995). Seasonal changes in microbial biomass and nutrient flush in forest. Soils. *Biol. Fert. Soil*s 19, 220-226.

[40] Marumoto, T., (1984) Mineralization of C and N from microbial biomass in paddy soil. *Plant Soil* 75, 165-173.

[41] Arunachalam, A., Pandey, H.N., (2003). Ecosystem restoration of Jhum fallows in northeast India: Microbial C and N along altitudinal and successional gradients. *Res. Ecol.* 11, 1-6.

[42] Haron, K., Brookes, P.C., Anderson, J.M., Zakaria, Z.Z., (1998). Microbial biomass and soil organic matter dynamics in oil palm (Elaeis guinensis Jacq.) plantations, West Malaysia. *Soil Biol. Biochem.* 30, 547-552.

[43] Smolander, A., Kitunen, V., (2002) Soil microbial activities and characteristics of dissolved organic C and N in relation to tree species. *Soil Biol. Biochem.* 34, 651-660.

[44] Steenwerth, K.L., Jackson, L.E., Calderon, F.J., Stromberg, M.R., Scow, K.M., (2002). Soil microbial community composition and land use history in cultivated and grassland ecosystems of coastal California. *Soil Biol. Biochem.* 34, 1599-1611.

[45] Zeller, V., Bardgett, R.D., Tappeiner, U., (2001). Site and management effects on soil microbial Properties of subalpine meadows: a study of land abandonment along a north-south gradient in the European Alps. *Soil Biol. Biochem.* 33, 637-649.

[46] Arunachalam, A., Tangiang, S., Deb, S., Melkania, U., Arunachalam, K., Shrivastava, K., (2003). Areca-nut tree: An eco-friendly species in traditional Home gardens of Assam, Northeast India. *ANJAC J. Sci.* 2, 9953-54.

[47] Martikainen, P.J., Polajarvi, A., (1990). Evaluation of the fumigation-extration method for the determination of microbial C and N in a range of forest soils. *Soil Biol. Biochem.* 22, 797-802.

[48] Diaz-Ravina, M., Carballas, T., Acea, M.J., (1988). Microbial biomass and metabolic activity in four acid soils. *Soil Biol. Biochem.* 20, 817-823.

[49] Upadhyaya, K., Arunachalam, A. and Arunachalam, K. (2004) Effect of bamboo foliage on soil respiration, microbial biomass and N mineralization. *J. Bamboo Rattan* 3, 169-184.

[50] Van Gestel, M., Ladd, J. N. and Amato, M. (1992) Microbial biomass response to seasonal change and imposed drying regimes at increasing depths of undisturbed topsoil profiles. *Soil Biol. Biochem.* 24, 103-111.

[51] Arunachalam, A., Arunachalam, K., (2002). Evaluation of bamboos in eco-restoration of jhum' fallows in Arunachal Pradesh ground vegetation, soil and microbial biomass. *Forest Ecol. Manag*, 59, 31-239.

INDEX

B

C

D

E

F

G

H

I

J

K

L

M

N

O

P

S

T

U

V

W

X

Y

Z